S. Hrg. 109–382

CHEMICAL FACILITY SECURITY: WHAT IS THE APPROPRIATE FEDERAL ROLE?

HEARINGS

BEFORE THE

COMMITTEE ON HOMELAND SECURITY AND GOVERNMENTAL AFFAIRS UNITED STATES SENATE

ONE HUNDRED NINTH CONGRESS

FIRST SESSION

JULY 13 AND 27, 2005

Printed for the use of the
Committee on Homeland Security and Governmental Affairs

U.S. GOVERNMENT PRINTING OFFICE
23–157 PDF WASHINGTON : 2006

For sale by the Superintendent of Documents, U.S. Government Printing Office
Internet: bookstore.gpo.gov Phone: toll free (866) 512–1800; DC area (202) 512–1800
Fax: (202) 512–2250 Mail: Stop SSOP, Washington, DC 20402–0001

COMMITTEE ON HOMELAND SECURITY AND GOVERNMENTAL AFFAIRS

SUSAN M. COLLINS, Maine, *Chairman*

TED STEVENS, Alaska
GEORGE V. VOINOVICH, Ohio
NORM COLEMAN, Minnesota
TOM COBURN, Oklahoma
LINCOLN D. CHAFEE, Rhode Island
ROBERT F. BENNETT, Utah
PETE V. DOMENICI, New Mexico
JOHN W. WARNER, Virginia

JOSEPH I. LIEBERMAN, Connecticut
CARL LEVIN, Michigan
DANIEL K. AKAKA, Hawaii
THOMAS R. CARPER, Delaware
MARK DAYTON, Minnesota
FRANK LAUTENBERG, New Jersey
MARK PRYOR, Arkansas

MICHAEL D. BOPP, *Staff Director and Chief Counsel*
ALLISON J. BOYD, *Counsel*
JOYCE A. RECHTSCHAFFEN, *Minority Staff Director and Counsel*
HOLLY A. IDELSON, *Minority Counsel*
TRINA DRIESSNACK TYRER, *Chief Clerk*

CONTENTS

Page

Opening statements:
- Senator Collins 1, 49
- Senator Lieberman 3, 50
- Senator Voinovich 4, 59
- Senator Lautenberg 6, 52
- Senator Carper 7, 62

WITNESSES

Wednesday, July 13, 2005

Martin J. Durbin, Managing Director, Security and Operations, American Chemistry Council 8

Matthew Barmasse, Environmental, Health, Safety, and Quality Director, ISOCHEM, Inc., on behalf of the Synthetic Organic Chemical Manufacturers Association 12

Bob Slaughter, President, National Petrochemical and Refiners Association 15

Gerald V. Poje, Ph.D., Former Board Member, U.S. Chemical Safety and Hazard Investigation Board 29

Glenn Erwin, Project Director, Triangle of Prevention Program, United Steelworkers International Union 33

Carol L. Andress, Economic Development Specialist, Environmental Defense .. 36

Wednesday, July 27, 2005

Rear Admiral Craig E. Bone, Director of Port Security, Marine Safety, Security, and Environmental Protection Directorate, U.S. Coast Guard 53

Beth Turner, Director, Global Operations Security, E.I. duPont de Nemours and Co., Inc., Wilmington, Delaware 64

Jim Schellhorn, Director of Environmental Health and Safety, Terra Industries, Inc., on behalf of the Fertilizer Institute 67

John P. Chamberlain, Security Manager, Asset Protection Services, Corporate Security, Shell Oil Company, on behalf of the Shell Oil Company and the American Petroleum Institute 70

Chief Robert A. Full, Fire Marshal/Emergency Management Coordinator, Allegheny County (PA) Department of Emergency Services 74

Alphabetical List of Witnesses

Andress, Carol L.:
- Testimony 36
- Prepared statement with attachments 209

Barmasse, Matthew:
- Testimony 12
- Prepared statement 102

Bone, Rear Admiral Craig E.:
- Testimony 53
- Prepared statement 233

Chamberlain, John P.:
- Testimony 70
- Prepared statement 264

Durbin, Martin J.:
- Testimony 8
- Prepared statement 91

Page

Erwin, Glenn:
- Testimony 33
- Prepared statement with an attachment 144

Full, Chief Robert A.:
- Testimony 74
- Prepared statement 272

Poje, Gerald V., Ph.D.:
- Testimony 29
- Prepared statement 130

Schellhorn, Jim:
- Testimony 67
- Prepared statement with an attachment 253

Slaughter, Bob:
- Testimony 15
- Prepared statement with attachments 119

Turner, Beth:
- Testimony 64
- Prepared statement 238

APPENDIX

Survey entitled "PACE International Union Survey: Workplace Incident Prevention and Response Since 9/11", October 2004, by Paper, Allied-Industrial, Chemical and Energy Workers International Union (PACE), submitted by Mr. Erwin 150

"Security Vulnerability Assessment Methodology for the Petroleum and Petrochemical Industries, Second Edition," October 2004, American Petroleum Institute, NPRA, submitted by Mr. Chamberlain 277

"Security Guidelines for the Petroleum Industry," American Petroleum Institute, April 2005, submitted by Mr. Chamberlain 428

Paul Orum, Working Group on Community Right-to-Know, July 2004, report entitled "Unnecessary Dangers: Emergency Chemical Release Hazards at Power Plants," submitted for the record 482

Jon P. DeVine, Jr., Senior Attorney, Natural Resources Defense Council, prepared statement, with an attachment entitled "Critical Infrastructure Security Series, New Strategies to Protect America: Securing our Nation's Chemical Facilities," by Dr. Linda Greer 522

Meghan Purvis, Environmental Health Advocate, U.S. Public Interest Research Group, prepared statement with attachments entitled "Needless Risk, Oil Refineries and Hazard Reduction," August 2005, U.S. PIRG Education Fund, and "Survey of Chemical Industry Hazard Reduction to Protect Public Safety, 2002 Survey Summary" 549

The National Association of Chemical Distributors, prepared statement 587

Agricultural Retailers Association, submitted by Richard Gupton, ARA Director of Legislative Policy and Counsel, prepared statement with attachments entitled "Guidelines to Help Ensure a Secure Agribusiness," and "Agricultural Retailers Association: Security Vulnerability Assessment Workshop" 595

CHEMICAL FACILITY SECURITY: WHAT IS THE APPROPRIATE FEDERAL ROLE?

WEDNESDAY, JULY 13, 2005

U.S. SENATE,
COMMITTEE ON HOMELAND SECURITY
AND GOVERNMENTAL AFFAIRS,
Washington, DC.

The Committee met, pursuant to notice, at 9:32 a.m., in room 562, Dirksen Senate Office Building, Hon. Susan M. Collins, Chairman of the Committee, presiding.

Present: Senators Collins, Lieberman, Voinovich, Lautenberg, and Carper.

OPENING STATEMENT OF CHAIRMAN COLLINS

Chairman COLLINS. The Committee will come to order. Good morning.

Today marks the third in this Committee's series of hearings on the issue of chemical security. At our first hearing, we heard from experts about the potentially catastrophic impact of a successful terrorist attack on a chemical facility and about how vulnerable many chemical sites are.

At the second hearing, the Department of Homeland Security and the Environmental Protection Agency testified that current laws are not sufficient, and the Administration pledged to work with this Committee in developing appropriate legislation.

At today's hearing, we will hear from a variety of witnesses who have a longstanding interest in the safety and security of chemical sites.

Let me take just a moment to describe the chemical industry. By economics alone, it is impressive. The total value of chemical shipments in the United States approaches half-a-trillion dollars annually. The chemical industry represents our largest export sector, with exports totaling $91.4 billion in 2003. More than 900,000 people work directly in the American chemical industry, with an additional 700,000 supplier jobs and millions more in indirect jobs.

Perhaps even more significant than the economic impact is the impact of chemicals on our daily lives. Chemicals are necessary for more than 70,000 products that help make life in our country what it is today and that have helped us to achieve the greatest standard of living the world has ever seen.

How many people have enough food to eat because fertilizers and other agricultural chemicals have helped to make America the breadbasket of the world? How many Americans would die of cholera and other diseases if we did not have chemicals to treat our water supply? How many children's lives are saved each year by the chemical compounds that make up prescription medicines? Where would we be without computers and other consumer electronics, which are not possible without chemicals?

It is an unfortunate fact of life that many things in this world that have the greatest capacity for good also have the greatest capacity to cause harm. Chemicals fall in that category. While of immense benefit to society, chemicals can also cause tremendous damage.

Since the first large-scale use of chemical weapons in World War I, chemicals have been the most used weapon of mass destruction by both governments and terrorists. As we learned in chilling detail in this Committee's first hearing, even necessary and legitimate chemicals have an immense capacity to cause death and destruction.

It is a further fact of life that we often fail to appreciate the significance of a threat until a catastrophe occurs. For example, many of our most important chemical safety measures were not established until after the tragic deaths of thousands following a chemical accident in Bhopal, India. The Chemical Safety Board, as well as the EPA's Risk Management Plan program, were both established in response to Bhopal.

Many companies have recognized the need for stronger security and have already taken strong steps to improve security at their chemical sites. Many in the industry have subscribed to well-regarded voluntary programs such as the Responsible Care® program. I applaud these efforts and strongly encourage the continuation of voluntary actions to improve security.

Unfortunately, as the Department of Homeland Security testified at our earlier hearing, not all companies abide by such codes of conduct. I look forward to hearing from our first panel of industry representatives today about their views on the need for mandatory measures to complement the voluntary efforts.

Our second panel consists of representatives from environmental, labor, and public advocacy groups. Environmental groups and other public advocates have long sought to increase public recognition of the risks inherent in operating large chemical facilities, particularly near large population centers. Similarly, labor representatives have long pushed for greater worker safety at chemical plants.

Given that the chemical industry presents both tremendous benefits as well as immense risks, it is critical that any legislation strike a carefully thought out balance. Terrorists seek to use our infrastructure and assets to cause maximum disruption to our society and harm to our economy. In our search for a solution to the threats that we face, we must be careful not to accomplish the terrorists' objectives for them by harming our economy.

I look forward to hearing from industry, labor, and environmental groups in today's hearing. Their different views and perspectives will be most helpful to this Committee as we continue our work on this critical issue.

Senator Lieberman.

OPENING STATEMENT OF SENATOR LIEBERMAN

Senator LIEBERMAN. Thanks very much, Chairman Collins. As you have noted, this is the third in a series of hearings that our Committee has held on chemical site security. Since there are not many subjects that get three hearings before the Committee in 3 months, it should be very clear that the Chairman and I and the Members of the Committee consider chemical security to be a particularly urgent challenge for our Nation and for this Committee.

This hearing, as we all know, comes just 1 week after terrorists in London demonstrated yet again their intention and capacity to attack and kill innocent civilians, to find and exploit weaknesses in our homeland defenses. And even though the most recent incident was an attack on a mass transit system, it was a very loud and painful warning that we need to continue to be alert, to be vigilant, to identify and close vulnerabilities in our own country.

By any measure, the chemical industry today is one of the sectors in American life that is most vulnerable to terrorist attack. At our first hearing, we heard compelling testimony about the potential risk posed by chemical sites across the Nation. We were told that chemical facilities represent potential weapons of mass destruction. If released through accident or terrorist attack, the chemicals stored or manufactured in these plants could kill thousands of people in surrounding communities.

At our second hearing, the Department of Homeland Security agreed that chemical facilities posed a serious risk from terrorist attack. While describing some significant initiatives taken by the chemical industry itself, the Department conceded that these voluntary measures are not enough. Rather, the Department said we need new legislation to ensure that all facilities that use or store significant amounts of hazardous chemicals, and therefore pose a terrorism risk, are subject to minimum security standards. I agree.

Today, we will hear from representatives of the chemical industry and other stakeholders, that is, those who work at chemical sites and also environmental and safety advocates who work on issues relating to the operation of chemical facilities. These witnesses really can help us on this Committee answer some of the most difficult questions that we will need to answer as we attempt to draft responsive and sensible legislation.

For example, one of our witnesses today, the American Chemistry Council, developed a security code for its members after September 11. I would like to learn more about what this code requires, what are its strengths and weaknesses, and how it might inform any Federal mandates, statutory mandates, for chemical facilities.

Another important question that I have relates to local preparedness and response. While some chemical facilities have clearly tried to improve security on their premises themselves, they also rely on local officials to secure the area outside their gates and respond in the event of an accident or an attack. Based on testimony at our earlier hearings and on interviews by our Committee staff, I am concerned that State and local officials will need more resources than they now have to carry out those responsibilities, and I hope today we can get some clarity about what is the best division of labor between the chemical industry and public authorities and what needs to be done by whom to ensure effective security and response capability.

Third, I am also concerned that there may be many citizens who live near chemical facilities who haven't been adequately prepared and informed about what to do if there is an accident or an attack at a chemical facility, and so I hope our witnesses can help us to decide how we can improve public readiness here.

And finally, and perhaps most difficult, we have to resolve critical questions about how to define and regulate the word "security." Some have argued that any legislation should be limited to physical security measures, such as gates, surveillance cameras, and access controls. Others say that these types of measures will never stop a determined terrorist and that we must instead figure out how to reduce potential damage from these sites. Some have said that this will and should require that the chemical industry look into alternative substances or technologies to reduce the amount of harmful chemicals it employs or configure them in ways that minimize the risk of hazardous release.

I know that there is great disagreement about whether these issues, all of them, should be addressed in chemical security legislation, but there should be no disagreement, and I don't believe there is, about the need to make our chemical industry and processes as safe as possible, indeed, safer than they are today, and the question is how to best get there.

A final word, Madam Chairman. Although the Administration is not testifying today, I am sure that they are listening, and so I want to reiterate my request made at our last hearing that the Administration and the relevant departments take a real leadership role in crafting chemical security legislation. I know you and I are prepared and eager to work with them. We need the benefit of the Administration's work on this issue and its recommendations on legislation it believes is needed, and we need that as soon as possible. Thank you very much.

Chairman COLLINS. Thank you. Senator Voinovich.

OPENING STATEMENT OF SENATOR VOINOVICH

Senator VOINOVICH. Thank you, Madam Chairman, for holding this series of hearings on chemical facility security. I compliment your diligence in examining the issue. I look forward to a bipartisan legislative effort to ensure that our Nation's chemical sector is secure from the threat of terrorist attack.

The chemical industry is a critical component of our Nation's infrastructure. It is massive, impacting every facet of our daily life. The scope and complexity of the chemical industry warrants careful consideration of any new security initiatives.

During the first hearing of this Committee on April 27, we heard alarming statistics that warned of a devastating loss of life in the event of a terrorist attack against a major chemical facility. Senator Lieberman, in your opening remarks, you made reference to the threat that is there.

In the hearing on June 15, we heard from the Environmental Protection Agency and the Department of Homeland Security regarding the safeguards that have already been implemented industry-wide. I think we must recognize that there has been a lot of legislation addressing safety at chemical facilities.

I recently hosted a round table discussion in Cincinnati and included local officials, law enforcement, and emergency response personnel. I was informed of the Community "Right-to-Know" laws, which require companies to disclose what is inside their facilities, assess the potential risk, and develop a response plan. So there has been a lot of work on the local level and by the industry that we should take into consideration when we pursue this legislation.

Today, we are going to hear differing views on how the Federal Government should best secure the chemical industry. I look forward to learning the perspectives of each party as we begin to debate the Federal role in securing this vital sector.

Though the risk of terrorism is serious, as last week's horrific attacks on London's transportation sector demonstrate, I must reiterate my belief that the Federal Government cannot protect against every potential threat that we can possibly conceive of in this country. Doing so would bankrupt the Nation. I would like to state publicly that one of the stated goals of the terrorists, the people who have announced that they would like to do us harm, is that they want to hurt our economy. We should learn the lessons of the Cold War. The Soviet Union bankrupted themselves trying to protect against whatever the United States might do to them.

So as we address the issue of chemical facility security, I think as a Nation, we need to take into consideration just how we are going to handle this. We must be wary of throwing money at this issue. Further, if we require that the industry incur the cost of enhanced security, it will have a horrific impact on the economy. I would like to emphasize the importance of a balanced approach between self-regulation by industry and more proactive Federal action.

Industry leaders like the American Chemistry Council and the National Petrochemical Association should be commended for building a strong foundation for chemical safety. It is my hope that the significant safety measures developed by industry will be incorporated into legislation and built upon. Likewise, we should carefully evaluate the laws already on the books and seek to enhance those relevant to chemical security.

As we further explore the issue, I would like to iterate four points. First, efforts to enhance the security of our facilities should be sharply focused on prevention, protection, and consequence management of potential terrorist attacks.

Second, Federal action to address chemical facility vulnerabilities must not be burdened with extraneous issues.

Third, critical information must be protected from unnecessary public disclosure, providing it only to responsible government authorities that need to have access to such information.

And fourth, Federal action should be based on risk and vulnerability. In other words, security considerations should be based on factors such as potential for adverse economic impact and serious loss of life. A one-size-fits-all approach will not work for chemical security.

Finally, Federal legislation should adhere to a comprehensive cost-benefit analysis so as not to place industry at a competitive disadvantage. As my colleagues may know, the chemical industry is experiencing economic hardship as a result of natural gas costs. In fact, we have gone from a Nation that exported chemical products to a Nation that is now importing chemical products because of the high cost of natural gas. The industry is already under economic stress.

I think we ought to take all these things into consideration when we are putting this legislation together. Thank you, Madam Chairman.

Chairman COLLINS. Thank you. Senator Lautenberg.

OPENING STATEMENT OF SENATOR LAUTENBERG

Senator LAUTENBERG. Thank you, Madam Chairman, for convening this hearing, yet another on chemical security. As I look at the witness table, I just left a Durbin and now we face another Durbin. Welcome. Part of the family, right?

Mr. DURBIN. Indeed.

Senator LAUTENBERG. But we know you are objective and we welcome you. [Laughter.]

My concern about the security of chemical plants dates back to the late 1990s, when I introduced the first bill in Congress to deal with the problem. And while the industry has made substantial investments in trying to improve the safety around these plants, more obviously needs to be done.

Now, 2005, we are well past the time to start acting to confront the terrible risks that have not diminished, but rather have increased since September 11, and I commend Chairman Collins for calling this hearing.

In view of the devastating attack in London last weekend, it is clear that we can't let down our guard. But as the 9/11 Commission cautioned, we must not focus so much on the last attack that we fail to continue to develop our own strategy.

Since September 11, we have focused on the security of our aviation system. But the London attacks remind us that there are many other potential targets in our country, particularly chemical facilities. With over 15,000 chemical plants, storage facilities in the country, we have quite an array of facilities that under attack, could be devastating. More than half of these are located in areas where an attack could claim thousands or even millions of lives.

In my State, New Jersey, we lost 700 of our friends, neighbors, and loved ones on September 11. We all hope that we can prevent something like that from ever happening again. But as horrible as the attacks on September 11 were, most of the victims were adults, but this wouldn't necessarily be the case in an attack on a chemical plant, since an incident there could kill or injure thousands of innocent children at home or school. The Congressional Research Service has calculated that more than 8,000 schools or hospitals are near a chemical facility.

Now, according to EPA, the largest zone of vulnerability to widespread death and destruction is in South Carney, New Jersey. You know that New Jersey has an industrial past, and we welcome the jobs and the industry in our State. But in this particular area, it is believed that an attack on this chemical facility could kill as many as 12 million people. It is a densely populated area, the New York-New Jersey region.

The threat is clear and our response deserves some acceleration. New Jersey has some 1,600 chemical facilities within our State borders. Not a single one of these facilities is legally required to take any of the risk-reduction steps identified by experts at our hearing a few months ago.

Ignoring the threat of a chemical plant attack won't make it go away. So I urge my colleagues on this Committee, who I know are very committed to the issue, to try to move forward from this hearing toward a legislative remedy. I am not sure that we can legislate everything that we want. Senator Voinovich was correct. I mean, we can't disrupt an industry that provides so much good, keep it from operating efficiently or at costs way beyond their capacity. But we do have to protect our citizens where we can, and I thank all the witnesses who are with us and look forward to hearing their views. Thank you, Madam Chairman.

Chairman COLLINS. Thank you. Senator Carper.

OPENING STATEMENT OF SENATOR CARPER

Senator CARPER. Thanks, Madam Chairman, and to our witnesses, welcome. We are glad that you are here today.

I was sitting here listening to Senator Lautenberg talk about all the chemical plants that they have in New Jersey. At one time, Delaware was known, among other things, as the chemical capital of the world, with companies like DuPont, Hercules both headquartered there with a number of facilities there, as well. I don't think we ever had 1,500. You may have bragging rights there.

We are known for a number of other things. We are also known as the First State, the State that started the Nation, as well as the Nation's summer capital, home of tax-free shopping, Small Wonder. I expect we could go around the Committee here and ask for each of us to tell what our States are known for or famous for, and we could all do that. And we may not be the chemical capital of the world, but we have a great deal of interest in the security of the chemical plants that we do have.

In Delaware, we have a bit of a reputation for being able to get things done, for being able to work across the aisle, for using common sense. It is one of those rare States where actually Democrats kind of like Republicans and vice versa. It reminds me a little bit of this Committee. This Committee has a reputation for getting things done, and with the leadership of Senator Collins and Senator Lieberman, we do work well across the aisle. I am told they like each other, and frankly, we like them, too.

This is an issue whose time has come. There are other Committees that have sought to deal with this without a great deal of success. The ball has been punted, if you will, in our direction, and we are on the receiving end, and I am pleased to see that we are going to receive that ball and take the kickoff and run with it, and I look forward to providing some of that upfield blocking and maybe a lateral from time to time, and let us see if we can't get this ball in the end zone and provide, whether it happens to be the chemical capital of the world in Delaware or our neighbors to the East, a little greater security not only for the folks who are really living around those plants, but also those who are working there, too. Thank you.

Chairman COLLINS. Thank you, Senator.

Our first panel of witnesses represents some of the largest chemical industry associations. Our first witness will be Martin Durbin, the Managing Director of Security and Operations and the Senior Director for Federal Relations at the American Chemistry Council. ACC member companies are responsible for approximately 90 percent of basic industrial chemical production in the United States. We welcome you, Mr. Durbin.

I would also like to welcome Matthew Barmasse, Director of Environmental, Health, Safety, and Quality at ISOCHEM, Incorporated. Today, however, he is here representing the Synthetic Organic Chemical Manufacturers Association. He has more than 25 years of experience in the chemical industry and will provide this Committee with the perspective of how a smaller company like ISOCHEM has improved security.

Last, I would like to welcome Bob Slaughter, the President of the National Petrochemical and Refiners Association. The NPRA has more than 450 member companies, including virtually all the refiners and petrochemical manufacturers in the United States. So we welcome you, as well.

Mr. Durbin, we are going to begin with you.

TESTIMONY OF MARTIN J. DURBIN,[1] MANAGING DIRECTOR, SECURITY AND OPERATIONS, AMERICAN CHEMISTRY COUNCIL

Mr. DURBIN. Madam Chairman, Senators, good morning. My name is Marty Durbin, and as the Managing Director for Security and Operations at the American Chemistry Council, I appreciate the opportunity to provide testimony on behalf of ACC.

Allow me to directly address the question posed by this hearing: "What is the appropriate Federal role for chemical facility security?" On behalf of ACC, I am here this morning to repeat and continue the call we have made for more than 2½ years, and that is the need for legislation to set mandatory national standards for security at chemical facilities and provide the necessary regulatory authority to the Department of Homeland Security to ensure this critical part of our national infrastructure is protected.

ACC represents more than 130 of the leading companies in the U.S. chemical manufacturing sector, and as noted, we are responsible for nearly 90 percent of basic industrial chemical production and are an essential part of our Nation's critical infrastructure. As many of you have noted, the products of chemistry are critical in many aspects of our lives, from cleaning our drinking water to supporting agriculture and spurring medical innovations to prevent and treat disease.

In my brief remarks, I would like to highlight the following. First, the leadership role that ACC members have taken to further ensure the safety and security of their products, their facilities, the supply chain, and the communities in which they operate, an in-

[1] The prepared statement of Mr. Durbin appears in the Appendix on page 91.

vestment to date of more than $2 billion in security since September 11.

Second, the great strides we believe have been made by the Federal Government and our industry, cooperatively, to secure the chemical sector.

Third, the real need for Federal legislation to provide nationwide assurances that all portions of the industry take the same aggressive actions that ACC members and others are taking.

And finally, our views on the important and often misunderstood subject of inherent safety.

Security isn't new to our members, but the tragedies of September 11, 2001, brought swift and decisive action from the industry leaders of our association. Without waiting for government direction, ACC quickly issued site and transportation security guidelines in October and November of that year, after which ACC's Board of Directors launched an aggressive effort to develop a new, Responsible Care® Security Code. Implementation of Responsible Care®, which is ACC's signature program of continuous improvement in environmental, health, safety, and now security performance, is mandatory for our members.

The Responsible Care® Security Code and ACC member security enhancements have been widely and uniformly acknowledged by government and security experts. State and local governments have used the code as a model for their own regulation of chemical facility security, and the U.S. Coast Guard, which regulates security for nearly 240 chemical facilities under the Maritime Transportation Security Act, recognized our Security Code as an alternative security program for ACC members.

The Security Code itself required each of our member companies to take the following four steps broadly. First, they had to prioritize every facility by risk.

Second, they had to assess the vulnerabilities using methodologies that were developed by Sandia National Laboratories and the Center for Chemical Process Safety, which is a program of the American Institute of Chemical Engineers.

Third, they then had to implement security enhancements commensurate with the risks that were identified by those assessments and taking into account inherently safer approaches, engineering, and administrative controls and other security prevention and mitigation measures.

And finally, they had to verify the implementation of those physical security measures using third parties that are credible in the local community, such as first responders and law enforcement officials. All ACC member company facilities have completed their vulnerability assessments, implemented security enhancements, and to date nearly all have had those enhancements verified.

The ACC Security Code also covers transportation and cyber security. It allows our members to extend the reach of the code throughout the physical and virtual value chain. Separate guidance documents were developed to assist members in implementing the code with those companies who transport our products, including rail, truck, and barge.

Specific to cyber, our members lead an industry-wide cyber security program that has developed guidance documents and a broad practices standards and technology initiative. We believe our members provide a model to other industries with similar automated systems. Some of our members' cutting edge facilities, in fact, have hosted visits by staff from DHS and this Committee, and we have received very positive reports.

All of the guidance materials I have mentioned addressing site, transportation, and cyber security, as well as the code itself, are publicly available through our website so they can have the broadest possible effect beyond our membership.

Now, turning to our partnership with the Federal Government, the Homeland Security Presidential Directive Number 7 specifically names DHS as the lead or sector-specific agency for the chemical sector. To achieve the infrastructure protection objectives of that directive, ACC and its members have worked in close partnership with DHS over the past years, facilitated site visits to our member facilities, and participated in their Buffer Zone Protection Program that provides support and resources to local governments.

We created, fund, and maintain the Chemical Sector Information Sharing and Analysis Center, a two-way 24/7 communications tool between DHS and the chemical sector, which we operate as a public service through our CHEMTREC program in cooperation with DHS.

We participate regularly in exercises and drills at all levels, from facility-based emergency preparedness and response drills to the recent national level TopOff 3 exercises.

We also facilitated development of the Chemical Sector Coordinating Council, a group of 16 leading trade associations that coordinates communication between DHS and our sector for purposes of infrastructure protection. In fact, all three organizations represented on this panel are members of that Council.

Along with others in the sector, we are working with DHS to develop tools and methods to help intelligently allocate protective resources on a risk basis. That is not to say everything is working perfectly in our relationship with DHS, but we are all learning together, and we have made great strides to improve the partnership between our sector and the agency, and we have established a constructive relationship that will allow for even better things as we move forward.

So why is Federal legislation necessary? Despite all the progress that has been made to date, there is no way to assure that all chemical facilities that need to be protected are taking the same kinds of aggressive steps that ACC members have taken to protect this critical sector. No doubt, many non-ACC members have also taken appropriate steps, and they should be commended. But as highlighted by DHS Assistant Secretary Stephan at last month's hearing, there are high-risk facilities that have not.

ACC has led the effort to ensure that all chemical facilities are secured against the threat of terrorism. We have worked continuously with Congress and the Administration for enactment of national security legislation that will first establish national standards for security of chemical facilities. We agree with DHS that those standards should be risk-based, reasonable, clear, and equitable, and that they be performance-oriented in a way that will provide flexibility to facilities.

Second, require those identified facilities to conduct vulnerability assessments and implement security plans.

Third, provide oversight, inspection, and enforcement authority to DHS.

Fourth, protect sensitive security information.

And finally, recognize responsible voluntary efforts. Naturally, we believe that any Federal legislation should enable DHS to give credit to ACC members for their substantial actions and investments to implement the Responsible Care® Security Code. As witnesses at your April hearing concurred, ACC members deserve a level playing field and a common set of expectations. But let me be clear. We are not asking for an exemption from the law, only that DHS be allowed to recognize our members' significant actions just as the Coast Guard has already done.

Without Federal action on this vital topic, State legislatures will fill the void. Both Maryland and New York have already enacted chemical facility security laws. And while ACC was able to support both of those statutes, we strongly believe that a national program, not an incomplete patchwork of potentially conflicting State efforts, is necessary.

Finally, Madam Chairman and Senators, in the debate over chemical security, no issue has proven more controversial than the role of inherent safety. Because of ACC members' deep investment in this issue, I want to spend the balance of my time explaining our views and why we feel so strongly about them.

In a nutshell, inherent safety means designing a process to minimize hazards in the first place rather than managing and controlling them with protective equipment or procedures. This concept was invented by the chemical engineering profession and our industry has long embraced it. Under the Responsible Care® initiative, inherent safety is a key element in the design and modification of facilities and job tasks. Our members continually conduct process hazard analyses of our facilities, and those analyses can lead us to change processes, modify procedures, or substitute materials to reduce and manage risks. And, as I noted earlier, the Responsible Care® Security Code mandates that our members take inherently safer approaches into account in assessing possible security measures.

I cannot overemphasize, however, that inherently safer chemical processing requires considering all the risks potentially associated with a process. Inherent safety typically involves making very challenging judgments to ensure that risks are not unwittingly shifted or substituted and that overall risks are reduced.

Many inherently safer approaches involve trading one risk against the potential of another.

For example, advocates of inherent safety frequently speak of reducing onsite inventories or reducing or eliminating storage of hazardous materials. While that may be appropriate, reducing inventories at a facility may also increase the number of truck shipments through a neighborhood. Similarly, replacing a low-temperature, low-pressure process that uses a toxic chemical with a process that uses a less-toxic chemical but operates at a higher temperature and pressure may increase the potential hazard to workers.

The challenge of trying to oversee inherent safety decisions is compounded by the complexity of chemical industry processes. Chemical companies make tens of thousands of products, and there are no standard processes for making them. To expect effective regulatory oversight in this area is unrealistic, at least without great difficulty, expense, and delay. In fact, in the Clean Air Act Risk Management Program rulemaking, EPA concluded that requiring and reviewing multiple process options at each regulated plant would not lead to greater advances in process safety.

Members and witnesses at April's hearing agreed on the importance of this legislation, and in Senator Voinovich's words at the time, any legislation must be sharply focused on security and not burdened with extraneous issues. We firmly believe that judgments about inherent safety are fundamentally process safety decisions that must ultimately be left to the process safety professionals. So mandating IST, we believe, should not be part of any security-focused legislation.

In closing, I just want to say that it has been nearly 4 years since September 11, and now is the time to act. So we welcome this hearing, and we are committed to continuing to work with this Committee and others to see that legislation is enacted in this session of Congress.

Thank you, and I would be happy to answer any questions.

Chairman COLLINS. Thank you very much. Mr. Barmasse.

TESTIMONY OF MATTHEW BARMASSE,[1] ENVIRONMENTAL, HEALTH, SAFETY, AND QUALITY DIRECTOR, ISOCHEM, INC., ON BEHALF OF THE SYNTHETIC ORGANIC CHEMICAL MANUFACTURERS ASSOCIATION

Mr. BARMASSE. Madam Chairman, Members of the Committee, my name is Matt Barmasse. I am the Director of Environmental, Health, Safety, and Quality for ISOCHEM, which is a small chemical manufacturer located in Western New York. My company mainly produces phosgene and phosgene derivatives, serving very diverse customers and markets, from pharmaceuticals to photographic products.

I am appearing today on behalf of the Synthetic Organic Chemical Manufacturers Association, also known as SOCMA. I appreciate the opportunity to speak with you about the appropriate Federal role in chemical site security. SOCMA is the leading trade association representing specialty and batch chemical producers, most of which are small companies. As a condition of membership to SOCMA, chemical companies must subscribe to Responsible Care® and its security code.

I will focus my remarks today on the nature of specialty chemicals and batch manufacturing, our relationship with DHS, EPA's Risk Management Program, and our perspective on Inherently Safer Technology.

Specialty chemicals are essential ingredients and building blocks for other products and perform very specific functions based largely on their molecular structures, which give them unique physical and chemical properties. Without these substances, nylon would not be

[1] The prepared statement of Mr. Barmasse appears in the Appendix on page 102.

strong enough to use for seat belts, medicine would revert back to what it was in the 1800s, and our Armed Forces would not have the modern equipment and supplies necessary to defend our country.

Because of their complex chemistries and narrowly focused applications, specialty chemicals are typically produced in small quantities, batch by batch. Most batch producers change products frequently, often on customer demand and short notice. This leads to frequent changes in the risk profile of the site. In many cases, batch producers are located in nondescript industrial or office parks with most of the processing equipment either indoors or out of view, making them difficult to recognize as chemical facilities.

Does this mean that my company and other SOCMA members feel that we should do nothing about security? Absolutely not. ISOCHEM conducted a security vulnerability analysis and accordingly enhanced its security policies and procedures. We spent over $750,000 to upgrade our physical and cyber security since September 11. And again, we are a small company.

I do believe, however, that a one-size-fits-all approach to security is neither appropriate nor feasible. Instead, SOCMA and its members support a tiered, risk-based approach.

SOCMA has established a strong working relationship with the Department of Homeland Security. DHS officials have met with SOCMA and its members on many occasions. SOCMA staff and member company experts are routinely consulted by DHS on technical issues and participate on DHS work groups, such as the team developing RAMCAP. SOCMA is a founding member of the Chemical Sector Coordinating Council, which also works closely with DHS.

DHS has also visited our site, providing valuable insight and constructive suggestions to enhance security. ISOCHEM has also been involved in our area Buffer Zone Protection Program, enabling our region to receive direct DHS funding for security upgrades. We are also participating in a RAMCAP pilot project which will be conducted over the summer. In addition, DHS is working with other Federal, State, and local agencies, trade groups, and individual companies to secure America's chemical facilities.

The Committee should be aware of other important efforts currently underway. State and local authorities are often in the best position to help secure our Nation's infrastructure, and there are many ongoing efforts to augment chemical site security. At the community level, we all have a mutual interest in mind. None of us want our communities to be attacked by terrorism.

In earlier hearings before this Committee, some have suggested that a number of RMP facilities are unwilling or unable to secure their facilities. While there may be some outliers, which are primarily small-scale chemical users rather than manufacturers, I am not easily convinced that they are very attractive terrorist targets. Simply put, the figures often cited by the press, 15,000 chemical facilities that put thousands or even millions of people at risk, are just not an accurate depiction of reality.

In fact, the RMP database, especially the worst-case scenarios, were never designed to be realistic. EPA and DHS officials have made this point repeatedly, and this has just been reaffirmed by the Congressional Research Service. Yet I repeatedly see RMP data used to scare people into thinking that the chemical industry is putting our communities at significant risk. This is both irresponsible and inaccurate. It is unfair to the chemical industry, DHS, and the local authorities with whom we work closely.

An important consideration missing from RMP methods include the safety systems in place at our facilities, our outstanding emergency response capabilities, residential and industrial building codes, and the realities of how hazardous materials behave when released, which will explain why we don't see Bhopal-like incidents occurring here in the United States.

That is not to say RMP data cannot be useful. While we believe that most facilities falling under the RMP program are not attractive terrorist targets, the list does provide a reasonable universe of sites to begin screening and prioritizing according to risk.

Inherently Safer Technology (IST) is probably the most misunderstood and controversial aspect of chemical site security. IST is a philosophy, it is not a technique, and it is certainly not a panacea for securing America's chemical facilities. Many non-scientists have been led to believe that the only way to achieve inherent safety is by substituting for the hazardous materials used in chemical manufacturing and processing. Application of IST, however, is bound by the laws of physics and nature. Physical laws place restrictions on what can and cannot be done when trying to make a chemical. In chemistry, reactive substances must be used to form new molecules and many reactive chemicals are, by their very nature, hazardous.

Where hazardous chemicals are used, they are highly regulated by EPA and OSHA and appropriately managed by chemists in universities, government, and industry. The fact of the matter is that scientists cannot produce the materials that make our standard of living possible without using very specific chemicals.

Making medicine is a good example. Phosgene is a key building block for an important starting material in a pharmaceutical application. The structure of phosgene allows for transfer of atoms that is clean, meaning that it does not allow side reactions to occur that would contaminate the compound with potentially toxic byproducts. Using phosgene helps secure the safety of medicines used to treat diseases, such as MS.

Another important factor is the potential for transferring risk from one area to another. For example, if the amount of a chemical stored onsite is reduced, the only way to maintain production schedule is to increase the number of shipments to the site, which increases the transportation and transfers the risk.

The very nature of hazardous chemicals provides important economic incentives for companies to use the safest and least hazardous chemicals possible, including reduced accidents, cheaper transportation and disposal costs, cheaper insurance rates, fewer government regulatory requirements, and avoidance of facility down time.

With all these incentives in place, the question becomes why do chemical companies still use hazardous materials? The simple fact is that the law of physics and nature are much larger drivers than anything else. No Federal program mandating IST will change the science of chemistry. Instead, such a program would result in nothing more than a burdensome paperwork exercise forced on companies just to justify their scientific methods and decisions while doing nothing at all to enhance security.

As noted earlier, chemical sites are extremely diverse as are the chemistries that take place within our facilities. Because of this, a one-size-fits-all approach to security of chemical facilities with prescriptive standards just will not work, nor will attempting to mandate Inherently Safer Technologies.

SOCMA and its members support a tiered risk-based approach to security that begins with a mechanism to screen and prioritize sites and concentrates further work on areas with the greatest degree of risk. Any Federal oversight of security in a chemical sector needs to account for the significant voluntary efforts already undertaken. It should also use performance-based fundamentals that provide the flexibility needed to implement effective site-specific programs.

Key elements of such a program include a clear definition of covered entities and any exceptions; recognition of past efforts and voluntary programs that are substantially equivalent to DHS requirements; flexibility in achieving compliance; compliance assistance for small companies; risk screening for prioritization across covered facilities; DHS approved security vulnerability assessments for higher-priority sites; Federal preemptive authority for DHS; retention of security plans containing critical infrastructure information with availability to DHS upon request; and finally, recognition of efforts by the regulated community under other security programs.

Madam Chairman, Members of the Committee, thank you for your consideration of SOCMA's perspective of these important issues, and I am happy to answer any questions you have about my testimony.

Chairman COLLINS. Thank you. Mr. Slaughter.

TESTIMONY OF BOB SLAUGHTER,[1] PRESIDENT, NATIONAL PETROCHEMICAL AND REFINERS ASSOCIATION

Mr. SLAUGHTER. Thank you very much. Madam Chairman, Senator Lieberman, and other Members of the Committee, my name is Bob Slaughter. I am President of the National Petrochemical and Refiners Association.

NPRA's member companies constitute an extremely broad representation across two industries, the petrochemical industry and the refining industry, as well as their suppliers and vendors. On behalf of our members, I do want to begin by thanking you for the opportunity to appear today and for holding this important hearing, as well as for the very balanced and fair opening statements.

We would like to offer the following summary of our complete testimony. Maintaining the security of our facilities has always been a priority at refineries and petrochemical plants. It is job one. It simply has to be that way. Our industries have long operated globally, often in unstable regions where security is an integral part of providing for the world's energy and petrochemical needs.

[1] The prepared statement of Mr. Slaughter with attachments appears in the Appendix on page 119.

After the occurrence of the tragic events of September 11, those industries realized, as did everyone else, that additional threats had to be taken into account to secure the critical assets that we own. Our members began implementing additional and far-reaching measures to address these new threats, and you have asked what are some of those steps.

We developed, along with our sister association, the American Petroleum Institute, a peer-reviewed Security Vulnerability Assessment methodology especially attuned to the needs of refining and petrochemical industries. The Department of Homeland Security has endorsed this methodology and, in fact, uses it in instances to train its own people.

Under that methodology, you analyze a facility to determine the vulnerabilities. You identify potential threats. You identify potential security vulnerabilities. You determine the risk by measuring the likelihood of an attack and the consequences, and you recommend appropriate incident mitigation and countermeasures. You identify the appropriate security measures and incorporate them in a security plan addressing the SVA findings, which is then implemented.

Our members have conducted security vulnerability assessments pursuant to these plans, and they have prepared and implemented facility security plans in response to the findings. In 2004, the SVA methodology was extended to transportation-related activities, including pipelines, rail, and truck transportation.

We developed an extremely close working relationship, as well, with key Federal agencies, as well as State and local law enforcement officials, to obtain and exchange critical information. We are actively partnering with DHS on many important security initiatives, including the development of the Risk Assessment Methodology for Critical Asset Protection, or RAMCAP, the Homeland Security Information Network, HSIN, and the Buffer Zone Protection Plan, among others. Other groups that we work with include the FBI, the Department of Transportation, DOE, the Department of Defense, the CIA, the Government Accountability Office, and, of course, the Department of Homeland Security and its various components, particularly the U.S. Secret Service, Transportation Security Agency, and Coast Guard.

We have held joint training exercises simulating terrorist attacks on numerous occasions with both Federal and State officials. We have developed training programs involving Federal and State Government officials. We have shared best security-related practices among large and small companies that constitute our diverse membership at NPRA meetings and conferences. We have held five national security conferences involving large numbers of companies in both industries since 2001. Again, they have shared best practices, they have heard from experts, they know what the state of the art is when it comes to security practices.

Our members, like others, have complied with the 2002 Maritime Transportation Security Act. The Coast Guard has jurisdiction over a majority of the 150 refineries and 200 petrochemical manufacturing facilities in the United States. SVAs and plans have been submitted to the Coast Guard. They have been reviewed and approved. Companies have designated Facility Security Officers to oversee implementation. Quarterly drills are required to test the elements of these plans.

Companies themselves have taken strong new security measures. They have reconfigured sites. They have set critical assets back from perimeters and installed electric intrusion detection systems, implemented card access controls using biometric technology. They have acquired enhanced security community systems, shared security response plans with local law enforcement and appropriate Federal agencies. They have conducted drills and exercises to test security and response plans, and hired additional security personnel. There is an even more complete list of this, which in itself is still partial, in the filed testimony.

You have asked for NPRA's position on legislation. We do not oppose reasonable chemical security legislation and regulation. However, the existing system, we believe, is working well and care must be taken to do no harm to current efforts in fashioning your ultimate product. Although we have not advocated legislation, we realize this Committee and DHS have both announced support for new regulatory authority, and in response, we have developed some principles that we hope the Committee will consider and adopt in Federal legislation, and we look forward to working with you on that.

Our principles are, you need to be prudent in fashioning what could amount to a significant additional and costly mandate on America's scarce refining and petrochemical facilities. There has been a lot said about how scarce our refining facilities are in the United States. We have not built a new refinery in the United States since 1976. So security needs to be maintained at these facilities, but we have to have an eye toward the impact on their survivability and the maintenance of these facilities in the United States.

The same with petrochemicals. As Senator Voinovich pointed out, the petrochemical industry has been under intense pressure on natural gas prices in recent years, so no one wants to compromise security, but requirements need to be reasonable. These are scarce assets and necessary to national security.

We hope you will try to maintain the close and highly productive relationship that currently exists between the Department of Homeland Security, other Federal, State, and local governmental bodies, and the refining and petrochemical industries. That relationship is largely responsible for the success of security programs in those industries thus far. We are concerned about the impact of new legislation on this productive relationship. The dynamics of the relationship could be affected and the current level of information sharing could be diminished and that would not be productive, and we hope you will keep that in mind as you fashion your legislative product.

We hope that you will use MTSA as the model for any new security legislation. It has clear performance-based requirements. Essentially here, we are talking about support for a tiered approach based on risk. We favor reliance on Security Vulnerability Assessments and responsive facility security plans with exercises, documentation, reporting procedures, and audits, protection, above all, for sensitive security information.

We think there should be self-assessment and auditing. We have had good experience with Coast Guard jurisdiction. We would assume that you would set up a Department of Homeland Security jurisdiction for facilities not subject to Coast Guard jurisdiction. We think that a facility that currently is partially covered by the Coast Guard should be able to opt in its whole facility if it chooses. We hope you will preempt other Federal and State programs so there aren't a lot of overlapping requirements that will make it difficult to comply and understand what the rules are.

We hope you will credit companies for security programs already implemented by companies. We have not developed and marketed a proprietary NPRA program for our members. We have tried to let them know what the state of the art is. We have some of the largest meetings in the world in the petrochemical and the refining industry, and we have invited folks to come in and talk about their programs, including ACC, so that our members will know what is available. We let them make their own choice.

We hope you will help companies with background checks, to define the criteria for denying access to facilities, and hopefully allow companies to access and utilize government resources and databases in making employment decisions.

Again, we hope you will require DHS to develop a tiered risk-based approach to regulate chemicals and facilities.

We were very much encouraged by the DHS statement before this Committee and others that they are developing core principles based on risk, reasonable, clear, equitable and enforceable security standards, ones that recognize investments and the progress that companies have made so far. We are committed to continuing that progress however we go forward.

So just to conclude, I want to underscore again that refiners and petrochemical manufacturers take very seriously the responsibility to maintain and strengthen security at facilities. We urge the Committee to fully consider the impact of legislation on existing programs and practices. Please use MTSA as the template for developing new chemical security requirements and embrace and support the core principles outlined by DHS at this Committee's June 15 hearing.

I am happy to answer any questions the Committee may have on our testimony. I want to thank you again for offering us the opportunity to be here today.

Chairman COLLINS. Thank you.

Mr. Durbin, you testified that compliance with the Responsible Care® Security Code is mandatory for ACC members. First, could you explain to us how ACC monitors compliance with the code, and second, what would your suggestions be for compliance measures to be included in the legislation that we will be drafting?

Mr. DURBIN. Senator, for the Responsible Care® program overall and the Security Code, we have set the guidelines for the companies to follow within the code and they self-assess. And in the case of the Security Code, they actually had to report to a third party that they had completed the steps that I outlined. And again, if they had not done that, they had not met those guidelines within the code, then we have a governance process that would first try to bring them into compliance, and if not, make clear that they are no longer eligible for membership.

With regard to compliance within legislation, again, I think, clearly, we have stated very clearly that there should be clear oversight, inspection, and enforcement authority for DHS. All that we asked, just as the other witnesses have, as well, is that we give DHS the ability to look at work that has been done through programs such as those that have been cited and determine whether or not they are essentially equivalent to those regulatory programs, and if so, let us not force companies to duplicate efforts that they have already made.

Chairman COLLINS. Mr. Slaughter, in your written testimony and again this morning, you have cited the Maritime Transportation Security Act as a model that this Committee could use in drafting chemical security legislation. Under that law, the Coast Guard has the authority to shut down a facility if the Coast Guard determines that the facility has not established sufficient security measures. In fact, the law prohibits a facility from operating unless it has submitted and is in compliance with a security plan approved by the Secretary of Homeland Security. Would you support giving the Department of Homeland Security similar authority to shut down chemical facilities that the Secretary determines have not taken the necessary steps or security measures that the Department deems necessary?

Mr. SLAUGHTER. Obviously, any regulatory entity, Madam Chairman, has got to have ultimate authority to enforce its requirements. I think you have to hope that any regulatory authority will use wisely whatever authority they are given, and I don't believe that anyone in the industry would be disinclined to grant that as the ultimate authority to the Coast Guard under MTSA. But again, one would hope there would be a number of steps and the good working relationship has been set up with the Coast Guard and DHS, but that is probably one aspect of that regulation, yes.

Chairman COLLINS. Mr. Barmasse, more than 3 years ago, the CIA first alerted us to the possibility of an al Qaeda attack on chemical facilities, and since that time, many experts both inside and outside of government have warned the industry that you are a potential target. That is different, however, from knowing the specifics, from knowing that there is a specific plant that is at risk or a specific plot against a particular sector.

I am curious about the flow of communication and information sharing between the Department and smaller companies like yours. I suspect that the Department has a very close communication and working relationship with the ACC and with larger industry players. But could you tell us how a threat that would involve plants that are smaller, like yours, would be conveyed and assess for us the extent of communication and information sharing between the Department and the smaller manufacturers?

Mr. BARMASSE. We have been very pleased with the flow of information from DHS and through the chemical sector, ISAC, which anybody can participate in to get that type of information on chemical site security. We signed up for that. We get notices and information on potential threats. And the Department of Homeland Security and their different offices within the Department have been very forthcoming with information and sharing information. They visited our site. They have assessed our security procedures that are in place. They provided valuable information on how to assess threats, and we have found that the information flow from them through the chemical sector, ISAC, has been very good, which all small chemical companies would be available to. So it has been a very good relationship to date and the information has flowed very well.

Chairman COLLINS. That is good to hear.

Mr. Durbin, one of the issues raised by witnesses at our previous hearings is that while 80 percent of the industry is complying with voluntary codes and has taken sometimes very expensive measures to improve security, there is a smaller percentage, possibly as much as 20 percent, according to the Department, that has not implemented the kinds of security measures that your members have embraced.

Are there competitive issues at play here? What I am thinking of is that a company that makes the investments, and they may well be expensive investments, to improve security may be put at a competitive disadvantage compared to a counterpart that does not make those investments.

Mr. DURBIN. Certainly. I think it is clear that we have—just speaking for ACC members, we can point to more than $2 billion worth of investment in security. That doesn't count what my counterpart organizations here at the table have also invested there. But while that is certainly a consideration, and something that I think from our members' standpoint, yes, we would like to see the playing field leveled and ensure that as we do move forward, we are not forced to make duplicative investments, the fact of the matter is our primary drive here is that you have a critical sector, critical part of this national infrastructure that has to be protected, and we have to have those nationwide assurances that the entire sector is acting in ways that it should.

Chairman COLLINS. Thank you.

Senator LAUTENBERG. Madam Chairman, may I make a request that questions be answered by the witnesses in writing? I have to go to another hearing.

Chairman COLLINS. Certainly. The hearing record will remain open for 15 days. Senator Lieberman.

Senator LIEBERMAN. Thanks, Madam Chairman. Thanks to the witnesses for their testimony this morning.

Let me ask this question. Despite some of the significant steps that the industry has taken, which you have testified to today, there have been media reports relatively recent that suggest an unacceptable level of access to some chemical facilities with dangerous materials. Most recently, the *New York Times* reported that the stretch of Northern New Jersey between the Newark Airport and Port Elizabeth, which has more than a dozen chemical plants and a lot of other potential targets—storage tanks, refineries, and pipelines—was very accessible to trucks. Apparently, you could drive within 100 feet of storage tanks. A *Times* reporter and photographer, and I quote here from the story, "found the plant only loosely guarded as they drove back and forth for 5 minutes, snapping photos."

This experience echoed previous incidences, which I am sure you are familiar with, including one highlighted on "60 Minutes" where reporters easily gained access to a chemical facility near Pittsburgh, which contained very toxic and explosive chemicals.

Given the work that the industry has done, how do you explain these incidences and what do they say to us about what more should be done? Mr. Durbin, do you want to start?

Mr. DURBIN. Sure. In the instances that were cited with "60 Minutes," if there is access to a facility, and certainly getting to the more sensitive areas of a facility, frankly, that is unacceptable, and I think that is why you have to have programs that are focused on making sure those things won't happen and why we as an organization have been calling for national legislation to make sure that we do have those kinds of standards set in place.

It is difficult to comment on other stories without knowing more details, but not all security preparations are obvious or visible. So I am reluctant to get into specifics on any one——

Senator LIEBERMAN. No, I understand——

Mr. DURBIN [Continuing]. And you are talking about public roadways and what have you. But in general, again, I think that this just points out why there needs to be a nationwide set of standards to be sure that all those facilities that ought to be taking these kinds of actions are doing so.

Senator LIEBERMAN. I appreciate that answer. Mr. Barmasse.

Mr. BARMASSE. And again, I am not familiar with the specifics of those, but we are also supportive of legislation that is reasonable and flexible for the risks associated with facilities. A facility like ours takes quite a few steps to make sure that our facility is adequately secured. We have gone through a lot of the risk assessments and worked with DHS to help identify those threats. And I think many of the small companies are doing similar-type things. So we would be very supportive of legislation that does provide those types of security.

Senator LIEBERMAN. Mr. Slaughter.

Mr. SLAUGHTER. Senator Lieberman, we work very closely through our NPRA Security Committee with our members, who go from the largest to the smallest of companies. I can tell you from what I have seen personally and what I have heard is that they are extremely sensitive to problems such as were discussed in this particular article, which I also have read.

Senator LIEBERMAN. Right.

Mr. SLAUGHTER. And we certainly have sent a very strong message, and they have heard it and they have heard it from others, that this is unacceptable behavior. So it is difficult to determine—the company names I have seen are people who are not our members, but sometimes you don't see them. But this is behavior which seems to be very different from what we are seeing in our members who are watching to see if anyone takes pictures of the facility or anything. So it is difficult to determine who the outliers are. All of us are united here in efforts to get the information about best practices out and to see that they are enforced.

Senator LIEBERMAN. OK. I appreciate the answers. I think you draw the same conclusion I do, which is that these stories, generally speaking, speak to the need for national standards and for legislation.

Mr. Durbin, let me ask you this. After September 11, I know that your organization added a security requirement to the Responsible Care® Security Code that requires facilities, and I applaud this, to conduct a vulnerability assessment and then prepare and implement a security plan. There is third-party verification of plan implementation. However, the third-party review consists of verifying that the chemical facility took the steps outlined in the security plan, but it doesn't conduct an independent assessment of whether the plan is adequate to the threat.

Is there a need for a truly independent assessment of the sufficiency of the security measures taken in our Nation's chemical facilities?

Mr. DURBIN. You are absolutely correct, Senator, in your explanation of the verification process, and that is how it was set out when the code was developed. At that time, the overall program was moving from one of a separate set of codes to what is now the Responsible Care® Management System. So we put the code in place and the verification piece that you described in place in the interim.

Now, as we move forward, we are moving to RCMS, modeled on ISO 14,000, where there actually will be third-party certifications and audits of companies that will encompass everything they have done in the environmental, health, safety, and security area. So moving forward, there will be those independent third-party auditors coming in to certify that they have taken appropriate actions.

Having said that, we were also working toward trying to get a government role that would help to assure that the actions taken were indeed up to the measure on whatever the national standards are that would be set.

Senator LIEBERMAN. OK. My time is up. Thank you for that answer.

Chairman COLLINS. Thank you. Senator Voinovich.

Senator VOINOVICH. I have been thinking about this from a perspective of a former governor and former mayor, and I am wondering how you get all of this done? Specifically, what percentage of the industry is covered by MTSA?

Mr. SLAUGHTER. For refining, it is over half of the refining facilities and probably over half of the petrochemical facilities, as well, Senator Voinovich. They tend to be located close to coasts and large navigable waterways.

Senator VOINOVICH. OK. How much different is the MTSA regulations as to the Responsible Care® Security Code? How similar are they?

Mr. DURBIN. Actually, Senator, they are very close. In fact, as I noted, the Coast Guard was given the authority to look at programs like Responsible Care® and determine whether or not they were substantially equivalent. We worked with them over about a 6- to 9-month period to walk through their regulations and our program, and at the end, the Coast Guard was willing to declare that the Responsible Care® Security Code was an alternative security plan for complying with MTSA. They did require each facility to provide some additional information on what they will do when we raise the alert levels in the port, but overall, our companies did not have to go back and redo vulnerability assessments——

Senator VOINOVICH. So from the Committee's point of view, if we looked at your Responsible Care® Security Code and looked at the MTSA regulations, that could give us a nice picture of what we should be doing in terms of regulation. Now, does the Coast Guard verify that MTSA is being carried out?

Mr. SLAUGHTER. Yes, Senator.

Mr. DURBIN. Yes, sir.

Senator VOINOVICH. From your perspective, is it pretty conscientious?

Mr. DURBIN. It is extremely conscientious. It is one of the, frankly, rare times that our members say that a Federal agency is extremely conscientious, and also, they have a very good relationship with that group.

Senator VOINOVICH. How do your recommendations differ from one another? If you read the testimony, you are almost all in sync about what you think the legislation should look like. You all agree that there should be national legislation. So how much different, in terms of your consensus of the legislation, is it from what the Department of Homeland Security has suggested as the kind of legislation that needs to be implemented? Is there a wide discrepancy? I have asked my staff to look at that, but from your perspective, how far off are you?

Mr. DURBIN. Again, just responding to what we have heard so far, what Assistant Secretary Stephan laid out in his testimony and from discussions with them, so from the broad context, I think we are very supportive of the approach that they are taking on this. Again, nothing specific to respond to yet, but very supportive of the structure they have laid out.

Senator VOINOVICH. It would be interesting to get from DHS their opinion about what the industry folks are recommending in terms of the legislation.

The other issue, then, is the bureaucracy. I understand that the Coast Guard is responsible for the facilities or navigable water. What bureauacy do you suggest should monitor the rest of the facilities?

Mr. SLAUGHTER. That is correct, and we suggested DHS outside of the Coast Guard.

Senator VOINOVICH. It appears that the Coast Guard is a good role model for them to follow.

Mr. SLAUGHTER. Right.

Senator VOINOVICH. The other issue, of course, is Inherently Safer Technologies. The concept that reduction or elimination of particular chemicals or alternative approaches will lessen the threat. What is your opinion on I.S.T?

Mr. SLAUGHTER. I would say we have concerns about an IST requirement, Senator Voinovich, because a lot of—there are great incentives to go to Inherently Safer Technologies if they are effective and practical today. But if you get into a situation where it is mandated and you get into an extensive review process as to why didn't you do A, B, C, and D instead of what you are doing, this whole program may be very difficult to implement and be very problematic for everyone and just be a papermaking exercise, as the SOCMA testimony pointed out.

Mr. DURBIN. I would echo those comments. I think the Inherently Safer Technology is clearly something that our member companies, this industry really drives toward, but it does not lend itself to a regulatory approach.

I believe one of the Senators in your opening statements talked about the dichotomy between those who just want physical and those who say you have to have this approach. I don't think it is that stark of a contrast here. If you are doing a meaningful vulnerability assessment that has a meaningful methodology behind it, that is going to point you in that direction toward process changes as well as other ways of managing risk.

For example, the GAO report responding to Senator Byrd that was provided in March, they visited ten ACC member company facilities. Seven of those facilities noted that they made process changes as part of their security enhancements.

Senator VOINOVICH. My time is up. Thank you.

Chairman COLLINS. Senator Carper.

Senator CARPER. Thanks very much.

I have a couple of questions. One, I find it helpful with a panel like this where there is a fair amount of consensus, before you wrap up, just to come back again and tell us where you think the consensus lies among the three of you. A follow-up question is going to be, where do you disagree?

And then I think I am going to ask you to sum it up by saying, again, the purpose of this hearing was what is the appropriate Federal role, and I am going to ask you to sum up again and say this is what we believe, each of you, this is what we believe the appropriate Federal role is.

So if you could, Mr. Durbin——

Mr. DURBIN. Sure. At the risk of speaking for my colleagues——

Senator CARPER. Where is the consensus, what are the differences, what is the appropriate Federal role?

Mr. DURBIN. The consensus I am hearing here this morning is that the Federal role that is put in place needs to be a risk-based tiered approach that will set national standards to ensure that everyone in the chemical sector that has been identified is taking the appropriate steps. But again, it needs to be a risk-based program that is reasonable, clear and measured, and provides some flexibility, and also recognizes the efforts that have already taken place within the industry.

Mr. BARMASSE. And I agree with that, and I would like to add a few things to that——

Senator CARPER. Go right ahead.

Mr. BARMASSE [continuing]. Especially for the smaller facilities and smaller chemical companies that may not have the expertise of the larger companies. I think Small Business assistance or compliance assistance is going to be a very important component of anything that is drafted, and so I believe that is the extent of my additional comment.

Senator CARPER. All right. Mr. Slaughter.

Mr. SLAUGHTER. We also would agree that it is very important that everything rely on a tiered, risk-based approach, which is what DHS apparently is talking about. I suspect where there may be a little bit of disagreement is that, I think the impression is left sometimes that industry has not focused on this issue and done a great deal of work. We have.

I would say at the same time there are competitive issues here. I think we need to have a flexible program that fits requirements to facilities and responds to the risk and threats at that particular facility. If large companies can make certain investments but they go beyond what is necessary to secure facilities that may be owned by someone with less capital, we don't want to lose facilities in the petrochemical and refining business unnecessarily. So rather than force everyone to do what the largest companies in the world are doing, we need to focus, as I think the MTSA does, on what does a facility really need to do rather than going beyond in any case. If there are competitive concerns, as the Chairman mentioned and questioned earlier, they run both ways, and I think a reasonable program will take care of both elements of competitive concerns.

And as I said before, we have not been advocates of Federal legislation. We have focused on working with our members to help them do everything they can do at their facilities. But given the position of the Committee, the position of DHS, we want to work with you to fashion reasonable requirements and look forward to working with you in that. And I agree with you, there is a substantial consensus at the table with just small differences and concerns.

Senator CARPER. Does anyone else want to mention differences, where you might differ?

[No response.]

OK. I will come back again to the issue of the appropriate Federal role with a specific focus on this Committee, if you will. Any closing thoughts?

Mr. DURBIN. Again, just to restate, the ACC believes there needs to be a Federal role. We believe DHS should play that role in coordinating the efforts of the Federal Government to protect this critical sector. They have worked very diligently with our sector. You have heard all three organizations talk about the good working relationship there, and I think that is absolutely the case.

Allow them to take that expertise that they have built over at the agency and that relationship and really put together and build a meaningful program that will also take advantage of not only the existing actions of the industry itself, but the existing actions of various Federal agencies that we all deal with on a day-to-day basis, not just EPA. We are talking about DEA and the Department of Commerce and Department of State, OSHA and what have you. Those are all the things that need to be coordinated.

Mr. BARMASSE. I would like to add that I have a legitimate concern that, being in New York State, there is New York State security legislation drafted, and if there are vast differences between Federal and State legislative activities, it could conceivably require us to spend a lot more time, effort, and money to comply with two totally different types of programs, and we would be supportive of Federal preemptive authority over the State programs so you don't have to do two totally different things.

Senator CARPER. Any last comment, Mr. Slaughter?

Mr. SLAUGHTER. Well, Senator Carper, I just say that the real trick in doing this will be not to harm the existing relationship that exists with DHS and industry. Particularly with DHS, the information flow is very good right now. There is a lot of understanding and it grows all the time—between the industries and DHS. If they become a regulator, you don't want to do too much harm to that relationship. The nature of it will change somewhat, but you want that information flow to be maintained and not to set up a purely adversarial relationship.

Senator CARPER. All right. One more real quick one. There are many times when safety and security actions mesh together well. There are some instances when security priorities have conflicted with safety. Are you aware of any times when we have had a conflict between the security priorities and the safety priorities?

Mr. DURBIN. One example that sticks out, more on the transportation side, was the use of placards for hazardous materials as they are being transported. The question raised is does that make it a target, or do you need to maintain that as the useful tool that it is for first responders and others that need that information in the event of an accident?

Our association very clearly agreed that placards should stay because they do play an important role for first responders, and the first responder community themselves said, until we come up with a better way of doing this, those need to stay on there. So that is the only kind of obvious conflict, but DHS clearly stepped in and resolved that, as well, and said they are staying on. We are not going to try to change that at this time.

Senator CARPER. Anybody else?

Mr. BARMASSE. The only thing I would add to that is that the protection of the information may be a conflict. The security-sensitive information and people's right to know what is going on at these facilities is a very important consideration. I think that information, it is very important that it is protected, kept within the chemical facilities and possibly with only DHS so that this information isn't publicly available beyond that and might pose another threat to the chemical facilities.

Senator CARPER. Gentlemen, thanks very much. Madam Chairman, thank you.

Chairman COLLINS. Thank you.

I want to give my colleagues the opportunity for one last question each to this panel before we go on to the second one. I understand from your testimony that each of you would oppose including in legislation a requirement involving Inherently Safer Technology, and Mr. Durbin, you have cited the complexity of the chemical processes. In addition, others have cited to me a fear of litigation resulting from the requirements.

But let me ask you a broader question. Do you think that the Department of Homeland Security should have any authority to regulate chemical processes, chemical use, or chemical storage? Mr. Durbin.

Mr. DURBIN. I believe that with regard to chemical processes, use, and storage, there are existing regulations in place. Our companies have to perform process hazard analyses as part of the PSM rule at OSHA and with RMP and——

Chairman COLLINS. If I could interject, just for a second. Those programs are not aimed at security. Those programs are aimed at enhancing worker safety or environmental health and safety. So they have a different justification. They may, in fact, help safety and security, but that is a different issue.

Mr. DURBIN. That is correct, but that is why it is important that you have a meaningful vulnerability assessment that would be required that would essentially point you toward and encourage the use of different technologies or things that you could put in place to change not only your process, but perhaps the way you distribute it and the way that your plan is configured. We have countless examples where our member companies have done just that to address security issues.

Chairman COLLINS. But should the Department be able to require a process change if the vulnerability study indicates that this is an issue for a particular facility?

Mr. DURBIN. I think I could only answer that by saying we would have strong concerns about the agency making those types of decisions, as to what process should or shouldn't be used or what material should or shouldn't be used. I think we should use that authority to really drive companies toward finding those solutions.

Chairman COLLINS. Thank you.

Mr. Barmasse, same question for you. Should the Department have any authority in this area?

Mr. BARMASSE. I think the Department's expertise is going to be in the area of security and not chemistry, and it is going to be very difficult for security experts to have the expertise to understand how to regulate what goes on in a process. Chemists and scientists spend a tremendous amount of time trying to understand their process, and they develop these processes in the safest manner they can. And security experts would have a difficult time understanding the intricacies and the complexities of a chemical process and be able to make any meaningful suggestions or recommendations on that. So I think it is outside the realm of their area of expertise.

Chairman COLLINS. Thank you.

Mr. Slaughter, what about a requirement that companies have to consider Inherently Safer Technology, which is different from having the Department mandate specific chemical processes?

Mr. SLAUGHTER. My answer, I am sorry, is somewhat hackneyed, is that the devil is in the details on that one because—the devil is in the details because the question is, how is that written? What is reviewable? I mean, you can end up in the exact same place just with that type of requirement as you can actually giving them authority to mandate changes in processes.

I agree that the SVA methodology and process will lead to information about potential problems and a dialogue with the regulator. But I think we would have significant concerns about either type of provision being included in the legislation.

Chairman COLLINS. Thank you. Senator Lieberman.

Senator LIEBERMAN. Thanks, Madam Chairman. Thanks again to the witnesses.

My question goes to the interaction of the chemical industry with local governments, and I would just ask each of you to respond briefly. Mr. Durbin, first, if you might, I am interested in knowing whether the Responsible Care® Security Code includes a requirement to conduct drills and exercises or interact in any way with first responders and local communities.

Mr. DURBIN. Absolutely. In fact, it was one of the founding principles within Responsible Care®, the original Care Code, was the community awareness and emergency response. Frankly, that is one of the good stories here, is that after September 11, this is an area where we didn't have to start from scratch. Our member companies generally had very well established and good relationships with first responders in their communities. In fact, in many cases, you will find that the first responders work at our facilities. The volunteer firemen—some of our security directors happen to be the deputy sheriff of the county or the fire chief of the neighboring community. So there is a very robust relationship that already existed there and drills that have been taking place all along. So this was just one more way of focusing our effort.

Senator LIEBERMAN. Mr. Barmasse and Mr. Slaughter, do you believe that the chemical facilities should have a role or a requirement to play in ensuring that the surrounding communities which they might impact have a well-functioning local emergency planning committee, and just briefly, because our time is going, what is your sense of the current relationship generally between the chemical facilities that you are involved with and the local surrounding communities?

Mr. BARMASSE. I would be happy to respond to that, and I would like to say that it is not just large companies that do those types of things. It is small companies, also. We work very actively with local emergency planning committees. Previously, it was always on response to chemical accidents, but now, we have even worked with them and broadened the local law enforcement to provide security and vulnerability assessments from a security perspective.

The Buffer Zone Protection Program brought in State, local, and county law enforcement agencies to perform buffer zone protection analysis. We have had drills and we have had meetings with our local and county emergency planning committees that discuss just response to terrorism activities.

So I believe that the integration has already occurred in a lot of cases, and not just at the larger LEPC levels. It is happening with smaller companies and at the smaller level.

Senator LIEBERMAN. Good. Mr. Slaughter.

Mr. SLAUGHTER. I would agree that is the case with large to small companies across our membership in both industries, Senator, and I would also say that the State and local law enforcement personnel plus also first responders have been active participants in all the exercises that we have been doing for several years with Federal and State agencies on terrorist-related events.

Senator LIEBERMAN. I thank the three of you.

I think, Madam Chairman, that the testimony of this panel has been significant. I, at least, have not heard up until today this kind of clarity of statement that, while some progress has been made voluntarily and in other ways, that the status quo with regard to chemical security of facilities in America today is no longer acceptable, that there is a larger necessary and appropriate Federal role.

Now, obviously the question is, what is that role, and there are going to be a lot of disagreements about that. But most encouraging from your testimony today, I think we are all at the same table. The Administration is. Obviously, we are. And I presume that the representatives of the stakeholders on the next panel are. Under your leadership, Madam Chairman, I am more encouraged after hearing this panel that we are going to get something done in this critical area in this session of Congress. Thank you.

Chairman COLLINS. Thank you. I, too, want to thank this panel for excellent and very constructive testimony. We look forward to continuing to work closely with you. Thank you.

I would now like to call up our second panel of witnesses today. Our first witness on the second panel is Dr. Gerald Poje. Dr. Poje is a toxicologist by training and has years of experience dealing with safety issues in the chemical industry. Dr. Poje recently completed his second term on the U.S. Chemical Safety and Hazard Investigation Board, where he earned the distinction of the longest-serving member of that Board. He currently is serving on the National Academies of Science Expert Committee assessing the vulnerabilities of the Nation's chemical infrastructure.

Our second witness on this panel will be Glenn Erwin, the Project Director of the Triangle of Prevention, or TOP Program, and the Catastrophic Accident Investigator for the United Steelworkers. Mr. Erwin has more than 30 years of experience in the petrochemical industry and in particular with health and safety issues. The Steelworkers Union recently merged with PACE, the largest chemical workers' union in the United States, and we welcome you, as well.

And finally, we will hear from Carol Andress, who is an Economic Development Specialist for the environmental organization known as Environmental Defense. She has led Environmental Defense's work to foster pollution prevention and improve the public's awareness of chemicals in the environment, and we thank you for coming today, as well.

We are going to start with Dr. Poje.

TESTIMONY OF GERALD V. POJE, PH.D.,[1] FORMER BOARD MEMBER, U.S. CHEMICAL SAFETY AND HAZARD INVESTIGATION BOARD

Dr. POJE. Thank you, Madam Chairman and Senator Lieberman, for the opportunity to testify before this Committee on strengthening the chemical sector's security.

With its history of catastrophic releases, the chemical sector has had too many unintentional incidents of public terror to leave unregulated the potential for intentional terror. As last Thursday's events in London and yesterday's blast at a Spanish power station tell us, terrorism, maybe home grown, is becoming an all too frightening global specter.

My written testimony focuses on a number of issues. However, my oral testimony today, I hope, will convey my passion and urgency for preventing these chemical disasters.

[1] The prepared statement of Dr. Poje appears in the Appendix on page 130.

While America's worst chemical disaster occurred in Texas City in 1947, my wake-up call came more than 20 years ago when I was a young toxicology professor. I vividly remember the world's worst chemical disaster. It began as a violent runaway reaction within a methyl isocyanate (MIC) storage tank in December 1984 at the Union Carbide pesticide plant in Bhopal, India. After about 1,500 pounds of water entered an MIC tank, possibly caused by a routine line washing procedure, an exothermic reaction ensued. Excessively heated and pressurized gases burst through a rupture disk and opened a pressure relief valve, allowing approximately 50,000 pounds of MIC and reactants to be released through an elevated scrubber vent system.

The cooling gas formed a dense, low-lying cloud in that early morning and slowly and quietly drifted through the adjacent housing and much of the central city. MIC is highly reactive, irritating, and a toxic gas that is soluble in the aqueous fluid membranes around eyes and lungs. Victims awoke gasping for painful breaths and stumbled bleary-eyed into the streets with no indication of which direction to seek relief. Immediate fatalities were estimated at 3,000, with an accumulation of almost 20,000 disaster-related deaths in subsequent years. Injuries estimates range from 200,000 to 500,000. Casualties overwhelmed the city's four hospitals and several clinics that supplied only a total of 1,800 hospital beds and 300 doctors. Now, how many American communities could triage such an event?

What made Union Carbide such a tool of mass destruction in Bhopal? Well, I think the root causes lie in the systemic problems at the facility and within the community.

Lack of awareness and knowledge of the hazards—MIC was produced and utilized as a high-volume intermediate chemical, and yet its hazards under specific process conditions were not well understood by the workers, or the management, or the emergency responders.

Deficient hazard assessments—the hazards associated with contamination of the MIC in the storage tanks and their operations under high temperatures and pressures were poorly assessed and, therefore, abnormal situations were not managed.

Inadequate operating procedures—procedures were just insufficient, poorly written, understood, and executed.

Insufficient staffing and preparedness for the abnormal situation—managing staff at that facility were relatively new, unfamiliar with its processes. Employee responsibilities were not clearly established. Staffing had been downsized and staff turnover was high.

Failure to maintain the essential design and safety equipment—major changes had occurred without them being assessed for their safety impact. A refrigeration unit was shut down and the refrigeration material drained. The flare tower had been shut off for maintenance and was inoperable. The scrubber system, which had the ability to detoxify smaller amounts of MIC, also was turned off at the time of the event.

Inadequate investigations and failure to implement audit recommendations—prior deadly incidents that caused fatalities, injuries, and evacuations and smaller releases at the facility were not fully investigated and their root and contributing causes not established.

The equipment mechanical integrity was not maintained. Valves, pipes, and other pieces of equipment were corroded and leaking and unable to contain the material.

And there was inadequate emergency planning and response. The community was not even alerted to the disaster that was impending in their midst.

And there was lack of public oversight and authority. The government of India did not have rules, regulations, and authorities to conduct the appropriate management of such facilities.

You might think that this incident was long ago and far away and off topic. However, the CSB observed every one of these deficiencies in our investigations during my tenure, and who among us could not imagine a terrorist scenario being successful at such an operation and location? In fact, a consultant to the company speculated that the real cause was sabotage.

Let us look at a tale of two countries. While most Americans remember the events of September 11, few recall the major chemical catastrophe that occurred just 10 days later. On September 21, a huge explosion tore through the AZF fertilizer factory in Toulouse, France. Nearly 400 tons of ammonium nitrate detonated with a force equivalent to 3.4 on the Richter scale. AZF is owned by Atofina, the chemicals unit of TotalFinaElf, one of the world's largest petrochemical and petroleum producers.

The blast created a crater 50 meters in diameter and 10 meters deep. Windows shattered in buildings throughout the city's center three kilometers away. Thirty people were killed, 10,000 injured, and a further 14,000 sought treatment for acute post-traumatic stress. Over 500 homes were rendered uninhabitable and 27,000 others were damaged. Alarm systems failed, telephone lines were severed, frustrating public communications of safety messages. Nearby businesses collapsed and others had long-term business interruptions.

Thousands of tons of liquified ammonium, ammonium nitrate, and solid fertilizers and other chemicals at nearby businesses prompted additional concerns about possible domino effects. Because so many windows and building structures were damaged, sheltering in place would have been impossible if toxic chemicals were released.

The event greatly exceeded the consequences of the scenarios that have been used for planning emergency response. More than 1,500 firemen, special emergency personnel, and 950 policemen responded to the event, yet the early responders arrived on scene lacking exposure assessment equipment and personal protective equipment to cope with the toxic cloud.

The facility had been inspected several times in 3 years by local authorities, but not for the inadequacies of the ammonium nitrate fertilizer management in a warehouse of that facility, a warehouse mostly operated by the subcontracting workers and not by the management itself.

The Toulouse disaster, as many others have, and you already know, prompted nationwide debate about acceptable risks in communities. The French legislature extensively reviewed policies and practices and new legislation has focused on strengthening safety management systems of technological risk, including enhanced worker training and roles in risk prevention, improved safety management coordination and roles for contract workers, expanded public information about the risks and involvement in prevention, and better land use planning and siting around these high-risk facilities.

Now, with 20/20 hindsight, could we imagine what would have happened if that event occurred in the United States on September 21, 2001? The same corporation had a facility in Michigan that just 2 months earlier had sent 2,000 people into an evacuation mode and killed three in using a chemical called methyl mercaptan.

If there is a silver lining in this cloud of terrorism, it is, I believe, the urgent motivation to reign in the risks posed by the chemical sector. I urge the Committee to see the development and maintenance of competent management systems for safety as essential underpinnings to enhanced security. These have to go together. We need to have U.S. policies that will force the marriage between these two domains such that we are not Balkanizing security into a Homeland Security Department that is completely ignorant of all of the essential security features that have to be part of a security paradigm.

I give you five—or six recommendations to consider. One, ensure that whoever has responsibility monitors the scope of the chemical sector problem. We know that we have 9,000 incidents occurring annually in just 15 States in this country. We don't have a nationwide surveillance system to tell us how many chemical events are occurring in America.

I ask that you also establish a Department of Homeland Security responsibility that promotes effective coordination with other agencies. If these agencies are only on bended knee to Homeland Security about security issues and there is no interdigitation of security's work with these other agencies functions, we will lose a golden opportunity for strengthening our whole system of safety and security.

Set requirements for a security management system. We heard on the previous panel the importance of the words "management systems." I believe that those are the critical underpinnings for us being able to have a much more effective approach. One where effectiveness is observed, in my particular unique safety portion of the world, by looking at exceptions. Yes, I know about good coordination between agencies. I know about good work of trade associations. But I have had to look at the safety exceptions, when good practice and oversight don't work. We have to make more abundant use of such features of the safety landscape of the chemical sector and force the study of the exceptions, the exceptions that are causing evacuations and injuries in communities right now and are showing us where those relationships aren't working. I think we have to keep a high focus on that.

I also believe that the ultimate solutions for security and safety will be found in reducing the volumes and the toxicity of the hazardous chemicals. We need to have a better way of making an attack on that problem.

And finally, we need to employ effective training approaches. An absolute critical step to improving security at the chemical plants is going to be to properly train the workers who respond to the disruptions. We have some good models, and I think they need to be built upon for enhancing security.

Thank you for this opportunity to testify, and I would be happy to answer any questions.

Chairman COLLINS. Thank you. Your testimony is a powerful reminder of why we are committed to passing legislation.

Mr. Erwin.

TESTIMONY OF GLENN ERWIN,[1] PROJECT DIRECTOR, TRIANGLE OF PREVENTION PROGRAM, UNITED STEELWORKERS INTERNATIONAL UNION

Mr. ERWIN. I would like to thank you, Chairman Collins and Senator Lieberman and the rest of the Committee. I would also like to thank the staff. Too often, the ones that do the work never get the recognition, so I would like to thank the staff.

As Dr. Poje said in his remarks, he said he wanted to share with you his passion for this concept here. He reminded me of an 86-year-old cowboy friend I have in Texas that always said, "Whatever you do, you do with all your heart, mind, and soul." So I guess you have my mind in the written agenda that I gave you, or the written comments. Like Dr. Poje, I would like to share my heart and soul. I promise you I won't sing. I am not Aretha Franklin. [Laughter.]

But I would like to talk about some things that are very near and dear to me. Just as recently as March 23, 2005, I lost a very good friend in an explosion in Texas City, one of the most wonderful, Godly men I had ever met. As a matter of fact, the last Christmas that I saw him, he was gathering up a pickup load of toys to take to the Texas State Penitentiary in Huntsville, Texas, to make sure that none of the children there had a Christmas without toys. He was killed in that explosion. Now, I know we are here to talk about intentional acts of sabotage, but whether it is an intentional act or an accidental act, his life was cut short and our community has really lost a wonderful person.

Now, I believe that we, in the oil and petrochemical industry, oil refineries and chemical, I believe that we will be a target. It is not "if" but "when." I am certain it is going to happen. I think one of the reasons for it is we are too easy, very easy to gain access.

We did a survey.[2] We have distributed that. We have also submitted that for your review. But only 3 percent of our people think that we have done an excellent job in preparing to prevent an intentional act of sabotage. So, we are too easy.

There is such a large vulnerability. There is such a potential on what they can do if they get access into certain chemical plants, and our industry is just too important. If we disrupt the flow of energy, the flow of gasoline, the flow of chemicals, as everybody said before, we are going to really impact our country.

---

[1] The prepared statement of Mr. Erwin with an attachment appears in the Appendix on page 144.

[2] The survey entitled "PACE International Union Survey: Workplace Incident Prevention and Response Since 9/11", October 2004, by Paper, Allied-Industrial, Chemical and Energy Workers International Union (PACE) appears in the Appendix on page 150.

Now, I want you to think layers of protection, and that is what we need to do, is we need to look at protection, some way to protect from this worst possible thing happening. And I guess I would ask you to visualize, I couldn't think of anything better, but maybe an onion. Let us make it a 10–15 layered onion that was developed by Texas A&M—— [Laughter.]

It goes great with barbecue and will give you something to pick on Senator Hutchison about, about somebody talking about onions from her State.

But anyway, I want you to visualize an onion and just kind of take the outer skin of it. The outer skin of it, the first layer is our security. It is the fence line. It is to keep the unauthorized people from being there, the gates to control the flow of who goes in and out in normal admission, and also to train and equip our guards. That is our first layer. That is the one that we need first to put in place, but it is not there yet.

I just stood at the front gate of a major multi-national oil company the other day right at lunch time, and I watched the flow through the front gate of one car going in after another, and I watched a pickup truck, and I will use this one for an example. There were two people in it, and they drove up to the gate, and they showed their badge, and they went right on through. Well, sitting in the back of that truck was five or six buckets, closed-top five-gallon containers, and I looked at the guy that was next to me, and I said, "What is in the buckets?" And he said, "I don't have any idea." And I said, "Well, does the guard?" And he said, "No." I said, "Why won't he check them?" He said, if he did, nobody would get back from lunch, and he would be in trouble for holding up the flow of traffic.

So I think that we are vulnerable there. I don't think we have control of our main gates yet. So that is the first layer.

The second layer of security is inside the plant. Once you are inside the plant, there are different areas. But our security is set up for perimeter. Our security is not set up for everything within it. We treat a kerosene tank, the accessibility to a kerosene tank, the same as we do to a hydrofluoric acid tank. In fact, the same plant, as we drove by, and we drove on a road, not 100 feet, maybe 150 feet from a hydrofluoric tank that contained probably 800,000 pounds of hydrofluoric acid.

That didn't bother me as much as to see 50 or 100 people with a flurry of activity going on around that. And I said, "Is that tank empty?" And he said, "Oh, no, that tank is full." They had heavy equipment operating within 20 to 25 feet of a line, the suction line to that tank. Now, had they have hit that, whether intentionally or unintentionally, knocked that suction line loose from that tank, I asked our guide, I said, what would have been the effect, and he said it would have been catastrophic. And I said, "Well, how bad?" I said, "Thousands?" And he said, "More like 10,000, maybe 100,000 if the wind direction is right," if that happened.

So that is the second layer. There needs to be added precaution once inside and not treat everything just the same.

Let us peel another layer. Let us go now to substitution, and we have talked—they have used some fancy words for it. I am not going to use that, but let us get rid of some hazards.

Just like the HF tank, it is used for an alkylation process that you can also use sulfuric acid for. Now, why does one company use one method that doesn't have the potential and another company use the other one? I can't answer that. There are lots of other examples of how we can eliminate, how we can substitute, how we can change. I guess economics is one reason, but if you start looking at the human toll if something happened to a sulfuric tank versus a hydrofluoric tank, there would be a tremendous incentive to try to move to the others.

Now, some companies may not want to hear that I feel, and our institution feels, that there should be some mandatory look at what you handle. Whether you use the HF, sulfuric, chlorine, or bleach, I think somebody has to do it, and it is not just economics. It should be based on the vulnerability.

Now, you may not get that law passed, but I will tell you the second best thing. Pass a law where the plant manager or the CEO has to live in that plant, and I will tell you what, they would look at it just a little different. You know what the dirty little underbelly is? It is that most of the people that manage our facilities don't even live in the same town. They move further away.

Let us peel another layer—reduction. Reduce the hazard. Look, there are things that we can do, and it has got to be mandatory to look at trying to reduce the hazard. My old cowboy friend would say, if you are going to raise cattle, you have to have a bull, but he doesn't have to have horns. Look at doing something to try to reduce the hazard in the materials that we work with.

We can store it in smaller amounts. They say you have to truck more in that way. If you use 1,000 pounds a year, I don't care if you store 100,000, that is what you have to use to get in and out. I don't see the math. So I think we need to look at trying to reduce it.

Let me peel another layer. Next is to minimize what we have other than just the amount in a tank. We had an 800,000-gallon, or pound tank of hydrofluoric acid. Wouldn't it be less hazardous to have four 200,000 if you have to do it? There are just some things like that that make common sense to me that I understand why we don't do it, the things that we have to look at.

Now, I want to emphasize that there are a couple hazards in the plant we have to look at. One of them is explosives. The other one is toxics. Nine-eleven was explosives, but Bhopal was toxic.

Now, I have a friend that drives a truck, and he drives a hydrofluoric acid truck—methyl mercaptan. He drives a methyl mercaptan truck. And I was talking to him and I said, "well, what would happen—what are you doing to prevent somebody from using your truck as a weapon of mass destruction? What is to prevent somebody from hijacking it?" He said, "Well, I have a Global Positioner Satellite on top of my truck." And I said, "Have you got one on the tank?" He said, "No." And I said, "Well, all they have to do is just to waylay you and take the tank, isn't it?"

Let me visualize, can you imagine what a tank of methyl mercaptan could do if they drove it into the right area and somebody knocked the belly cap off that thing and just released all the contents of that highly-toxic material at the right place, at the right time? It would be devastating. We need to put the positioning satellites on the trailer, not just the truck. We need to see where the actual shipment is going.

Now, look, I lost a friend, I said, to that explosion. I have had other people that have been hurt in fires. I have walked into Ben Taub Hospital and walked into the burns institute. There were four people in there and I was trying to find my friend, and I couldn't tell the four people apart. I couldn't even identify him. His own mother didn't even know which one he was.

The incidental act and the intentional act still have the same effect, but if we can prevent the intentional and really prepare ourselves to prevent for those along at the same time that we are looking for the intentional acts, I think we are going to gain so much more.

Let me give you a personal example. I am running out of time, but I will tell you what—on Halloween night, 1987, it was Friday night in Texas. We had a football game. And on Friday night in Texas, what is the most important thing that goes on? I have two kids, a 17- and 15-year-old that were already down at the stadium, and I was preparing to go, and as I was sitting there, I came across the eyewitness news that we had a leak in town, shelter in place, stay off your phones and behave yourself. Don't get out of the house. There I sat, with two kids at the football field. They told over the TV where the spill was occurring. It was occurring at a Marathon facility. Well, I could just draw a beeline from my house to there and right in the middle of it was where that stadium was.

I know what it would be like if the leak that occurred was a contractor dropped and hit the vapor line of that tank. Now, had he hit the liquid line of that tank, it would have killed both my kids. Both of them were exposed, but it was minor because the vapors were coming up, not the liquid being left off.

Look, the hazards are out there, the potential in our communities. We have to do some things. We have to look at layer protection. We have to work together. We have done our survey. We said there is more that can be done. Our members say that there is more that can be done. It is not just me sitting here. It is 125 sites that were surveyed. It says we are not ready enough. They are not involving the people. We have not involved the actual workers to the extent that we can.

Now, we support legislation. I am out of time and I am going to shut this off, but we support it. It is in my written comments. We can do better. I think we can do better. And I appreciate your effort for convening this and attempting to try to make our workplaces and our communities safer. Thank you.

Chairman COLLINS. Thank you very much. Ms. Andress.

TESTIMONY OF CAROL L. ANDRESS,[1] ECONOMIC DEVELOPMENT SPECIALIST, ENVIRONMENTAL DEFENSE

Ms. ANDRESS. Good morning, and thank you for the opportunity to testify today. I represent Environmental Defense, a national environmental group where I work on pollution prevention issues. I

[1] The prepared statement of Ms. Andress with attachments appears in the Appendix on page 209.

will summarize my written statement, but I ask that my full statement and the attachments be entered into the record.

Chairman COLLINS. Without objection.

Ms. ANDRESS. On the issue of chemical security, I want to describe an example that I believe is illustrative of the challenges and the opportunities before you. It is about an actual chemical plant in Baltimore, Maryland, that was subject to three separate but overlapping security programs. It was covered by ACC's Responsible Care® Security Code, and, in fact, the facility had already passed the company's mandatory third-party verification process. The facility was also covered by the Maritime Transportation Security Act because it is located on a navigable waterway. The Coast Guard approved the security plan that the facility developed under ACC's voluntary program. The facility is also covered by a Baltimore ordinance on mandatory security plans. Despite these requirements, a reporter was able to enter the facility, enter an unguarded gate, reach two fully loaded chlorine tank cars, and then leave without ever being challenged.

This is not an isolated example. Investigative reporters have documented lacks and inadequate security at many facilities storing and using extremely dangerous substances. An enterprising reporter, or more troubling, a determined terrorist could likely gain access to most if not all of the several thousand facilities that use or store large quantities of dangerous substances. This includes about 2,800 facilities, all of which have 10,000 people or more living within a projected danger zone. These very high-risk facilities are located in almost every State.

So the problem is significant, pervasive, and yet unaddressed. This is why your commitment to a strong chemical security program is so important.

I want to return to the example of the plant in Baltimore. What lessons can we learn from this? First and most importantly is that a sole reliance on a strategy of guards, gates, and guns is simply inadequate and bound to fail. Physical security alone cannot prevent a determined terrorist.

Second, current security programs which, frankly, are largely voluntary, are not effective. This suggests that the accountability mechanisms in the existing laws are not enough.

So what should we do? The most effective and economical way to achieve security is to design the products and processes that reduce the use of these extremely dangerous chemicals. Reducing the source of the problem, the chemicals and processes, makes a facility less attractive as a terrorist target. It cuts the needs and costs of security measures. And it minimizes the likelihood of a major chemical accident. This is classic pollution prevention. But more importantly, this is how you get real, lasting, cost-effective security.

My written statement provides examples of some high-hazard industries that have eliminated or significantly reduced their vulnerabilities to terrorist attack. This includes refineries, power plants, sewage treatment, and water treatment facilities.

The challenge then is not how many guards, gates, and guns are needed but how to foster more widespread risk reduction. Several State laws and one local law provide a road map for how to achieve that risk reduction. These include New Jersey's Toxic Catastrophe

Prevention Act, Massachusetts's Toxic Use Reduction Act, California's Accidental Release Prevention Act, and Contra Costa County's Industrial Safety Ordinance. These laws are aimed at spurring facilities to cut their use of certain toxic chemicals and the results are impressive.

At the start of New Jersey's program, 575 facilities reported having chlorine tanks on site. Now that number is 26. Contra Costa County, California, experienced a 36 percent reduction in acutely hazardous substances between 1990 and 1994.

Lessons from these programs suggest three key principles for a Federal chemical security program. First, Congress should mandate the most effective, most efficient, and safest option. This means establishing requirements that all facilities conduct a thorough evaluation of ways to switch to safer chemicals or processes, reduce the amount of dangerous chemicals used, or reduce the amount stored onsite. When those options are practicable, the facility should be required to implement them. High-risk facilities, especially, should be expected to make significant investments in reducing the quantity and nature of the hazardous chemicals onsite.

I realize not every facility will be able to eliminate or significantly reduce the hazards. When a facility finds that there is no safer option that is technologically feasible, or where the alternatives are prohibitively expensive, particularly when compared to the potential damages, or when the available alternatives would create an equal or greater hazard to public health or the environment, then they should provide a justification for why an alternative approach is not practicable.

Safety cannot be voluntary. The issue is too important and the market mechanisms are simply inadequate. Facilities that are facing daily questions about operational efficiency and financial performance have little interest in dealing with catastrophic hazards that seem remote. For that reason, Congress needs to mandate that a reasonable process be put in place for getting safer approaches in place. The complexity of the industry should not be an obstacle to action.

A second key principle is accountability. I trust that most facilities will make a good faith effort to implement safer approaches. However, this is far too important to rely solely on good intentions. Facility owners and operators must be accountable to Federal authorities and the public for reducing hazards. I believe accountability measures should include government oversight and intervention, especially when facilities do not perform; public disclosure of the reasons why they were unable to implement alternative approaches; and linking public funding with safer operations.

This is especially applicable at sewage and water treatment plants that receive substantial public money and yet continue to use chlorine gas in populated areas. It frankly makes no sense to me to have taxpayer money going to basically pre-position a deadly and unnecessary chemical in a populated area and then spend Homeland Security money to try to protect the chemical. Taxpayer money should not be spent at facilities that pose an unnecessary risk to the American public.

The third principle is that Federal legislation should avoid creating loopholes for voluntary programs. We commend ACC's,

SOCMA's, and MPRA's early efforts to protect their facilities. But as we have seen with many news reports, voluntary programs alone are wholly inadequate. Creating special conditions for facilities that participate in these voluntary programs will undermine your efforts to safeguard facilities. Allowing facilities to follow their own standards has not been deemed acceptable for airports or nuclear plants and should not be acceptable for chemical plants.

We agree that companies should not have to reinvent work done previously. Congress should allow them to submit prior documents with supplements, as needed. For example, vulnerability assessments done by drinking water facilities under the Bioterrorism Act should be considered as part of meeting their obligations under a chemical security program.

However, it is particularly important that work done as part of a voluntary industry program be strictly scrutinized. It is one thing to recognize the security efforts performed under Federal statutes. However, it is completely unacceptable to rubber stamp voluntary measures that have not been evaluated or enforced by a Federal agency.

My written statement elaborates on some additional issues to include in chemical security legislation, including requiring buffer zones and simulating community evacuation drills with the community and coordinated by local emergency responders.

Efforts to protect Americans from terrorist attacks are often costly and complicated. Instances when protection of the public can be achieved in a cost-effective manner should be aggressively pursued. That some of these options have side benefits, such as eliminating the potential for chemical accidents, makes them all the more appealing, and I do not consider these to be extraneous issues. Safety and security cannot be separated.

Congress should insist that facilities take all reasonable steps to reduce risks of catastrophic chemical release. Thank you.

Chairman COLLINS. Thank you for your testimony.

Each of you has argued for mandating a reduction in the use of dangerous chemicals or the substitution of less-hazardous chemicals wherever possible. But if we draft legislation so that it is truly risk-based, so that the level of regulation is ratcheted up depending on the hazards at a particular facility, wouldn't the companies have an inherent incentive to use less-dangerous chemicals or smaller amounts of hazardous chemicals in order to get into a lower-risk category with fewer regulations imposed upon them? Dr. Poje.

Dr. POJE. Actually, Senator, that is a very good point. The earlier mentioning of the Toxic Catastrophe Prevention Act in New Jersey, I think, has given us quite a few examples to look at for how a regulatory regimen over time has caused the mobilization of the industry to change its pattern and practice of the use of chemicals.

A certain portion of chlorine-using facilities, particularly in the water and wastewater treatment arena, have migrated out of chlorine gas usage for biocidal treatment, and that has come in part because of a higher degree of oversight and a ranking of high hazardness for that particular chemical in that regulatory regimen.

Now, to be fair to the previous panel, there is enormous complexity in the diversity of processes being used throughout the chemical producing and using sector. However, I think it is abundantly clear to me that there are some processes whose moment has come for inherently safer approaches and we need to be able to challenge the usage of those chemicals in ways that embrace clear alternatives available. I think Ms. Andress has given us a pretty clear example with chlorine in the water-treatment industry.

Do we taxpayers want to pay both for the development of a wastewater treatment facility using the most highly hazardous form of biocidal treatment and then a second payment for using Homeland Security protection measures to be imposed over that? I think that is just foolish, and we clearly don't have the resources to perpetuate such a poorly thought out system.

Chairman COLLINS. Mr. Erwin, wouldn't there be an inherent incentive for companies to change to less-hazardous processes if we draft the legislation so there is a different level of regulation depending on the risks involved?

Mr. ERWIN. That might be very true. The more hazardous it is, the larger the problem. It is a very complex issue. There are some things that are hazardous they can't get rid of. And a lot of the companies have done a lot of work, and I don't want to sound like they haven't because they have done a lot of work to try to substitute, when they can reduce. But not everybody has.

There are some forward-thinking companies. There are some companies that are very responsible. And then there are some that are not. There are some that keep the books right and some that don't. We know that for a fact, too, and it is the same thing here. But you are right. That may be true.

Chairman COLLINS. Ms. Andress.

Ms. ANDRESS. Well, I think implicit in that kind of risk-tiering approach is that safer approaches are, frankly, the best option. And so from that standpoint, I find that appealing. I think, however, I am concerned that it would leave—it potentially leaves some fairly high-risk facilities to simply adopt a physical security approach and that, I don't think, is enough, to just rely on physical security.

Chairman COLLINS. Dr. Poje.

Dr. POJE. If I could just make one additional comment on that. My experience for 7 years has been to look at safety tragedies in the chemical sector, so I have a very myopic view of seeing where failures occur. Having said that, though, I also see that is the place where Phoenix-like, we can rise up out of the ashes to do a much better job.

Bhopal changed policies in the United States to be more aggressive. There was a chemical facility in the State of Texas right after the terrible Bhopal tragedy in India, a DuPont facility, that within an 8-month period switched dramatically out of methyl isocyanate usage. It actually had plans already developed. Now, the acceleration of the implementation of those plans took the terrible Bhopal tragedy to make it happen.

The Chemical Safety Board under my tenure conducted 33 investigations. Only less than 10 percent of those investigations involved chemical processes covered under the RMP system. Now, in one way of thinking, RMP is an appropriate approach for risk ranking systems, the conceptual basis of what is worst, highest toxicity, highest amounts. Those are all very rational designs that we have to employ.

But there is one other piece of the equation to consider. What happens when failure tells us there are other management processes that are having terrible problems. In fact, every one of those 9,000 incidents that I mentioned occurring in those 15 States is an enormous red flag to everybody—a red flag to those who want to do harm that we have management problems here and harm can be had in this fashion. If we don't embed the responsibility within DHS to have to hold them up for an example and examine them in a detailed way, we are going to lose the advantage of those disasters to strengthen the whole of the system of safety and security.

And I would argue if there is a pattern within these safety incidents that identifies particular chemicals and processes having the most frequent problems, we had better figure out solutions for them quickly. And, there should be governmental resources, if there isn't private sector resources, to help make that happen.

Chairman COLLINS. Mr. Erwin, before I turn to Senator Lautenberg, I am very interested in the results of the study that you were involved with. It prompts in my mind a question about whether the Department of Homeland Security involves the workers, goes to the head of the local union if there is one, when it does an assessment of the security of a chemical plant. Do you happen to know? DHS has pointed to these site visits that it has undertaken. Do you happen to know whether workers, union representatives, are interviewed by DHS officials when they do these site visits?

Mr. ERWIN. We are not party to when they come in like we are when OSHA comes in or when the CSB comes in, and we are not included in the conversations, to my knowledge. I don't know of any union leaders or employee representatives that have been included in this area. I think it is nonexistent.

Chairman COLLINS. That is very helpful and something I will follow up with the Department on, because I think, judging from your testimony and experience, that they could learn a lot from talking to the employees of these facilities.

Let me ask just one other related question. Has the Steelworkers Union or PACE shared its survey with the Department of Homeland Security, do you know?

Mr. ERWIN. Well, we have copies here. I would be glad to give them a copy. But what we did when we prepared the report, we shared this with the governmental agency that we were working with, the National Institute of Environmental Health Scientists. Now, it is our understanding that they have shared that in the report with other agencies with whom they are working, and I guess they work with DHS, too.

Chairman COLLINS. Thank you. Senator Lautenberg.

Senator LAUTENBERG. Thank you, Madam Chairman. Thank you all for your testimony. I am sorry that I wasn't here, Madam Chairman, when the first panel was still up because I was struck by some things that were said, and one of them related to Mr. Barmasse's testimony about IST.

You talked, Ms. Andress, about how much use has been reduced of chlorine, and we know that here in this district, the wastewater treatment had enormous reductions in threat as a result of transfer from chlorine to another material that appeared to be substantially safer. I don't like to ask questions that Mr. Barmasse could have answered, but I am compelled by the structure to ask you.

I think in Mr. Barmasse's testimony for SOCMA, he made the point that Inherently Safer Technology is probably the most misunderstood and controversial aspect of chemical site security. While it seems self-explanatory, the term as used in chemical and engineering may be misleading to non-scientists. It is an approach to chemical processing that considers procedures, equipment, and the use of hazardous substances.

Don't we have data that refutes the fact that IST can be seriously employed with a lot less expense or risk than the other massive changes that have to be made? Are we out of reliable changes of one material for another that can make us safer?

Ms. ANDRESS. I am not sure I understand the question.

Senator LAUTENBERG. Well, the question is whether or not we have exhausted the opportunity to make substitutions of materials. Dr. Poje may want to say something about that. Have we run the gamut on substitutable materials?

Ms. ANDRESS. Well, you are correct in that we have—there is quite a bit of knowledge and expertise out there, and, in fact, there is quite a bit of knowledge about these issues at the various State institutions in New Jersey, Massachusetts, and elsewhere. So there probably—I don't think we have exhausted it. I think there is still ground to be tilled. I think from my standpoint, the most important point is that we haven't exhausted the adoption of the safer chemicals. There still are several wastewater treatment plants using chlorine gas in heavily urban areas. That is simply outrageous.

Senator LAUTENBERG. Would that transfer be relatively transparent with regard to costs? Dr. Poje, do you have any knowledge about this?

Dr. POJE. Yes. I think as I said earlier and I think as the first panel reflected, there is a great complexity to certain aspects of the chemical industry. But we have already heard from two panelists here about the use of hydrofluoric acid in alkylation within the oil refining industry for which there are two alternatives. One, hydrofluoric acid, has a much higher risk than the other sulfuric acid. In the chlorination and alternative biocidal treatment of water and wastewater, there are also clear examples.

In other chemical process areas, it takes specific research and analysis to make processes inherently safer. Now, within the chemical industry, there are some that are leaders in doing this R&D work, and it gives them the competitive advantage of new materials. In fact, the greatness of our chemical industry comes in large measure by very innovative chemistry and R&D to help get a competitive advantage in the global market over those who are producing things in a less efficient way.

The term that has grown of art recently is green chemistry. That is the most vibrant aspect of the chemical industry's development and the chemical sciences development. How do we optimize across 12 different principles for making a better chemical science that will be of advantage to us for our lives and lifestyles in the future? There are aggressive programs in universities all over this country and across the globe to promote that end.

One of the Green Chemistry principles is to design things so that they are inherently safer and so that we prevent chemical accidents. I would argue in the post-September 11 world, also to prevent terroristic disasters.

Senator LAUTENBERG. Dr. Poje, you heard the testimony of our first panel, and yet it is clear that you believe that improvements in chemical safety and security beyond the industry's Responsible Care® program are needed. How do you draw those conclusions because I think they are quite different from the idea we heard earlier.

Dr. POJE. I certainly draw those conclusions from my more intimate experience, having studied the pattern of safety tragedies that occur when chemicals aren't appropriately managed. And when you see after the incident that safer alternatives could have been available, you are forced to ask the question, what are the barriers that prevented people from either knowing about those alternative approaches or for economically employing them?

Clearly, there are two different worlds that we have to be concerned about. Greenfields development, in which we should have the best and most cutting-edge technologies applied as we develop new facilities. Then there is the brownfields, the facilities that already have tanks and concrete and "hardened" facilities for which making changes is going to have to come out of someone's capital budget.

And I think that is where the artfulness of business decisionmaking is coming into this debate. How much can you mandate of that to existing facilities before you wind up mandating those facilities to leave the country and go offshore. We have to be concerned about that. I think we do need this domestic industry and its jobs and its opportunities here in the United States. But how do we avoid expanding the risks that we see?

Last April, I had the terrible experience of having to lead a team from the Chemical Safety Board to a place in Dalton, Georgia, that was a SOCMA member that for the first time had been using a chemical called allyl alcohol. And while the investigation is still ongoing at the Chemical Safety Board, rudimentary aspects of safety and emergency response just were not operational in that fairly sizeable community of Dalton, Georgia.

When that chemical was being used for the first time, there were poor plans on how to deal with abnormal situations. The reaction got out of control and it bubbled out of the reactor. There was no secondary containment available. A bucket, a small plastic bucket, was being used to capture what was coming out of this reactor, and that poor containment allowed toxic gas to emanate into the surrounding community.

Emergency responders turned out to deal with it. Police went door to door. Police without any kind of protective equipment went gasping into this cloud of toxic allyl alcohol, trying to get community members out of harm's way. There was no awareness within this community about those hazards.

In fact, I was quite shocked. When I gave a press briefing the next day after I had arrived and announced to people what chemicals were involved, they didn't already know that information at that time. They didn't know it at the hospital. They didn't know it in the broad community. You shouldn't wait for somebody from Washington to come and investigate and find out what people might have been exposed to. You need to know that at the hospital.

Those 154 people went to the hospital on a cold mountain Georgia evening to be stripped naked, hosed down before they were allowed to go in and be examined for any possible impact. But if the medical system did not know what they were exposed to, how would you possibly be able to deal with the hazards that those people had?

So we here have a reason to start asking more serious questions about who is using chemicals, when they are using them, how they are using them, and make sure that we are adhering to even the minimal standards that currently operate for risk management and process safety management. There are a whole bunch of chemicals that are outside of the RMP system, and the Chemical Safety Board has had to investigate numbers that are not currently covered by Federal safety standards.

I would not want Homeland Security to think that somehow it can pull out of another agency the named list of chemicals, talk to the industry and thereby say that these are the only ones we are going to worry about, and consequently blindly miss other risks that are around us. And those risks, the ones that I see, have seen, are ones that announce themselves through mismanagement as releases into communities.

Senator LAUTENBERG. Madam Chairman, if I might, just another question.

Chairman COLLINS. Certainly.

Senator LAUTENBERG. It was said by Mr. Barmasse in the first panel that, in response to a memo to Congressman Markey of Massachusetts, the population potentially affected under an EPA worst case scenario release is calculated in a circle around the facility. It is unlikely that this entire population would be affected by any single chemical release even if it is a worst case accident. So this challenging to the data that are being used to describe the threat.

All three of you have had occasion to look at these. Would you agree that the figures that are used are under suspicion in terms of their accuracy?

Ms. ANDRESS. Well, he is correct when he says that everybody within the vulnerability zone would not be affected in the event of a release. The idea is there is a circle drawn around the facility. It shows where the potential could be. But on any given day, an incident is only going—depending on prevailing wind conditions, it is only going to move in one direction or another. But it does get to this issue of how do we determine what are the risky facilities, and I am aware that, for example, the Department of Homeland Security has its own system for evaluating risk.

At Environmental Defense, we recognize we need to prioritize. We are not talking about a rigorous government oversight of rural facilities that have minimal, if any, offsite consequences. We do believe that every facility that poses an offsite risk needs to evaluate safer options, but where we need to focus government resources is on the high-risk facilities.

But I do think the EPA numbers are useful in that they are transparent. We know how those numbers got arrived at, whereas the DHS numbers, it is largely a secret methodology, and it is predictable, and we think both the public and the agencies need—and the companies need that kind of clear basis for knowing what the priorities are.

Senator LAUTENBERG. Well, the "Right-to-Know" law that I helped coauthor in Federal statute derived from a similar law that was developed in New Jersey. The thing that triggered the Right-to-Know law in New Jersey was when the firemen approaching a chemical fire had their protective gear virtually melt in front of them. What happened is there was an incredible amount of participation by industry on a voluntary basis to reduce the emissions and to identify these hazardous chemical facilities that were located in lots of places in New Jersey. So it was a good start.

But as we look now, there wasn't an interest then by some terror group that was looking for a way to really do us a lot of damage, and so there are things that we can do on a voluntary basis, but there are also things that we have to do.

Mr. ERWIN. Can I comment on this issue right here?

Chairman COLLINS. Absolutely.

Mr. ERWIN. Any institution is just like a body. The head of it is the only one that gets to dream. The ones down in the rest of the body live in reality. [Laughter.]

Having said that, the worst case scenario, we only look at single worst case scenarios, and if we are going to deal with terrorists, don't you believe that they are smart enough to hit more than one? So when we look at a worst case scenario, we are not dealing with a single incident. We are going to look at multiple things. I mean, if I was going to do it, I would knock the HF tank, I would hit your power supply, I would also do some other things all at one time, and I am not a terrorist, so I don't think like that. But just imagine what they could do when they hit a lot.

So I think we need to go back and reassess what our worst case is now in the light of terrorism because I think it has changed. I don't think that our worst case that we looked at now is truly our worst case. I think we need to go back and do a reassessment on that.

Senator LAUTENBERG. Thank you, Madam Chairman. Thank you to the witness table.

Chairman COLLINS. You are welcome.

I just have a couple of final questions for this panel. One of the most important tasks that this Committee will face will be to define in legislation the universe of chemical facilities that DHS should be regulating for security. Do you have any advice for the Committee on how we should define the universe? Dr. Poje.

Dr. POJE. Yes. I think it is clear that the usage of chemicals is widespread in our society. One could go to an individual consumer who goes to a Home Depot and picks up a can of pesticide for use on their lawn and that person is handling a chemical. Could we possibly reach down and touch such persons for the way that they are securely and safely managing it? You can't do that. So there has to be a scale that moves in some direction toward those that are using the highest and the worst chemicals.

I think there has been an awful lot of work done in this country in the chemical safety arena to examine and reexamine that question. I think we can build off of that platform to define what are the highest risks in a measurable fashion above the next tier. The Risk Management Planning program obviously establishes three different tiers of program responsiveness for dealing with that kind of work. I think that should be examined and looked at, and I would hope you would get the assistance that you called for from Homeland Security and from OSHA and EPA and those who have had that kind of responsibility to work together to come up with such a proposal.

Again, my written statement, though, asks also that we be prepared for the exceptions. Do not put the blinders up that says that listed chemicals in regulated amounts are the only thing we worry about. Force yourself and DHS to have to confront the annual reality of chemical releases and cross compare. Are the chemical incident events reflective of the reality that we have chosen for our regulatory programs?

Chairman COLLINS. Mr. Erwin, do you have any advice for us on the scope of our legislation?

Mr. ERWIN. I think the scope of your legislation should be based on potential, the potential risk, the potential vulnerability, the amount of who could be harmed, and if you based it on that, it would be very inclusive.

Chairman COLLINS. Ms. Andress.

Ms. ANDRESS. Well, I would start with the Risk Management Program. It is, as I mentioned earlier, a transparent system. We know how those numbers are derived at. They are imperfect on both sides. As the industry panel noted, they may exaggerate the risks in some respects, but then in others, as Mr. Erwin noted, they don't take into account, what if all of the inventory were to be released at one time. But I actually think maybe that makes them the best option because they are kind of in the middle.

And then in terms of—I know there have been proposals, for example, to say that all facilities above a certain vulnerability should do X, Y, and Z, and I think that is—we recognize again the need to prioritize, and there are a number of facilities in the Risk Management Program that don't need heavy regulation and oversight. But as I said earlier, I think everybody needs to have—all facilities in the program that pose a potential risk to communities and workers nearby need to do an evaluation of the safer alternatives. And then you focus government resources on the high-risk facilities.

Chairman COLLINS. Thank you. Yes, Mr. Erwin?

Mr. ERWIN. Because I have never been here before, I may not understand the procedure, but do I have to ask to have my written comments and the survey added into the record?

Chairman COLLINS. All of the statements, surveys, and anything else that you wish to submit will be included in the record.

Mr. ERWIN. Thank you.

Chairman COLLINS. Just one final question, Dr. Poje. In response to an earlier question, you made a statement along the lines that we need to be able to challenge the chemicals and chemical processes employed by chemical facilities. This raises a question that I posed to the first panel.

You have already testified that you would support imposing some sort of IST requirements directly on facilities. But could you also support an alternative approach whereby we would not impose those requirements across the board, but give DHS the explicit authority to require changes if specific vulnerability studies for particular sites indicated a problem that could be addressed that way?

Dr. POJE. Certainly, I would, if I understand you correctly, I would see the advantage of having a Federal entity have some oversight in this area, particularly if it could identify common problems across the country and for which there should be some mandated inherently safer approaches. However, I would also have to say, my experience on the Chemical Safety Board is that the knowledge as to where a particular process could best change oftentimes is dependent upon the best process engineering competency within that facility itself or within that corporation.

I don't think that we are going to be able to guarantee that any Federal agency is going to become the best repository for that intimate process information. The agency should be the coordinator, convener, collaborator for drawing that information into a public arena so that more of the public would be able to see what are the opportunities. And thereby, more of the facilities that might not have access to getting to a professional American Institute of Chemical Engineering meeting would find out what inherently safer techniques are being used through the services provided by a Federal entity required to make sure that information gets out to the public. And I think it would also help this Committee do an effective job of oversight on whether we are making the progress as rapidly as we could.

Chairman COLLINS. Thank you. I want to thank all of our witnesses today for truly excellent testimony that will be very valuable to this Committee as we undertake the Herculean task of weighing all these arguments and drafting legislation.

We will be having a final hearing in this series of four hearings focusing on chemical security. That hearing is tentatively scheduled for July 27.

Again, I want to thank you all. We look forward to working closely with you.

The hearing record will remain open for 15 days for the submission of any additional questions for our witnesses as well as other materials.

This hearing is now adjourned. Thank you.

[Whereupon, at 12:10 p.m., the Committee was adjourned.]

CHEMICAL FACILITY SECURITY: WHAT IS THE APPROPRIATE FEDERAL ROLE?

WEDNESDAY, JULY 27, 2005

U.S. SENATE,
COMMITTEE ON HOMELAND SECURITY
AND GOVERNMENTAL AFFAIRS,
Washington, DC.

The Committee met, pursuant to notice, at 10:03 a.m., in room SD–562, Dirksen Senate Office Building, Hon. Susan M. Collins, Chairman of the Committee, presiding.

Present: Senators Collins, Voinovich, Lieberman, Carper, and Lautenberg.

OPENING STATEMENT OF CHAIRMAN COLLINS

Chairman COLLINS. The Committee will come to order.

Good morning. Today, the Committee is holding its fourth and final hearing on the security of our Nation's chemical industry against terrorist attack. The goal of these hearings has been to help this Committee develop comprehensive, bipartisan legislation to address what is clearly one of our Nation's greatest homeland security vulnerabilities.

Throughout this series of hearings, we have learned that the United States is home to an astonishing number of facilities that manufacture, use, or store chemicals for legitimate purposes that could cause devastation if turned against us as weapons.

The Environmental Protection Agency has listed some 15,000 chemical facilities that produce, use, or store large quantities of hazardous chemicals. The Department of Homeland Security uses a different methodology and has identified 3,400 facilities that could potentially affect more than 1,000 people if attacked and nearly 300 chemical facilities where a toxic release could potentially affect at least 50,000 people.

We have heard expert testimony regarding recent chemical accidents in our country that have also resulted in injury and death. We have learned that the chemical industry is enormous, diverse, and vital to the American economy. The U.S. chemical manufacturing industry approaches half a trillion dollars annually in sales. The chemical industry represents our largest export sector, with exports totaling $91.4 billion in 2003. More than 900,000 people work directly in the U.S. chemical industry, which supports an additional 700,000 supplier jobs and millions more indirectly.

From national defense and high-tech to agriculture and health care, the chemical industry produces more than 70,000 products that improve the well-being of the American people. And these hearings have reminded us that the terrorist enemy we face has a clear strategy of turning the tools of free and productive societies into weapons. They did it on September 11, 2001. They did it in Madrid last year. And they have done it in London, not once but twice this month. Given the chance, they will surely do so again.

Currently, the Federal Government's regulation of the security of chemical facilities is limited. The Department of Homeland Security's representative and many other witnesses have testified that new legislation is required to strengthen the security of chemical sites. The Department points out that approximately 20 percent of the overall chemical industry sector that it believes to be at high risk does not subscribe to voluntary industry security standards. While I applaud those many companies that have taken voluntary measures, an unacceptable number have not. Moreover, given the severity of the threat, I believe that it is a mistake to rely on voluntary measures alone.

To date, we have heard from witnesses representing industry, labor and environmental associations, as well as chemical safety professionals, homeland security experts, and the Department of Homeland Security and the Environmental Protection Agency. Today, we will hear from company security chiefs who will describe the day-to-day challenges of securing these sites. A local emergency manager with decades of experience in responding to chemical incidents will also testify. And we will begin our hearing today by hearing from the U.S. Coast Guard, which is responsible for implementing the Maritime Transportation Security Act of 2002.

Throughout these hearings, the results-based cooperative approach of MTSA has been described as a security success story. Maritime commerce is no less diverse or vital to our economy than is our chemical industry, and the security issues are no less challenging.

I will be very interested to hear the Coast Guard's views on the extent to which MTSA could be used as the template for the chemical security legislation we will begin drafting next month.

I look forward to hearing from all of our expert witnesses today. Senator Lieberman, welcome.

OPENING STATEMENT OF SENATOR LIEBERMAN

Senator LIEBERMAN. Thank you, Madam Chairman. This is, as you have said, the fourth—and I believe the last for now—in a series of hearings that the Committee has held on chemical security. I think this series has really served to inform us of the vulnerabilities we face as well as the various responses that we can take together to strengthen our defenses against a potential terrorist attack or a chemical accident. I hope it is clear that Chairman Collins and I and the Members of this Committee consider securing our most exposed chemical storage and manufacturing sites a top priority for this Committee and, indeed, for our country.

I am heartened that we have agreement with the Administration and a large portion of the chemical industry itself that legislation is necessary. The fact is that the recent news from Sharm el-Sheikh and London reminds us again that the war of Islamist terrorists against us is continuing and it is global; that terrorists will exploit weaknesses in our homeland defenses wherever they find them; and that they aim to kill as many innocent men, women, and children as possible to spread fear and panic throughout our countries to bring about the political changes that they desire.

By any measure, the chemical industry today is one sector where a successful attack could have catastrophic consequences for our people and our country, and that is why we must and will continue to work with haste to do everything we possibly can to prevent such an attack.

At our first hearing in April, we learned of the potential risk posed by thousands of chemical sites across the Nation. One witness described chemical facilities as potential weapons of mass destruction. At our second hearing, in June, the Department of Homeland Security testified that voluntary safety measures taken by the chemical industry, commendable as they were, were not enough and that the Administration supports legislation to secure the most hazardous facilities by imposing minimum security standards.

Earlier this month, industry representatives told us that legislation was, in fact, needed in their opinion to establish Federal security standards. The largest chemical trade associations—which is to say the American Chemistry Council, the Synthetic Organic Chemical Manufacturers Association, and the National Petroleum and Refiners Association—all agreed that Federal standards would improve security, although they opposed Federal mandates requiring companies to implement the so-called inherently safer technologies.

Today, we are going to hear from a variety of stakeholders, both public and private. Some of our witnesses will argue that Federal controls should be limited to—and have argued, in fact, that Federal controls should be limited to standards for physical security measures such as gates and surveillance cameras. But I must say that I am impressed by the arguments of most of the security experts that we have heard that physical measures alone will not stop a determined terrorist attack. Knowing that, I believe we must look long and hard and thoughtfully at what can be done to reduce the inherent hazards at chemical sites by finding alternative substances or technologies to reduce the risks or configure plants in ways that minimize the possibility of a hazardous release. In other words, how can we ensure that the chemical companies are doing all they can to achieve better safety and security through such measures?

I am also concerned that too many local preparedness and response teams may not be able to respond effectively to an attack on a chemical plant, and I believe that State and local officials, who are also the first preventers, need more resources than they now have if they are expected to protect the areas just outside a chemical facility's fence, as now seems to be the case.

And, finally, I want to be sure that the people who live near chemical facilities have been informed and prepared about what to do if there is a breach at a chemical facility. Today, in too many places, I conclude that is not the case, leaving the public uninformed and unnecessarily at risk.

I know that there are still disagreements about details, and they are not insignificant disagreements. But I must say, as we come to this fourth hearing, I am very encouraged that we all are walking along the same road, which will lead us to an agreement that will make our chemical plants safer and that will guarantee that they pose as few risks as possible to the American people.

Thank you, Madam Chairman.

Chairman COLLINS. Thank you, Senator.

Senator LAUTENBERG. Madam Chairman, are we going to be permitted to make opening statements?

Chairman COLLINS. As the staff had informed all Members, we just did opening statements at the first day of this series of hearings. But if you and Senator Voinovich——

Senator LAUTENBERG. I just think this subject is such an important one, and in particular, the area I come from is highly vulnerable to terrible destruction if an attack is placed against any of the chemical facilities, and I don't want to upset the routine, but I would hope, Madam Chairman——

Chairman COLLINS. Senator, if you would like to make a few comments, that is certainly fine with me. We do have a number of witnesses this morning, so I would ask that they be brief.

Senator LAUTENBERG. Sure.

Chairman COLLINS. I will call on Senator Voinovich and offer him the same courtesy, if he would like to make any comments.

Senator VOINOVICH. The only comment I would like to make is I have been working on this subject for about 4 years.

Chairman COLLINS. You are very knowledgeable.

Senator VOINOVICH. In that amount of time, the issue was before the Environment and Public Works Committee, and now it is over at Homeland Security and Governmental Affairs. I am really pleased that you and our Ranking Member have taken it upon yourselves to have these extensive hearings on this issue. I hope that as a result of them we can come up with some legislation that is fair.

Chairman COLLINS. Thank you, Senator. You have worked long and hard on this issue, and we are very fortunate to have your expertise to help guide us as we draft legislation jointly on this issue. So thank you for your participation.

Senator Lautenberg.

OPENING STATEMENT OF SENATOR LAUTENBERG

Senator LAUTENBERG. I won't extend the courtesy, Madam Chairman, but I would hope that in the future there is an opportunity to lay out a point of view. And I, too, started on this a long time ago. As a matter of fact, before I took my sabbatical, I had proposed a chemical hazards structure so that we could identify these things. But I look forward to the hearing, and I commend you, Madam Chairman, for having held these hearings. But the change in procedure is one that I would hope would change.

Chairman COLLINS. Senator Lautenberg, we will discuss that further after the hearing, but the procedure was made very clear.

Our first witness today is Admiral Craig Bone of the U.S. Coast Guard. Admiral Bone is the Coast Guard's Director of Port Security and brings to this job more than 28 years of service to our country. It is noteworthy that he was the Deputy Commander of Activities-New York on September 11 and later served as the Commanding Officer and Captain of the Port Activities-New York, where he laid the groundwork with the maritime industry for the implementation of MTSA. We look forward to hearing his testimony.

Admiral Bone, thank you for being with us, and you may proceed.

TESTIMONY OF REAR ADMIRAL CRAIG E. BONE,[1] DIRECTOR OF PORT SECURITY, MARINE SAFETY, SECURITY, AND ENVIRONMENTAL PROTECTION DIRECTORATE, U.S. COAST GUARD

Admiral BONE. Good morning, Madam Chairman, Senator Lieberman, and distinguished Members of the Committee. I am Rear Admiral Craig Bone, Director of Port Security in the Coast Guard's Marine Safety, Security, and Environmental Protection Directorate. Today, I intend to discuss the Coast Guard's role to secure chemical facilities on the waterways of the United States.

A terrorist incident against a facility in our marine transportation system could have a disastrous impact on public safety, our Nation's economy, and international trade. Such an incident, if it were to occur in a strategic port, could also threaten our military mobilization capabilities. Clearly, the security of the chemical sector is vital and important to the protection of the public from harm.

Of more than 3,000 facility security plans that the Coast Guard has reviewed and approved under the Maritime Transportation Security Act, commonly known as MTSA, we have approved 300 for chemical facilities. This represents about 2 percent of the approximate 15,000 chemical facilities in the United States that use or store chemicals. The Coast Guard also approved an alternative security program for the American Chemistry Council. An alternative security program is an option afforded to facility operators under MTSA. Instead of creating their own facility plan, operators of facilities required to meet Title 33, Code of Federal Regulations, Parts 101 through 106, may meet an alternative security program that has been approved by the Coast Guard. Approximately 50 chemical facility operators have chosen to use the American Chemistry Council's alternative security program rather than create their own individual plans.

The Coast Guard has completed compliance inspections of all facilities that currently have facility security plans or the alternative security program to verify that they are operating within their respective plans. Since the July 1, 2004, implementation date for MTSA, the Coast Guard has taken control actions, which include restrictions to or suspension of operations, against 45 facilities. Three of those facilities were from the chemical industry.

The Coast Guard's work in implementing MTSA for waterfront facilities has been a collaborative effort with other Federal, State, and local agencies as well as the private industry partners. We have worked in conjunction with the Information Analysis and Infrastructure Protection Directorate within the Department of Homeland Security to ensure that all MTSA plans are consistent with their buffer zone protection plans.

The Area Maritime Security Committees, led by the local Coast Guard Captain of the Port, have identified their port's specific

[1] The prepared statement of Admiral Bone appears in the Appendix on page 233.

vulnerabilities and created a plan to address those vulnerabilities. The Area Maritime Security Committees, which include representatives from the oil and chemical sector, developed the Area Maritime Security Plans to address the risks specific to their ports. These area plans focus on critical port operations and infrastructure, which include regulated chemical facilities under MTSA as well as those facilities merely located in close proximity to the navigable waterways but do not engage in marine transfer operations. Such facilities would not be regulated under MTSA. These plans address how local, State, and Federal resources will be deployed to prevent terrorist attacks and protect critical infrastructure in our ports, waterways, and coastal areas.

We have developed a security matrix under Operation Neptune Shield, which is our internal plan to identify highest-risk threats and conduct operations which prevent and protect the public, facilities, and vessels from a terrorist attack. The matrix establishes a protocol of risk awareness and surveillance to include vessel traffic, air patrols, cutter presence, security zones, vessel escorts, security boardings of vessels, and positive control measures used to mitigate the vulnerabilities inherent in the ports, waterways, and maritime domain.

We continue to address highest-risk maritime operations. As such, we have contracted for a special assessment of inland barges which carry certain dangerous cargos, evaluating their vulnerabilities and identifying the blast consequence analysis.

The Coast Guard will continue to perform facility security plan compliance examinations and spot checks on waterfront facilities that are regulated under Title 33, Code of Federal Regulations, Part 105, including facilities identified as chemical, production, and storage operations. Those facilities will continue to be held to a standard commensurate with the vulnerabilities of the facility, the threat to the facility, and the consequences of a successful attack.

Since September 11, the Coast Guard has worked closely with Federal, State, and local agencies and members of the chemical industry to enhance the security of the chemical sector and the marine transportation system. We have established a robust strategy to enhance public safety from potential threats to chemical facilities located in the maritime region. We have conducted vulnerability assessments, implemented comprehensive security plans, and worked again with the Federal, State, and local agencies and industry to exercise those plans against realistic scenarios.

The MTSA has provided the foundation piece for chemical facility security in our ports. Our work is far from complete. We will build upon this foundation using a program of regular training and exercises, an annual review of plans.

The Coast Guard, in concert with the other Federal agencies, State and local authorities, and partners in industry will continue to refine the tools and analysis that aid senior leaders and first responders alike in their ability to protect, prevent, and rapidly respond to maritime transportation security incidents. We want to minimize the damage in such an incident and aid in recovery operations.

I thank you for the opportunity to testify today. I will be pleased to answer any questions at the appropriate time.

Chairman COLLINS. Thank you very much, Admiral. I very much appreciate your testimony and the expertise that you bring.

Many of our witnesses, including representatives of the Department of Homeland Security, have indicated that the framework under the Maritime Transportation Security Act might be one that we could use in drafting a chemical security bill. So I would like to ask you more questions about the specifics of the implementation of the MTSA.

First of all, how does the Coast Guard verify and enforce compliance with MTSA regulations?

Admiral BONE. Well, first off, the plans have to be approved in accordance with the regulations, so that is the starting point, which we have already completed. But then there is annual compliance examinations. We have already, again, examined each facility, and our inspectors go out with a checklist that includes the performance dimensions required of the facility to deal with their particular risk environment. In other words, not every facility is exactly the same; each facility has different types of chemicals, has different vulnerabilities. So facilities conducted—we confirmed that the facilities, in fact, conducted their actual facility risk assessment or their vulnerability assessment and then have put into place the actions necessary to protect that facility. And, again, the checklist includes such things as access controls, training of individuals, looking at realistic scenarios for threats to that facility or attempts to basically threaten that facility and cause a transportation security incident.

Again, we do not look at everything that could happen. We look at those things that would have significant consequences if it was, in fact—if someone entered improperly or took actions.

Chairman COLLINS. Let me pick upon a point that you made that not all facilities are the same. This is a point that the chemical industry representatives have made to us repeatedly, that they are not in favor of a one-size-fits-all regulatory scheme, and they have pointed to the performance-based regulations of MTSA as a possible model.

Could you explain for us the differences between performance-based and prescriptive regulations and how you have implemented a performance-based framework?

Admiral BONE. An example might be someone could prescribe 20-foot-size fences plus perimeter guards outside that have to be located 20 feet or 100 feet from the facility to address incoming traffic. That may be one standard in one highly populated area or high risk, particular high risk, but what if the facility, for example, you are worried about the cargo that is there being taken, or being used, which is a byproduct on the facility, versus the product being the target itself to cause it to explode or blow up at the location.

You may be able to do that in a different way in a different location, say if you are in a rural area on the inland rivers versus if you are sitting, as Senator Lautenberg mentioned, in New Jersey, in the port of New Jersey.

The requirements, however, that are in place are access control. Establish effective access control that will not allow someone who is not properly identified and is not supposed to be there to do business from entering your facility. There are multiple ways of doing that, both within your facility but also with the help of State and local agencies. In other words, you may hire additional security contractors and maybe people within your own. You may actually have the local authorities assisting you in establishing those access controls.

Chairman COLLINS. The focus is on the goal, not how to reach the goal.

Admiral BONE. Yes.

Chairman COLLINS. Under MTSA, the Coast Guard has the authority, I am told, to shut down a facility that is not in compliance with MTSA regulations. Has this actually happened? Has the Coast Guard shut down facilities for noncompliance?

Admiral BONE. Yes. Again, since July 2004, there have been 32 cases where we have actually shut down a facility, these facilities—not all of them chemical. When we are talking about MTSA, three of which were chemical facilities. But the majority of those, the very beginning when the program started up, some of them had not submitted their Federal security plans and as such they were not allowed to continue to operate until they had submitted them early in the process.

Chairman COLLINS. Do you think that authority is important for us to give the Department of Homeland Security?

Admiral BONE. Yes. If you have a significant violation of security such as access, illegal access or breach of the facility, and there are not proper procedures in place, then you have compromised that security, safety, and the well-being of the public, and I think that it is imperative.

Chairman COLLINS. Does the Coast Guard have authority under the MTSA to mandate changes in the storage of chemicals at facilities if you deem the security plan to be inadequate?

Admiral BONE. Well, we start off with an adequate security plan, so if they decided to move something to a different location, then we would—the Captain of the Port has the authority to seek an amendment to their security plan or actually require a modification of, again, their protective measures or their performance, or it may be a determination that they make to move it in order to continue operations.

Chairman COLLINS. Thank you. Senator Lieberman.

Senator LIEBERMAN. Thanks, Madam Chairman.

Admiral, thanks very much for your extremely helpful testimony. Some of the experts in this area that I have talked to and we have heard from have said, given the thousands of chemical facilities in this country, that they worry whether the Department of Homeland Security has the kind of infrastructure or capacity to adequately monitor and oversee implementation of security measures in this sector. In contrast, I have heard that one reason why the Coast Guard has been able to effectively implement the provisions of MTSA is because the agency does have an existing infrastructure at the ports, obviously, where the Captains of the Ports are clearly in charge of ensuring that security is improved in their areas of jurisdiction. The Coast Guard also received substantial additional resources to implement MTSA.

I wanted to ask you to reflect a little bit on the existing DHS infrastructure, as you understand it, and also how the Coast Guard determined how much in the way of additional resources it needed to adequately oversee implementation of MTSA.

Admiral BONE. Well, first off, as you stated, we were fortunate because we already had experience working in the safety and environmental arenas with these facilities and, in fact, had area committees similar to the area security committee, both for environmental and port safety operations, two separate committees.

However, we did not have the additional bodies, as you have said, in order to do that and actually called up people on Title 10 in order to do that until the Administration and the Senators and Congressmen provided the additional resources for us, which in turn was about $101 million and approximately 800 people to carry this out.

Now, when we look at what differential, what do we have to put in place in order to establish and actually execute the plans and the review of the plans and approval of the plans is separate from the execution of the compliance. But, again, part of the regulatory process, which we follow, includes looking at those alternatives, looking at the approach in order to do this and identifying what is required in order to execute it. So it is a very deliberate process, identifying—and if you are going to be having requirements that are annual, quarterly, semi-annually, that will drive your resource requirements as well.

Additionally, we look at what we are currently doing in security and what our controls were already able to capture. And the fact that we are monitoring these from a safety standpoint as well as security, our visits are more frequent to these facilities in any event.

Senator LIEBERMAN. As you probably know, Robert Stephan, the Acting Assistant Secretary for Information Analysis and Infrastructure Protection, came before the Committee, and he is the one who said on behalf of the DHS that voluntary measures were no longer enough and that the Department was working on legislation. I wanted to know whether the Coast Guard is currently involved in those efforts within DHS to try to flesh out proposals for broader chemical security safety legislation.

Admiral BONE. We are not involved in security regulation drafting, but we have, in fact, been working closely with IAIP on looking at comprehensive chemical reviews of facilities, of chemical facilities, and that process. We have also been involved with them in looking at the liquefied natural gas facilities and comprehensive security assessments of those to follow the current assessments that have already been conducted.

Senator LIEBERMAN. So beyond your answers to my first question, what lessons have you learned from implementing MTSA that could be important for Congress to keep in mind as we consider legislation to broaden the requirements for chemical safety in the country?

Admiral BONE. First, I think, is that you have an industry that is a mature industry and that is a risk-based industry that has been engaged in safety and environmental protection and actually understands risk probably better than any other group, if they are professional in what they undertake. Risk management is not a new thing. The threat vectors may be different for this group. So it is key that you engage, as we do, actually, in almost all of our rulemakings, with the industry component as you go forward and you continue that process because they have expertise that you should use.

The other is that you have to have compliance. I think that it is not just—there has to be a mandated set of requirements. A voluntary system, as we learned in our environment and in our safety system, for those that are already respectable and the best companies, they are not just going to meet what you have, they are going to exceed it. And that is true also in the security arena. And we learned that also with MTSA.

Senator LIEBERMAN. But they are not all the best companies, are they? In other words, the best companies will exceed the minimum that we set, but others need that minimum.

Admiral BONE. But we can learn a lot from those companies that set those examples, that have been doing it—actually, many of them, this is not a tremendous change to their way of doing business because identification and access control and worrying about threats, even internal individuals doing harm to their facilities because they have certain dangerous cargos that, in fact, are such high risk, many of them actually have gone down this, what I will call a decision tree process to make sure the critical links in the chain are removed so that something cannot happen.

I mean, when I think back on my career, I have worked with this industry in the past to try to prevent catastrophic incidents from a safety arena. Actually, in our experience with MTSA, they, in fact, were one of the leaders in this along with the passenger vessel, cruise ship industry, who were already involved with security operations.

Senator LIEBERMAN. Any other lessons?

Admiral BONE. I think you need to make sure you exercise your plans. I think that if you do not have exercises, if it is not drilled, people are not trained to do this. And if you do not have exercises that involve not just the people in the facilities themselves, but the vessel that may be located there, the emergency response—I heard mention that one of the emergency response agencies, the local authorities that are quite often the people that are providing that layer of defense for the security of this facility, if they are not built into it, it would be a big mistake to not include them in the drills and exercises, particularly the exercises, because that is where you find out your gaps. You put the framework together, which is what MTSA does. It puts the security framework together, hardens the targets, and allows for the entities to engage. Then the question is: How do you buy down the risk in that system? How do you collectively use that? In my experience, things have changed drastically from just response. In today's environment, if you have an incident, you not only are responding to that incident, but at the same time you are heightening security in and around the facilities.

So some of the same people that were engaged before in responding and controlling traffic now may also be tapped to go provide increased security. So you may build a plan, but until you actually exercise it, you will not find all your communications. You will not really clearly know your true resource requirements. And you may find some things that you have more than enough of and other things that you need.

Senator LIEBERMAN. Thanks, Admiral. My time is up. You have been very helpful. Thank you.

Chairman COLLINS. Thank you. Senator Voinovich.

OPENING STATEMENT OF SENATOR VOINOVICH

Senator VOINOVICH. Again, Madam Chairman, thank you for these hearings.

In the previous hearings, we have learned of the vulnerabilities of this sector and the need to adequately secure it from the threat of terrorist attacks. We have also begun to understand the patchwork of safety and security measures, both public and private, that begin to address both the safety and the security of the chemical industry in the absence of a comprehensive Federal approach to chemical security.

Before I ask my questions, I would like to express my thoughts, Madam Chairman, on this issue. Risk is inherent in business. While it is possible to manage risk and mitigate its impact, elimination of risk is impossible. Unfortunately, we continue to see the brutal nature of terrorism, and we know that the possibility of a terrorist attack is very real. That threat must be addressed by enhanced security.

That said, I want to reiterate my belief that the Federal Government cannot protect itself against every single threat, and I think that is what Osama bin Laden would like to see us try to do because in the process of doing so, we will bankrupt this country.

Therefore, I want to emphasize the importance of a balanced approach between self-regulation by industry and more proactive Federal action. Admiral Bone, I have been very much impressed with your testimony here today. It seems to me that the MTSA-approach to chemical facility security could be the benchmark for the way we go about handling this. You have come up with your standards; if necessary, the industry has come up with their alternative security system. You have approved it. You supervised it. I am very impressed with the way you are doing things. Madam Chairman, if we are thinking about who is going to run the show after we pass the legislation, it seems to me that maybe we ought to suggest to Michael Chertoff that the person should be Admiral Bone. [Laughter.]

Admiral BONE. Thank you, Senator.

Senator VOINOVICH. One of the things that I would like to know is who takes care of the facilities that you do not oversee? I understand that you are responsible for facilities or navigable water ways; but who takes care of the portion that is not in your jurisdiction?

Admiral BONE. Well, maybe I need to clarify something. Actually, this was different than the regulation that applied to transfer operations under environmental or safety in that it actually takes you to the gate. So if the facility is, in fact, a single structure or a single perimeter, then we do, in fact, when it comes to security and access control, have authority under MTSA to regulate that facility. However, if portions of a facility owned by the same company are located, as you say, on another location, for example, if one company has this plot and area and then there is another one that is completely separate and has separate access controls, separate processing, not a transfer of cargos that has a maritime nexus to it, then you are right, those facilities would be inland facilities and would not have a maritime nexus.

The key is we need some type, from water, either—for MTSA, we need from water access for transportation of goods. But from the Port and Waterway Security Act, we have authority over a facility if it is adjacent to the waterway and it presents a risk or a threat. In those cases, MTSA does not apply, but if there is a threat vector directed, say, at the chemical sector, we can, in fact, impose requirements on that facility regarding access controls and assist in that with our own assets, and that is working with the State and local.

Senator VOINOVICH. So we could look at the proximity or the nexus of facilities with that water facility perhaps, to look at who would handle those that are not in your jurisdiction.

Admiral BONE. What I would tell you is that each plan identifies the exact perimeter and the layout of that facility, so that it is clear if it is a MTSA facility or not.

Senator VOINOVICH. OK. But what I am saying is that for the facilities that are not subject to MTSA, what entity oversees them?

Admiral BONE. Right now I don't know anyone that is actually applying any standards along these lines other than the States themselves or the local communities that may have input particular requirements or safety requirements, and then did it under the guise of safety and protection of the public.

Senator VOINOVICH. What do you think about using MTSA and the alternative security system as a prototype for expanding the method to a nationwide chemical security effort.

Admiral BONE. I think it is a good model and it is a good framework, and I do not see why it would not work. I think that you have to look at the organizational constructs. You create an area maritime security committee and things like that, you may need to look at some other organizational construct further inland just because of the nature of the relationship or the entities and how DHS could best manage.

Senator VOINOVICH. And you are confident that the MTSA regulations and the alternative security system that the industry has come up with, in terms of regulations, gets the job done in terms of securing these facilities?

Admiral BONE. Yes, it has definitely improved the security of the facilities. Again, I want to make sure it is clear that this is a systemic approach. The hardening and the protective actions by the industry of the facilities is one piece of securing that facility from a terrorist event. It has to be layered, no different than from a vessel that is coming to the facility. When we look at our work overseas, if you are looking at a terrorist with intent to do harm——

Senator VOINOVICH. But what I am really interested in is that in terms of the regulations, that they get the job done. Do you feel comfortable that we do not have to come back and add another 50 pages of regulations. Do you feel that the regulations you have get the job done?

Admiral BONE. Right, I believe so.

Senator VOINOVICH. Thank you.

Chairman COLLINS. Thank you. Senator Lautenberg.

Senator LAUTENBERG. Admiral, you know that I have great respect and I would say friendship with the Coast Guard and so much appreciate how you get things done, typically with ever reduced resources to do it but more assignments. That is really an anomaly, I must tell you. But you carry on in a form that makes us all proud.

When you do an assessment of risk, do you do risk assessments throughout the ports that you have jurisdiction over or are involved in?

Admiral BONE. Yes, sir. We have completed 55 risk assessments of what we believe to be the 55 most critical economic and military strategic ports in the United States. We have completed those.

Senator LAUTENBERG. So when you do a risk assessment, you try to measure the damage that might occur if an attack takes place, and that deals with things like volatility of material. How about the density of population nearby? Does that figure into it?

Admiral BONE. We look at the threat vector, we multiple that by the vulnerability, and then by the consequence. And part of that consequence could be public safety. It could also be the economic harm. If you are looking at a port, if you are looking at a facility, then a facility—again, you will use the same assessment only if it is a microcosm of the port. But we looked at the combined port system and the vulnerabilities of that system, not just of an entity. And we also had threat assessments that were conducted as well so that you have some validity of what that threat—what is the true risk that you are trying to address.

Senator LAUTENBERG. Do you have a file, a list, a database that lists the most vulnerable and a scale that defines where all of these places stand in terms of one risk in one place compared to the other sites?

Admiral BONE. Yes, Senator. We have, in fact, identified that——

Senator LAUTENBERG. Are you familiar with the——

Admiral BONE [CONTINUING]. For ports. And then we work with DHS to look at critical infrastructure and critical assets.

Senator LAUTENBERG. So when you look at the port of New York, Newark, Elizabeth, you are, I assume, familiar with the identification of the 2-mile stretch from Newark Liberty Airport to the harbor. And that is described as the most vulnerable, most damage-susceptible place in the country.

Admiral BONE. I would hope it is not the most vulnerable now along the waterfront. But I would say that does present a very high risk. It is a high-risk environment that you have to have countermeasures for. I agree.

Senator LAUTENBERG. So you support applying resources based on risk assessments?

Admiral BONE. Yes, sir.

Senator LAUTENBERG. Because we do that on the maritime side, the port security side. And that is different than our grant programs that we have otherwise. And you know that Secretary Chertoff, the Administration, and the Chairmen of the 9/11 Commission all suggest focus on the risk and that is how we should distribute our resources. That is quite logical.

What is the difference between a risk-based security decision and threat-based? Is there a difference or is that just terminology?

Admiral BONE. It sounds to me like terminology. I think that risk includes the threat vector as well as the vulnerability or the probability of the event and the consequence when you say risk-based. If you say threat-based, then usually that is counterterrorism direct, meaning here is the threat, I go after the threat. I know exactly where it is, I counter the threat. So that may be the differential.

Senator LAUTENBERG. Admiral, I thank you for your service and the Coast Guard.

Madam Chairman, I would ask that the record be kept open because I have to go, and I have other questions. I will not be able to hear the other panelists.

Chairman COLLINS. Without objection.

Senator LAUTENBERG. Thank you.

Chairman COLLINS. Senator Carper.

OPENING STATEMENT OF SENATOR CARPER

Senator CARPER. Thanks, Madam Chairman. Admiral, welcome. I love your name. [Laughter.]

I have to slip out, in just a minute, and go to the Capitol to participate in an event with one of our colleagues, Blanche Lincoln, and some others, on a matter. As a result, I cannot ask you but maybe one question. I am going to miss the beginning of the testimonies from the panelists who follow, and I want to especially welcome Beth Turner from DuPont, and I look forward to hopefully returning to ask some questions.

My staff was good enough to prepare some real good questions, and my colleagues have asked them all. So I am just going to ask you one.

Let's say you are sitting on this side of the table, and we are sitting out there, and you are thinking about what do I do now. The hearing is over, time to craft the legislation and to introduce a bill. What would it look like?

Admiral BONE. I think that it would start with the end in mind, meaning, again, what is the risk, the loss that is unacceptable. And I would frame it around that.

We framed MTSA around a transportation security incident that looked at significant loss of life or direct impact on the transportation—significant impact on the transportation system as the baseline. You have to decide that, I think, again, inland for those facilities. Then I would craft legislation very similar to the protective measures when it comes to industry's responsibility to execute security around their facilities. And I think that you may have some nuances in that maybe trucks go to one location where we have fleeting barges in a location, maybe trucks or railroad—trains have places that they may come together different than we do for barges. But I think you are going to have to look at the nuances between the transportation systems and the storage systems and develop it from there.

But I think the framework is in place, and I think it is something, too, that we have seen that industry and a portion of the States and the local enforcement entities understand. So why would you want to create something that is so new or so different than now if I am over here on this side of the street I work this way, if I am on this side of the street, I have to do something completely different.

Senator CARPER. That sounds like pretty good advice. Thank you. Thanks for your service as well and being with us today.

Admiral BONE. Thank you, sir.

Senator CARPER. Thank you, Madam Chairman.

Chairman COLLINS. Thank you.

Thank you very much, Admiral. I think now we will be consulting with you as we begin to draft the legislation over the August recess, and I hope we can call upon you for advice.

Admiral BONE. Yes.

Chairman COLLINS. Thank you.

I would now like to call forward our second panel, which consists of three security chiefs from different parts of the chemical sector as well as a local emergency manager. Our first witness will be Beth Turner. Ms. Turner is the Director of Global Operations Security for DuPont and is responsible for the security of DuPont's operating assets around the world. Ms. Turner led the American Chemistry Council team that developed the original Responsible Care® Security Code, about which we have heard so much, and the team also reassessed the code in 2004. She is currently serving as Chairman of the Chemical Sector Coordinating Council. Welcome.

Our second witness is Jim Schellhorn, the Director of Environmental Health and Safety for Terra Industries. Mr. Schellhorn is responsible for security for Terra's North American operations. In addition to providing testimony about his own experience with security for Terra's fertilizer facilities, he will be representing the views of the Fertilizer Institute, and we thank you for being here today.

Third, we will hear from John Chamberlain, who is the Corporate Security Manager for Shell Oil Company. Mr. Chamberlain has years of experience working with Shell's refineries, chemical plants, and distribution terminals, as well as more than 30 years of law enforcement experience. He also serves as the Vice Chairman of the Security Committee for the American Petroleum Institute and will be representing both API and Shell today.

And last, but certainly not least, we will hear from Chief Robert Full, who is the Fire Marshall and Emergency Management Coordinator for Allegheny County in Pennsylvania, an area that encompasses the city of Pittsburgh. Chief Full has more than three decades' experience with hazardous materials and chemical safety. He has been a volunteer firefighter for 34 years and the county's emergency manager for the past 7 years. He chairs or has chaired the local emergency planning committee for the last 6 years.

I would also note that he has had firsthand experience with terrorism. On September 11, when Flight 93 crashed in Somerset County, Chief Full was part of the team that responded to that tragedy. Chief Full, we welcome you as well.

Ms. Turner, we will start with you. Thank you.

TESTIMONY OF BETH TURNER,[1] DIRECTOR, GLOBAL OPERATIONS SECURITY, E.I. DUPONT DE NEMOURS AND CO., INC., WILMINGTON, DELAWARE

Ms. TURNER. Thank you very much, Madam Chairman and Senator Lieberman. Distinguished Members of the Committee, it is my pleasure to be here with you today. My name is Beth Turner, and as the Chairman indicated, I am Director of Global Operations Security for DuPont. In that responsibility and that role, I have responsibility for the security of our operating assets around the world, and the Chairman has indicated some of the other roles that I have so I will not repeat those. But it is a pleasure to be here, and thank you for the opportunity.

My testimony will first address the actions that DuPont has taken to protect our employees, our communities, and our facilities, so first I will cover that; second, our views regarding the critical security legislation that we are here to discuss; and third, some brief comments on our activities and working with industry programs.

For over 200 years, DuPont has focused on safety. The founders of our company established an uncompromising commitment to safety when we opened our first gunpowder operation in Delaware, and that safety commitment continues today. Our focus on safety is driven by what we, in DuPont, know as our core value commitment to our employees and the communities in which we operate.

So, in that context, the world-changing events of September 11, 2001, compelled us to view security in a different light. Quickly after the 2001 attacks, senior corporate leadership made security a high priority by integrating it into the company's safety core value, and this sent a very strong and powerful message across the company about the importance of security. The bottom line of that change is hardening and heightening of security at our facilities across the company. We assessed over 500 locations worldwide, and we used a risk-based approach to sort these facilities into categories, and we called the highest category Category 1 facilities. A security leader was designated at each of these locations to become a focal point for security. These site security leaders have worked tirelessly since the events of 2001, and it is their outstanding work that I am so pleased to recognize today to this Committee.

These security professionals partnered with process safety professionals in our company and conducted security vulnerability assessments of our Category 1 facilities, looking at equipment, staffing, procedures, the practices we have in place, and our preparedness. We accelerated the timing for the overall vulnerability assessment process and completed our upgrades and our verification of all Category 1 sites 9 to 12 months ahead of the American Chemistry Council deadline for that work.

While I cannot speak publicly about specific measures that we took, I can describe in general terms the types of upgrades that we have implemented at our U.S. Category 1 facilities so you get an idea of the kinds of things we have done.

Equipment upgrades include fencing, motorized gates, turnstiles, signage, access control systems, video surveillance, additional light-

[1] The prepared statement of Ms. Turner appears in the Appendix on page 238.

ing, fence-associated electronic intrusion detection systems and alarm monitoring, crash gates, and barricades. We have since implemented a special maintenance program to ensure that this new equipment remains functional and reliable, and we have another round of upgrades currently underway.

Other measures that we implemented include increased patrols of site perimeters, significant reductions in traffic coming on site, more stringent identification checks, and increased inspections of rail cars, vehicles, and other trucks and other vehicles on site.

In addition, our entire workforce is very alert to suspicious activities, and I will talk more about that in just a few minutes.

Security officer staffing has been significantly increased. These officers received additional training, and they are continually retrained.

Strong process safety management is a key part of our DuPont safety culture, and it is a very important means to protect our employees and our contractors. Process safety analyses are performed to identify ongoing improvements, and they consistently include inherently safer evaluations.

We require extensive criminal background checks for all employees upon hiring and all contractors that seek access to a DuPont U.S. site.

We have long-standing relationships with local law enforcement and emergency planners, and these relationships have been reinforced. Together, we train, we drill, we exercise, we work together on investigation of suspicious activities. We are active in local emergency planning committees and mutual aid groups and, in fact, we offer our own DuPont transportation emergency response teams to assist other companies in transportation incidents.

We work with a range of trade associations and Federal Government agencies such as DHS, the Coast Guard, the FBI, and the Joint Terrorism Task Forces, and we have found government to be a very willing and helpful partner in our efforts to secure our sites.

When the national threat level is elevated, security measures are immediately reassessed by headquarters and by individual sites, even if there is no connection to the chemical industry or DuPont. Additional measures that we might implement are determined based on the specific threat environment at the time. Each DuPont Category 1 site has carefully planned for security actions that might be required in extreme circumstances, and we have an automated crisis notification system that can contact all of these sites within 10 minutes or less.

Perhaps the most powerful security measure activated since September 11 is the involvement of our employees and our contractors. They have been trained to be alert and to report anything unusual, and believe me, they do.

In summary, DuPont and our employees have done a lot, and our security enhancements are continuing. We recognize that an effective security program is a journey. It requires constant vigilance and continual improvement, and we are committed to that.

So now I would like to turn to the Federal legislation. While many security measures have been implemented voluntarily, we believe that there is an important role in government to ensure that all chemical sites are taking appropriate action. Accordingly,

DuPont supports meaningful and effective legislation and believes that ten important elements should be addressed, and I now will go through those briefly.

First, we believe the legislation must have a clear security focus so that we get the job done in a timely and effective manner.

Second, legislation should be risk-based so that government and the private sector resources can focus where they can provide the greatest benefit.

Third, we believe that regulatory authority for the chemical sector should reside with DHS. DHS and the sector are already working together and also DHS regulates portions of our sector through the Maritime Transportation Security Act, as has been discussed.

Fourth, we believe that chemical security legislation should be guided by a clear Federal program rather than a patchwork of State and local programs.

Fifth, it is important to recognize the different yet complementary roles for government and the private sector in security matters. The private sector can and should take reasonable steps to secure its facilities against threats, but it is the role of government to defend the Nation's infrastructure.

Sixth, flexibility is important. Our sector is very diverse. In DuPont alone, we operate thousands of chemical processes, employing a wide range of raw materials in both rural and urban locations. Chemical security legislation should be risk-based and allow DHS to tailor its regulations with the diversity of the sector in mind.

Seventh, is the Maritime Transportation Security Act. It has proven, in my opinion, to be a very effective security regulation for DuPont facilities, and I suggest that it be a model for regulating the highest-priority facilities.

Eighth, the work already done under programs such as Responsible Care® and the Maritime Transportation Security Act has materially enhanced security, and these prior efforts must be credited.

Ninth is the protection of sensitive security information, and protection of that information is critical. We must obtain strong protection for information that we need to ensure does not get into the hands of the wrong people.

The final issue is inherent safety, commonly referred to as inherently safer technology, or IST. As the Committee knows, IST is a process safety matter, and we believe that it should stay with the safety arena and not be mandated in the chemical security context. DuPont believes that inherently safer technology is a mainstream component of process safety and that it has an important role to play in security. And inherently safer has not only been an integral part of our process safety system for many decades. In addition, it is now part of the security vulnerability assessment process that we all ran and the teams that conducted those assessments included both security and safety professionals at the table, the safety professionals being the ones that understand IST.

Each chemical process is complex and unique, a complex array of piping and pressure vessels, tanks, pumps, valves, raw materials, and operating conditions at a variety of temperatures and pressures. So given the complex and unique nature of each process, safety evaluations do require special expertise and consideration of a wide range of possibilities for inherently safer operation. Therefore, companies must have the flexibility to assess and decide upon options.

I was also asked to comment on the Chemical Sector Coordinating Council in my role as Chair of the group. I am pleased to report that the council is strong, and after only one year of existence has tackled a number of substantive issues. I can speak further about the council during the question-and-answer session.

In closing, Madam Chairman, I want to thank you and the Members of the Committee for allowing me to share what DuPont has done to build a strong security system and process in place in our operations. We have very successfully integrated security, engagement, and responsibility into our culture, and we know there is more to do. We take this responsibility very seriously, Madam Chairman, and we appreciate the trust and the confidence that has been placed in us by the public and government. Therefore, we are taking the necessary actions to appropriately harden and heighten security across the company.

Our corporate leadership is very committed to continually strengthening security. Security and safety of our operations are critical to our employees and neighbors and, in fact, are essential to the future of the company.

Thank you very much for the opportunity to testify. We appreciate the important work of this Committee, and we have enjoyed working with you to date and hope we can do that in the future.

Chairman COLLINS. Thank you. Mr. Schellhorn.

TESTIMONY OF JIM SCHELLHORN,[1] DIRECTOR OF ENVIRONMENTAL HEALTH AND SAFETY, TERRA INDUSTRIES, INC., ON BEHALF OF THE FERTILIZER INSTITUTE

Mr. SCHELLHORN. Thank you. Madam Chairman and Members of the Committee, I am Jim Schellhorn. I am the Director of Environmental Health and Safety for Terra Industries and am responsible for security for Terra's North American operations. I am here today to testify on behalf of The Fertilizer Institute. TFI is the leading voice of the Nation's fertilizer industry, representing the public policy, communication, and statistical needs of manufacturers, producers, retailers, and transporters of fertilizer. I appreciate the opportunity to appear here today.

Terra is headquartered in Sioux City, Iowa. We are a leading international producer of nitrogen fertilizers. Our primary products are anhydrous ammonia, ammonium nitrate, urea, and urea ammonium nitrate solution. Our facilities operate 24 hours a day, 7 days a week, and Terra employs approximately 1,200 people in North America and the United Kingdom. We are proud of the vital role the fertilizer industry plays in modern agriculture.

Fertilizer is essential to food production. Without the contribution of our fertilizers to crop production, roughly one-third of the world's population would be without food. Because food production depletes soil nutrient supplies, farmers rely on fertilizers to keep the soil productive. With the help of commercial fertilizer, North

[1] The prepared statement of Mr. Schellhorn with an attachment appears in the Appendix on page 253.

American farmers are able to produce the most abundant and affordable food in the world.

The fertilizer industry is very diverse. Companies such as Terra produce and sell fertilizer into the retail distribution system, which in turn sells it to farmer customers. Most of our production and storage facilities, like many others in the industry, are located in rural communities. For instance, Terra's Verdigris plant, where I work, is located in a rural area of northeast Oklahoma near the Tulsa port of Catoosa. Because we produce and store anhydrous ammonia and because our operations include a waterfront facility, the Verdigris plant is subject to many Federal safety, security, and environmental regulations, including OSHA's process safety management standard, the U.S. Coast Guard's facilities security regulations under the Maritime Transportation Security Act, or MTSA, and EPA's risk management program requirements. Companywide, in the United States, Terra has five locations subject to MTSA and nine locations subject to PSM and RMP requirements.

Shortly after the events of September 11, TFI formed a security task force, of which Terra is a member. In September 2002, TFI's security task force developed and the board of directors adopted an industry Security Code of Management Practices designed to help the fertilizer industry secure the manufacture and transport of its products.

The voluntary code calls on the industry to use methodologies developed by the Center for Chemical Process Safety, the Synthetic Organic Chemical Manufacturers Association, or an equivalent methodology when conducting security vulnerability assessments, or SVAs, and when making security-related improvements.

The code establishes benchmarks for conducting SVAs and implementing security measures, for conducting employee training and drills, for communicating with law enforcement, conducting audits, and verifying physical site security measures through a third party, and the code provides timelines for these activities by ranking facilities at high, medium, and low risk levels.

I would like to take a moment and discuss specific measures Terra has taken and continues to undertake to secure our facilities and the products we produce.

After TFI developed the security code, we immediately began to conduct security vulnerability assessments and audits at all of our facilities. We used both outside law enforcement experts and internal resources to identify vulnerabilities, implement countermeasures, and develop security plans. The process we utilized ranked both our facilities and the vulnerabilities we identified based upon risk. Using those rankings, we began to address the highest risks first.

Since September 11, Terra has installed many physical security improvements, including additional lighting, fences, physical barricades, and video monitors at strategic locations. All gates are locked when unattended, and facility access is tightly controlled by security or Terra employees 24 hours a day, 7 days a week. All of our product carriers and drivers are pre-approved. All deliveries to our facilities are checked at the gate prior to authorizing access. And criminal background checks are required for all contractors and all Terra employees.

We have also recently implemented a system to ensure delivery receipts for all truck shipments of ammonium nitrate from Terra-owned facilities. All of our facilities now have active security plans, and our waterfront facilities are in compliance with the Coast Guard facility security regulations.

Terra Industries and other members of TFI have undertaken tremendous efforts to ensure that criminals intent on harming our country cannot purchase and misuse fertilizer products. For example, after the tragedy in Oklahoma City in 1995, the fertilizer industry partnered with the Bureau of Alcohol, Tobacco, Firearms, and Explosives in outreach programs called "Be Aware for America" and "Be Secure for America," which were aimed at protecting our products and our places of business.

After the terrorist attack on September 11, the fertilizer industry launched "America's Security Begins with You," a new program, which has been endorsed by ATF, the Department of Homeland Security, and the Association of American Plant Food Control Officials, who regulate fertilizer at the State level. The campaign urges that security plans be developed and implemented, records of sales be maintained, and that law enforcement be alerted to any suspicious activity.

These programs have primarily focused on ammonium nitrate, the fertilizer used in the Oklahoma City bombing. Recognizing the changing nature of the Nation's security, Senators Cochran, Pryor, Roberts, and Chambliss recently introduced the Secure Handling of Ammonium Nitrate Act of 2005. The bill directs the Department of Homeland Security to promulgate regulations requiring all facilities that handle ammonium fertilizer to register at the State level and maintain records for all purchases of ammonium nitrate. The fertilizer industry's support of the Senate legislation—and parallel legislation introduced in the House—takes the industry's voluntary programs to the next level through the creation of a uniform Federal set of rules for sellers and purchasers of ammonium nitrate.

We believe that chemical facilities will most effectively address security when given the flexibility to use measures that will address the risks specific to each facility. Quite simply, we at Terra and others in the industry have not employed a one-size-fits-all approach at our facilities, and we believe that legislation requiring us to do so would be counterproductive.

Equally important, Congress must recognize the security measures already taken and facilities covered under other Federal regulations, such as the Coast Guard's facility security requirements, to avoid duplicate regulations.

There has also been considerable debate over whether Congress should mandate the use of inherently safer technologies, or IST. IST is not a security measure. It is a safety concept that has been misapplied by some groups in a way that we fear could lead to the ban or restricted use of basic nitrogen fertilizers. For instance, if anhydrous ammonia manufacture was banned in the United States as a result of an IST mandate, there would be no nitrogen fertilizer manufacturing in the United States because ammonia is the basic feedstock for all other nitrogen fertilizer. U.S. farmers would have to rely on imported fertilizer to grow their crops, and indirectly, the

American public would have to rely on foreign fertilizer for their food supply.

Terra and the fertilizer industry are not opposed to evaluating process safety of our operations and considering potential safety improvements. On the contrary, process hazard analyses and risk assessments we have conducted as part of our PSM and RMP programs and the security vulnerability assessments we have performed include consideration of ways to minimize hazards. However, this type of hazard assessment can only work when applied by a site owner's engineers and safety professionals who truly understand the facility's operations.

Madam Chairman and Members of the Committee, American farmers, fertilizer producers, and retailers are committed to security. We have demonstrated that commitment through the significant number of voluntary security steps we have taken and will continue to take. Without question, we very much want to help Congress in its endeavors to shield this country from acts of terrorism. We support Department of Homeland Security Secretary Chertoff's efforts to evaluate the Nation's vulnerabilities and prioritize the Federal Government's response based on risk assessment.

As the Federal Government proposes its suggestions for chemical facility security legislation, we recommend such proposals be based on reasonable, clear, and equitable performance standards. TFI and its members believe that to be effective, fair, realistic, and feasible to implement, the legislation must: Provide for the varying levels of risk posed by different kinds of chemical facilities; recognize the security measures our industry has already taken and complement Federal regulations with which we already comply; and reject attempts to mandate inherently safer technology.

Furthermore, we urge that the Federal regulations preempt any such action by State or local governments. Layering Federal regulation upon a patchwork of State regulations is, at best, inefficient and, at its worst, an impediment to efficient compliance.

I thank you for the opportunity to testify and look forward to answering any questions you might have.

TESTIMONY OF JOHN P. CHAMBERLAIN,[1] SECURITY MANAGER, ASSET PROTECTION SERVICES, CORPORATE SECURITY, SHELL OIL COMPANY, ON BEHALF OF SHELL OIL COMPANY AND THE AMERICAN PETROLEUM INSTITUTE

Mr. CHAMBERLAIN. Chairman Collins, Ranking Member Lieberman and Members of the Committee, my name is John Chamberlain. I am a Manager with Corporate Security for Shell Oil Company. I also serve as the Vice Chairman of the Security Committee for the American Petroleum Institute. I have many years of experience working with Shell's energy operations, and also 30 years of law enforcement experience.

I am pleased to appear before you today to testify on the issue of chemical security, representing both Shell Oil and the American Petroleum Institute, API.

---

[1] The prepared statement of Mr. Chamberlain appears in the Appendix on page 264.

The U.S. oil and natural gas industry is committed to protecting the reliable supply network of fuels and products to keep our economy growing. Our industry has long operated globally, and often in unstable regions overseas, where security is an integral part of providing for the world's energy needs.

After September 11, 2001, the industry partnered with Federal, State, and local authorities to reevaluate and strengthen our domestic security. Within months of the attack, the industry developed security measures for all segments of the oil and gas network, including pipelines, refineries, terminals, and others.

One reason I believe the industry was able to move so quickly is that we have high caliber security professionals with both military and law enforcement backgrounds on our staff. These former FBI, Secret Service, and Delta Force personnel are experts in physical security, and they are employed protecting our industry's assets. A large number of security personnel in the oil and gas industry, including myself, have security clearances necessary for classified briefings we have with the Federal intelligence community, and that is important.

I want to speak briefly about two areas: one, the numerous broad actions to address security in the energy sector that we support, including industry actions, Federal security laws, and public/private sector partnerships; and two, I want to talk about specific proposals that we think would be counterproductive to security.

Although it is rarely reported on, the oil and natural gas industry, in partnership with government agencies, has taken quite thorough and painstaking actions to improve security. We have operated under new Federal security law, Federal security partnerships, industrial practices, and enhanced intelligence sharing networks, and we support these ongoing efforts. Little has been communicated about the actions that Congress, industry, government agencies, State and local first responders have taken. Examples of these actions are—what we heard today—the Maritime Transportation Security Act, the TSA background check requirements under the PATRIOT Act, and the Department of Transportation's security requirements for hazardous materials, all security laws that we operate under and support.

The industry collectively created industry-wide methods to address two stages of security, first finding the weaknesses and then protecting them. First API and the National Petrochemical Refiners Association produced the methodology for SVA or Security Vulnerability Assessment. This is a method for managers to identify security vulnerabilities in the wide range of oil and natural gas operations. This SVA methodology is sophisticated. It is a risk-based tool used to identify the security hazards, threats and vulnerabilities. We co-wrote this with the Department of Energy's security personnel, and DHS today is using this methodology to train their field inspectors.

I would like to submit a copy of this document for the record.[1]

Chairman COLLINS. Without objection, Mr. Chamberlain. Thank you.

---

[1] The document entitled "Security Vulnerability Assessment Methodology for the Petroleum and Petrochemical Industries, Second Edition," October 2004, American Petroleum Institute, NPRA, submitted by Mr. Chamberlain appears in the Appendix on page 277.

Mr. CHAMBERLAIN. In addition, this security tool is accepted by the American Chemistry Council's Responsible Care Code and is an example of the government-recognized industry practices that are now in operation in this business.

API and Federal security personnel next completed the Security Guidelines for the Petroleum Industry. This booklet instructs operators and plant managers in how to protect facilities and respond to changes in the threat level. The third edition was completed earlier this summer. These are working methods and countermeasures the oil sector uses to protect all segments of industries, and I would like to also submit this after testimony.[1]

Chairman COLLINS. Without objection, Mr. Chamberlain. Thank you.

Mr. CHAMBERLAIN. Some legislators may be tempted to treat security as a concern to be addressed with proscriptive inflexible regulations. This would result in a one-size-fits-all approach that provides a roadmap for terrorists in my opinion. We ask that you recognize that a terrorist, unlike a pollutant or physical workplace environment, is clever, deliberate, and has the ability to adapt against a checklist of arbitrary rules. This is one reason we value our close professional partnership with government, industries, and the intelligence community.

Let me give you an example of a more risk-based approach. Like other integrated oil companies, Shell and other API members have joined with the Department of Homeland Security in developing a common system for comparing security risks across the Nation's very critical infrastructure. The system is called Risk Assessment and Management of Critical Asset protection and has the acronym of RAMCAP. It will give Congress and the Executive Branch, through the Department of Homeland Security, the tools they need to make decisions and allocate resources for security. We support the risk-based concept being adopted in the RAMCAP program.

Overall, we hope that you would avoid provisions that would be counterproductive to the gains that we have made in security since September 11. There are specific proposals that we have concern would be disruptive to our industrial security operations. Although we are in the energy business, some proposals to address the security of chemical sites could affect the energy industry, as well as agricultural, water treatment, food, dairy processing, and other small businesses. These U.S. industries and farms are essential for our national security and economic vitality and are not traditionally thought of as chemical industry facilities.

Concerning inherently safer technology, we strongly oppose any environmental mandates for inherently safer technology pursued under the guise of security. It would be counterproductive to protecting our infrastructure. Security law covering companies should be risk-based and not seek to legislate out the elimination of all risk, which quite frankly is impossible. Private farms and company facilities that need to use dangerous substances intensify their security plans based on the risk level.

[1] "Security Guidelines for the Petroleum Industry," American Petroleum Institute, April 2005, submitted by Mr. Chamberlain appears in the Appendix on page 428.

Infrastructure security laws already passed by Congress, such as the Maritime Transportation Security Act, the Bioterrorism Act, require vulnerability assessments and security plans for private facilities and vessels, but they do not create a new requirement for IST. In fact, no other security law requires IST and that is for good reason.

First, creating an inherently safer technology requirement for farms and businesses and others in the name of national security could actually increase risk. For example, in reducing volumes of hazardous chemicals stored at a facility, you may reduce the on-site risk, but consequently you could increase the transportation risk where the material has to be transported by rail, truck, or barge traffic to the site that used to keep it on site, and this could potentially increase risk to the overall system.

Under new IST authority, a government order for a change to materials or processes could very well result in accidental or intentional harm and create a new liability for complying with the law. Process safety concepts are already incorporated under existing Federal health and safety requirements. They are both in the Occupational Safety and Health Administration's Process Safety Management Program and the EPA's Risk Management Program.

American farms and companies will continue to comply with Federal and State and local requirements as they are today. Farms and company facilities, through self-interest, consider the safest, most innovative and cost-effective technology as they do business. New government mandates for IST could require bureaucrats without expertise and courts to determine the best technology of businesses. Creating a new security IST authority will allow government micro-management in mandating substitutions for all processes and substances, and this would greatly inhibit and limit operational flexibility and innovation.

I want to mention information protection, too. It has been mentioned earlier. But in addition to FOIA exemption, I believe information protection is extremely important in anything to do with security legislation. I would like to see additional protections made to prevent the leak of vulnerability information which could provide a roadmap to terrorists or other criminals. Any information developed in regard to this security legislation should be protected from civil discovery.

I want to mention, too, MTSA. We have heard a lot about it already. Should the Committee conclude that new legislation is needed, we would suggest that it not apply to facilities already covered under the existing MTSA legislation. We would also suggest that sites that contain areas only partially covered by MTSA have the option for the entire facility to be covered by MTSA instead of a new law, something the Senator questioned the Admiral about earlier. We would support that as it would avoid conflicting regulations in a single facility, to have part of it under MTSA and part of it under some other requirement when you have a common management for the site.

Examining the MTSA security law, I would like to highlight a few characteristics for your consideration. In implementing a broad new security law, the Coast Guard has overall done a successful job without impeding the commerce it protects. This is a credit to the

Coast Guard century-long experience in protecting onshore and offshore commerce, as well as the existing relationships of local stakeholders and the respective captains of the port. Without this security expertise and these relationships with private sector operations, the MTSA would not have been able to be successful. Many agencies do not have the security expertise of the Coast Guard and should not have responsibility for counterterrorism.

Like the MTSA, other Federal security laws have protected and strengthened our infrastructure, instead of having a Federal bureaucracy attempt to redraw or micromanage how private operators function. In other words, we believe that a security rule or law has to be a risk-based philosophy. The required security protections need to meet the risk under which the facility is operating.

In conclusion, oil and natural gas operations are safer now and more secure as a result of the public/private partnerships and numerous new Federal security requirements. We urge the Committee to carefully consider the effect any new Federal law would have upon existing security laws, industry practices, and the partnerships that have been developed with government thus far.

The oil and gas industry is committed to protecting the reliable supply, supply network of fuels and products to keep our economy growing, and whether or not new security legislation is passed, we are going to continue to work with the government to consistently reevaluate and improve security of U.S. oil and gas operations.

I thank you.

Chairman COLLINS. Thank you. Chief Full, welcome.

TESTIMONY OF CHIEF ROBERT A. FULL,[1] FIRE MARSHAL/ EMERGENCY MANAGEMENT COORDINATOR, ALLEGHENY COUNTY (PA) DEPARTMENT OF EMERGENCY SERVICES

Mr. FULL. Good morning, Chairman Collins and Senators. It is a distinct honor and privilege to be invited here to testify on behalf of chemical facility security today and its impact to the local and county level of government. My County Chief Executive, Dan Onorato, extends his appreciation for this opportunity as well.

I speak this morning as a 30-year first responder as a firefighter, paramedic, and a hazardous materials technician, as both a career professional and a volunteer firefighter from Allegheny County in Southwestern Pennsylvania. I serve as my county's local Emergency Management Coordinator, and also the local Emergency Planning Committee Chairperson. I also have had the privilege to serve as the chairman of one of our Regional Counterterrorism Task Forces in Pennsylvania, representing 13 counties, a population of 3.1 million people, which would also include the city of Pittsburgh.

Allegheny County in Pennsylvania has the city of Pittsburgh as its county seat and is famous for Three Rivers, steel making, research centers, world class medical systems, education institutions such as the University of Pittsburgh, Duquesne, and Carnegie Mellon, major transportation systems, and the Pittsburgh Steelers and Pirates. The county covers some 730 square miles with a popu-

[1] The prepared statement of Mr. Full appears in the Appendix on page 272.

lation of 1.3 million residents, and most unusual, with 130 separate local municipalities.

This morning as I awoke early to fly here, I took a shower and made my coffee with crystal clean and safe water. My clothes have synthetics in them. The breakfast fruits that I enjoyed were free from bacteria and were hearty from the vine. The fuel in both my car and the airplane I flew in worked extremely well today. As I look around here I see so much of the positives and the need for a strong and safe chemical industry. It has been said and reinforced that one of the main reasons the United States enjoys the highest standard of living is through the use of our chemicals in all aspects of our daily lives.

On behalf of those that I represent, the first responder community and local government, we could not agree more in the need to support and protect our chemical industry. I am humbled to be with such fine representatives of the chemical industry. I know personally firsthand the representatives from these various organizations have done an outstanding job in working with us at times at the local level to provide us training and resources so we can better serve the public.

As a first responder, paramount to the success of doing your job is to be able to protect lives and property during emergencies. An individual comes into public safety as a first responder and he/she is primarily trained to deal with the aftermath of an incident which was caused by perhaps an accident, an act of God, or an intentional act.

Every day in this country the men and women of our public safety departments, police, fire, emergency medical services, 911, demonstrate great courage and conviction to be the best they can be. These folks plan, train, exercise, and respond to any emergency no matter what the case. No matter how good a public safety organization is, there will be times that their training, skills, knowledge, capabilities will be overwhelmed, or they may not have the expertise to deal effectively with the situation.

To minimize this scenario, having a strong emergency plan and relationships with pertinent persons in advance pays dividends each day at the local level across America. It is cliché, but it is not the time or place to exchange business cards during the time of an emergency.

I would like to focus now on chemical safety. In 1986, the Federal Government enacted the SARA Title III, Emergency Planning and Community Right to Know Act. The overall success of this law cannot be overstated and can be measured in my county and throughout the country by the reduction in chemical spill emergencies, better informed employees and responders during emergencies, Federal, State, and local government input and coordination, and so much more.

In my career I have had an opportunity to specialize in hazardous materials response emergencies. I was the first city of Pittsburgh Hazardous Materials Chief and served in that capacity for 13 years, and today I oversee five hazardous materials teams in my county, and I have logged in excess of 2,000 responses to hazardous material emergencies.

I have come to see firsthand the potential life-threatening situations that are involved when chemicals are accidentally or intentionally released from their containers and processes. The chemicals and materials are found in fixed facilities during production, transfer, storage, and along with the transportation to and from market via highway, railroad, water, air, and pipeline. Responding to chemical spills requires quick informed decisionmaking along with specialized tools and equipment. Incidents of vapor clouds, running liquid spills, unidentified products, and fires severely complicate local response actions, many times to the point that a community may not be able to react fast enough to save its residents. Transportation accidents involving chemicals provide even a greater challenge as they move in and out of our neighborhoods, by our schools, homes, and places of business.

The SARA Title III law targeting fixed chemical facilities, followed by similar legislation in my Commonwealth of Pennsylvania, has directly contributed to saving lives, property, and the environment. The SARA law has allowed us to be proactive through planning, training, and networking versus reacting and always responding to the unknown and not knowing the players when you get there. The Federal Government has served us all well with this law, but we need to update some of the provisions to meet the needs of today.

I believe we all knew it would come some day or another, and never did any of us expect it to come in a manner so coordinated with such devastating results as it did on September 11. It did, and we should have learned from it and should not forget. I was always told by my father that mistakes and accidents can and will happen. Most importantly you learn and work to make sure that you do not make the same mistake twice. We may have missed it the first time to a degree, but let us do everything to prevent it from happening a second time. The next time when it comes, we are told by the top security minds in our government, it may be greater in magnitude with even more loss of life and property, utilizing weapons of mass destruction, involving chemicals, biological, nuclear, radioactive materials, and explosive devices. We need to get and be ready now.

At the local government and first responder levels we are concerned that our residents believe that we can protect them effectively against the threat of WMD and chemical releases from a terrorist act, which could easily be one of our own chemical facilities in our neighborhoods.

Our men and women on the front lines in our communities have been working hard in getting some of the special training and have begun to reap the benefit of some of the generous homeland funding that has been provided by this Congress and the President by putting new specialized equipment in the hands of first responders and local governments and extra training. The sharing of intelligence between the levels of government has not been better. However, we are not where we need to be as of yet and have a long way to go, but we are better off today than we were yesterday.

Terrorism threat assessments and uniform strategies that deal with them are a common requirement and a need at all levels of government. In looking at all the potential hazards and threats to our communities, chemical facilities and their transportation rise with just a few others to the very top. It is not that we do not know what is in the plant or what is being transported in most cases. We do, through the impact of the Federal and State laws. But we do not know for sure what safety and security measures are in place to keep something or someone from getting to them. Can the bad guys use them against us? The fact is that there are some chemicals and materials, if released from their containers for whatever reason or by a terrorist, that can greatly cause injury and death to our unprotected public. We have to make sure that we do everything in our collective powers to make sure that we understand and make chemical facilities and their transportation safe.

Madam Chairman and Members of this Committee, today you are hearing from some of the most notable and responsible chemical companies in our country. I have had the opportunity to work with these folks and their people in safety and response personnel. The communities are top notch, well trained. They have excellent plans. They are in good financial condition and have in most cases good security systems.

Unfortunately, that is not the case around the country for many of us at the local level. There are so many other companies that are in our neighborhoods that are less fortunate that really concern us and pose a unique risk. These companies will not do anything unless there is some force of law to cause them to do it.

The American Chemistry Council has done a good job in stepping up to the plate with providing a voluntary program with materials and training on chemical plant security. A problem exists that it is voluntary, and second, not all companies belong to the Council, especially in my county.

Today we have an opportunity to be proactive versus reactive. Chemical plant safety and transportation is an issue that needs to be and should be addressed on a national level to ensure uniformity, and not at the State level, even though my State government has a fabulous State law that was enacted utilizing the SARA law and additional legislation from the Federal Government as a template.

I do not have a political or legislative expertise on whether or not a new law or tweaking an old one is the best way to go. I leave that, and the people that I represent, we leave that to you.

I was around in the 1980s when there was a great outcry from the chemical industry about how the SARA Title III law was unnecessary and that the industry voluntary program for planning and response was more than adequate. The law almost was not enacted. It took a real wake-up call. It took several thousands of folks to die in Bhopal, India, coupled with an incident in West Virginia that was just on the brink of catastrophe to raise enough concern that our Congress enacted the law.

Today we hear some of the same in different forms, or you have heard some of the same perhaps from other folks testifying before you in the past. Security, trade secrets, plans, products, we have heard it all before. What if it gets out, etc.? Together we can work it out.

The local governments and the people that are going to be responding to these incidents need to be a part of the process and be part of the solution because we are the ones on the front lines who are going to be out there responding when that emergency comes in, no matter what. We speak to international terrorism, but we know that we have grown some phenomenal terrorists at home as well. That is not to say that even a domestic terrorism event cannot be superseded by somebody who is mentally deranged, whether he is an employee of the company or not.

I do not know of many trade secrets that have been given up or critical information that has been given out. If so, then that information and those folks that made that available inappropriately should be held accountable and sanctions for doing so should be applied.

LEPCs have been a great tool to ensure effective planning and community safety. We can have experienced security people look over the plans as necessary. I do not advocate LEPCs as a policing agency by any stretch of the imagination for security. We can utilize the JTTFs, which are in place around the country. We have great relationships with U.S. Coast Guard, Department of Homeland Security, and our local law enforcement, but I do advocate that we cannot appreciate or effectively plan for incidents within our jurisdictions without the full benefit of all aspects of the hazards, the risks, and the vulnerabilities that we face.

The public is counting on us. I know my residents are counting on me and the 10,000 first responders in my county. Shame on all of us if we wait until it is too late. We can do something now, and we should move forward. Thank you.

Chairman COLLINS. Thank you, Chief. I think your local Chamber of Commerce ought to give you a special award for getting in all of the advantages of the county in which you live, and some beyond them as well.

At the end of your testimony you talked about an issue that we are going to have to deal with as we draft legislation, and that has to do with information sharing and the protection of sensitive security information. Under MTSA, the vulnerability assessments and security plans for individual facilities are maintained by the Captain of the Port, and a copy is also kept at the Coast Guard headquarters in Washington, DC. Do you believe that local law enforcement ought to also have access to or a copy of the vulnerability assessment? Where do we draw the line?

I will tell you that one chemical company told me that the Coast Guard actually lost its security plan. And I have great respect and admiration for the Coast Guard, but it seems to me if we are concerned about information that that was not a good indicator. But who should have access? Where should these vulnerability assessments, which obviously contain very sensitive information, and security plans be kept and who should have access to them?

Mr. FULL. Well, clearly, we have heard today, Senator, that there is an outstanding program that goes on where the Coast Guard does deal with the maritime issues of chemical plant security. I believe that those files, they are kept with the Captain of the Port, is just that, they are kept with the Captain of the Port right now.

I would argue the fact that a good bit of that information that has probably been developed has been developed without any local input or any knowledge of the local responders that may be involved with that in concert with local law enforcement or anybody that is familiar with security aspects from anywhere other than the maritime folks.

Chairman COLLINS. Mr. Chamberlain, what is your answer to that? What is your advice to the Committee on how can we strike the right balance between ensuring the security of this very sensitive information, and yet making sure that if someone like Chief Full, who is going to be called upon to respond, understands what security issues or vulnerabilities may exist at a plant?

Mr. CHAMBERLAIN. The facility security plan goes into tremendous amount of detail on single-point sources that could shut down your facility, basically your Achilles heel, and that is what you are going to identify and then protect against. Those types of things I think need to be kept classified, as they are today.

We work closely, and our facilities, wherever we operate, have close relationships with local first responder groups. We usually have various law enforcement and safety committees that we are active on, so we are not trying to surprise anybody in the types of issues that they may need to be responding to. The response is typically going to be after the fact, after something has occurred. Part of the plan is to try to prevent something from occurring. I think what you have under MTSA is a very workable approach. It has worked so far so well.

Chairman COLLINS. Mr. Schellhorn, should chemical facilities maintain vulnerability assessments and security plans on their site or on file with the Department of Homeland Security in Washington? What are your views on this as we are drafting legislation?

Mr. SCHELLHORN. They certainly should maintain a copy of the vulnerability assessment on site, and we do that now under MTSA, and a copy of that vulnerability assessment and plan, I would think, would be submitted to the regulatory authority, like we do now under MTSA. I do not think a copy should be submitted to the local fire department or emergency management authority. My personal opinion is you want to limit the distribution of those plans and vulnerability assessments.

However, what we have done is we invite the local authorities, the local emergency management agency, the LEPC chairman, the local law enforcement authorities to our facility. We share the details of our security plan and our vulnerability assessment with those individuals at our site so that they are familiar with what we are doing and familiar with the details of our security program.

Chairman COLLINS. Ms. Turner, what is the right balance here from your perspective? How do we ensure that this very sensitive information does not fall into the wrong hands, and yet make sure that first responder groups or those who would be called upon to act in the event of a terrorist attack on a chemical facility or an accidental spill do have the information they need? What is your advice to us?

Ms. TURNER. I think it is extremely important that the first responders have access to the information they need in order to know what to expect from the hazards that they are going to be responding to, and that information is freely shared today so that our first responders know the hazards they could encounter, what kind of equipment they need to have with them, and that is very important to keep that information there so they can access that.

We might think about that information as different than the vulnerabilities that are associated with getting into a chemical facility, as you just said, whether it is accidental or intentional, the nature of the chemical information is what you need, different than the vulnerability and separate from the vulnerability of the facility from a security standpoint.

Now, on that latter information we do, as was just said, we keep our vulnerability assessments on site, and then we vet the person who wants information, and we are pretty free with showing it to people that have a need-to-know basis, and I think that is the right thing to do. But it is very sensitive information that we want to be certain is properly secured, and in fact, that is why in my testimony I indicated that beyond MTSA we do need a framework for protecting that vulnerability information.

Chairman COLLINS. Thank you. Senator Voinovich.

Senator VOINOVICH. I would like to pursue your line of questioning.

Chief Full, you are the Emergency Management Coordinator, Allegheny County; is that right?

Mr. FULL. Yes, Senator.

Senator VOINOVICH. Do you have a list somewhere in your office of the chemical facilities that you have in Allegheny County?

Mr. FULL. Yes, sir, we do. In Allegheny County we have 235 chemical facilities which are required under the SARA law to have emergency plans and have reported the amount of chemicals and so forth within the facilities.

Senator VOINOVICH. Have the chemical facilities in the county for the most part done their threat assessment?

Mr. FULL. We believe that the majority of them, but there are some of them that are on the threshold of reporting now through the process of even some of the outreach of the SARA Title III law and the reporting and so forth. We find that there is more and more chemical companies that are reducing their amount of stored materials, putting them in the transportation stream, and falling out of the need for them to report.

Senator VOINOVICH. Is it mandatory that a fertilizer company share with you their vulnerability?

Mr. FULL. No, sir, not whatsoever. That is why the suggestion is——

Senator VOINOVICH. Do you think it should be mandatory?

Mr. FULL. I believe that we can go into exactly—the Federal Government address in the SARA Title III law—first off, again, our experience has been we have held close trade secrets. We are familiar with that. We are certainly not going to give up the ship here. We are just as interested as the corporate chemical facilities to make sure that it does not get into the wrong hands, but at the same time we believe also that we are responsible and in the law it already addressed that fire and local fire folks can get the fire information, medical——

Senator VOINOVICH. How about the "Right-to-Know" laws? Has there been pretty good compliance with those?

Mr. FULL. Yes.

Senator VOINOVICH. So to clarify, your fire department has on file what chemicals are on the premises.

Mr. FULL. Right.

Senator VOINOVICH. So with the "Right-to-Know" law, the community has access to knowing what is on file there, correct?

Mr. FULL. There is nothing on security though, sir. There is nothing of them to share it with us at all other than——

Senator VOINOVICH. But the fact is that you would not want them to share that information with, say, the community. You would like a provision that provides the necessary information to those that will be responding, and that allows you to have a good idea of the vulnerabilities, so that you have a better idea of how you would coordinate with them to respond if something happened.

Mr. FULL. That is what we are asking for right now, right.

Senator VOINOVICH. Mr. Chamberlain, how do you feel about that, and Mr. Schellhorn and Ms. Turner, how do you feel about that?

Mr. CHAMBERLAIN. I would like to make a distinction between responding to an emergency—which certainly the Chief and the first responders do—an emergency has usually already occurred, and the security plans, the security vulnerability assessment, and the facility security plan also address prevention, what you are doing to prevent an emergency. There are no chemicals or products on site that our first responders do not know are there. We are not trying to hide anything at all. It is merely the sensitivity of giving somebody a roadmap on how to shut you down or how to do damage that you want to carefully control.

And certainly, I think, MTSA does that today. I would encourage any future legislation would have that sensitivity in there. We want people to know what they are going to be coming into if they are coming out to assist with an emergency.

Senator VOINOVICH. I would be interested to get your best thoughts on how you would get that done.

Mr. Schellhorn—fertilizers—how much more has your product gone up because of natural gas costs? [Laughter.]

Thank you for being in business.

Mr. SCHELLHORN. It has not gone up as much as the natural gas price has gone up, I assure you.

I would like to add something if I may to what Mr. Chamberlain just said.

Senator VOINOVICH. Sure.

Mr. SCHELLHORN. Additionally, communicating with neighbors about what to do in the event that there is a release is extremely important, and Senator Lieberman touched on this in his opening statement. It is very important that neighbors know what to do, that they know when there is an incident, they know how they are going to learn if there is an incident, and then they know what to do to protect themselves, and the fertilizer industry has been very involved in that kind of community outreach program, as I know others in the chemical industry have been. Community awareness and emergency response programs have addressed that.

I have brought some information. I spoke to some of the staff about this earlier. We have an outreach program that has been in place for more than 10 years, where we visited with our neighbors to talk about shelter-in-place programs, and we have telephone notification systems that call our neighbors within a very short period of time if we have an accident. I know DuPont has that system in place, and so do many of the other chemical plants. I would like to share this with the Committee if I may. That is a very important part. These programs are coordinated with LEPCs and the local fire departments. So that is also, I think, an important part of this whole effort.

Senator VOINOVICH. Can I just ask one more question?

Chairman COLLINS. Absolutely.

Senator VOINOVICH. You represent the Fertilizer Institute.

Mr. SCHELLHORN. Yes, sir.

Senator VOINOVICH. Does the Fertilizer Institute also belong to the American Chemistry Council?

Mr. SCHELLHORN. No, sir. We are not a member of the American Chemistry Council.

Senator VOINOVICH. How about API companies, are you part of the American Chemistry Council or do you have a separate——

Mr. CHAMBERLAIN. No. API is a separate manufacturing group.

Senator VOINOVICH. So when we talk about 150 companies that are in the American Chemistry Council that are working with the Coast Guard, that does not include any oil companies?

Mr. CHAMBERLAIN. No, that is not correct. Shell is a member of the American Chemistry Council. When you asked if API was a member, those are two——

Senator VOINOVICH. OK. But that is what I meant.

Mr. CHAMBERLAIN. Yes, my company is a member——

Senator VOINOVICH. They belong to API and they belong to the American Chemistry Council?

Mr. CHAMBERLAIN. Yes.

Senator VOINOVICH. And you are part of the 150 companies that are in that organization?

Mr. CHAMBERLAIN. Yes.

Senator VOINOVICH. Mr. Schellhorn, my last question is regarding Senator Cochran's legislation, "Secure Handling of Ammonium Nitrate." How does what you are requiring in that legislation differ from what is in MTSA or what is being done by ACC?

Mr. SCHELLHORN. Yes, sir. The Cochran bill is specific to ammonium nitrate manufacturing, distribution, and retail sales of ammonium nitrate specifically. It is a registration.

Senator VOINOVICH. So, when considering legislation, we ought to be aware of the differences through the industry.

Mr. SCHELLHORN. Yes, sir.

Senator VOINOVICH. Thank you.

Chairman COLLINS. Thank you, Senator.

Senator Carper, you did miss excellent testimony from Ms. Turner, but I know you made a great effort to get back here in order to ask questions, and I am pleased to call upon you.

Senator CARPER. I apologize for leaving. Senators Lincoln, Lieberman, and myself, and a few others have just unveiled legislation to address the issue of children having access to pornography on the Internet, to create almost like a step that some would have to go through to register their age, to be able to identify their age, so that if you are under the age of 18 you cannot get on; to impose a 25 percent tax on the profits for the Internet, and to use those monies to develop new technologies to help keep kids clear of that kind of temptation. I apologize. It is an important issue. Not to say that this is not important as well, but that is why I have been away.

Thank you all for coming in. I especially wanted to welcome Beth Turner to our hearing today, and if I may I would just like to ask the first question of you, Beth.

We are proud of DuPont and we are proud of DuPont's reputation as a good steward of the environment, and my wife who worked there for 28 years, just retired last summer, and in a number of her jobs she was in charge of safety with the people in her workforce around her. She not only was that way at work, she was that way at home. I tell the story about how we would go on family vacations or be staying at a hotel, and get the kids to bed in their room, and we were getting ready for bed. My wife was probably one of the few—I do not know what other spouses talk about just before they go to sleep, but my wife is going through, out loud, just making out the escape routes from the hotel. Which door do we go out? Which direction do we go? Which stairs do we go down? So it was a company that puts a whole lot of emphasis on safety, and we are proud of them and respectful for that.

I would ask of Ms. Turner, if I could, could you describe DuPont's experience with the Maritime—and you may have addressed this, and if you have I apologize—but with the Maritime Transportation Security Act, and how you and DuPont believe that law can inform our Committee's work in a broader chemical facility security bill? Specifically I would like to hear how DuPont implemented the requirements and about your ongoing compliance assurance.

Ms. TURNER. Good morning, Senator Carper, it is nice to see you.

Senator CARPER. My pleasure.

Ms. TURNER. I only spoke briefly about the Maritime Transportation Security Act. In my comments I indicated that the regulations have very effectively secured our sites and that we would view them as a model for security and higher priority sites.

In terms of how we approach the regulation, we identified our facilities that are impacted. There are some very specific criteria in the regulations about facilities that have wharves on navigable bodies of water and that unload certain dangerous goods. So I went through that analysis of which facilities fell into that classification. And there are a number of very clear requirements on what you have to do once you are in, one of which is identify an individual who is a formal facility security officer, and put them through some very specific training. So our approach was to identify the facilities and then take them through the whole process as a group.

So we centralized, did our training. Much of what we had done for the Responsible Care® Security Code in terms of our vulnerability assessments, the things we had done for DuPont all fed into that very nicely so we were able to integrate it all, which is an important thing, so that the sites could see an integrated effort, and not, "I have to do this for responsible care and that for MTSA, and have to do this for corporate headquarters." So we made sure it all fits together.

On the ground with the Captain of the Port and their staff we have had a tremendous working relationship. I am very impressed at how grass roots oriented the Coast Guard is in deploying itself to work with sites. They do not let us get lax. They may show up at 2 a.m. in the morning when we least expect it, or run a boat down the channel and see if our cameras can pick it up. Our impacted sites have really tried to incent our security officers to see the Coast Guard before they think we can see them. So we have given out prizes and awards for sort of detecting the Coast Guard, and it has generated a lot of energy.

We have been successfully inspected by the Coast Guard at all of our regulated facilities.

Senator CARPER. I think in your testimony you described how DuPont categorized its sites. I think you may have just alluded to it. Category 1 sites, I am told, are your highest priority group.

Ms. TURNER. That is right.

Senator CARPER. Category 2 sites have no potential for off-site release or theft of materials. Is that correct?

Ms. TURNER. Yes, that is correct.

Senator CARPER. Many folks have advocated—I think even here today—for a risk-based tiered approach to regulating facilities. Let me just ask what criteria and what methods did DuPont use in categorizing your sites and your facilities? Do you think that the categorization that DuPont used is a sufficient approach, or do you think some additional steps or categories might be appropriate as we try to develop a risk-based tiered approach?

Ms. TURNER. I think that the categorization was absolutely critical. I find that—and let us not talk for a minute about whether you have two categories or four or however many. The fact that you can spread facilities out over certain categories is absolutely critical deploying resources. I treat and work with and defend and protect a Category 1 site very differently than I do a Category 2 site because the potential consequence is so very different.

From my standpoint I think that we have to have—and I think we have all been in agreement—that risk-based approach is very necessary here. I might just mention why we had a Category 1 or 2. It is really an internal thing. The American Chemistry Council had four tiers. The first three would have been equivalent to our Category 1, and the only difference was a 6-month delay that you could spread out. So the Tier 1 had to be done first, then Tier 2 6 months later, Tier 3 6 months later, and then Tier 4 after that. I simply made an internal decision that I wanted to treat everything as Tier 1.

So we identified all of our facilities, and we also tried to make it simple by saying it does not matter whether a facility is an RMP facility or not. If it can create an off-site consequence, then I put it in Category 1. And then we just took those Category 1s, again, just like we did MTSA, right through the process as a group. And for our company—and I am only speaking for our company—that created some efficiencies. For other companies, obviously having more tiers was a helpful thing and you could spread the effort out.

Senator CARPER. My time has expired. I appreciate those responses, and again your presence here.

Ms. TURNER. Thank you.

Senator CARPER. How did these guys do? Did they do a pretty good job in their testimony?

Ms. TURNER. They did great.

Senator CARPER. I wish I could ask them a few questions, but I am afraid time does not allow. Thank you again for joining us and for your valued input. Thank you, Madam Chairman.

Chairman COLLINS. Thank you very much, Senator.

Ms. Turner, I want to follow up on the issue that Senator Carper just raised. I do believe that we need a tiered approach. The security for a local fertilizer dealer may not be the same level that is needed for a large chemical plant on that two-mile stretch in New Jersey that Senator Lautenberg has referred to. One of our challenges is defining the scope of the chemical industry for regulation by the legislation that we are drafting. Each of the three of you, each of your respective companies has chemical facilities listed under the EPA RMP program, and you just were referring to that. Each of you also have facilities that are covered under the Maritime Transportation Security Act and are regulated for security by the Coast Guard.

Of your companies' chemical facilities that are not covered by the MTSA regulations, how would you identify which ones you think should be covered under a new chemical facility security regime? In other words, I am trying to pick up where Senator Carper left off, on his categorization. As we do this tiered approach, there are going to be some facilities, perhaps a local potato farm in Northern Maine, that should not come under the law at all. There may be others that need some coverage but at a lower level, etc. How should we define the scope of facilities that should be covered? Ms. Turner.

Ms. TURNER. Thank you. Speaking from DuPont's standpoint, the criteria that I used was the ability to create consequence off site. I think that is a very important discriminator, and I would recommend that as a consideration for the Committee. It is in our self-interest as a company not to create off-site consequence. We want the safest communities. We want our employees to be safe, and so the whole concept that we want to be able to contain our chemicals in the vessels where they belong, and focus on those facilities that have the potential to go beyond our fence line is the internal criteria I have used.

My view is that it does not matter how far the off-site consequence goes. If it goes off site then it needs to be in the highest priority category, and that is the approach DuPont has used.

Chairman COLLINS. That is helpful. Mr. Schellhorn.

Mr. SCHELLHORN. I agree with what Ms. Turner has said. One thing that I would add, however, is the four categories of risk is pretty helpful for dividing that group of facilities into highest, medium, and lowest risk facilities based on the significance of the off-site impact. I certainly understand why DuPont did what they did and just grouped everything that had off-site impact into Tier 1. But when you are looking at a universe of all facilities, breaking it down into Tier 1, 2, 3, and 4 is, I think, helpful because that helps to focus attention on the very highest risk facilities, and then down from there.

That criteria is established criteria. American Chemistry Council has a methodology for doing that.

Chairman COLLINS. Thank you. Mr. Chamberlain.

Mr. CHAMBERLAIN. Yes. Let me just mention all of our major chemical facilities do happen to fall under MTSA, but if there were other—and then I also wanted to just make sure that the Chairman and the other Senators realize that a number of chemical facilities are co-located on sites with refineries. In my case, my two biggest chemical facilities share a property with a refinery that you would not know where one stops and the other begins. So the co-located chemical facilities are another aspect that you need to have in the planning and mapping of future issues that you deal with.

Certainly off-site consequences is something that should be considered in trying to determine the severity. You also need to look at what is off site? If the closest population is 15 miles away and you are surrounded by a sugar cane farm, the consequence of an off-site release is not the same as if you are in a major metropolitan area with neighbors living on your fence line. So you have to look at the entire picture, look at the vulnerabilities that you have and the consequences of a worst-case scenario.

Chairman COLLINS. Chief Full, do you have any thoughts on this issue?

Mr. FULL. Senator, what is interesting to me here right now is the fact that we receive in our emergency management agency, emergency plans from companies that are just eloquently put together by consultants and so forth. We will get plans that are 100 pages thick. They will answer all kinds of questions in there about the vulnerabilities to the community and different things like this. Then we will find some other companies, that they will send us a three- or four-page report as well. Many of those folks, especially the ones that come from the biggest companies, have never consulted with us at the local level.

There is a disconnect right now between what we hear right now from the table here, and what goes on at the local level at times. How can folks really sit and say what is going on there without consulting with the local folks to see what vulnerabilities there are out there before the plans are done, and quite frankly, that is more the exception than the rule.

We come upon plans. We review the plans. It will say if people are injured here, they are going to go to XYZ Hospital. You tell the hospital that this particular chemical company has identified their hospital to take the injured, and they say, we do not know anything about it. They never talked to us about it.

I mean, my crusade here today on behalf of the folks at the local level on that is, again, just to ensure that whatever comes about—and we certainly need chemical plant and transportation security—additional security. In whatever form it comes from, we do have to have a strong input in coordination with us at the local level.

We are going to be there to handle the aftermath, and all too often it is sort of like they say, well, we will call the first responders and they will come. But frankly, we need to be involved in preventing too. We do not want to have to respond to these things because we know that going in that we are going to have very little or no impact, positive impact, and we are going to lose a lot of our folks as well as a lot of residents if we are not involved. We should know what the risks are, and so forth. I think there can be a happy balance between sensitive information and Achilles heel scenarios and so forth along those lines, but clearly we need to be involved, and we see all too often that we are not.

Chairman COLLINS. Ms. Turner, I am going to ask you my final question of the day. It has to do with, perhaps, the most controversial issue that we will have to wrestle with in this bill, and that is the inherently safer technology issue. It is clear from the four hearings that I have chaired that some people want this bill to be a hazards reduction bill. There are others who want this bill to be strictly limited to the physical security of chemical plants.

You have testified this morning that DuPont believes that inherently safer technology and chemicals are mainstream components of process safety and have a role to play as companies evaluate security. But you have also said that DuPont does not believe that inherent safety could or should be mandated by regulation, and you have called that unworkable. Similarly, and I think it was Mr. Schellhorn who pointed out that inherently safer technology is a safety process, it is not a security measure.

Is there a middle ground here? What I am wondering is whether it makes sense in our legislation to require companies to evaluate inherently safer technology as they do their safety plans, but in their vulnerability assessments, but not have the Federal Government mandate specific processes or get involved in second guessing, if you will, the safety processes used in the plants. It seems to me that it does make sense for companies to be required to look at whether safer chemicals or processes could be used to help make their plants less vulnerable to an attack. What are your thoughts on this?

Ms. TURNER. First, let me speak briefly about what is behind the testimony in terms of those things. We are saying IST has a role in security, but we are saying do not mandate it in security regulation, and in some respects that could appear to be sort of a contradiction.

I think that when we look back over the history of inherently safer technologies, at least in our company, we have been pursuing this for 40 years through—first the safety systems are engineering designs for our plants. We have very mature infrastructure for managing process safety management, inherently safer technology. We also have two codes of Responsible Care®, the process safety code and the security code that focus on inherently safer.

Then the Sandia methodology for conducting vulnerability assessments, which is a site that just because it is the one that we chose has a very structured approach for going through inherently safer from the very first step of the methodology when you form your team, characterize your facility. You have both process safety experts and security experts at the table because they bring separate expertise. So in the greater context we have these drivers for inherently safer in a very mature safety system that has IST embedded in it.

In my view, the place where we do not have something complementary is in the pure security side of the house, and that is why we recommend the passage of legislation so that we then can essentially bolt or marry these two together.

Now, in terms of how do we feel about what you said at the end about some view of requiring consideration, I think that my response is it depends on what is in the language. We would like very much to work with you if the Committee decides to go in that direction, using a phrase, the devil is in the details.

I think the thought I want to leave with you is—I will speak broadly—responsible chemical companies have many incentives for a look at inherently safer. A big one is keeping our facility safe, keeping our employees safe, keeping our employees' families safe. We cannot run a company if we are not doing that. But the other driver is, as you said at the last hearing, the incentive to a chemical company to pull its risk down through any tool is there both because it is good business and because it helps us bring down the risk category in the face of regulation and other drivers.

Chairman COLLINS. Your points are very well taken, and DuPont, of course, is renowned for its commitment to safety. It is difficult for me to imagine that the Department of Homeland Security could teach DuPont anything about the process of inherently safer technology. We might, however, be able to help you improve your security in general I would hope. But that is not going to be true of every chemical facility. I am thinking of the ones that Chief Full has talked about in his county, not all of whom are members of the American Chemistry Council or comply with the Responsible Care® Security Code or even have the sophistication perhaps of a DuPont.

Then we get into the dilemma of what if the Department of Homeland Security, in reviewing a plant, perhaps doing an audit of its vulnerability assessment, and comes across improper storage of chemicals, where there clearly is an increased security risk because of a lack of a secondary containment, for example, or some other measure. So should the Department, in such a case, be able to step in and mandate an improvement in the storage of the chemicals as the price of approving the security plant? How do we draw the lines here?

Ms. TURNER. I think that is going to require some analysis of the roles of the different regulatory agencies that are at play in the chemical industry. Right now I would not see Department of Homeland Security as having the kind of expertise to look at how a tank is built. That does not mean we could not embed it there, and I am not so sure that would not divert DHS from the security mission they need. It is possible to build it there, but I think the better approach might be to look at what resides in OSHA and what resides in EPA for driving the safety part that has been in place before September 11 ever came.

So it is certainly unacceptable if a chemical company is doing something that is blatantly unsafe, and somewhere in the regulatory regime we need to have an agency that has the capability to enforce its regulations. I am just not sure in my mind that is going to be a great focus for the security piece, which we need to stand up very quickly in a very thoughtful manner.

So it is a very important issue, and I appreciate how hard the Committee is working to figure out where the right place is on that issue, and we want to work with you on it. It is very important.

Chairman COLLINS. Thank you. I very much look forward to working with all of you. We are going to use the August recess to draft what I hope will be a comprehensive, effective, bipartisan, and reasonable bill on chemical security. We do not have that many of those around here that meet all of those criteria, and that is why we have spent so much time on this issue. This is our fourth hearing. There are not very many issues that Congress debates that have this many hearings and this kind of consideration, but I think this is enormously complex and enormously important. I really appreciate all of you sharing your expertise today.

I also want to thank the Committee staff, which has worked very hard to put together this series of hearings. You noticed that they all groaned when I said we would be spending the August recess drafting the bill. [Laughter.]

But I am very committed to introducing a bill in September, and we are going to try to adhere to that timeline.

This hearing record will remain open for 15 days for the submission of additional questions and other materials. I thank you all for your cooperation and advice to the Committee.

This hearing is now adjourned.

[Whereupon, at 12:23 p.m., the Committee was adjourned.]

APPENDIX

Testimony of

Martin J. Durbin

Managing Director, Security & Operations
American Chemistry Council
1300 Wilson Blvd.
Arlington, VA 22209
703-741-5575
marty_durbin@americanchemistry.com

Before the Senate
Committee on Homeland Security
& Governmental Affairs

"Chemical Facility Security: What is the Appropriate Federal Role?"

July 13, 2005

Madam Chair and Members of the Committee, my name is Marty Durbin, and I am the Managing Director for Security & Operations for the American Chemistry Council (ACC). I thank you for this opportunity to speak today on behalf of the Council's members on the important subject of security in the business of chemistry, a critical sector of America's infrastructure. We thank you as well for devoting so much of your time and energy to this important subject. We agree with you, Madam Chair, that "this issue is simply too important . . . to accept inaction."

The 128 members of the ACC manufacture essential life-saving products critical to homeland security, and life-enhancing everyday items that keep the economy moving. Our products are critical to daily life and crucial to the war on terrorism. We are essential to making bullet-resistant vests, night vision goggles and stealth aircraft. The products we manufacture are essential to the things that make modern life possible, from plastics to pharmaceuticals, from cars to clothing. And the products of chemistry are critical in many aspects of American life, including keeping our drinking water safe, supporting agriculture, and spurring medical innovations to prevent and treat disease.

ACC represents the leading companies in the U.S. chemical manufacturing sector, an industry which is the largest exporting sector in the economy ($91 billion), and employs nearly one million people in America alone, with $460 billion in sales. Our members are responsible for approximately 90% of basic industrial chemical production. In addition, the U.S. chemical industry has the largest share of knowledge workers of any industry, and it is the largest private industry investor in research and development.

Madam Chair, I welcome the opportunity to highlight four things for you and the Committee:

1. The leadership role ACC members have taken – at a cost of over $2 billion since 9/11 – to further enhance the safety and security of their products, their facilities, their supply chain and the communities in which they operate;

2. The great strides the federal government has taken, in cooperation with the chemical sector, to secure the industry;

3. The need for national legislation to provide an appropriate federal regulatory and oversight role in chemical facility security; and

4. Our views on the important and frequently misunderstood subject of inherent safety.

I. ACC Has Taken a Leadership Role in Enhancing Chemical Security

Even before September 11, 2001, Council members had begun to address the challenge of terrorist threats to our operations, by developing site security guidelines for chemical companies. Our Board of Directors was actually meeting that sad day, and their reaction to those events was swift and decisive. We quickly completed and issued our security guidelines, and a companion set of transportation security guidelines, in October and November of that year.

In those uncertain months, we shared those guidelines with state and federal agencies, and we and OSHA posted them on our public websites to make them as broadly available as possible. We also partnered with EPA to hold regional security briefings for our members and other chemical companies, state and local government officials, and first responders.

In January 2002, our Board launched an aggressive effort to develop a new Responsible Care® Security Code. Now in its 17th year, Responsible Care® is ACC's signature program of ethical principles and management systems designed to continuously improve our members' safety, health and environmental performance -- and now, their security performance as well. Implementation of Responsible Care® is mandatory for all members of the American Chemistry Council, as well as Responsible Care Partner companies, who represent chemical carriers, warehouses, logistics planners and others along the supply/value chain. In developing the Security Code, we consulted closely with plant-level Community Advisory Panels, and with first responders and government agencies at all levels. In June 2002, the Board adopted the Security Code.

Former Homeland Security Secretary Ridge has referred to the Security Code as a "model program," and at this Committee's June 15 hearing, Acting Under Secretary Robert Stephan recounted the "very legitimate, very real, and very qualitative improvements in security across the board" that DHS has observed at ACC members' facilities. At the April 27 hearing, John Stephenson of the Government Accountability Office also focused on these accomplishments, adding that "ACC is very good." Indeed, Madam Chairman and Senator Lieberman, our members were very gratified by your statements of commendation and appreciation last month for the work that companies like ACC's members have done to voluntarily to secure their facilities. Moving to the state and local level, New Jersey has accepted the Code as a "best practice" for chemical facility security, the City of Baltimore has adopted a security ordinance that recognizes the Code as an alternative means of compliance, and Maryland has enacted legislation mirroring the Code. In published reports, the Security Code, and ACC members' security enhancements, have been widely and uniformly recognized, from the *Washington Post* editorial page[1] to GAO reports.[2]

The Security Code requires member companies to:

- Prioritize their sites by degree of risk, sorting them into four tiers. This process was begun before the Code was adopted, and every ACC member company completed it on schedule in June 2002.

---

[1] "Some of the biggest security gains have been made cheaply, sometimes thanks to unobtrusive, even private-sector initiatives. The 140 large companies that form the American Chemistry Council, for example -- a group with both financial and practical interests in not having their chemical plants blown up -- have created their own security code, internal communications system and inspectorate." *The Washington Post*, p. A26 (May 27, 2005).

[2] "To its credit, the chemical industry, led by its industry associations, has undertaken a number of voluntary initiatives to increase security at facilities. For example, the ACC, whose members own or operate 1,000, or about 7 percent, of the facilities [handling large quantities of hazardous materials in the country] requires its members to conduct vulnerability assessments and implement security improvements." GAO, "Homeland Security: Voluntary Initiatives Are Under Way at Chemical Facilities, but the Extent of Security Preparedness is Unknown" (GAO-03-439, March 2003), at "Highlights."

- Thoroughly assess vulnerabilities, using rigorous methodologies developed by Sandia National Labs and the Center for Chemical Process Safety (CCPS), a program of the American Institute of Chemical Engineers (AIChE).
- Implement security enhancements commensurate with risks, and taking into account inherently safer approaches, engineering and administrative controls, and other security, prevention and mitigation measures.
- Verify the implementation of these physical security measures, using third parties that are credible with the local community, such as first responders or law enforcement officials.

All 2,040 ACC member company facilities have completed their vulnerability assessments, and virtually all have completed their enhancement verifications. Progress in implementing the Code was verified by GAO in its most recent report on chemical facility security.[3]

Our Security Code is not just limited to physical plant security. It covers the complete "value chain" for chemicals, from suppliers to customers, including transportation. Value chain management is an area where we have a long and successful history of partnering with and supporting federal agencies to safely steward our products and to prevent their diversion and misuse, such as for making illegal drugs or chemical weapons. In fall 2002, the Council issued a detailed value chain guidance document to enhance the security of our products outside the fence line. Our members who also belong to the Chlorine Institute have, together with the Association of American Railroads, implemented a chlorine rail car security plan.

The Security Code also covers cyber security, to protect our highly automated operations from being attacked electronically. Here again, the efforts of ACC members provide a model to other industries employing similar automated systems. Our members lead a broad Chemical Sector Cybersecurity Program to promote cybersecurity in our industry. In spring 2003 the Program issued a cybersecurity guidance document. The Program also launched a broad cybersecurity practices, standards and technology initiative through CIDX, the Chemical Industry Data Exchange. All of these guidance materials, and the Security Code, are available through our websites (www.americanchemistry.com and www.rctoolkit.com) so that they can have the broadest possible effect beyond our membership. Information about the Chemical Sector Cybersecurity Program can be accessed at www.chemicalcybersecurity.com. The CIDX materials are similarly available at www.cidx.org/CyberSecurity/default.asp.

II. The Federal Government, Working with ACC, Has Greatly Enhanced the Security of the Chemical Sector

ACC and its members have worked closely with the Department of Homeland Security during its first two and a half years of existence. We concurred with GAO's recommendations in 2003 that the federal government should develop "a comprehensive national chemical security strategy that is both practical and cost effective," and that should:

---

[3] Based on work conducted between October 2004 and March 2005, GAO stated: "All 10 of the chemical facilities we visited reported making significant progress in fulfilling the requirements of the security code." GAO, "Protection of Chemical and Water Infrastructure: Federal Requirements, Actions of Selected Facilities, and Remaining Challenges" (GAO-05-327, March 2005), at 5, 37. ACC members' implementation of the Code is discussed in detail at pages 17-21.

- "Identify high-risk facilities based on factors including the level of threat and collect information on industry security preparedness;
- Specify the roles and responsibilities of each federal agency partnering with the chemical industry;
- Develop appropriate information sharing mechanisms; and

Develop a legislative proposal, in consultation with industry and other appropriate groups, to require these chemical facilities to expeditiously assess their vulnerability to terrorist attacks and, where necessary, require these facilities to take corrective action."[4]

A. Identify High Risk Facilities

Starting in March 2003, DHS partnered with ACC to facilitate visits to our members' facilities. ACC also worked with DHS to develop methods for evaluating facilities based on potential physical and economic consequences. And even before the creation of DHS, the Coast Guard and state offices of homeland security or counterterrorism visited facilities to offer advice on enhancing facility security.

Today, DHS' Protective Security Division (PSD) and the Coast Guard are actively visiting chemical facilities, reviewing vulnerability assessments and security plans, understanding common vulnerabilities and developing plans, in conjunction with local law enforcement and responders, to protect facilities and their communities. Information gained from these visits supports the development of DHS's "Buffer Zone Protection Program" to provide support and resources to local governments in plant communities. ACC is also working closely with PSD to develop, refine and publicize its "Risk Analysis and Management for Critical Asset Protection" (RAMCAP), which allows DHS to compare the vulnerabilities of disparate assets and resources against a series of benchmark threat scenarios. RAMCAP will enable DHS to allocate protective resources rationally, on the basis of risk.

B. Specify the Roles and Responsibilities of Federal Agencies

In December 2003, the President issued Homeland Security Presidential Directive/HSPD-7, which clearly defines roles for various federal agencies in protecting the nation's critical infrastructure and key resources, and specifically names DHS as the lead or "sector-specific" agency for the chemical sector. With DHS's blessing, ACC organized the Chemical Sector Coordinating Council -- a group of 16 national chemical trade associations that coordinates communications between DHS and our sector for purposes of infrastructure protection. ACC serves as the administrative secretariat for the Sector Coordinating Council.

The federal Maritime Transportation Security Act (MTSA), which was enacted in late 2002, puts the Coast Guard in charge of regulating security within ports, on vessels, and at facilities that have the potential to be involved in a transportation security incident. Roughly 240 chemical plants in the United States -- including most of the largest facilities nationally -- are currently subject to rigorous Coast Guard oversight under the MTSA. These facilities have all conducted security

[4] *See* "Homeland Security," *supra* note 2, at 27.

vulnerability assessments, have implemented facility security plans, and have been inspected by the Coast Guard. Facility security plans specify actions the facility will take at different MARSEC (threat) levels regarding access control, restricted areas, handling cargo, delivery of vessel stores and bunkers, monitoring, security incident procedures, and barge fleeting facilities. They also include schedules for employee security training and response drills and exercises. Even more facilities are covered by area (i.e., port) security plans.

ACC supported the MTSA throughout the legislative process and we have worked closely with the Coast Guard to make the law a success. In particular, the U.S. Coast Guard recognized the Responsible Care® Security Code as an Alternative Security Program ("RCSC–ASP") for purposes of fulfilling facility security regulatory requirements under the MTSA. The RCSC—ASP was the first alternative security program the Coast Guard approved for facilities.

Cyber security is one area where needed progress will require DHS to better focus and prioritize its efforts. The chemical industry views physical security and cyber security as tightly coupled issues. Protection of our physical and cyber assets is critical to our security and ACC members have taken great initiative to secure their cyber assets. We do not believe that the National Cyber Security Division (NCSD) has been focused enough to help us in this effort.

We believe that better management of cyber issues at DHS is an important component to reaching overall security goals. However, we have not been able to have a strategic discussion with NCSD. Instead NCSD appears to be offering tools to solve a problem, before the strategic dialogue has taken place. Lack of continuity in leadership and staffing at NCSD contributes to the lack of progress. For example, nearly one year after his resignation, the Director of DHS's NCSD has not been officially replaced.

C. Develop Appropriate Information Sharing Mechanisms

Effectively securing privately-held infrastructure -- like the business of chemistry -- requires a partnership between the private sector and the government. Within seven months of 9/11, ACC and the FBI created a Chemical Sector Information Sharing and Analysis Center (ISAC) to share security information daily between the federal government and companies that make and use chemicals. The Chemical Sector ISAC provides 24-7 capability for DHS's Homeland Security Operations Center (HSOC) to contact the chemical sector as well as for individual members of the ISAC to convey incident or threat information to DHS. Members of the ISAC receive daily intelligence reports from DHS as well as episodic alerts and warnings. Open to any chemical sector business, whether or not it is a Council member, the ISAC has almost 600 participants. The Council runs the ISAC for free as a public service through its CHEMTREC service,[5] in cooperation with Department of Homeland Security (DHS). It is located at http://chemicalisac.chemtrec.com. ACC is also one of the first critical infrastructure sectors to be piloting DHS's new Homeland

[5] CHEMTREC® is a 24-hour-a-day emergency communications center that ACC has operated as a public service since 1971. CHEMTREC® provides emergency responders with round-the-clock resources for information and assistance for spills, leaks, fires, explosions and other emergencies involving chemicals and other hazardous materials. CHEMTREC has provided critical information to emergency service workers for incidents ranging from the attacks at both the World Trade Center and the Pentagon to the Columbia space shuttle disaster.

Security Information Network – Critical Sectors (HSIN-CS), a set of secure communications and collaboration capabilities. ACC anticipates that the Chemical Sector ISAC will eventually be integrated into HSIN.

On behalf of the chemical sector, ACC recently participated in TopOff 3, the third in a series of congressionally mandated emergency response exercises. TopOff 3 was the first such exercise to involve the private sector. ACC's involvement in TopOff 3 helped generate ideas for further improving the Chemical ISAC and added significant value to other signature parts of the exercise. The success of the public – private sector cooperation and coordination during TopOff 3 clearly underscored the value of private sector involvement, not only for providing expertise but ensuring that the business impacts of terrorist events and official reactions (or inaction) to such events are considered in both short and long term emergency management planning. ACC is now actively supporting development of lessons learned from TopOff 3 and the design of TopOff 4.

Information sharing between DHS and critical private infrastructure sectors like chemicals is a relatively new and complex challenge, and there are, understandably, still a number of ways in which it can be improved – a topic on which a subgroup of the Homeland Security Advisory Council has just made recommendations to Secretary Chertoff. Those recommendations, which ACC supports, include:

- ***Regular, detailed threat briefings between DHS and each sector.*** ACC believes that senior corporate security officials with security clearances should be able to meet regularly with DHS intelligence analysts to discuss threat information. The current semiannual briefings should be more frequent, be limited to single sectors, be more interactive, and focus on classified information.
- ***Revising rules and policies to promote information sharin***g. ACC feels that DHS has been slow to roll out its "Protected Critical Infrastructure Information" program for voluntarily-submitted information on threats, vulnerabilities and countermeasures. It has also been unclear regarding its ability to protect such information under other exemptions from the Freedom of Information Act. Finally, it has been indecisive regarding applicability of the Federal Advisory Committee Act to the activities called for by HSPD-7 -- sector coordination and information sharing. Public right-to-know is an important value, but operational communications about security simply must remain protected.

III. The Need for Federal Legislation

ACC recognizes that not all chemical facilities are currently regulated under the MTSA. We also recognize that not all chemical facilities belong to ACC. No doubt many non-ACC members have taken steps comparable to those our members have taken, but as Under Secretary Stephan has estimated, something like 20% of high risk facilities have not.

As a result, ACC has been taking a leadership role at the federal level to ensure that all chemical facilities are secured against the threat of terrorism. We have worked continuously with Congress and the Administration to secure enactment of national security legislation that will:

Establish national standards for security of chemical facilities. We agree with Under Secretary Stephan that these standards should be:

- *Risk based, reasonable, clear, and equitable*. The only sensible way to address the risks posed by terrorist attacks on our homeland is to adopt a risk-based system of prevention and preparedness. Different chemical facilities pose different risks, based on their differing vulnerabilities and consequences, and any regulatory system must reflect those differences and require security measures commensurate with those risks.
- *Performance-oriented*. Facilities need flexibility to select among appropriate security measures that will effectively address risks. Under Secretary Stephan noted that an overly prescriptive system could, by its predictability, actually assist terrorists in targeting their attacks.

Require identified facilities to conduct vulnerability assessments and implement security plans. Vulnerability assessments should be based on rigorous methodologies like those accepted under the Responsible Care Security Code.

Recognize responsible voluntary efforts. Based upon their substantial and verifiable efforts to date, ACC members strongly believe that federal legislation should enable DHS to give credit for their substantial voluntary, at-risk expenditures implementing the Responsible Care® Security Code. Under Secretary Stephan testified that "[w]e should recognize the progress that responsible companies have made to date." GAO's John Stephenson likewise stated: "I would expect that any federal system would give them credit for – indeed, recognize" ACC members' efforts. Mr. Richard Falkenrath concurred that ACC member companies deserved "a level playing field" and "a common set of expectations" that all chemical facilities would be required to meet. We are not asking for anything less stringent than everybody else, only that DHS be allowed to recognize our members' significant actions, just as the Coast Guard has done.

Provide oversight, inspection, and enforcement authority to DHS. DHS must have the legal authority to police compliance with its standards and to take enforcement action if necessary.

Protect sensitive information. Information about the vulnerabilities of facilities, and the measures they have taken to reduce them, is literally a roadmap for terrorists. A law that required such information to be created, but then permitted it to be released publicly, would be worse than the status quo. Senator Voinovich, Mr. Falkenrath and others have emphasized the overriding importance of ensuring that this information is protected from public release in any fashion.

In the absence of Congressional action on chemical security, state legislatures are beginning to fill the vacuum. Both Maryland and New York have enacted chemical facility security laws. ACC supported both of these statutes, and is working with the two states' offices of homeland security on their implementation. However, we strongly believe a national program, not a patchwork of potentially conflicting state efforts, is necessary.

IV. ACC's Views on Inherent Safety

In legislative and policy debates over chemical security, no issue has proven more controversial than the concept of "inherent safety" and what role it should play. Because of ACC members' deep investment in this issue, I would like to spend the balance of my time explaining our views and why we feel so strongly about them.

The concept of inherent safety was invented by the chemical engineering profession. In fact, it is no exaggeration to say that the business of chemistry, and indeed ACC members, wrote the book on inherent safety. The leading reference on the subject -- *Inherently Safer Chemical Processes: A Life Cycle Approach*, also known as the "Gold Book" -- was written by nine process safety experts, every one of whom worked for an ACC member company at the time.[6] The concept of inherent safety has been well understood within the process safety community for many years. Basically, it means designing a process to minimize hazards in the first place, rather than managing and controlling them with protective equipment or procedures.

The business of chemistry has long embraced inherently safer approaches. For over a decade and a half, our Responsible Care® initiative has required ACC members to have mechanisms for reviewing the design and modification of facilities and job tasks, with inherently safer design and material substitution at the top of the hierarchy of controls. This drives our members continually to develop and implement safer processes. We conduct process hazard analyses of our facilities, and those analyses can lead us to change processes, modify procedures, or substitute materials to reduce and manage risks. As I noted earlier, the Responsible Care Security Code mandates that our members take inherently safer approaches into account in assessing possible security measures. It is also in companies' best interest to implement inherent safety when that technology is effective. Such changes not only reduce risks for employees and surrounding communities, but typically reduce long-term costs associated with maintaining other protection systems or regulatory compliance oversight that would otherwise be required. In fact, the GAO documented that seven out of the 10 ACC member facilities it visited had included process changes as a part of their security enhancements.[7]

I cannot overemphasize, however, that inherently safer chemical processing requires considering *all* the risks potentially associated with a process. Inherent safety typically involves making very challenging judgments to ensure that risks are not unwittingly shifted or substituted, and that overall risks are reduced. Many inherently safer approaches involve trading one risk against the potential of another. For example, advocates of inherent safety frequently speak of reducing onsite inventories, or reducing or eliminating storage, of hazardous materials. By reducing inventories, though, a facility may increase the number of truck shipments through the plant's neighborhood. Similarly, replacing a low temperature, low pressure process that uses a toxic chemical with a process that uses a less toxic chemical, but operates at higher temperatures and pressure, increases the potential hazard to its workers.

Fundamentally, ACC has been dubious of any regulatory initiative that involves government agencies or other third parties reviewing and approving -- or disapproving -- facilities' decisions regarding inherent safety, whether in the context of security or otherwise. The history of "inherently safer" approaches is full of examples of unintended consequences: chlorofluorocarbons, underground storage tanks and PCBs were all originally regarded as inherently

---

[6] *Inherently Safer Chemical Processes: A Life Cycle Approach* (1996), published by the Center for Chemical Process Safety of the American Institute of Chemical Engineers.
[7] *See* "Protection of Chemical and Water Infrastructure," *supra* note 3, at 21.

safer, from the perspective of fire or explosion. Their possible effects on stratospheric ozone, groundwater or health, however, were not fully appreciated until later.

The challenge to regulators is compounded by the complexity of chemical industry processes. There are no "standard processes" for making chemicals, and "[c]omplex process systems, especially those with a long history of safe performance, should not suddenly be changed without careful thought and consideration."[8] To expect effective regulatory oversight in this area is unrealistic, at least without great difficulty, expense and delay. In fact, in the Clean Air Act Risk Management Program rulemaking, EPA concluded that requiring and reviewing multiple process options at each regulated plant would not lead to greater advances in process safety.[9] In doing so, it recognized that no small, central group of people can be so omniscient as to be able to understand the huge range of issues involved at so many unique facilities.

The challenge facing regulators – and even businesses – is further heightened by that fact that, while the concept of inherent safety is generally agreed upon, "a systematic methodology to measure inherent safety does not exist and it is not currently possible to know how inherently safe a plant or equipment item is because it is not possible to evaluate the principles that have been applied."[10] Another leading process safety expert concurs: given "the lack of formal and agreed inherent safety approaches . . . [e]xperience has shown that regulators and industry have a difficult time interpreting inherent safety and agreeing on adequacy of efforts."[11] This is not to say that such methodologies cannot be developed – they should, and ACC supports efforts to do so. But even if agreement on methods is achieved, leading process safety experts discount the feasibility of using them in a regulatory system: "[T]he complexity of process plants essentially prevents any prescriptive rules that would be widely applicable."[12]

Members and witnesses at April's hearing agreed on the importance of legislation, in Senator Voinovich's words, being "sharply focused" on security and not "burdened with extraneous issues." Dr. Falkenrath clearly stated that chemical security legislation should not be used as a "back door" for addressing environmental or safety issues, and maintained that the government should not have the power to order hazard reduction measures to be taken. Mr. Stephenson agreed, adding that many types of chemicals and chemical processes do not lend themselves to such approaches without massive capital expenditures, and that, in general, facilities using or storing such chemicals can make such changes more easily than manufacturing facilities.

---

[8] David Moore, "Judging Effectiveness of Inherent Safety for Safety and Security of Chemical Facilities," presented at the 20P[th] Annual CCPS International Conference (April 11-13, 2005), at 3.
[9] *See* 61 Fed. Reg. 31699 (June 20, 1996). EPA has also concluded, as Thomas Dunne stated last month, that attempting to use this Clean Air Act authority to regulate security "would subject the agency to significant legal vulnerability and protracted litigation." Dr. Falkenrath similarly testified that he "disagrees" with the Clean Air Act approach, adding that it would be "politically imprudent" to accomplish such a significant intervention in the economy via such an indirect and imprecise mechanism.
[10] Sam Mannan, White Paper, "Challenges in Implementing Inherent Safety Principles in New and Existing Chemical Processes" (2002). Dr. Mannan is Director of the Mary Kay O'Connor Process Safety Center at Texas A&M University.
[11] David Moore, *supra* note 8, at 1.
[12] Mannan White Paper, *supra* note 10, at 6.

In the final analysis, ACC firmly believes that judgments about inherent safety are fundamentally process safety decisions that must ultimately be left to the process safety professionals. We will remain concerned about legislation that would enable government officials focused on security to second-guess process safety decisions.

IV. Conclusion

In closing, I want to reiterate our commitments. Our member companies are committed to doing all they reasonably can to enhance the security of their operations and products against those who would do us harm. But we know that our nation will not be safe until all chemical facilities that need to be protected have taken steps equivalent to those taken by our members.

Madam Chair, it has been almost four years since 9/11. The attacks last week in London confirm that our enemies, and their determination to harm us, are still very real and present. Now is the time to act. We welcome this series of hearings and the Members' stated willingness to work in a bipartisan way with the Administration. We are committed to working with you, and others, to see that legislation is enacted by this Congress. Thank you and I'd be happy to answer any questions.

SOCMA

SYNTHETIC ORGANIC CHEMICAL MANUFACTURERS ASSOCIATION

Testimony
of
Matthew Barmasse

Environmental, Health, Safety and Quality Director
ISOCHEM Inc.

On behalf of the

Synthetic Organic Chemical Manufacturers Association

Before the

Senate Committee on
Homeland Security and Governmental Affairs

On

Chemical Facility Security: What is the Appropriate Federal Role?"

July 13, 2005

1850 M Street, NW • Suite 700 • Washington, DC 20036
(202)721- 4100 • Fax (202) 296 - 8120

I. Introductory Comments

Madam Chair, members of the Committee, my name is Matt Barmasse. I am the Director of Environmental, Health & Safety and Quality for ISOCHEM in Lockport, New York. I am appearing today on behalf of the Synthetic Organic Chemical Manufacturers Association, known as "SOCMA".

I appreciate the opportunity to speak with you regarding the appropriate federal role in the security of America's chemical facilities. My goal is to share with you some of the activities of SOCMA and its members with respect to chemical risk and security. I will also describe the unique nature of the batch and specialty chemical manufacturing sector of the U.S. chemical industry and our efforts working with the Department of Homeland Security (DHS) to ensure appropriate communication and information-sharing between the federal government and the chemical sector.

SOCMA is the leading trade association representing specialty and batch chemical producers. Approximately 90 percent of SOCMA's members are small businesses, according to SBA definitions. While commodity chemicals make up most of the production volume in the global marketplace, specialty chemicals make up most of the diversity (or number of different chemicals) in commerce. As a condition of membership to SOCMA, chemical companies must subscribe to Responsible Care® and its security code. This self-imposed program requires conducting a security vulnerability assessment, developing a plan to reduce vulnerabilities and enhance security, and obtaining third-party verification that all of the actions in the plan have been carried out. My company, ISOCHEM Inc., has been an active SOCMA member for many years and I have been active in SOCMA's Responsible Care Committee and Employee, Process & Safety Committee..

ISOCHEM Inc. is a small facility located in western New York, north of Buffalo, with 93 employees and approximately 25 million dollars in sales. ISOCHEM Inc.

manufactures mainly phosgene and phosgene derivatives serving many markets including: pharmaceutical, agrochemical, plastics, cosmetics, dyes, paints and coatings, sealants, photographic, and flame retardants.

I will focus my remarks today on four specific areas. First, I will explain the nature of batch manufacturing, the contributions of our industry sector, and the unique circumstances that demonstrate why a cookie-cutter approach will not achieve our nation's security goals. Second, I will provide information on how ISOCHEM, and SOCMA more generally, are addressing security and working with the Department of Homeland Security, as well as with local and state officials. Third, I will discuss the nature of the EPA Risk Management Program. And fourth, I will explain SOCMA's perspective on the engineering concept of Inherently Safer Technology, known as "IST," and why attempting to legislate this philosophical approach is not a panacea for securing America's chemical facilities.

II. The Unique Nature and Role of the Batch and Specialty Chemical Manufacturing Sector

Specialty chemicals are essential ingredients and building blocks for the manufacture of almost everything made in the United States. Specialty chemicals perform very specific functions, based largely on their molecular structures, which give them unique physical and chemical properties. Without these substances, nylon would not be strong enough to use for seatbelts, medicine would revert back to what it was in the 1800s, and our armed forces would not have the equipment and supplies necessary to defend our country.

Because of their complex chemistries and narrowly focused applications, specialty chemicals are typically produced batch-by-batch in reaction vessels. Batch processes are very different from the 24 hours a day, 7 days a week continuous operations that produce commodity chemicals. Since continuous processes employ continuous feeds and yields, the production volume is usually far greater than for batch processes. The main difference, however, is that a batch process and the chemical reaction (which yields the desired product) has a distinct beginning and end for each batch. In addition to processes having variable risk, the products that are stored onsite also change on a continual basis.

In addition to differences in processing, another distinct feature among specialty chemical producers is the variability of risk at production and storage sites. Batch producers are necessarily flexible and they can make many different products during any given production year. Their business is driven by customer demand, and many chemicals are made on short notice. As a result, the types of products onsite at a specialty chemical facility often change from week to week or even day to day, leading to similarly frequent changes in the risk profile of the facility. This ever-changing risk profile makes planning a successful attack difficult. This fact must be accounted for when looking at security and vulnerability.

Batch and specialty chemical producers also vary widely in appearance, which often makes them difficult to recognize as a chemical facility at all. In many cases, batch processing equipment is located either inside of building structures or contained in areas out of view from the road. Often, the sites are located in non-descript industrial or office parks and contain few features that would make them stand out as having anything to do with chemicals. Recognizability and location of equipment can greatly hamper surveillance efforts and make the facilities less attractive as targets. This attractiveness concept also must be considered when looking at vulnerabilities and potential countermeasures, and it is a fundamental part of the current DHS approach to the chemical sector..

Does this mean that my company and other SOCMA members feel that we do not have to consider the security of our facilities? Absolutely not. However, it does mean that when we apply our limited resources to security, we do it in ways that make sense and that actually reduce vulnerability or enhance security. Based upon the unique differences between large continuous chemical manufacturing facilities and small batch manufacturing facilities, it should be clear that a cookie cutter approach or one-size-fits-all approach is neither appropriate nor feasible for the variety of sites that make up the chemical manufacturing sector. Instead, SOCMA and its membership support an approach to security that focuses on actual risks identified by a vulnerability analysis. These risks should be addressed in a written, site-specific security plan that is kept onsite and made available to DHS upon request. This approach also should be tiered so that it imposes escalating requirements on those sites that pose higher levels of risk.

III. <u>How SOCMA and Specialty Chemical Firms are Addressing Security</u>

SOCMA's security activities started long before there was a Department of Homeland Security, even before September 11, 2001. In February 2001, SOCMA formed a partnership with the American Chemistry Council and The Chlorine Institute to proactively address site security. Together, we co-authored a guidance manual on site security for the chemical industry and distributed it to members and non-members alike. After the terrorist attacks in September 2001, SOCMA and its members cooperated with multiple federal agencies and began to develop additional tools and approaches for companies to identify their particular vulnerabilities and enhance security. SOCMA co-hosted a series of workshops throughout the country in late 2001 and early 2002. These included a workshop in Arlington to teach the fundamentals of vulnerability analysis in October 2002.

In addition, we created the concept of a Chemical Security Summit and partnered with the American Chemistry Council to develop what is now is now an annual conference attended by hundreds of chemical industry representatives. SOCMA also developed a unique vulnerability analysis model that is geared for variable risk facilities; the model has been downloaded from the SOCMA web site by more than a thousand different entities. All of SOCMA's security products and services are available to any firm that manufactures, handles or stores chemicals.

ISOCHEM and other individual SOCMA members are taking aggressive steps to secure their facilities. Since 9/11 ISOCHEM has instituted an integrated approach to security management using voluntary programs, such as the Responsible Care Security Code, as a guideline. Our security management plan includes enhancements to physical security, personnel security, surveillance, communication and threat assessment, vulnerability assessments, emergency response planning, transportation, supply chain and customer security, and cyber security.

To expand upon this without specifically identifying sensitive security information, ISOCHEM has spent over $750,000 since 9/11 on fencing, surveillance systems, access control, background checks, security guards and infrastructure, transportation security enhancements, community alert siren, and closure of a local road. Our security plan now includes response to terrorism, homeland security threat level changes, cyber security, communications and threat information, local law enforcement coordination, analysis of threats and vulnerabilities, and third-party verifications of security plans.

SOCMA supports DHS's request for additional authority to enhance security in the chemical sector, and we recommend that any such authority require covered facilities to take the following steps:

- Perform a risk screen based on potential consequences of an attack and attractiveness as a target
- If found to be at risk, perform a detailed vulnerability analysis
- Develop plans to enhance security, according to the risks and vulnerabilities that have been identified
- Develop a site security plan that contains the plans for enhancements and includes standard operating procedures and policies pertaining to security

IV. DHS Efforts to Secure Chemical Facilities

Since the inception of the Department of Homeland Security, SOCMA has forged a strong working relationship with DHS. DHS provides regular security briefings and outreach to the chemical sector, has addressed our Board of Governors and member committee meetings, and is generally available whenever we have questions or need its perspective. It is not a one-way relationship, however. Because the specialty chemical industry is unique and diverse, SOCMA staff and member company experts are routinely consulted by DHS on various topics related to chemicals. SOCMA staff also participates on DHS work and issue groups, such as Risk Assessment and Management for Critical Asset Protection (RAMCAP) and the Chemical Sector Coordinating Council, which have been discussed in earlier testimony before this committee..

I have been very impressed from my personal experiences working with DHS by their efforts to develop a framework for enhancing security at U.S. chemical facilities. SOCMA staff and several member company representatives and I attended the DHS tabletop security exercise in February at the Maritime Training Institute in Jessup, Maryland. This conference facilitated networking with DHS personnel and other industry experts on the perceived threats and best practices in security at chemical facilities, making it easier to enhance security at our sites. DHS also conducted a site assistance visit at our facility, which led to an outstanding third-party assessment of our security practices. The auditors provided extremely helpful suggestions for improvement of our security plans and practices. In addition, state and local law enforcement conducted a buffer zone protection assessment at our facility. that looked at potential vulnerabilities outside the facility boundaries, allowing the local authorities to apply for federal grants to enhance the security outside facility boundaries. And finally, we are participating in a pilot project being conducted by New York's Office of Homeland Security in August to assist in testing the DHS RAMCAP methodology for comparing security risks across the chemical sector.

DHS does not just work with the chemical industry, however. The Department has met with leaders from other critical infrastructure and business sectors that handle or store hazardous materials. They also coordinate closely with other federal agencies and experts. For example, DHS has been coordinating with EPA to use existing EPA data on chemicals and facilities, rather than trying to reinvent the wheel. Not all of the data are particularly well-suited for security purposes, but EPA's data have provided a rational starting point to help identify potentially vulnerable sites. Also, several representatives from U.S. national laboratories are members of the DHS RAMCAP team, which is developing a standard approach to screening and prioritizing critical infrastructure according to risk. Additionally, DHS has consulted with the FBI on theft and diversion issues and how RAMCAP could be modified to help identify potential vulnerabilities in those areas.

As you can see, DHS is working with other federal agencies, trade groups and individual companies to secure America's chemical facilities. However, I think there are other efforts underway that are equally pertinent here and should not be discounted. State and local authorities, from law enforcement to fire departments and other emergency services, are often in the best position to ensure the security of our nation's infrastructure. There are many efforts underway within the chemical sector that have enhanced security and this coordination will continue because at the local level, we all have a mutual interest in mind. None of us wants our communities to be affected by terrorism. I am active in my community and have family, friends and neighbors who I care deeply about. My children go to local schools. The same holds true for others who work at chemical facilities around the country. At the local level, we have incentives that are stronger than just complying with regulations.

V. <u>RMP Facilities</u>

Under the Clean Air Act Amendments of 1990, EPA requires facilities possessing certain listed chemicals above threshold amounts to develop risk management plans that include an assessment of the worst-case scenario in the event of a release from a single chemical process. These worst-case scenarios include estimates of the population potentially at risk, based on the application of very conservative EPA criteria and guidelines. Roughly 15,000 facilities submitted plans under this RMP program, and the data they submitted have been routinely misinterpreted ever since. Nevertheless, the RMP list provides a legitimate starting point for any discussion of facilities that should be covered by an expanded DHS program.

Recent witnesses before this committee have suggested that not all RMP facilities are covered by the 16 associations that make up the Chemical Sector Coordinating Council and that there may be outliers that are unwilling or unable to secure their facilities. While there may be some outliers, I will not be easily convinced that those outliers, which are primarily small-scale chemical users rather than manufacturers, are very attractive targets to terrorists. In fact, if you study the RMP list closely, you will find that only about 10 to 12 percent of facilities on the entire list are even involved in chemical manufacturing. Simply put, the figures often cited by the press—15,000 chemical facilities that put thousands or even millions of people at risk—are just not an accurate depiction of reality. In fact, the RMP database, especially the worst-case scenarios under RMP, were never designed to be realistic.

EPA and DHS officials have made this point repeatedly, and this has just been reaffirmed by the Congressional Research Service, which noted in a June 27 memo to Rep. Edward Markey of Massachusetts that "[s]ince the population potentially affected under an EPA worst-case scenario release is calculated in a circle around the facility, it is unlikely that this entire population would be affected by any single chemical release, even if it is a worst-case accident.". In spite of these frequent caveats and clarifications, I repeatedly see RMP data used to scare people into thinking that the chemical industry is putting the nation at risk. This is both irresponsible and inaccurate, and it is unfair to both the chemical industry and to DHS and the local authorities with whom we work closely.

Consideration of the RMP numbers demands more perspective than the media and other alarmists give them. The EPA models used to estimate affected populations under worst-case scenarios for RMP assume that gases will spread out in a perfect circle. In reality, gases usually form plumes that drift in a specific direction. In effect, this reduces the potentially affected population to a small fraction of what the RMP data tell us. That is why historically, when catastrophic releases have occurred in this country, you do not see the kinds of numbers that the media are claiming for injuries and fatalities.

Another factor to consider that greatly diminishes injuries and fatalities is our nation's emergency response system. The United States has what are arguably the best systems in the world to handle chemical emergencies. For instance, there are national-level mutual aid networks for specific chemicals that can provide on-site experts and equipment to help mitigate emergency situations. Areas with concentrations of chemical facilities have Local Emergency Planning Committees, known as LEPCs, which conduct exercises and drills to test response capabilities. We have community-wide procedures for sheltering in place and, when necessary, evacuation.

Our emergency response capabilities, residential and industrial building codes and the realities of how hazardous materials behave when released, explain why we don't see Bhopal-like incidents occurring in the United States. That is not to say that RMP data cannot be useful. While we believe that most facilities falling under the Risk Management Program are not attractive terrorist targets, the list does provide a reasonable universe of sites to begin screening and prioritizing according to risk.

VI. <u>The Philosophy of Inherently Safer Technology (IST)</u>

Inherently safer technology or IST is probably the most misunderstood and controversial aspect of chemical site security. While it seems self-explanatory, the term as used in chemistry and engineering may be misleading to non-scientists. IST is an approach to chemical processing that considers procedures, equipment and the use of less hazardous substances in these processes.

Many non-scientists have been led to believe that the only way to achieve inherent chemical safety is by reducing the amount of hazardous substances used in chemical manufacturing and processing. Application of IST, however, is bound by the laws of physics and nature; a simple reduction in the use of hazardous chemicals is often not possible within the confines of a particular reaction or process. Such reductions often result in transferring risk to other points in a chemical process or the supply chain, without actually reducing it. To place the current IST debate in context, I will begin with an illustration of the limitations of chemical substitutes, then discuss the difference between a hazard and a risk, and finish with an explanation of why reducing a hazard in a process does not necessarily reduce the overall risk.

Most natural processes involve chemical reactions in one form or another, but chemistry is bound by the laws of physics and nature. These physical laws place restrictions on what can and cannot be done when trying to make a chemical. For instance, a molecule (i.e., a chemical) is made up of atoms (e.g., sodium, carbon, chlorine, etc.) that are in specific locations or positions on the molecule. In organic chemistry, the goal is to take the atoms from one molecule and move them to locations on another, different molecule so that it takes on a specific function or behavior.

The laws of physics and nature dictate if, how and when those atoms can be moved. To achieve certain critical structural changes, reactive chemicals must be used, and many are by their very nature hazardous, i.e., toxic, flammable, etc. In light of these constraints, scientists seeking to achieve certain chemical changes are often left with few alternatives. Where hazardous chemicals are used, they are highly regulated by EPA and appropriately managed by chemists in universities, government and industry. The fact of the matter is that scientists usually cannot produce the materials that make our standard of living possible without using very specific chemicals. Making medicine is a good example.

Often, to make medicine it takes multiple steps. Each step in the process carefully moves atoms from one molecule to locations on another molecule. Eventually, the scientist will obtain the desired chemical that performs a precise medicinal function. The movement of these atoms, from one molecule to another, is a chemical reaction and can only take place using certain materials. The chlorine atom, for instance, when it is located on a specific part of a molecule, allows these steps to take place. One common misconception, though, is that any chlorine atom will do. That is not the case. Chlorine atoms take on different behaviors, or physical properties, depending on the atoms to which they are attached.

Table salt consists of the sodium (Na) and chlorine (Cl) atoms, which make up the chemical sodium chloride (NaCl). The chlorine atom used to make medicine, on the other hand, often comes from phosphorous trichloride (PCl_3). PCl_3 has one phosphorous and three chlorine atoms. The sodium atom that is attached to the chlorine atom in table salt gives the chlorine a different nature than the one attached to the phosphorous atom in PCl_3. The very specific nature of the chlorine atom in PCl_3 is critical to its fundamental role it pharmaceutical manufacturing. By contrast, to use the chlorine in table salt in the drug manufacturing process would require the application of electric energy to the salt, resulting in the formation of chlorine gas, which is corrosive and poisonous by inhalation. At that point, it is no longer table salt; it has been converted into a compound with similar hazards to the PCl_3. The complex chemistry associated with making medicine has well-defined physical boundaries and requires the use of reactive chemicals. That is why, generally, medicine is not made from table salt.

For several years, people have debated the hazards and risks of certain chemicals. Part of the length and intensity of these debates may be due to how people define hazard and risk. In the sciences, hazard and risk take on different meanings than the typical dictionary definitions. Before a coherent discussion of IST can take place, it is important to understand the definitions used by scientists so that chemical information is not misinterpreted. In essence, a hazard is part of a certain chemical's nature, while risk depends on the circumstances in which the chemical is stored, used or handled.

When discussing chemicals, a hazard is a characteristic of a substance that gives it the *potential* to produce an undesirable consequence *under certain conditions.* The inherent hazard of a chemical does not change and does not depend on circumstance. Risk, on the other hand, can vary with conditions. It is related to the *likelihood* that an undesirable event could take place and the consequences the event can produce; in other words, the likelihood that a hazardous thing would cause harm.

For instance, a car has hazardous properties (i.e., heavy weight, flammable fuel) that *under certain conditions*—high speed, bad road conditions, driver intoxication, etc.—can produce serious damage. The weight of the car and the flammability of the fuel that propels it—two of its hazards—do not change. Operated under proper speeds and conditions, however, cars are considered to be at a reduced (and acceptable) degree of risk because they are less likely to be involved in an accident. Furthermore, we as society accept the risks inherent in automobile use because they are outweighed by the benefits.

Chemicals can also have hazardous characteristics. Just as conditions affect the risk posed by operating a car, the risk a specific chemical presents depends upon the conditions of how and where the chemical is stored, used or handled. These conditions are as important as the chemical's hazardous properties when trying to determine its degree of risk. For example, household oven cleaners and drain openers are corrosive—a hazard—and can cause severe burns on skin and permanent blindness if splashed into the eyes and not treated immediately. Despite these hazardous characteristics, they are used in most households because of their grease-cutting properties. When these products are clearly labeled, which is required by law, and used with adequate precautions, they do not pose a significant risk. In fact, anything can be handled safely with the right precautions. Consumers accept that and use hazardous products accordingly.

As noted earlier, IST is a conceptual and often complex framework that covers procedures, equipment, protection and, when feasible, the use of less hazardous chemicals. Its premise is that if a particular *hazard* can be reduced, the *overall risk* associated with a chemical process will also be reduced. In its simplicity, it is an elegant concept; however, reality is not always simple. A reduction in hazard will reduce overall risk if, and only if, that hazard is not displaced to another time or location, or does not magnify another hazard. If the hazard is displaced, then the risk will be transferred or increased, not reduced. Here are several examples of how factors related to likelihood affect overall risk when attempts are made to reduce hazard:

Reducing the amount of a chemical stored on site

A manufacturing plant is considering a reduction in the volume of a particular chemical stored on site. The chemical is used to manufacture a critical nylon additive, which is sold to another company and used to make seat belts stronger. Because it is a critical component for nylon strength and seatbelt production cannot be disrupted, the production schedule cannot change. If the amount stored on site is reduced, the only way to maintain the production schedule is to increase the number of shipments to the site. This leads to more deliveries (an increase in transportation risk), more transfers of chemical from one container to another (an increase in transfer risk) and, since there is now a greater chance that production could be disrupted by a late shipment, there is an increase in economic risk. This analysis only accounts for the risk to the manufacturer and does not include the risk to the customer making the seat belts or those using seat belts.

Substituting a Reactant in a Chemical Reaction

Phosgene is a key building block for an important starting material in pharmaceuticals. The structure of phosgene allows for a transfer of atoms that is clean, meaning that it does not allow side reactions to take place that would contaminate the compound with potentially toxic by-products. Using phosgene helps ensure the safety of medicines used to treat diseases such as multiple sclerosis .

Substituting Sodium Hypochlorite for Chlorine

Some people point to the Blue Plains water treatment plant in Washington, DC, as a prime example of how easy it is to substitute sodium hypochlorite solution for chlorine gas as a wastewater disinfectant. Unfortunately, several important facts are usually missing from these explanations. First, the conversion was not an overnight process; in fact, the substitution began prior to September 11 and included retrofitting the plant to accommodate the substitution. Second, the District of Columbia is in a different situation financially than other municipalities, in that it often receives federal funding to make such expensive changes possible. Also, it takes a large amount of sodium hypochlorite to achieve the same sanitizing effects as chlorine. But the most important fact that is missing from this story is that it takes chlorine to make sodium hypochlorite. The facilities producing the hypochlorite must now use and store vast quantities of chlorine in very few locations to keep up with the increased demand for hypochlorite. There are only a handful of sodium hypochlorite producers in the United States, which means that more and more chlorine will have to be concentrated in a few locations to keep up with demand. The ultimate result of this is a huge increase in risk at chemical facilities that produce hypochlorite, but a modest reduction in risk at the water treatment plants, which typically use 1-ton cylinders of chlorine.

In science, risk is dependent on the circumstances and surroundings of a hazard. A simple reduction in hazard will not necessarily result in a reduction of overall risk. IST decisions, therefore, are and should be based on risk, not simply on inherent hazards.

Scientists support the concept of using inherently safer technologies whenever possible. They have one major motivating factor: their own safety. Scientists spend hours each day in laboratories and manufacturing facilities that use and produce chemicals. It is difficult to imagine that any scientist would not want to work under the safest conditions possible. In addition, at most chemical companies, executive offices are in the same buildings, or very close to the same buildings that contain the processing, storage and laboratory areas.

There are also important economic incentives for companies to use the safest and least hazardous chemicals possible. These incentives include reduced accidents among laboratory and processing workers, cheaper transportation and disposal costs, cheaper insurance rates and fewer government regulatory requirements. In addition, the lost productivity caused by a system that is out of operation or by the absence of lost raw materials can put a company out of business.

With all of these incentives in place, the question becomes: Why do chemical companies still use hazardous materials? The simple fact is that the laws of physics and nature are a much larger determining factor in selecting process materials than anything else. No federal program mandating IST will change how these processes are run in any significant way. Instead, such a program would result in government micromanagement of the decision making process at individual facilities, would impose burdensome paperwork requirements on the regulated community, would duplicate certain key requirements of other federal and state regulatory programs, could slow chemical production activities, and could lead to manufacturers moving production overseas.

VII. Conclusion

As you can see, chemical facilities are extremely diverse, as are the chemistries that take place within the manufacturing plants. Because of this diversity, a one-size-fits-all approach to security with prescriptive standards will not work, nor will attempting to mandate inherently safer technology.

SOCMA supports programs that promote enhanced security in the chemical sector based on an evaluation and prioritization of risks, threats and vulnerabilities. Given the broad range of processes and operations that are part of the chemical sector, these programs should focus first on facilities most likely to present the highest risks. Any federal oversight of security in the chemical sector needs to account for the significant voluntary efforts already undertaken, factor in the diversity in operations and risks presented, and use performance-based fundamentals that provide the flexibility needed to implement effective, site-specific programs.

SOCMA supports DHS in its push for greater authority over our sector, assuming that the any future program adopts a tiered, risk-based approach to security at America's chemical facilities. Key elements of such a program include:

- A clear definition of covered entities and any exemptions
- Recognition of past efforts and voluntary programs that are substantially equivalent to DHS requirements
- Flexibility in achieving compliance
- Compliance assistance for small facilities
- Risk screening for prioritization across covered facilities
- DHS-approved security vulnerability assessments for higher-priority sites
- Federal preemption authority for DHS
- Retention of security plans onsite, with availability to DHS upon request
- Recognition of efforts by the regulated community under other security programs

Madam Chair, members of the Committee, thank you for your consideration of SOCMA's perspective on these important issues. I am happy to answer any questions you may have about my testimony.

NPRA

WRITTEN STATEMENT OF

BOB SLAUGHTER

PRESIDENT

NATIONAL PETROCHEMICAL & REFINERS ASSOCIATION (NPRA)

BEFORE THE

SENATE COMMITTEE ON HOMELAND SECURITY AND GOVERNMENTAL OPERATIONS

HEARING ON CHEMICAL FACILITY SECURITY: WHAT IS THE APPROPRIATE FEDERAL ROLE?

July 13, 2005

Introduction

Good morning, Madam Chairman, Ranking Member Lieberman, and Members of the Committee. I want to thank the Committee for holding this important hearing today. I look forward to discussing how the refining and petrochemical industries are performing the critical task of maintaining and strengthening the security of our national energy and petrochemical infrastructure. I will also discuss principles for chemical security that we hope the Committee will consider and adopt as it moves forward to develop legislation regarding chemical facility security.

NPRA, the National Petrochemical & Refiners Association, has more than 450 member companies, including virtually all U.S. refiners and petrochemical manufacturers, their suppliers and vendors. Petrochemical companies use manufacturing processes similar to those in a refinery. NPRA companies supply consumers with a wide variety of products used daily in their homes and businesses. These products include gasoline, diesel fuel, home heating oil, jet fuel, lubricants, and the chemicals that serve as building blocks for everything from plastics to clothing, medicine and computers.

Overview/Summary of Statement

Maintaining the security of our facilities has always been a priority at refineries and petrochemical plants. Refiners and petrochemical manufacturers are heavily engaged in maintaining and enhancing security – and were so before September 11. These industries have long operated globally, often in unstable regions overseas where security is an integral part of providing for the world's energy and petrochemical needs. When the tragic events of September 11, 2001, occurred, the nation realized immediately that additional threats had to be taken into consideration in order to protect our homeland. The refining and petrochemical industries drew the same conclusion. Industry – and I say this with special emphasis – did not wait for new government regulations before implementing additional and far-reaching facility security measures to address these new threats.

What are some of the steps our industry has taken to strengthen security? Industry has conducted security vulnerability assessments, prepared and implemented facility security plans, and developed close, working relationships with key federal agencies and state and local law enforcement offices to obtain and exchange information critical to maintaining infrastructure security. Industry has held joint training exercises simulating actual terrorist attacks and developed educational programs involving federal and state government officials with security expertise. Industry personnel from the largest companies to the smallest have shared best practices at NPRA meetings and conferences. With this strong evidence of our commitment to facility security as background, NPRA urges the Committee to consider the following facts:

- The refining and petrochemical industry will continue to maintain and improve our security operations to protect the vital network that provides a reliable supply of fuels and other petroleum and petrochemical products needed to keep our nation strong and our economy growing.

- Industry, in cooperation with government security agencies, has reassessed security vulnerabilities and implemented strong and effective security measures since September 11, 2001.

- Essential working relationships and information networks have been established between government security agencies and the refining and petrochemical industry to exchange "real-time" intelligence data on security issues to allow them to respond rapidly to terrorist threats.

- Industry has partnered with the Department of Homeland Security (DHS) on many important security initiatives and programs, including the Risk Assessment Methodology for Critical Asset Protection, or RAMCAP, the Homeland Security Information Network (HSIN), and Buffer Zone Protection Plans.

- Industry complies with the security requirements under the Maritime Transportation Security Act (MTSA) which is administered by the U.S. Coast Guard. The Coast Guard and industry are working together closely to achieve the security goals of the Act.

- MTSA has been an effective security regulation. It enjoyed broad bipartisan support in Congress. For these reasons, NPRA recommends that the Committee use MTSA as a model as it develops new DHS regulatory authority to address chemical security issues.

- Any new legislation should recognize and give credit to companies for the security programs they have already implemented.

Industry has Conducted Facility Security Vulnerability Assessments

In 2003, NPRA, working with the American Petroleum Institute (API), DHS and the Department of Energy (DOE), developed and provided industry a peer-reviewed security vulnerability assessment (SVA) methodology for our industry. In 2004, industry expanded that methodology to include transportation-related activities, including pipelines and rail and truck transportation. DHS has endorsed the vulnerability assessment methodology and uses it to train its employees.

The security vulnerability assessment methodology is a sophisticated and effective tool used to identify the security hazards, threats and vulnerabilities of a facility, and to evaluate the best measures to provide safe facility operations to protect employees and the public. The methodology provides the framework for a complete security analysis of the facility and its operations. Depending on the type and size of the facility, the assessment utilizes expertise in physical and cyber security, process safety, facility and process design and operations, emergency response, management, law enforcement, and other disciplines as necessary.

Differences in geographic location, type of operations, and on-site quantities of hazardous substances all play a role in determining the approach taken. Security vulnerability assessments typically include the following types of activities:

- Analyzing the facility to determine what critical assets need to be secured, their importance and their interdependencies and supporting infrastructure;

- Identifying and characterizing potential threats against those facilities and assessing their attractiveness as targets;

- Identifying potential security vulnerabilities that threaten the asset's service or integrity;

- Determining the risk represented by these events or conditions by evaluating the likelihood of a successful event and the consequences if it were to occur; and

- Making specific recommendations for incident mitigation and countermeasures appropriate to the risk level.

Based on the results of the security vulnerability assessment, companies identify appropriate security measures and incorporate them in security plans which are then implemented. Individual facilities have spent many millions of dollars in upgrading their security posture to assess and address risk and other related factors outlined here. A small facility in a remote location may have to spend hundreds of thousand dollars; larger ones, in more populous areas, have spent many millions.

The Maritime Transportation Security Act of 2002 Serves the Nation Well

A majority of the almost 150 refineries and 200 petrochemical manufacturing facilities in the United States are subject to the jurisdiction of the U.S. Coast Guard, and are therefore regulated pursuant to the security requirements of MTSA. (See attached map of U.S. refineries.) The Act requires that these facilities conduct security vulnerability assessments and submit comprehensive security plans to the U.S. Coast Guard. These security plans were submitted by facilities in December 2003. They have been reviewed and approved by the Coast Guard. MTSA also requires companies to designate facility security officers who oversee the implementation of their security plans. This officer is required to conduct drills on a quarterly basis to test elements of the facility's security plan. We understand that the Coast Guard has been pleased with the petroleum and petrochemical industry's implementation of the Act.

Industry has Implemented Strong, New Security Measures since September 11

Media reports sometimes leave the impression that the industry has not taken new security initiatives since September 11. That simply is not true. With the critical information gained from conducting their security vulnerability assessments, facilities have taken the following specific measures to enhance security:

- Reconfigured sites allowing critical assets to be set back from the perimeter.

- Installed sophisticated, state-of-the-art electronic intrusion detection systems around our perimeters and on buildings.

- Implemented card-access controls with new biometric technology readers, such as retina or thumbprint scanners.

- Acquired enhanced security communication systems.

- Shared security response plans with local law enforcement and appropriate federal agencies.

- Conducted drills and exercises to test security and response plans.

- Hired additional security personnel to assist in our security efforts, which are an around the clock, seven days per week priority.

This is just a partial list. A longer list of measures taken by our industry is included as an attachment to this statement, but it, too, is only a partial list of measures already taken as a result of a dynamic process.

Industry is Working with DHS to Improve Risk Assessment

NPRA members are working with DHS on the RAMCAP, or Risk Assessment Methodology for Critical Asset Protection, project. This approach to risk assessment and management will provide a consistent framework for the assessment, reporting and management of terrorism risks across the nation's critical infrastructure and to other key resources. This will be accomplished by developing a common risk-based method for comparing security risks, thereby giving Congress and the executive branch the tools they need to make decisions and allocate resources based on risk. In short, RAMCAP aims to put all infrastructures and key resources, including refineries and petrochemical plants, on a common risk platform.

Industry is Working with DHS to Develop Buffer Zone Protection Plans

Our members are also working with DHS, states, and local officials to protect and secure areas surrounding our facilities, which they neither own nor control, by developing buffer zone protection plans. These plans will identify specific threats and vulnerabilities within the buffer zone, analyze and categorize the level of risk, and recommend corrective measures to local law enforcement to reduce the risk of a terrorist attack.

Industry Participates in Private and Public Information Networks to Enhance Security

As stated earlier, information sharing is a vital part of our industry's security efforts. NPRA members serve on several security-related public and private sector boards and task forces. These include participation on the Boards of the Energy Information Sharing & Analysis Center, or ISAC; the Oil & Natural Gas Sector Homeland Security Coordinating Council; and the Chemical Sector Coordinating

Council. NPRA also serves on a working group of the Homeland Security Advisory Council (HSAC), helping to resolve legal impediments that hinder the submission of private sector information to government officials. NPRA members have also agreed to serve on a working group of the President's National Infrastructure Advisory Council.

One particularly important initiative underway – once again, as a cooperative effort between DHS and industry - is the creation and implementation of the Homeland Security Information Network, or HSIN, for the petroleum and chemical industries. HSIN is an information sharing system facilitated by the DHS in partnership with the critical sector organizations. It links owners and operators with each other and with DHS and FBI to enable collaboration in protecting critical resources and to address physical and cyber threats, vulnerabilities, and incidents, and to share information about potential protective measures and best practices.

Industry Sponsors Educational Programs and Holds Training Exercises with DHS and Other Government Officials to Enhance Security at Facilities

NPRA has established a standing committee on security which has held or co-sponsored more than a dozen national facility security conferences and workshops. The agenda has featured federal and state policymakers, security and counterterrorism experts, and the sharing of best practices to afford participant companies the opportunity to learn which new approaches have worked for others. In February of this year, for example, NPRA conducted an intensive training workshop for persons designated as Facility Security Officers which helped them to better fulfill their responsibilities under MTSA. NPRA has held two training exercises in cooperation with Texas Homeland Security. The exercises were conducted by Texas A&M University's National Emergency Response and Rescue Training Center and Texas Engineering Extension Service. The most recent training exercise, "Safe Horizon," was held in March of this year. This exercise was focused on incident deterrence and prevention of a presumed terrorist attack. These training exercises and educational programs provide information that allows companies to better assess the effectiveness of their own security policies, plans, and procedures, and make modifications as necessary.

Industry Relationships with Federal, State and Local Officials Enhance Facility Security and should not be Impeded

The success of security programs in the refining and petrochemical industries is due in large part to the excellent working relationship industry has established with various federal, state, and local governmental bodies. NPRA and its member companies work with more than a dozen federal agencies, as well as state and local law enforcement agencies and emergency responders throughout the nation to share critical infrastructure information and obtain updates on the latest intelligence concerning terrorist focus and targets. Agencies we work with include the FBI, the Department of Transportation, the Department of Energy, the Department of Defense, the CIA, the Government Accountability Office, and, of course, the Department of Homeland Security and its various components, including the U.S. Secret Service, the Transportation Security Agency, and the U.S. Coast Guard.

Industry's relationship with DHS and other security agencies allows immediate access for both government and industry to rapidly changing information vital to maintaining facility security. Frankly, we are concerned about the impact of new legislation on this cooperative relationship. If DHS becomes an industry regulator through enactment of federal security legislation, the dynamics of the relationship will certainly change and this level of information sharing could be diminished. Our homeland security posture, in other words, could be significantly impacted depending on the content and scope of federal legislation. We ask that you keep these concerns in mind as you develop your proposals.

NPRA does not oppose reasonable chemical security regulation; however, the existing system is working well and care must be taken to do no harm to current efforts in fashioning your ultimate product. Although we do not advocate legislation, we realize that this Committee and DHS have both announced support for new regulatory authority to address chemical security. In response, we have developed some principles that we hope the Committee will consider and adopt in federal legislation.

NPRA's Principles for Chemical Security

Our first principle concerns the general construct of any chemical security legislation or regulation. Given the success of Maritime Transportation Security Act, it is NPRA's strong recommendation that MTSA be used as a model for any new security legislation. MTSA has a proven, successful track record and provides all of the essential tools needed to maintain and strengthen security. A MTSA-type regulatory program would include clear performance-based requirements, security vulnerability assessments, facility security plans, exercises, documentation, reporting procedures, audits, and protection for Sensitive Security Information, or SSI. Such a regulatory program should also provide for self-assessment and auditing, possibly to include a program similar to OSHA's Voluntary Protection Program or EPA's Performance Track.

Federal legislation should continue existing U.S. Coast Guard jurisdiction over facility security, and authorize DHS to promulgate MTSA-type security requirements for chemical facilities not regulated by the Coast Guard. Legislation should avoid overlapping jurisdiction with other federal agencies by giving this federal program preemption over other federal or state programs. In addition, some facilities are only partially covered by MTSA. In these cases, we would suggest that they be given the option of submitting security plans to the Coast Guard where logistically appropriate. Legislation or subsequent regulation should allow this type of "opt in" activity to occur.

As previously mentioned, after 9/11 industry did not wait for new government regulations before implementing enhanced facility security measures. Refiners and petrochemical manufacturers have conducted security vulnerability assessments and adopted facility security plans. Any new legislation should recognize and give credit to these companies for the security programs they have already implemented.

An important part of any facility security plan is making sure that the workforce is trained, qualified, and dependable. If background checks of employees and contract employees are required, we hope the Committee will direct DHS to define specific criteria for denying workers access to a facility. Companies conducting background checks should also be authorized to access and utilize government resources and databases, as is done now for the financial sector.

Federal legislation should require that DHS develop a risk-based approach to regulating both chemicals and facilities. We would suggest that DHS use Section 112(r) of the Clean Air Act Amendments of 1990 (pertaining to risk management plans) as the starting point to define the chemical sector. DHS should then, by regulation, develop a list of chemicals of interest based on security risk as the qualifier for a chemical site to be regulated. The RAMCAP project will be one tool for DHS to use to assess security risk. DHS should also be given flexibility to set the appropriate chemical thresholds based on risk.

NPRA was encouraged by the core principles for chemical security announced by DHS. Those principles for addressing chemical security are based on risk and provide reasonable, clear, equitable and enforceable security standards, while recognizing investments and progress that companies have made to date. We concur with these principles and look forward to working with both the Committee and DHS as legislation is developed.

Conclusions

To conclude, Madam Chairman, refiners and petrochemical manufacturers take very seriously their responsibilities for maintaining and strengthening security at their facilities. Our industry has complied with modernized, post 9-11 federal security requirements. We have utilized expert engineers who understand our facilities better than anyone else to conduct vulnerability assessments and implement new measures to protect against new threats. We have called upon experts throughout all of industry, government agencies, and the security industry to determine the best practices to protect our facilities. And perhaps most importantly, the industry has created an outstanding working relationship with government security agencies to receive rapidly the critical information needed to fight terrorism. This working partnership has been very effective in encouraging the exchange of information to allow the industry to focus on the security threats that exist today and are most relevant. NPRA and its members look forward to continuing this security partnership.

In closing, I urge the Committee to fully consider the impact of federal legislation on existing security programs and practices, to use MTSA as the template for developing new chemical security requirements, and to embrace and support the core principles outlined by DHS at the Committee's June 15th hearing. I will be happy to answer any questions the Committee may have on our testimony.

NPRA

July 2005

FACILITY SECURITY MEASURES TAKEN BY PETROLEUM REFINERS & PETROCHEMICAL MANUFACTURERS

NPRA, the National Petrochemical & Refiners Association, has more than 450 members, including virtually all U.S. refiners and petrochemical manufacturers. Our members supply consumers with a wide variety of products and services that are used daily in homes and businesses and contribute to the nation's quality of life and security. NPRA is proud of the accomplishments refiners and petrochemical manufacturers have achieved in maintaining and strengthening facility security.

NPRA members report they have conducted comprehensive facility security vulnerability assessments and have identified and evaluated critical assets and infrastructure, such as dock facilities, high value production units, power stations, and other equipment which, if attacked by terrorists, could result in significant off-site consequences. Each individual facility is expected to determine what is most important for that particular facility. With this information, facilities have taken the following kinds of specific measures to enhance security:

Formalized information sharing networks with area businesses and local, state, and federal law enforcement and homeland security (such as membership in the Energy Information Sharing and Analysis Center, or ISAC, and the Homeland Security Information Network, or HSIN).

Shared security response plans with local law enforcement and appropriate federal agencies.

Conducted drills & exercises to test response plans.

Hired security personnel, some of which are used around the clock, seven days per week.

Conducted contractor background checks.

Installed perimeter fencing, ditches, berms, and jersey barriers.

Reconfigured roadways and installed speed devices to delay vehicular movement.

Installed a variety of fence-line intrusion detection devices, to include security lighting and area cameras.

Reconfigured sites, allowing critical assets to be set back from perimeters.

Acquired enhanced security communication systems.

Instituted perimeter patrols and surveillance, conducted by both company personnel and local law enforcement.

Installed electronic intrusion detection on buildings (e.g., infrared, motion detectors, door and window sensors).

Implemented card-access controls, with new technology access readers (e.g., biometrics, retina scan).

Required remote parking for employees or contractors, and contractor/visitor vehicles marked with identification (signs/cones).

Required ID badges to be displayed at all times, and instituted procedures for lost ID card and requiring parking decals.

Adopted shipments/deliveries verification process (e.g., close examination of shipping papers, driver's identity).

Identified restricted areas within facilities.

Monitored railroad traffic to and through facility.

Required all visitors to produce identification.

Restricted visitors from driving within the facility.

Prohibited any unannounced visitors.

Rotated access gates on random basis.

Conducted security officer training.

Installed secure mail handling procedures.

Reported suspicious activities (e.g., photo taking, vehicles parked unusually, aircraft over facility).

Conducted vehicle searches (interior & exterior).

Instituted sophisticated processes for collecting and evaluating intelligence/threat information.

Protected computer infrastructure.

Questions about this document may be directed to Maurice McBride, NPRA Director for Security, 202-457-0480, or mcbride@npra.org.

Location of U.S. Refineries 2005

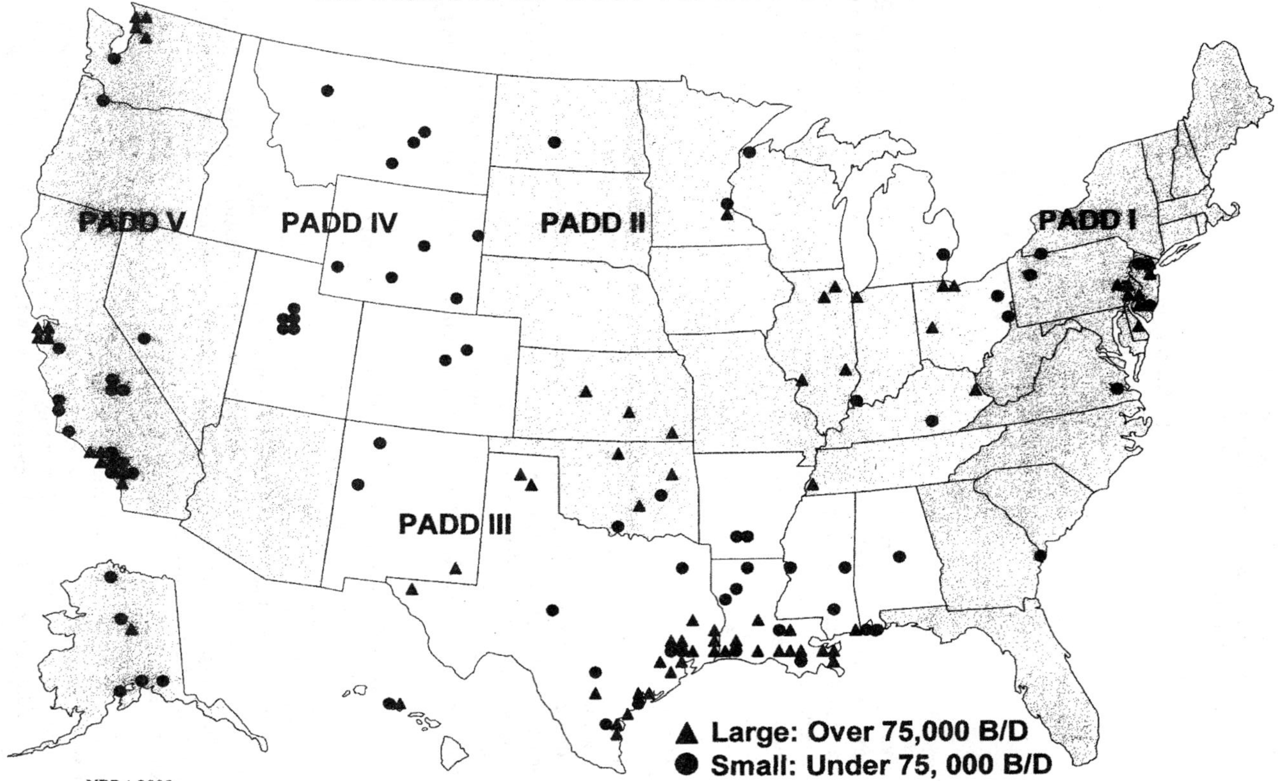

NPRA 2005

Testimony of

Gerald Poje, Ph.D.

Before the Senate Committee on Homeland Security and Government Affairs

July 13, 2005

Thank you Madam Chairman, Senator Lieberman, and members of the Committee, for the opportunity to testify before this committee. I commend you for your leadership in convening a series of hearings, as a prelude to considering new federal policies to strengthen the security of the chemical sector. As repeated Iraqi incidents and last Thursday's events in London tell us, terrorism is an all too frequent, emergent global hazard and must be addressed in the next generation of risk assessment and mitigation in all sectors of U.S. society. The chemical sector bears special attention given its history of catastrophic fires, explosions and toxic releases whose outcome can precipitate a sense of public terror.

My testimony focuses upon the chemical risks to communities, the need for new policies to consider the interface between safety and security and the recommendation to promote coordination across facilities, emergency responders, at-risk communities with and among federal agencies. Although known to me, others can provide specific illustrations of security weaknesses that support the call for a national approach to regulations. Much effort has already been expended in developing and using model vulnerability assessments and in implementing security programs that also should be considered by the Committee.

My professional competency is in the field of toxicology and chemical safety policy. Until last November, I served as a board member of the U.S. Chemical Safety and Hazard Investigation Board (the Board or the CSB). My tenure began with the agency's inception and remains the longest duration of any board member. The CSB is an independent federal agency whose primary mission is to investigate and promote the prevention of unintentional, major chemical incidents at industrial facilities. In addition to conducting root cause investigations and reporting on findings, the Board has been directed by Congress to conduct special studies that encompass analyses of policy, guidelines, regulations and laws governing chemical safety.

Prior to joining the CSB, I directed international programs and public health for the National Institute of Environmental Health Sciences, an institution that also has lead responsibility for the National Toxicology Program, the premier governmental approach for elucidating chemical hazards, and the Worker Education and Training Program, a leading peer-reviewed, competitive grants program for ensuring training of emergency responders to manage hazardous material incidents.

Safety and Security Risks Surrounding the Chemical Industry

The chemical sector is an important component of the American economy and fundamental to our current quality of life. Less than a year ago, The American Chemistry Council provided a detailed economic analysis of the chemical sector, estimating its business value as $459 billion, providing 900,000 direct jobs, supporting employment for nearly 700,000 suppliers and contributing nearly $30 billion in income and property taxes.[1]

However, as painful experiences have taught us, special risks are associated with this sector. Many American communities have suffered localized chemical releases from routine chemical processing, distribution, product usage or waste disposal that, in limited ways, contaminate air, water, or soil. Much larger societal use of specific chemicals over longer periods of time have resulted in releases with widespread regional and global impacts, such as food chain contamination by persistent, bioaccumulative, toxic chemicals and even holes in the stratospheric ozone layer.

Germane to the thrust of this hearing are catastrophic chemical risks that have proved costly in lives lost and livelihoods and property destroyed. This class of problems include major episodic explosions, fires and toxic releases that are generally characterized as low probability – high consequence (LP-HC) events. Low probability does not mean no probability, just very infrequent events at any single facility and within any given process at that facility. However, given the great diversity of facilities and processes across America, the aggregate annual, number of events are nationally quite significant.

U.S Chemical Accident Patterns and Costs

Despite valuable surveillance efforts among some states and federal agencies,[2] the true number, severity and trends of U.S. chemical incidents is not known.[3] Nationally, among 14,500 high hazard chemical-handling facilities required to file risk management plans with the U.S. Environmental Protection Agency (EPA) in 1999, more than 1100 of these facilities reported approximately 1,900 incidents over the five-year period from 1994 through 1999 – more than one incident per day. These incidents resulted in a total of 33 deaths and 1,897 injuries, to workers/employees and evacuation or sheltering in place of over 200,000 members of the public.[4,5]

---

[1] American Chemistry Council. 2004 Guide to the business of chemistry. Arlington, VA

[2] Horton, DK et al., "Surveillance of hazardous materials events in 17 states, 1993-2001: a report from the Hazardous Substances Emergency Events Surveillance (HSEES) System." Am J Ind Med 2004, 45:539-548.

[3] Mannan, S. et al, "National Chemical Safety Program, Annual Assessment Report – 2001" Publication of the Mary Kay O'Connor Process Safety Center at Texas A&M University.

[4] See, Kleindorfer, P. et al., Center for Risk Management and Decision Processes, The Wharton School, University of Pennsylvania, http://opim.wharton.upenn.edu/risk/downloads/00-1-15.pdf

In similar fashion, the Hazardous Substances Emergency Events Surveillance (HSEES) system established by the Agency for Toxic Substances and Disease Registry (ATSDR) within the Centers for Disease Control and Prevention collects and analyzes information about acute releases of hazardous substances that need to be cleaned up or neutralized according to federal, state, or local law, as well as threatened releases that result in a public health action such as an evacuation.[6,7] HSEES events are defined as any release or threatened release of at least one hazardous substance.[8]

For a five year period (1996-2001) surveillance systems from 13 state recorded 39,766 incidents (29, 994 at fixed facilities) of which 2,964 involved evacuations of up to 11,000 people. HSEES captures data on approximately 9,000 events annually – nearly 25 per day, however it is not a comprehensive tally of U.S. incidents.[9] Over the years the ATSDR aggregate data has remained fairly consistent, while individual states vary.

Direct losses from chemical releases have been estimated as about $1 billion dollars per year.[10] Taking into account indirect losses and other losses not covered by insurance companies, the losses would be conservatively estimated as three to four times larger, or additionally three to four billion dollars annually.

Role of Management Systems in Incident Prevention

---

[5] Note: during my tenure as a board member, CSB was involved in 33 investigations from 1998 through 2004 that resulted in 58 deaths and 199 injuries. Fewer than 10 percent of incidents investigated by the CSB involve RMP-covered processes (3 RMP covered incidents).

[6] See, http://www.atsdr.cdc.gov/HS/HSEES/hsees.html

[7] Data collected include: time, date, and day of the week; geographic location and place within the facility where the event occurred; event type (fixed-facility or transportation-related event); factors contributing to the release; environmental sampling and follow-up health activities; specific information on injured persons: age, sex, type and extent of injuries, distance from spill, population group (employee, general public, responder, student), and type of protective equipment used ; information about decontaminations, orders to evacuate or shelter-in-place ; land use and population information to estimate the number of persons at home who were potentially exposed; whether a contingency plan was followed and which plan.

[8] Unlike the EPA RMP program with a defined list of covered chemicals, HSEES program considers a substance hazardous if it might reasonably be expected to cause adverse human health effects. It also has a major exception in rejecting incidents involving releases of petroleum products.

[9] Funding limitations allow only fifteen state health departments currently to have cooperative agreements with ATSDR to participate in HSEES: Colorado, Florida, Iowa, Louisiana, Michigan, Minnesota, Missouri, New Jersey, New York, North Carolina, Oregon, Texas, Utah, Washington, and Wisconsin. Many of these states have contributed independently to support this program.

[10] "Economic Analysis in Support of Final Rule on Risk Management Program Regulations for Chemical Accident Release Prevention, As Required by Section 112 r of the Clean Air Act", CEPPO, US EPA, Section 6-p. 21, Exhibit 6-10, June 1996.

The avoidance of safety problems requires management's demonstration of commitment, a well trained, educated and knowledgeable workforce, effective supervisory process, and employee involvement and commitment. Since the early 1980s private practice, professional engineering guidance and governmental policy have evolved to address LP-HC problems from a simple system of technical requirements to control hazards into a newer management systems paradigm of prevention.

Whether by the Chemical Safety Board, by a major governmental safety agency or by a leading corporation, the best investigations of LP-HC events examine specific safety management systems for the root causes underlying chemical process incidents, since rectifying these causes will do the most to prevent recurrence of the incident.

Terrorism has added another risk factor to LP-HC events. In response, many practitioners of process safety have incorporated the new hazard into the existing hazard assessment approach that must be addressed as part of a larger management system to prevent chemical releases.

Special features of terrorist risks demand closer coordination with governmental security expertise about threat potential and additional capacity for on-site physical security assuredness. However, chemical security is linked inextricably to chemical safety. I urge the committee to see the development and maintenance of competent management systems for safety as essential underpinnings to enhance security.

Why Lessons Learned need to be considered from major chemical incidents
Unfortunately, major LP-HC incidents have happened in America. They have occurred with extremely deadly consequences in premier multinational corporations. And, they have occurred recently. Three incidents bear specific consideration from this committee about causes, consequences and coordination needs: the ammonium nitrate explosion in Texas City in 1947; the methyl isocyanate release at Bhopal, India in 1984 and the fertilizer factory explosion in Toulouse, France in September 2001.

With 20/20 hindsight and an understanding of current terrorist threat potential, each of these incidents could easily be considered as realistic scenarios for security incidents (in fact, each has had to bear allegations of intentional human causation). Furthermore, each incident provides details about infrastructural issues that must be addressed if we hope to manage effectively the consequences of either chemical security or process safety incidents.

1. Texas City, Texas - April 16-17, 1947
Anchored in the harbor of Texas City on the bright spring Tuesday morning of April 16, 1947 was a liberty class cargo ship, the "Grand Camp." During the previous few days it had been loaded with tons of an ammonium nitrate fertilizer, a cargo destined for European post war redevelopment as part of the Marshall Plan. Texas City was a boom

town, having rapidly developed as a major port, petroleum refiner and petrochemical producer during the war.[11]

For several possible reasons the warm fertilizer began to smolder, emitting a reddish-orange 'pretty' smoke, mobilizing the under-trained and under-equipped fire department,[12] and engendering a crowd of school children and adult spectators. Rather that douse the cargo with water, emergency responders were directed to close the hatches and the hot cargo was subject to ships steam heat, in a misperception that such action would starve the fire of available oxygen and preserve the economic value of the cargo. Shortly thereafter the fertilizer exploded, destroying the ship, the entire volunteer fire department and all arrayed alongside the dock.[13]

The detonation was heard in Houston and 150 miles away. A smoke plume 2000 feet high was observed from Galveston and Shrapnel rained upon the nearby petrochemical complex. Like falling dominos, pipelines broke and storage tanks were breached, triggering fires and secondary explosions in numerous businesses, and multiplying the fatalities and injuries. The casualties swamped the response capacity of Texas City Hospital, a small 20-bed clinic, serving a city of 18,000.

The carnage reigned throughout the day and into the night, culminating in a smoldering fire in the cargo hold of a second liberty ship, the High Flyer, a vessel that also contained ammonium nitrate fertilizer. Damaged and unable to be towed away from the dockside, the High Flyer exploded in the early morning of April 17, killing and injuring others, including emergency responders that had recently arrived from throughout the surrounding area. Fear deepened and Texas City fires burned for a week.

When the dust finally had settled, the toll was tallied at nearly 600 killed,[14] 3500 injured, homes and schools extensively damaged, making the Texas City event America's largest chemical disaster. Subsequent analyses and investigations demonstrated that the emergency response infrastructure was under prepared and quickly overwhelmed. Hazards were neither assessed, nor understood by all who could have demanded

[11] The majority of the very large petrochemical complex was located in an unincorporated area and not subject to local taxes. The residential population had grown so rapidly that the under-resourced elementary school operated in split sessions.

[12] Shortly before the event the town sold its only fire boat in a cost cutting measure.

[13] For more detailed analyses of the Texas City incident, see: Minituglia, Bill. 2003. City on Fire: The Forgotten Disaster that Devistated a Town and Launched a Landmark Legal Battle, HarperCollins Press, NY; Stephens, Hugh W., 1996. The Texas City Disaster, 1947, University of Texas Press.; http://www.chron.com/content/chronicle/metropolitan/txcity/ and http://sdsd.essortment.com/texascityexplo_rkvi.htm

[14] Scores of victims were never identified, having been burned beyond the detection capacities of the forensic technologies of that era.

operations with a greater sense of precaution.[15] Private practice and public regulations were woefully deficient to manage the hazards and respond to the emergency. The U.S. Coast Guard that had established and enforced much stronger safety precautions with ammonium nitrate when it was shipped as explosive material during WWII, had relaxed its vigilance when the same material from the same factories was shipped as fertilizer.

2. Bhopal, India – December 2-3, 1984

Safely conducting chemical reactions is a core competency of the chemical manufacturing industry. Reactivity is not necessarily an intrinsic property of a chemical substance. The hazards associated with reactivity are related to process-specific factors, such as operating temperatures, pressures, quantities handled, concentrations, the presence of other substances, and impurities with catalytic effects. Chemical reactions can rapidly release large quantities of heat, energy, and gaseous byproducts. Uncontrolled chemical reactions have led to serious explosions, fires, and toxic emissions, that kill and injure, damage property and threaten the environment.

The world's worst chemical disaster began as a violent runaway reaction within a methyl isocyanate (MIC) storage tank in the late Sunday evening of December 2,1984 at the Bhopal Union Carbide pesticide plant in Madhya Pradesh, India. After ~ 1,500 lbs of water entered the MIC tank, possibly caused by a routine line washing procedure, an exothermic reaction ensued. Excessively heated and pressurized gases burst through a rupture disk and opened a pressure relief valve, allowing ~ 54,000 lbs of MIC and reactants to be released through an elevated scrubber vent system. Cooling gases formed a dense, low lying cloud that in the early morning of December 3 slowly and quietly drifted through adjacent housing and circulated throughout much of the central city, including the railway station.

MIC is a highly reactive, irritating and toxic gas that is soluble in the aqueous fluid of membranes surrounding eyes and lungs. Victims awoke gasping for painful breathes and stumbled bleary eyed into the darkened streets with no indication of which direction to seek relief. The government of India estimated 1754 immediate fatalities. Others estimate initial fatalities as high as 3000 and an accumulation of 15-20,000 disaster related deaths in subsequent years, based upon elevated mortality rates among hundreds of thousands of injured people.[16]

Injuries have been estimated to range from 200,000 to 500,000, with the Bhopal Directorate of Claims having registered medical folders for 361,966 exposed persons by 1990. These casualties overwhelmed the city's four hospitals and several clinics that supplied a total of 1800 hospital beds and 300 doctors. Mitigation of the damages from the toxic chemical exposures were exacerbated by the city's inability to provide water to

---

[15] Unlike Texas City, the city of Houston had refused to accept the high volume of dangerous , ammonium nitrate fertilizer for loading as their docks.

[16] Dhara, V. R., and Dhara, R. 2002. The Union Carbide Disaster in Bhopal: A Review of Health Effects. Archives of Environmental Health 57(5): 391-404.

residential taps for more than a few hours per day, and the meager water supplied had quality problems.

Underlying systemic problems at the Bhopal facility and community included the following management system issues noted by several reports and analyses: [17]

- **Lack of awareness and knowledge of hazards.** MIC was produced and utilized as a high volume intermediate chemical, and yet its hazards under specific process conditions were not well understood by workers and emergency responders. Company personnel, nearby inhabitants and emergency responders were unaware of MIC toxicity. Medical and toxicological professionals debated appropriate treatment for months following the crippling exposures of Dec. 3. Citizen watchdog groups were lacking prior to the incident.
- **Deficient process hazard assessment.** The hazards associated with contamination of MIC storage tanks and their operations under higher temperatures and pressures were poorly assessed, and therefore abnormal situations were not managed safely.
- **Inadequacy of operating procedures.** Operating procedures were insufficient, poorly written, understood and executed. MIC tanks at the facility were filled above their recommended volume levels. A spare storage tank, intended to be empty for emergency dumping, instead contained high hazard intermediate chemicals.
- **Staffing insufficiency and lack of preparedness for abnormal situation.** Managers and staff were relatively new to the facility and unfamiliar with all the systems and personnel. Responsibilities of various employees were not clearly established. The facility staffing had been downsized. Staff turnover was high, and critical functions were severely undermanned. Staff training was not maintained.
- **Failure to maintain essential design and safety equipment.** Significant facility changes were not assessed for their safety impact and therefore not managed appropriately. The refrigeration unit designed to stabilize the pressure and temperature of the MIC in the storage tank was shutdown and the coolant was drained months earlier. The flare tower had been shut off for maintenance and was not operational at the time of the event. The scrubber system, which had the ability to detoxify smaller amounts of the MIC, was also turned off at the time of the event. Regardless, the system was not capable of neutralizing the quantity of MIC that escaped.
- **Investigation inadequacy and failure to implement audit recommendations.** Prior deadly incidents that caused fatalities, injuries and evacuations and smaller

---

[17] For more detailed analyses of the Bhopal incident, see: Kharbhanda, O., and Stallworthy, E. 1988. Safety in the Chemical Industry. Townbridge, Wiltshire: Redwood Burn Ltd.; Shrivastava, P. 1992. Bhopal: Anatomy of a Crisis (2nd ed). London: Paul Chapman Publishing Ltd.; Lees, F. 1996. Loss Prevention in the Process Industries (2nd ed: Vol. 2&3). Great Britain: Reed Educational and Professional Publishing. Kletz, T. 1999. What Went Wrong: Case Histories of Process Plant Disasters (4nd ed). Houston: Gulf Publishing Company.

MIC releases at the facility were not fully investigated and root and contributing causes established.[18] Significant safety audit recommendations had not been enacted.

- **Failure to maintain equipment mechanical integrity**. Valves and pipes were corroded and leaking. Many of the instruments and gauges such as pressure indicators were defective to the extent that workers did not trust them, thereby exacerbating problems of operating procedure adherence.
- **Inadequacy of emergency planning and response**. The scrubber system was not designed to handle the amount of MIC that breached containment. The water curtain system was not positioned high enough to contain escaping gas. Staff was confused as to whether or not to turn on the public emergency evacuation siren, and during the leak the alarm remained off for a matter of hours. No clear method of evacuation was established to manage such a release. Local zoning permitted dense, shanty dwellings to be close to the Union Carbide facility thereby increasing the population at risk.
- **Lack of Public Authority and Oversight**. As a emergent industrial nation, the government of India did not have laws, regulations and trained staff to ensure compliance with appropriate safety practice.

The Bhopal disaster prompted various assessments of causation, including one that speculated sabotage[19] and serious questions about the adequacy of international legal systems regarding responsibilities of multinational corporations.[20]

Union Carbide Corporation (UCC), a major multinational chemical corporation, headquartered in Danbury, CT had multiple U.S. production facilities, including those handling large amounts of MIC. Concerned about domestic chemical safety, Congress held hearings on chemical safety. UCC and the Occupational Safety and Health Administration (OSHA) conducted safety assessments of MIC operations at UCC's Institute, WV facility in late 1984 and early 1985 with generally favorable accounts of safety management. However, an aldicarb oxime release from the same Institute, WV facility in August 1985 sent over 130 people to the hospital, fueled widespread public doubts about the adequacy of high hazard chemical management by large corporations,

---

[18] In Dec. 1981 3 workers were exposed to phosgene, 1 died; 2 weeks later 24 workers were overcome by another phosgene leak. In February 1982 18 workers were affected by an MIC leak. In October 1982 3 workers and nearby residents were affected by a leak of hydrochloric acid and chloroform.

[19] See, for example: 1985 Report of International Confederation of Free Trade Unions International Federation of Chemical, Energy, and General Workers Unions (ICFTU-ICEF) mission to study the causes and effects of the methyl isocyanate gas leak at the Union Carbide pesticide plant in Bhopal, India on December 2-3, 1984, at: http://bhopal.net/oldsite/documentlibrary/unionreport1985.html; and Ashok S. Kalelkar, Investigation of Large-Magnitude Incidents: Bhopal as a Case Study, Arthur D. Little, Inc., Cambridge Massachusetts, USA, May 1988, at http://www.bhopal.com/pdfs/casestdy.pdf

[20] Despite its magnitude, the full circumstances and consequences of the Bhopal incident have not been deliberated in a court of law. For a fuller examination of the legal dilemma, see: Cassels, J. 1993. The Uncertain Promise of Law. Toronto: University of Toronto Press Inc.

oversight competency of federal agencies and precipitated significant changes in domestic policy regarding high hazard chemicals.[21]

3. Toulouse, France – September 21, 2001

While most Americans vividly remember the events of 9/11/2001, few recall the major chemical catastrophe that occurred just 10 days later. Mid-Thursday morning on September 21, a huge explosion tore through the AZF (Azote de France) fertilizer factory in Toulouse, France.[22] Nearly 400 tons of off specification granular ammonium nitrate (and perhaps contaminated with a reactive agent) stored in a warehouse detonated with the force of 20-40 tons of TNT and equivalent to an earthquake measuring 3.4 on the Richter scale.. AZF is owned by Atofina, the chemicals unit of TotalFinaElf one of the world's largest petroleum and petrochemical producers.[23]

The blast created a crater 50 meters in diameter and 10 meters deep. Windows shattered in buildings throughout the city center three kilometers away. Thirty people were killed: 22 on the site, 8 members of the public. National and local authorities estimated that 10,000 people were physically injured, and a further 14,000 sought treatment for acute post-traumatic stress for months following the explosion. Over 500 homes were rendered uninhabitable, some 27,000 other dwellings were damaged, and almost 11,000 pupils had their educations interrupted since 85 schools and colleges sustained damage. Insurers estimated the costs at 1.5 billion euros.

Alarm systems were rendered inoperable and telephone lines were severed, frustrating the public communications of safety messages. Telecommunications were affected as far as

---

[21] Most prominent policy changes were reflected in the Emergency Planning and Community Right-To-Know Act of the Superfund Amendments and Reauthorization Act (1986) and the chemical accident prevention provisions of the Clean Air Act (1990).

[22] For more information about the Toulouse incident see: Dechy, N., T. Bourdeaux, N. Ayrault,, M-A. Kordek, J-C. Le Coze. 2004. First lessons of the Toulouse ammonium nitrate disaster, 21st September 2001, AZF plant, France. J. Hazardous Materials 111 (2004) 131–138; Dechy, N. and Y. Mouilleau , 2004. Damages of the Toulouse Disaster, 21st September 2001. In Loss Prevention and Safety Promotion in the Process Industries, 11th International Symposium - Loss Prevention 2004, Praha Congress Center, Prague, Czech Republic. 31 May – 3 June, 2004.; also, http://www.uneptie.org/pc/apell/disasters/toulouse/home.html; http://www.environmenttimes.net/article.cfm?pageID=131; http://www.icem.org/update/upd2001/upd01-68.html;

[23] Just two months earlier the CSB directed an investigation team to assess an event on July 14, 2001, at the ATOFINA Chemicals, Inc., (ATOFINA) plant in Riverview, Michigan. Ultimately the National Transportation Safety Board found that a pipe attached to a fitting on the unloading line of a railroad tank car fractured and separated, causing the release of methyl mercaptan, a poisonous and flammable gas. The gas ignited, engulfing the tank car in flames and sending a fireball about 200 feet into the air. Fire damage to cargo transfer hoses on an adjacent tank car resulted in the release of chlorine, a poisonous gas that is also an oxidizer. Three plant employees were killed in the accident, several were seriously injured and nearly 2,000 residents were evacuated in Michigan and into Ontario. See: http://www.ntsb.gov/publictn/2002/HZM0201.pdf and http://www.semcosh.org/atofina_explosion.htm.

100 km away. Air traffic was rerouted away from Toulouse. A nearby business collapsed 45 minutes after the explosion and others were subjected significant, long term business interruptions.

Thousands of tons of liquefied ammonia, solid ammonium nitrate and solid fertilizer were stored in other portions of the AZF facility, and nearby chemical businesses stored others toxic and hazardous chemicals, prompting additional concerns about domino effects throughout the industrial park. Because so many windows and building structure were damaged, sheltering in place would not have been possible if toxic chemicals were released.

The event greatly exceeded the consequences of the scenarios that had been used for emergency planning. More than 1500 fireman and special emergency personnel and 950 policemen responded to the event. Yet early responders arrived lacking exposure assessment equipment and the personal protective equipment to cope with a toxic cloud. Communications among responders suffered because of severed land lines and saturated cellular networks.

The AZF facility had been inspected seven times in three years by local authorities, but not for the adequacy of ammonium nitrate fertilizer management. Within a few weeks of the incident, the European Parliament issued a resolution calling for member states to provide themselves with sufficient numbers of competent inspectors trained to the specific technological hazards of the regulated facilities.[24,25]

The Toulouse disaster engaged the highest levels of French governmental leaders and prompted nationwide debate through many formal dialogues in communities near the 1200+ high hazard French facilities. The French legislature conducted an extensive review and deliberations on policies and practices. New legislation[26] has focused on strengthening the safety management systems of technological risks, including:

- Enhanced participation of employees in risk prevention and enhanced training of those working at at-risk sites.
- Improved safety management, coordination and roles/responsibilities of contract workers.[27]
- Expanded requirements to inform the public and to involve it more closely in the prevention of industrial risks

[24] See: http://europa.eu.int/abc/doc/off/bull/en/200110/p104028.htm

[25] Some experts called for a doubling of the French inspectors, and the French Administration plans to have 1400 inspectors by 2007, up from 800 at the time of the incident.

[26] See: http://mahbsrv.jrc.it/downloads/frenchlegisEN/30july_law_on_risk_prevention.pdf

[27] At AZF, 250 regular employees worked alongside 100 subcontractors who were drawn from 25 different companies. Three different subcontractors worked in the warehouse where the explosion occurred . Some characterized AZF as having 'lost control' of the work of the warehouse contractors.

- Better control over land use planning and urbanization around the at-risk sites

The Toulouse disaster also influenced policies in a larger European context by stimulating amendments to the Seveso directive that governs each member country's approach towards chemical incident prevention, preparedness and response.[28] Among other amended provisions, facilities handling the forms of ammonium nitrate and ammonium nitrate fertilizer involved in the AZF event were made subject to the Seveso II requirements [29].

Summary: U.S. policy needs to establish and define a new federal role in chemical security that is consonant with the management systems approach in chemical safety.

Philosopher, poet, literary and cultural critic, George Santayana speaks to our current situation in his often quoted statement: "Progress, far from consisting in change, depends on retentiveness. Those who cannot remember the past are condemned to repeat it."[30]. As a CSB Board member, I was intimately involved in 33 field investigations and eight safety studies, many of which illustrated the systemic problems of Texas City, Bhopal and Toulouse. I urge the committee to seek progress in formulating new federal chemical security policy, but by building upon experiences in chemical safety.

While much more remains to be accomplished in setting, strengthening and enforcing standards, existing laws and regulations that govern the occupational and environmental safety of highly hazardous chemicals provide a good framework for considering federal role in chemical security. OSHA and EPA establish general duties for employers to safely manage specified hazards and the specific elements of process safety and risk management for regulated facilities to comply. Existing training and information access requirements for emergency preparedness and response provide a road map for new needs to enhance security.

Recommendations for Consideration in Federal Chemical Security Legislation

1. Monitor the Scope of Chemical Sector Problems

Thankfully, in the wake of 9/11 America has not become the victim of a terrorist initiated catastrophe in the chemical sector. However, our vulnerabilities are manifest. As noted above, approximately 9000 incidents occur annually in just 15 states, but a nationwide surveillance system is lacking.

[28] See: http://europa.eu.int/comm/environment/seveso/#2.14

[29] See: http://europa.eu.int/eur-lex/pri/en/oj/dat/2003/l_345/l_34520031231en00970105.pdf

[30] From The Life of Reason (1905)

At a minimum. comprehensive surveillance of chemical incidents whether due to safety or security management system failures would help inform policy makers and the public about sector vulnerabilities , such as which chemicals, processes, facilities and companies are involved in releases, what competencies and capacities are needed to respond to emergencies and what are the changing patterns of incidents. Armed with this perspective, policymakers could better set priorities for improving federal, state and local resource allocations.

2. Establish Department of Homeland Security Responsibility and Promote Coordination with Other Agencies
In a time of large budget deficits I urge Congress not to rob from Peter to pay Paul. Under-investing in programs for public health, occupational safety and environmental protection to resource a narrowly defined chemical security need, will backfire. The quickest return on investment will come from building upon existing strengths and promoting accountability for effective collaboration and coordination.

The Department of Homeland Security should have primary federal expertise is in assessing and addressing chemical security. However, the lessons learned from chemical incidents show that many other agencies have essential roles and responsibilities that need to be employed if we hope to protect the chemical sector comprehensively. OSHA and EPA have set standards for occupational process safety and risk management with the community at large. The Department of Health and Human Services has responsibility for public health protection and promotion. The CSB sets the standards for investigating and gathers important information about process incidents and their community impacts.

3. Set Requirements for a Security Management System
Much work has been accomplished on defining hazards of concern, developing vulnerability assessments tools, implementing security plans; auditing, testing and response exercises; employing inherently safer chemicals and processes, and coordinating with local response agencies and mutual response entities. Establishing strong policy that defines primary and secondary federal responsibilities for security management systems that complements safety management systems is needed.

4 Evaluate Security Management Systems Effectiveness When Failures Occur
While all stakeholders hope for effective assessment planning and management to avoid LP-HC events, experience tells us that some entities will not succeed on their own. Investigating the root causes of chemical incidents has proven quite valuable for strengthening the management systems to prevent recurrences. When wielded effectively by public agencies, such investigations have proven extremely valuable for educating the agency and the larger community about preventable causes of incidents.

Effective programs set standards and routinely audit for compliance on schedules designed to maximize responsiveness from the regulated community. However, more can be done to promote security vigilance. For the Department of Homeland Security to wait for a verified terrorist incident before thoroughly investigating management system competencies at a chemical facility would be a strategic mistake, since chemical incidents occur frequently and these incidents manifest systemic problems that need to be solved for both safety and security. The CSB has had significant success in promoting prevention by widely publicizing the results of a few well selected, noteworthy incidents, and has had much success in collaborating with other relevant entities during the course of an investigation.

5. Support Research, Development and Technology Transfer for Safer Chemicals and Processes

The ultimate solutions to security and safety risks will be found in reducing the volume and toxicity of the chemical hazards, an inherently safer approach. Following the Bhopal tragedy a few major corporations developed aggressive programs to evaluate their storage and use of extremely toxic chemicals, resulting in important process changes that reduced the volume and use of high hazard intermediate chemicals. The American Institute of Chemical Engineers produced good guidance documents on inherently safer chemical processes.

Some chemical processes are overdue for implementing inherently safer technologies. However, if America is to maintain its leadership role in field of chemistry broader support is needed for Green Chemistry Principles that include inherently safer chemistry for incident prevention.[31] The Congress should seek to involve the Department of Homeland Security with the National Science Foundation, the National Institute of Science and Technology, the Department of Energy, and the Environmental Protection Agency National Research Council to enhance research, development and technology transfer whose outcome will enhance safety, security and economic prospects for the chemical sector.

6. Employ Effective Training Approaches

An absolutely critical step to improve the security at chemical plants is to properly train the workers who respond to plant disruptions – both external responders like fire fighters, emergency medical personnel and police, but also workers inside the plant whose immediate reaction to a crisis can make an enormous difference in whether the crisis is controlled quickly with a minimum number of injuries and damage to the facility. The possibility of a plant suffering an unintentional mishap currently is much more realistic than a terrorist attack. Whether a mishap at a plant results from an intentional versus unintentional act, the release consequences are generally the same.

This country - through the private sector and public organizations like the National Institute of Environmental Health Sciences and the National Fire Academy - has trained millions of workers to safely handle uncontrolled hazardous waste sites as well as

[31] Anastas, P. T.; Warner, J. C. 1998. *Green Chemistry: Theory and Practice*, Oxford University Press: New York, and see: http://www.epa.gov/greenchemistry/principles.html

hazardous materials emergencies, in transportation and in fixed facilities.[32] Most of this training has been done under the OSHA Hazardous Waste Worker and Emergency Response standard (HAZWOPER, 29 CFR 1910.120), which was promulgated in 1989.[33] Workers trained under this standard represent a potent force already in place in fire houses, on trains hauling chemicals, in chemical plants, in waste water treatment, and in the nation's nuclear weapons facilities.

Other key consensus standards that have served this nation well and must not be relegated a lesser status through any new efforts to legislate greater chemical security . Firefighters have relied upon standards from the National Fire Protection Association, particularly NFPA 472, 473, and 1600. FEMA, through the National Response Team, has developed a set of training guidelines that have been recognized as definitive among emergency response experts.

Thank you again for the opportunity to testify before you.

---

[32] NIEHS has successfully supported twenty primary awardees, representing over one hundred different institutions that have trained more than 1.2 million workers across the country and presented 69,000 classroom and hands-on training courses, which have accounted for nearly 18 million contact hours of actual training. Awardees developed the official safety and health training for site workers at the cleanup of the World Trade Center, and first reported on site health and safety issues.

[33] See: http://www.osha.gov/Publications/OSHA3114/osha3114.html

Testimony of Glenn Erwin
July 13,2005

Madam Chairman, Senator Lieberman, and members of the Committee, thank you for allowing me to testify this morning on behalf of the 850,000 members of the United Steel, Paper and Forestry, Rubber, Manufacturing, Energy, Allied Industrial and Service Workers International Union or the USW for short.

The comments I make this morning reflect my 35 years of experience within the petrochemical industry. I currently serve as the Program Director for USW's Triangle of Prevention (TOP) Program. This program is a system based, union led, company supported method for finding and fixing the potential failures within a facility. We have trained over 20,000 union and management employees to understand and use the TOP approach to identify problems and make recommendations to correct them.

My comments here this morning are both professional and personal, as I am sure this topic is with the group I sit before.

My invitation letter asks me to discuss the risks posed by the chemical industry to the security of both the workers inside the chemical facilities and the communities that surround these facilities. When I look at the potential effects of a catastrophic failure within a facility, I can see little difference between intentional and unintentional releases. Although the causes are very different, they both have the same tragic effects.

I would like to begin by stating that the USW stands ready to work with the Congress, the Administration and the oil, chemical, paper, steel, nuclear and any other industries where we represent workers. Our goal is for workers and other members of the community to reap the benefits of a safe and secure place to work and live.

In the spring of 2004, we conducted a survey of 125 sites where USW represents workers. 1 These sites were those designated by the EPA as Risk Management Program (RMP) sites. The respondents to our survey reported that each of these sites had quantities of chemicals or other hazardous materials large enough to cause a catastrophic event onsite if those materials were involved in a fire, explosion or other release. Importantly, this study was a process of participatory research carried out among professional researchers, union staff and rank and file members from some of the same plants for whom chemical plant security is a central and vital issue.

My goal in appearing here today is to bring to light some serious gaps between the ideal we desire, and the reality with which we live. We will look today at some of the issues those gaps represent, as identified by our members. First, I will address issues of security. Second, I want

---

[1] Paper, Allied Industrial, Chemical and Energy Workers International Union (PACE). 2004. *PACE International Union Survey: Workplace Incident Prevention and Response Since 9/11.* Nashville, TN. PACE International Union recently merged with the United Steelworkers of America to form the United Steel, Paper and Forestry, Rubber, Manufacturing, Energy, Allied Industrial and Service Workers Intl. Union (USW).

to address issues related to prevention. Finally, I will present the USW's positions on legislative action.

1. Security

One of the gaps identified by our members in our survey was security. I think we can all agree that the reassessment of worksite security in the face of new terrorist threats has been a paramount issue since 9/11. Yet, in our survey, only four out of five high-risk facilities had conducted a reassessment of worksite security since September 11—twenty percent had not2 (p. 16). Similarly, only three in four sites reported that the company at their site had improved the systems to guard and secure the facilities (p. 16). These findings are consistent with news reports of the ease with which reporters have been able to get unfettered access to chemical plant sites that should have been secured.3 If the patterns in these data were to hold more broadly among the population of RMP sites, approximately 3,000 sites would still be without reassessment of worksite security and a similar number would have failed to act to improve security.

Let me illustrate from my personal experiences the ease of access to a facility, which we believe is a failure of security.

As recently as this summer, I stood at the main entrance to one of the nation's major oil refineries and watched pick-up trucks only slow down as guards waved them through. Sitting in the back of the trucks were several closed-topped buckets. When I asked the employee standing with me who they were, he said they were temporary workers employed by contractors. When I asked him what was in the buckets, he said, "I have no idea." I wanted to know why the guards had not screened the trucks' occupants and examined its contents. He said there was so much traffic it would be impossible to check them all. Following the September 11 attacks, this volume of unsecured traffic in and out the gates of our facilities is astounding.

This same facility had a storage tank containing 800,000 pounds of hydrofluoric acid. A release of this much hydrofluoric acid would create an enormous catastrophe. A lethal vapor cloud of hydrofluoric acid would extend for miles downwind and reach into one of the most heavily populated metropolitan areas in the country. As we drove past the tank, I watched approximately 50 people working in the area using heavy equipment less than 50 feet from the exposed liquid line leading to the hydrofluoric acid tank. My tour guide explained that the site was engaged in a "turn-around" and that these people were temporary contract workers. A "turn-around" is the term that describes the periodic shutdown of processing units for major maintenance. I asked if he knew any of these people. He replied, "No, they are just here for three to four weeks." As we drove, we discussed what the result would be if by accident, or on purpose, the bulldozer was driven into the liquid line of this tank. His reply was that thousands maybe tens of thousands would be killed.

---

[2] PACE. 2004. *PACE International Union Survey: Workplace Incident Prevention and Response Since 9/11.* Nashville, TN.

[3] Prine, C. 2002. "Lax Security Exposes Lethal Chemical Supplies," Pittsburgh Tribune-Review. Sunday, April 7, 2002; and CBS News, http://www.csbnews.com/stories/2003/11/13/60minutes/main583528.html.

There is no silver bullet or sole solution to having a safe and secure facility with large volumes of highly hazardous materials. The facilities in which our members and many others work are complex and closely coupled. They require layers of protection.

The concept of "layered protection" has not been applied to the access of critical areas within many of our plants.

First, the access to office buildings should be one layer. Stronger, more in-depth measures should be applied for access to other areas with higher hazard potential. Something must be done to control the "Front Gate Freeway" that exists in far too many facilities.

Second, as the potential for devastation increases, so should the security. A tank of hydrofluoric acid is of far greater concern than a tank of gasoline, but typically, there is no additional layer of internal security for those processes or vessels with the most serious hazard potential.

Think multi-layered protection.

After the access is controlled and limited to known, trusted and trained people, there are more important steps that we need to take to make our plants inherently safer. If we rely too much on security and deemphasize measures that will make sites inherently safer we will do so at our own peril.

2. Prevention

The most foolproof way to prevent our facilities from being turned into weapons of mass destruction by terrorists is to eliminate the very substances, that if released into the environment, could kill or harm workers and the people in surrounding communities. This has been done in many cases such as replacing chlorine with sodium hypochlorite in water treatment facilities.

At the time of the 9/11 attack on the Pentagon there were seven tank cars with a combined 550 tons of chlorine and sulfur dioxide at the Blue Plains Sewage Treatment Plant just across the river. This was enough to have killed thousands of neighbors, including those in Congress and the White House. Al Qaeda computers recovered in Afghanistan contained maps of similar plants. Eight weeks after the attack, engineers at Blue Plains were directed to get rid of the chlorine. Today D.C. treats its sewage with a much safer chemical, sodium hypochlorite, strong household bleach.

In one of our steel plants, the site employed a process using chlorine to treat certain waste streams. The contractor doing the work found it convenient to have as many as fourteen chlorine tanks cars on the site at one time. This quantity of chlorine could have put a major metropolitan area at risk. There was never a need for more than one tank car of chorine at a time. The union fought successfully first to reduce the amount of chlorine stored and later in persuading the company to use a different and safer process that eliminated the use of chlorine all together.

Earlier I spoke about a plant with large volumes of hydrofluoric acid. Hydrofluoric acid is used as a catalyst in a process called alkylation that chemically joins refining compounds. Alkylation can be carried out with the much more dangerous hydrofluoric acid or with the less dangerous sulfuric acid. Some facilities have become inherently safer by replacing hydrofluoric acid with sulfuric acid. Others have not.

There are other examples safer chemical substitution. Whenever they are possible, these types of substitutions provide the first and best layer of protection by eliminating the hazard.

Reducing the hazards that remain provides the second, and next best layer of prevention. Here we ask:
Are the quantities of hazardous materials stored and energy used as small as possible?
Recall, that the reduction in the volume of chemicals was the first step in the steel mill example just cited. Substitution with a safer process came later. Another example comes from Bhopal where the release of methyl isocyanate (MIC) killed and injured thousands. Union Carbide used MIC as an intermediate in the production of the pesticide Sevin. In a similar process used by Mitsubishi in Japan, the process was designed such that MIC was consumed immediately as it was being produced. Union Carbide could have produced Sevin without any MIC storage. No storage, no massive release.

We also ask:
Are reactive materials adequately isolated from each other?
Are the least hazardous conditions and least hazardous forms of materials being used?
Have systems been designed so that they are hardened against possible failures and forgiving of potential errors?

Each of these safeguards reduces the likelihood of a catastrophic release.

The next layers of protection is provided by mitigation and containment should vessels be breached. Here we ask:
Can hazardous materials be stored in smaller, separate containers?
Are systems sufficient to suppress, neutralize and contain a release if it occurs?
Will these key systems operate if the power supply is interrupted?

Moving to yet another layer, we must have in place preparedness, warning and response capabilities commensurate with the disaster potential at chemical facilities. I find that we are always too slow in sounding the alarm and that communication equipment is seldom, if ever, sufficient. Lives can be saved if automatic notification devices are installed to detect and trigger evacuation alarms when toxic or explosive material is released.

I have just covered:
Substitute for safer materials. Effective substitution requires less dependence on other prevention and response systems.
Further minimizing risks using other secondary prevention methods.
Mitigating the effects of a release should systems be breached.
Being prepared to carry out an effective response.

We are stressing these forms or prevention here because the overriding focus since 9/11 has been limited to security. Prevention has been bypassed as a priority.

Now here is the problem. In our survey4 (p. 39), 90% of respondents stated their facility had not worked with the local union, or hourly workers about plans or actions to prevent or respond to a possible terrorist attack. The people who know the most about these facilities are the full-time workers who run and maintain them. We are astounded that in the vast majority of cases these people have not been included in addressing chemical plant security and safety issues related to a possible terrorist attack. If workers are neither informed nor involved before an incident happens, how can there possibly be effective preventative systems in place? How could workers possibly contribute their vast knowledge, experience and skills to prevention, preparedness or response? We firmly believe that the lack of union or worker involvement in preventing terrorist attacks means that the systems are broken and in desperate need of repair.

In the 1990 Clean Air Act Amendments, Congress required OSHA to promulgate a process safety management standard addressing the risks of catastrophic chemical accidents. This same legislation required EPA to institute its complementary risk management program. Of course, the concern then was the accidental release of highly hazardous chemicals. It is time these programs be adapted and applied to our post-9/11 world. For example, the government should mandate that all facilities covered by the OSHA Process Safety Management Standard conduct a "Process Hazard/Terrorist Analysis." Sites would conducted this analysis in accordance with the present requirements for unintentional events, but would now include an analysis of terrorist potential at each point of review for each covered process. Furthermore, in because terrorist are capable of striking multiple targets simultaneously, we recommend that all "worst case scenarios" under OSHA and EPA now include multiple failures.

Finally, on the issue of the necessity of federal legislation …

I would take this opportunity to express my sincere thanks to those forward thinking facilities that are striving to achieve the excellence necessary to provide a safe workplace and be a good neighbor.

But if you review the results of USW's national survey you will see that in the days following 9/11 there has been some improvement in some areas by some companies. But, we ask, is some improvement in some areas by some companies enough? With what is at stake, we all know the answer is an emphatic no. Workers and members of our communities should not be placed at risk because some companies either have other priorities or choose to ignore the possibility of an attack. The phrase, "this will never happen to us," should be erased from our vocabulary.

Responsible companies should not be placed at an economic disadvantage because they allocate resources to address the threats we face.

To insure that not just some, but all prepare:

We support legislation that would dovetail with the OSHA Process Safety Management Standard and the EPA Risk Management Program with a focus on potential terrorist attacks.

---

[4] PACE. 2004. *PACE International Union Survey: Workplace Incident Prevention and Response Since 9/11.* Nashville, TN.

This legislation should certainly mandate high-level security measures—fences, guards, etc.—but its main focus should be on forging inherently safer processes and minimizing the storage of highly hazardous chemicals. EPA should enforce this part of the legislation, even if DHS enforces the fences and guards.

This legislation should include provisions for prevention, preparedness, emergency response and remediation. Our experience, coupled with our national study, shows that voluntary measures are not enough. The country needs strong legislation that will ensure that companies take all possible measures to protect our communities, our workers and our industries.

This legislation should strike an appropriate balance between, on the one hand, the need to keep critical information out of the hands of terrorists, and, on the other, the information needs of local responders, and the public's right to know. It is hard to imagine how we can win true protection without providing vital information to workers and communities. If they are kept in the dark, not only will opportunities for prevention and preparedness be lost, chaos will rein if an incident does occur.

Legislation should mandate the participation of workers and their unions as major contributors to both security and inherent safety.

There should be government funding for both research on, and promotion of inherently safer systems. Funding should also be provided for training and education for chemical site workers, emergency responders and remediation workers. A model program for this type of training already exists at the National Institute of Environmental Health Sciences' (NIEHS') Worker Education and Training Program (WETP). That highly successful program should be strengthened and expanded.

Every day hundreds of thousands of workers stand on the front lines, working skillfully and diligently to ensure the safety of our nation's chemical-related facilities. Since September 11, these workers have stood ready to make an additional contribution to workplace prevention, preparedness and response related to possible terrorist attacks. Neither the union nor its members want to stand idly by. Enlist us in the fight to keep our plants safe for our members as well as those across the fence-lines. Serving on the front lines, we know that the job of protecting our facilities and our country cannot be accomplished without us.

Thank you again for the opportunity to speak to you.

Paper, Allied-Industrial, Chemical and Energy Workers International Union (PACE)

PACE International Union Survey: Workplace Incident Prevention and Response Since 9/11

October 2004

A cooperative effort of New Perspectives Consulting Group, Inc. and the PACE Evaluation Team

PACE International Union ◆ Health and Safety Department◆P.O. Box 1475◆ Nashville, TN 37202
Voice: (615) 834-8590◆ Fax: (615) 833-9332◆ dortlieb@steelworkers-usw.org

Acknowledgments

This report was produced by New Perspectives Consulting Group, Inc.* with in-depth consultation from the PACE International Union[†] Evaluation Team. The Team is comprised of: 1) worker trainers from across the country who work in different industries represented by PACE; 2) PACE staff members; and 3) staff from the Labor Institute, the labor education organization that writes and develops educational programs for PACE union. The Team helped develop the project, design the survey and interviews, carry out the data collection, analyze the data, and review the report.

Many thanks to the Team members for their efforts:

Linda Cook
Mike Gill
Donna Howard
Michael Kaufman
Tom McQuiston
Tom Seymour
Kathy Smith
Doug Stephens
Brian Williams

In addition, we would like to thank:

- PACE International Union President, Boyd Young for supporting this study and recognizing its importance to PACE's membership
- Participating PACE local union leaders for their support and contribution
- Other PACE staff members who supported this effort and provided key input along the way: Steve Cable and Glenn Erwin

This survey and report was funded by grant number U45ES06175-13S1 from the National Institute of Environmental Health Sciences (NIEHS), NIH. Its contents are solely the responsibility of the authors and do not necessarily represent the official views of the NIEHS, NIH.

Tobi Mae Lippin and Toshiba Burns-Johnson
New Perspectives Consulting Group, Inc., Durham, NC

*New Perspectives Consulting Group, Inc., a Durham, NC based consulting firm, provides evaluation consulting services for PACE Union's health and safety programs.

[†] PACE International Union recently merged with the United Steelworkers of America to form the United Steel, Paper and Forestry, Rubber, Manufacturing, Energy, Allied Industrial and Service Workers International Union (USW).

PACE International Union Survey:
Workplace Incident Prevention and Response Since 9/11
October 2004

Executive Summary

Introduction

Background. The U.S. Clean Air Act Amendments of 1990 required the U.S. Environmental Protection Agency (EPA) to enact regulations establishing a Risk Management Program (RMP). Facilities that produce or store large quantities of 140 highly hazardous chemicals[1] must develop a Risk Management Program. The 15,000 RMP sites regulated by the EPA across the U.S. have been identified as possible targets for terrorist attacks. The Paper, Allied Industrial, Chemical and Energy Workers International Union (PACE) identified 189 RMP sites where 50,437 PACE members work. PACE-represented industries -- paper mills, petroleum refineries, chemical manufacturing and nuclear materials facilities may be targets. The communities surrounding these facilities are also at-risk.

Study Overview. PACE sought to gain a better understanding of issues related to prevention of and preparedness for possible intentional incidents (i.e., terrorist attacks) at sites represented by its local unions. In March 2004, PACE launched a self-administered mail-back survey questionnaire that asked respondents from high vulnerability PACE-represented facilities about issues and activities since the attacks of 9/11. Questions covered issues of: vulnerability assessment, prevention, emergency response, training, and involvement of the local union, hourly workers and the community.

Survey Population, Administration, and Response. PACE developed a target list of potentially high hazard sites to include in the survey based on the intersection of a list of EPA Risk Management Program (RMP) sites and a listing of PACE local unions/company sites. A packet of information including a letter from PACE International Union President Boyd Young was sent to the local union president and recording secretary of each PACE represented RMP facility identified. The local union president was asked to designate someone who was knowledgeable about what the company and the local union might be doing to lower the vulnerability of their site to intentional (terrorist attacks) and unintentional incidents. Survey data was collected between March and June 2004. The survey response rate of 70% was calculated based on the number of PACE represented facilities to which PACE mailed survey questionnaires (189), and the number of returned surveys (133).

[1] Quantities greater than thresholds listed by the EPA.

Of the 133 sites that returned questionnaires, this report's findings are limited to those 125 sites (95%) that responded *yes* when asked whether their worksite had quantities of chemicals or other hazardous materials large enough to cause a catastrophic event on-site if those materials were involved in a fire, explosion or other release. The findings for this report are limited to these 125 sites because they represent the PACE members at greatest risk. Of the 125 sites included, 100 also said that they faced the potential of a catastrophic event to the areas surrounding their site.

About the Respondents. The majority (82%) of the responding worksites were *chemical* plants (32%), *primary paper* mills (26%), or *oil refineries* (24%). The remaining 18% of the worksites were *other* types of industries. *Other* industries included the following: cement, automotive, nuclear, paper converting, wet milling, and synthetic rubber.

Limitations of the Data. It is important for you to remember the following limitations when you review these findings:

- This survey looked at perceptions only. It did not include an independent assessment of, for example, which employees actually received training since September 11, 2001, or which actions companies actually took.

- The survey respondents were selected from a list of Risk Management Program (RMP) sites. However, due to security limitations imposed since 9/11, the most accurate lists of RMP sites are not readily available. Therefore, some sites who did respond may not actually be RMP sites any longer, and some sites who were not surveyed may actually be RMP sites at this time. Readers should be careful not to assume that the findings can be generalized broadly to represent all PACE represented workplaces, all PACE represented sites from a specific industrial sector, or RMP sites in general.

Findings

Possibility and Likelihood of A Catastrophic Event

Ninety-five percent (95%) of the respondents reported that their sites have large enough quantities of chemicals to cause a catastrophic event if those materials were involved in a fire, explosion or release. Over half of the sites indicated that they face a *high* or *medium* likelihood of a catastrophic event due to a **terrorist attack** (54%) or an **unintentional incident** (59%).

What Companies Are Doing

Company Preventative Actions. In response to these vulnerabilities respondents' reports suggest that most employers assessed their sites vulnerabilities (66%) and worksite security (64%). Company actions appeared to focus more frequently on security, with almost three-quarters (73%) of the respondents reporting improved systems to guard and secure the plant.

All other company actions were reported to be taken at less than half of the study sites. These actions included improved communication systems (43%), improved training and procedures to prevent possible terrorist attacks (38%), updated warning systems (38%), improved containment of potential hazardous releases (34%), and improved quality and availability of personal protective equipment (30%). Some preventative actions, that could directly reduce the likelihood of a catastrophic event, were reportedly taken with the least frequency, such as: reduced volumes of hazardous substances (17%), strengthened plant vessels, tanks, piping or other structures (17%), and improved the siting of hazardous substances or processes (14%).

Company Actions To Prepare To Respond. When preparing to respond to an event caused by a **terrorist attack**, 68% of the companies provided emergency response training to employees in the past 12 months, and 59% conducted emergency response drills for the plant site. About half (47%) of the respondents reported that the companies at their worksites had updated *facility* emergency response plans since 9/11. Other company actions to prepare for responding to an event included: 46% informed local fire and police departments, HazMat teams, etc. about specific plant hazards; 42% put additional procedures in place to inform employees of emergencies; and 30% updated shutdown procedures.

Respondents used the *don't know* choice considerably more frequently in the set of questions about actions to inform local community services, or nearby residents or update the **community** Emergency Response Plan than when responding about actions at their facility. While, 23% knew their employers had informed local hospitals, health departments and emergency medical personnel about potential health threats from plant-specific exposures, 20% said these community health services were not informed, and 57% reported *don't know.*

Effectiveness Of Company Prevention and Response Actions

Effectiveness of Prevention Actions. Less than half (44%) of the respondents indicated that their company's preventative actions, including security efforts, were effective (includes: *very effective, moderately effective*, and *slightly effective)* in reducing the vulnerabilities of their site to a catastrophic event caused by a **terrorist attack**. Over one-third (36%) were *neutral* about the effectiveness, and one-fifth (21%) said the actions were ineffective (includes: *very ineffective, moderately ineffective*, and *slightly ineffective)*.

When considering the effectiveness of actions to *prevent* an event caused by an **unintentional incident**, one-third (33%) said the company's actions were effective. Forty-six percent (46%) were *neutral* about the effectiveness, and one-fifth (21%) said the actions were ineffective to reduce their sites' vulnerabilities to an event caused by an **unintentional incident**. On average, respondents rated the effectiveness of company actions to *prevent* a catastrophic event only slightly above *neutral* (**terrorist attack** = 4.2 and **unintentional incident** = 4.1) on a 7-point scale.

Respondent assessment of the effectiveness of the company actions *to prevent* a catastrophic event were also examined considering perceptions of a site's vulnerability to a catastrophic event (*high, medium, low*). Forty-five percent (45%) of the respondents who rated their sites with a *high* vulnerability level also rated their company's actions to *prevent* an event caused by a **terrorist attack** as ineffective. This ineffective rating is notably higher than ratings given by respondents from *medium* or *low* vulnerability sites who rated their companies' actions regarding an event caused by a **terrorist attack** as ineffective (medium vulnerability sites = 18% ineffective, *low* vulnerability sites = 11% ineffective). Overall, respondents rated the effectiveness of company actions to *prevent* an event caused by a **terrorist attack** (44%) higher than one caused by an *unintentional incident* (33%).

Effectiveness of Response Actions. Thirty-eight percent (38%) of the respondents indicated that their company's actions *in preparing to respond* to an event caused by a **terrorist attack** were effective (includes: *very effective, moderately effective*, and *slightly effective)*. As many were *neutral* (38%) about the effectiveness of actions *in preparing to respond* to an event caused by a **terrorist attack**, while almost one quarter (23%) said the actions were ineffective (includes: *very ineffective, moderately ineffective*, and *slightly ineffective)*. When considering the effectiveness of company actions in *preparing to respond* to an event caused by an **unintentional incident**, 44% said the company's actions were effective. The same percentage (38%) were *neutral* regarding the effectiveness of *preparing to respond* to an **unintentional incident** as they were to an event caused by a **terrorist attack**. Eighteen percent (18%) said the company's actions were ineffective. On average, respondents rated the effectiveness of company actions to *respond* to a catastrophic event caused by a **terrorist attack** only slightly above *neutral* (4.1) on a 7-point scale. Respondents' perceptions of the effectiveness of employers' actions *in preparing to respond* to an event caused by an

unintentional incident was slightly higher at 4.4, midway between *neutral* and *slightly effective*.

When rating the effectiveness of the company actions *in preparing to respond*, respondents from sites rated as having a *high* likelihood of a catastrophic event reported considerable differences from *medium* or *low* likelihood sites. When considering responding to an event caused by a **terrorist attack**, 44% of respondents who characterized their sites as *high* risk found their company's actions ineffective. This rating is considerably higher than the ineffectiveness ratings given by respondents at sites with a *medium* or *low* likelihood of an event (*medium* likelihood = 27% ineffective, *low* likelihood = 11% ineffective). However, most notable is that when considering the effectiveness of company actions *in preparing to respond* to an **unintentional incident**, the *high*est risk respondents rated their employers' actions with the highest levels of effectiveness in the survey, with 62% indicating that their company's actions were effective.

Training

About one-third of respondents reported that no employees at their sites received training about *preventing* (34%) or *responding* (28%) to a catastrophic event caused by a *terrorist attack* since 9/11. At sites where some training occurred, 38% reported that half or fewer employees received *response preparedness* training, and 27% reported that half or fewer employees received *prevention* training. Notably, a sizeable percent of respondents reported not knowing about training to *prevent* (25%), or *respond* (21%) to catastrophic events at their sites. Seventy-four percent (74%) reported that additional training was needed for members of their bargaining unit.

Involvement Of Hourly Workers, the Local Union Or Community

A strong majority of respondents reported no action had been initiated by the companies at their sites to involve the local union or hourly workers in company plans or actions *to prevent* or *respond* to a catastrophic event caused by a possible **terrorist attack**. About one-quarter reported involvement by the local union and hourly workers in making recommendations (local union = 25%, hourly workers = 22%), and being informed by the company (local union = 21%, hourly workers = 28%). Ten percent (10%) of respondents reported that their local unions had taken action to improve the company's plans or actions regarding prevention of or response to a catastrophic event. However, 83% reported no action had been initiated by their local union. Those respondents who indicated actions taken by the local union, described efforts to ask the company for additional employee training, and offers for the local union to work with the company on these issues.

Involvement of the community regarding company plans or actions was even lower. In addition, almost two-thirds of respondents selected the *don't know* choice regarding community involvement.

Recommendations for the Future

A number of action-oriented opportunities for PACE Union's Health and Safety Department and local unions emerge from this examination of the survey findings.

The PACE Evaluation Team Incident Prevention and Response Since 9/11 Work Group recommends that local unions examine this report's findings and consider the following questions:

1. What does this data mean for your local and for your site?
2. What actions do you want the company at your site to take regarding the following: preventing catastrophic events; preparing to respond to potential catastrophic events or emergencies; and involving your local union, hourly workers and the communities surrounding your facility?
3. What role should your local union take to initiate or advocate for the highest levels of prevention for your members, the facility, and the communities surrounding your facility?
4. How can your site work more closely in coordination with local emergency responders and health providers who would respond in an emergency?
5. Can your local union organize a training for your members about these issues, using the PACE Health and Safety Department curriculum?

Furthermore, the Evaluation Team Work Group recommends that the PACE Health and Safety Department take the following actions:

A. Educate and train PACE members about more effective actions companies could take to prevent catastrophic events using higher levels of prevention, rather than solely focusing on increased security measures.

B. Develop expanded training opportunities for PACE members about: 1) prevention and response to hazardous materials emergencies, and 2) the variety of roles local unions, hourly workers, and communities can play in prevention and response activities.

C. Increase the level of awareness about these issues within PACE Union.

Preventing and preparing to respond to potential catastrophic events whether caused by terrorist attacks or unintentional incidents are important issues facing PACE's membership. The PACE Evaluation Team hopes this assessment and report contribute to the dialogue and to effective action to meet these serious challenges.

Paper, Allied-Industrial, Chemical and Energy Workers International Union

PACE International Union Survey:
Workplace Incident Prevention and Response Since 9/11
October 2004

TABLE OF CONTENTS

ACKNOWLEDGMENTS i

EXECUTIVE SUMMARY ii
- Introduction ii
- Findings iv
- Recommendations for the Future vii

TABLE OF FIGURES ix

INTRODUCTION 1
- Background 1
- Study Overview 5
- Target Study Population 6
- Survey Administration and Response 7
- About the Respondents 9
- Report Lay-out 11
- Tips on Interpreting Charts, Tables, and Data Overall 12

KEY FINDINGS 14
- Likelihood of a Catastrophic Event 14
- Preventing a Catastrophic Event 15
- Plant Security 16
- Effectiveness of Prevention Actions 17
- Preparing to Respond 25
- Effectiveness of Actions in Preparing to Respond to a Catastrophic Event 27
- Training: Quality, Scope, and Need 35
- Involvement in Incident Prevention and Response by Local Unions, Hourly Workers, or Communities 38

STUDY LIMITATIONS 41

DISCUSSION AND CONCLUSIONS 42

Paper, Allied-Industrial, Chemical and Energy
Workers International Union
PACE International Union Survey:
Workplace Incident Prevention and Response Since 9/11
October 2004

Table Of Figures

CHARTS

Chart 1: Survey Response Rate........7
Chart 2: Possibility of a Catastrophic Event On-site........8
Chart 3: Type of Industry........9
Chart 4: Size of Workforce........10
Chart 5: Plant Security........16
Chart 6: Effectiveness of Prevention Actions to LESSEN VULNERABILITY to a Catastrophic Event Caused by a TERRORIST ATTACK........18
Chart 7: Effectiveness of Actions to LESSEN VULNERABILITY by High, Medium, or Low Likelihood of a Catastrophic Event due to a TERRORIST ATTACK........20
Chart 8: Effectiveness of Company Actions to LESSEN VULNERABILITY to a Catastrophic Event Caused by an UNINTENTIONAL INCIDENT........22
Chart 9: Effectiveness of Actions to LESSEN VULNERABILITY by High, Medium, or Low Likelihood of a Catastrophic Event due to an UNINTENTIONAL INCIDENT........24
Chart 10: Effectiveness of Actions in PREPARING TO RESPOND to a Catastrophic Event Caused by a TERRORIST ATTACK........28
Chart 11: Effectiveness of Actions in PREPARING TO RESPOND by High, Medium, or Low Likelihood of a Catastrophic Event due to a TERRORIST ATTACK........30
Chart 12: Effectiveness of Actions in PREPARING TO RESPOND to a Catastrophic Event Caused by an UNINTENTIONAL INCIDENT........32
Chart 13: Effectiveness of Actions in PREPARING TO RESPOND by High, Medium, or Low Likelihood of a Catastrophic Event Caused by an UNINTENTIONAL INCIDENT........34
Chart 14: Percent of Employees Trained to Prevent a Catastrophic Event Caused by a Terrorist Attack........36
Chart 15: Percent of Employees Trained to Respond to a Catastrophic Event Caused by a Terrorist Attack........36
Chart 16: Bargaining Unit Needs Additional Training Related to Terrorist Attacks........37
Chart 17: Local Union Taken Action........40

TABLES

Table 1: Site Covered by Standards and Regulations 10
Table 2: Likelihood of a Catastrophic Event 14
Table 3: Possible Actions to Prevent a Catastrophic Event 15
Table 4: Effectiveness of Prevention Actions to LESSEN VULNERABILITY to a Catastrophic Event Caused by a TERRORIST ATTACK 18
Table 5: Effectiveness of Actions to LESSEN VULNERABILITY by High, Medium, or Low Likelihood of a Catastrophic Event due to a TERRORIST ATTACK 19
Table 6: Effectiveness of Prevention Actions to LESSEN VULNERABILITY to a Catastrophic Event Caused by an UNINTENTIONAL INCIDENT 22
Table 7: Effectiveness of Actions to LESSEN VULNERABILITY by High, Medium, or Low Likelihood of a Catastrophic Event due to an UNINTENTIONAL INCIDENT 23
Table 8: Possible Actions to Be Prepared to Respond to a Catastrophic Event 26
Table 9: Effectiveness of Actions in PREPARING TO RESPOND to a Catastrophic Event Caused by a TERRORIST ATTACK 28
Table 10: Effectiveness of Actions in PREPARING TO RESPOND by High, Medium, or Low Likelihood of a Catastrophic Event due to a TERRORIST ATTACK 29
Table 11: Effectiveness of Actions in PREPARING TO RESPOND to a Catastrophic Event Caused by an UNINTENTIONAL INCIDENT 32
Table 12: Effectiveness of Actions in PREPARING TO RESPOND by High, Medium, or Low Likelihood of a Catastrophic Event Caused by an UNINTENTIONAL INCIDENT 33
Table 13: Possible Actions Taken by the Company to Involve Others 39
Table 14: Effectiveness of Prevention and Response Actions 46

Paper, Allied-Industrial, Chemical and Energy
Workers International Union
PACE International Union Survey:
Workplace Incident Prevention and Response Since 9/11
October 2004

Introduction

Background

EPA's Risk Management Program
The U.S. Clean Air Act Amendments of 1990 required the U.S. Environmental Protection Agency (EPA) to enact regulations establishing a Risk Management Program (RMP). Each facility that produces or stores large quantities of 140 highly hazardous chemicals[2] must develop a Risk Management Program. Facility operators at RMP sites are required to undertake hazardous materials accident prevention activities and to make reports to the EPA. The RMP reporting process includes an analysis of possible consequences of a major chemical incident to surrounding communities. There are 15,000 RMP sites regulated by the EPA across the U.S.

RMP Sites Are Potential Sources of "Weapons of Mass Destruction"
In 2000, The Department of Justice linked RMP sites to the issue of terrorist threats (or Weapons of Mass Destruction, WMD) when it stated:

> In recent years, criminals have with increasing frequency attempted to obtain or produce WMD precisely because such weapons are engineered to cause wide-scale damage to life and property. However, traditional means of creating or obtaining WMD are generally difficult to execute. In contrast, breaching a containment vessel of an industrial facility with an explosive or otherwise causing a chemical release may appear relatively simple to such a terrorist.[3]

RMP-Related Risk Estimates Limited for Assessing Terrorist Threats
While looking at the number, location and type of RMP sites may offer important insights into assessing possible terrorist threats, this vantage point is limited in that:

1. RMP data and analyses assess risks related to accidental rather than intentional incidents. Intentional acts that create hazardous chemical disasters may differ from and be more severe than accidental releases in a number of important ways. For example, the Government Accounting Office (GAO) reports that the RMP regulation requires facilities to estimate the effects of a toxic chemical

---

[2] Quantities greater than thresholds listed by the EPA.

[3] Source: United States Department of Justice Assessment of the Increased Risk of Terrorist or Other Criminal Activity Associated with Posting Off-Site Consequence Analysis Information on the Internet, April 18, 2000.

release involving the greatest amount of the toxic chemical held in a single vessel or pipe -- not the entire quantity on site. Therefore, for some facilities it is conceivable that an attack, where multiple chemical vessels were breached simultaneously, could result in an even larger release Such releases would involve more severe consequences, than those estimated in the RMP "worst-case" scenarios.[4]

2. RMP "Off-Site Consequence Analyses" only consider releases of a single hazardous chemical from a single source. However, the risks are potentially greater because releases in one system can trigger releases in adjacent systems involving other chemicals.[5]

3. Planning conducted as part of the RMP process primarily involves assessment of scenarios and possible consequences for off-site, rather than on-site populations. It is likely that any terrorist attack at an RMP site would put the entire on-site population at extreme risk.

Amplifying the potential severity of these possibilities, a 2001 U.S. Army analysis estimated that up to 2.4 million people could need medical treatment as a result of a major chemical disaster.[6]

Chemical, Refinery, and Other Site Risks

Shortly before the World Trade Center disaster, the EPA published a study of hazardous chemical accidents at RMP sites and reported:

- Among the 15,000 RMP sites considered to be at risk of a terrorist attack, 11% were petroleum refineries (1,609 sites) and 13% were chemical or petrochemical related manufacturing (1,945 sites).
- Petroleum refineries ranked first in the number of hazardous chemical accidents at RMP sites between 1994 and 1999. This accounted for 10% of all such accidents and was nearly double the number for any other single industry.

[4] United States General Accounting Office. 2003. Homeland Security: Voluntary Initiatives Are Under Way at Chemical Facilities, but the Extent of Security Preparedness Is Unknown. GAO-03-439, March 2003.

[5] Sources: Belke, J. 2000. U.S. Environmental Protection Agency. "Chemical accident risks in U.S. industry: A preliminary analysis of accident risk data form U.S. hazardous facilities." September 25, 2000; and United States General Accounting Office. 2003. Homeland Security: Voluntary Initiatives Are Under Way at Chemical Facilities, but the Extent of Security Preparedness Is Unknown. GAO-03-439, March 2003; and National Transportation Safety Board. 2002. Hazardous Materials Accident Report: Hazardous Materials Release From Railroad Tank Car With Subsequent Fire at Riverview, Michigan, July 14, 2001. Washington, D.C.: National Transportation Safety Board.

[6] United States Army, Draft Medical NBC Hazard Analysis of Chemical-Biological-Radiological-Nuclear-High Explosive Threat, Possible Scenarios & Planning Requirements, Army Office of the Surgeon General. Cited in: United States General Accounting Office. 2003. Homeland Security: Voluntary Initiatives Are Under Way at Chemical Facilities, but the Extent of Security Preparedness Is Unknown. GAO-03-439, March 2003.

- While the paper industry has far fewer RMP sites than refineries or chemical manufacturing facilities, two classifications of Paper Mills ranked 2nd and 4th in the rate of hazardous chemical accidents.[7]
- Sites classified as "chemical manufacturing" accounted for one in four of all RMP site hazardous chemical accidents.

While the EPA study focused on unintentional rather than intentional incidents, the knowledge that RMP sites are considered possible targets for terrorist attacks makes the findings of the study even more sobering. It is especially sobering for those who work at or live near refineries, chemical plants, paper mills and nuclear facilities.

The gravity of this situation was made more evident by the issuing of alerts in early 2003:

- On February 7, 2003 the Homeland Security Advisory System issued a "High" (Orange) state of alert. First on the list of potential targets was "the energy sector, including tank farms, refinery facilities, and oil tankers."[8]
- On February 12, another alert was issued warning of possible "conventional attacks against the U.S. nuclear/chemical-industrial infrastructure to cause contamination, disruption, and terror. Based on information, nuclear power plants and industrial chemical plants remain viable targets."[9]

As well, in its recent study of vulnerability and security preparedness at U.S. chemical facilities, the GAO stated:

> Chemical facilities may be attractive targets for terrorists intent on causing economic harm and loss of life. Many facilities exist in populated areas where a chemical release could threaten thousands. EPA reports that 123 chemical facilities located throughout the nation have toxic "worst-case" scenarios where more than a million people in the surrounding area could be at risk of exposure to a cloud of toxic gas if a release occurred.[10]

---

[7] Number of Accidents per Process per Year

[8] Sources: National Infrastructure Protection Center, Homeland Security Information Update, Information Bulleting 02-001, February 7, 2003. http://www.nipc.gov/publications/infobulletings/2003/ib03-001.htm

[9] National Infrastructure Protection Center, Homeland Security Information Update, Information Bulletin 03-003, February 12, 2003. http://www.nipc.gov/publications/infobulletings/2003/ib03-003.htm

[10] United States General Accounting Office. 2003. Homeland Security: Voluntary Initiatives Are Under Way at Chemical Facilities, but the Extent of Security Preparedness Is Unknown. GAO-03-439, March 2003.

In addition to these 123 chemical facilities, there are approximately 700 sites that could put 100,000 or more persons in the surrounding areas at risk from a chemical release, and approximately 3,000 sites could put at least 10,000 or more persons at risk. This adds up to nearly 4,000 sites and tens of millions of people at risk.

PACE Members at Risk
PACE identified 189 RMP sites where 50,437 PACE members work. PACE-represented RMP sites include:

- 26,696 workers at 47 primary paper mills
- 12,003 workers at 44 petroleum refineries
- 8,461 workers at 77 chemical manufacturing facilities
- 3,277 workers at 22 facilities with other classifications.

There are an additional 58,987 workers at 190 PACE-represented chemical plants, paper mills, petroleum refineries, and petroleum-product manufacturing facilities that use high volumes of highly hazardous chemicals.

In summary, PACE-represented industries -- paper mills, petroleum refineries, chemical manufacturing and nuclear materials facilities -- are some of the most at-risk sites for a terrorist incident in the United States. For PACE members and their fellow employees, merely the status of working at an RMP site or a site that uses highly hazardous chemicals puts them on the front lines in battle against both unintentional (accidental) and intentional (terrorist) incidents. In addition, hundreds of thousands – perhaps millions – of citizens who reside in nearby communities face similar threats.

Study Overview

In the 2003-2004 grant year, the Paper, Allied Industrial, Chemical and Energy Workers International Union (PACE) sought to gain a better understanding of issues related to prevention of and preparedness for possible intentional incidents (i.e., terrorist attacks) at sites represented by its local unions. The assessment addressed vulnerability to catastrophic hazardous materials incidents that could have effects either on- or off-site.

The purposes of this study were to:

- ▶ Learn what actions companies are taking to:
  - Prevent and respond to a catastrophic event caused by a potential terrorist attack or an unintentional incident.
  - Involve local union leaders, hourly workers, and the community in these efforts.
- ▶ Use the survey information to develop programs for PACE local unions to protect the workforce and surrounding communities from a potential catastrophic event caused by a terrorist attack or an unintentional incident.

The survey design, administration, analysis and report writing were conducted by the PACE Evaluation Team which is comprised of worker trainers and staff, with facilitation and guidance provided by New Perspectives Consulting Group, Inc., a Durham, North Carolina based evaluation consulting firm that has worked with PACE Union to evaluate its programs for over 10 years. The evaluation team developed a self-administered mail-back survey questionnaire that asked respondents about issues and activities since the attacks of 9/11. These related to potential catastrophic incidents including vulnerability assessment, prevention, emergency response, training, and involvement of the local union, hourly workers and the community.

Target Study Population

PACE developed a target list of sites to include in the survey based on the intersection of a list of RMP sites included in the Right-to-Know Network (RTK Net) database[11] and a listing of PACE local unions/company sites. In this process PACE identified 189 RMP sites where it is the collective bargaining agent for workers. This listing of 189 RMP sites includes 12,003 workers at 44 petroleum refineries, 26,696 workers at 47 primary paper mills, 8,461 workers at 77 chemical manufacturing facilities, and 3,277 workers at 22 facilities with other industrial classifications (see map below).

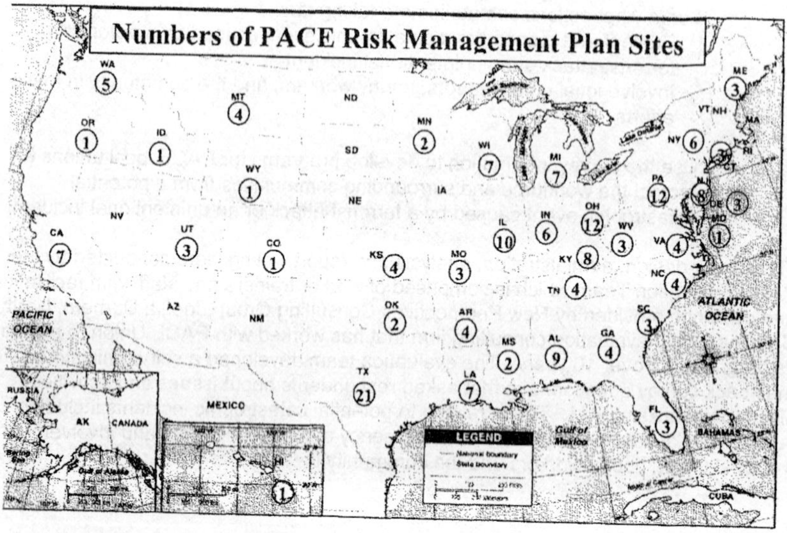

[11] Available at http://www.rtknet.org/rmp

Survey Administration and Response

A packet of information including a letter from PACE International Union President Boyd Young, instructions for completing the survey, the survey itself, and a return envelope was sent to the local union president and recording secretary of each PACE represented RMP facility identified through the RTK Net database. PACE requested that the local union president designate a local union member or group of members to complete the survey on behalf of the facility targeted by the survey. PACE asked that the person or people completing the survey be knowledgeable about what the company and the local union might be doing to lower the vulnerability of their site to intentional (terrorist attacks) and unintentional incidents. Suggested people for this task included: the local union president, secretary-treasurer, chair or member of the Health and Safety Committee, Health and Safety or TOP Representative, or other health and safety activist. Once completed, the surveys were returned, by mail, to PACE headquarters. After all of the surveys were collected, the surveys were forwarded to New Perspectives Consulting Group, Inc. for data entry, analysis, and reporting.

Survey data was collected between March and June 2004. The survey response rate of 70% was calculated based on the number of PACE represented facilities to which PACE mailed survey questionnaires (189), and the number of returned surveys (133). (See Chart 1 below.)

Chart 1: Survey Response Rate

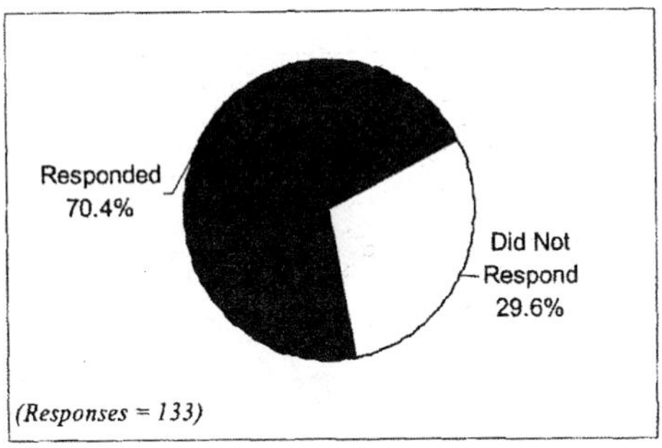

Of the 133 sites that returned questionnaires, this report's findings are limited to those 125 sites (95%) that responded *yes* when they were asked whether their worksite had quantities of chemicals or other hazardous materials large enough to cause a catastrophic event on-site if those materials were involved in a fire, explosion or other release. (See Chart 2 below.) We limited the findings for this report to these 125 sites because they represent the PACE members at greatest risk. Of the 125 sites included, 100 also said that they faced the potential of a catastrophic event to the areas surrounding their site. One site (that was not included in these findings) indicated that they did not have the potential for a catastrophic event on site, but they did off site.

Chart 2: Possibility of a Catastrophic Event On-site

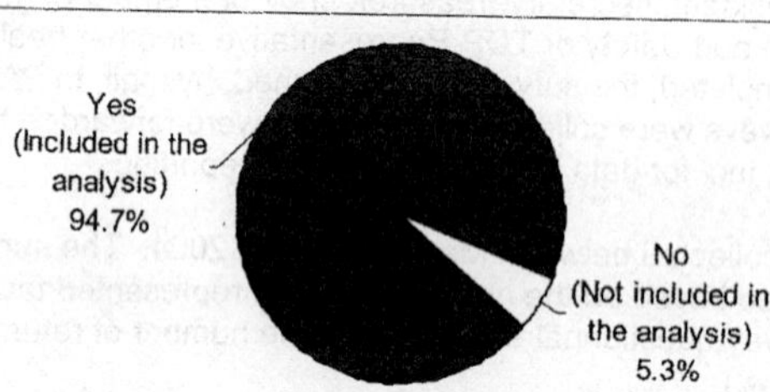

Q1. Think about possibilities on your plant site. Does your facility have quantities of chemicals or other hazardous materials on-site large enough to cause a catastrophic event within the plant site if those materials were involved in a fire, explosion, or other release? (Responses = 132, Missing = 0.8%)

About the Respondents

Type of Industry. The majority (82%) of the responding worksites were *chemical* plants, *primary paper* mills, or *oil refineries.* The remaining 18% of the worksites were *other* types of industries. (See Chart 3 below.) Respondents who indicated that they represented *other* industries included the following: cement, automotive, nuclear, paper converting, wet milling, and synthetic rubber.

Chart 3: Type of Industry

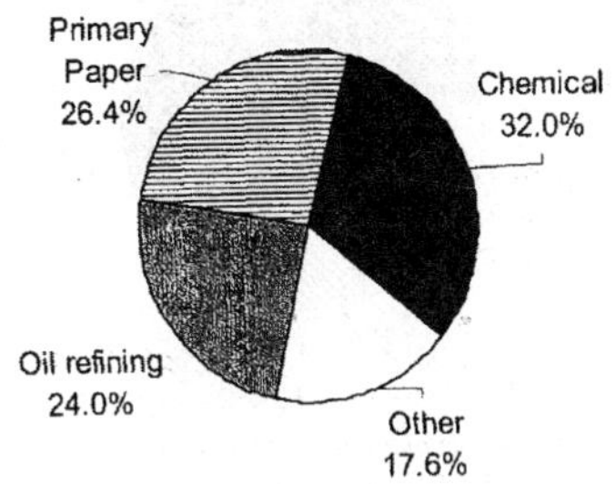

Q20. What type of industry is at your worksite? (Responses = 125, Missing = 0.0%)

Size of Workforce. The majority (44%) of the responding worksites had *100-499* employees. Twenty-five percent (25%) had from *500-999* employees and 21% had from *0-99* employees. About 10% of the worksites were relatively large, employing *1000* or more persons. (See Chart 4 below.)

Chart 4: Size of Workforce

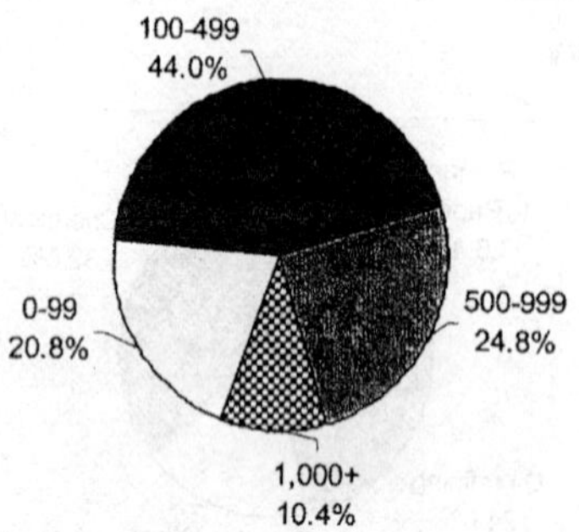

Q15. What is the size of the workforce at your worksite? (Responses = 125, Missing = 0.0%)

Coverage by Standards / Regulations. Respondents were also asked about regulations applicable to their worksites. These included: OSHA's Hazardous Waste Operations and Emergency Response Standard (29 CFR 1910.120, HAZWOPER), OSHA's Process Safety Management of Highly Hazardous Chemicals Standard (29 CFR 1910.119, PSM), and EPA's Risk Management Program (RMP). Seventy-six percent (76%) reported that they were covered by HAZWOPER, 79% reported that they were covered by the Process Safety Standard, and 50% reported that they were RMP sites. A notable percentage indicated that they *did not know* if their site was covered by these standards or regulations (22% for HAZWOPER, 20% PSM, 48% RMP). (See Table 1 below.)

Table 1: Site Covered by Standards and Regulations

Standard / Regulation	Yes	No	Don't know
HAZWOPER (Q 18)	76.4%	1.6%	22.0%
PSM (Q 16)	79.0%	0.8%	20.2%
RMP (Q 17)	50.0%	2.5%	47.5%

Q16. Is your site covered by OSHA's standard "Process Safety Management of Highly Hazardous Chemicals" (1910.119)? (Responses = 124, Missing = 0.8%); Q17. Is your site a Risk Management Program (RMP) site according to the Environmental Protection Agency? (Responses = 122, Missing = 2.4%); Q18. Is your site covered by OSHA's standard "Hazardous Waste Operations and Emergency Response (HAZWOPER)" (1910.120)? (Responses = 123, Missing = 1.6%)

Report Lay-out

The report begins by providing some guidance on interpreting the report's data, including the charts and tables. Following this, the findings are reported in the following sections:

- Likelihood of a Catastrophic Event
- Preventing a Catastrophic Event
- Plant Security
- Effectiveness of Prevention Actions
- Preparing to Respond
- Effectiveness of Actions in Preparing to Respond to a Catastrophic Event
- Training: Quality, Scope, and Need
- Involvement in Incident Prevention and Response by Local Unions, Hourly Workers or Communities

After these sections is a list of this study's limitations and the Discussion and Conclusions section. The Discussion and Conclusions section summarizes and interprets some of the main findings and links some of the findings together to provide a broad, cross-cutting view of the findings gathered from the RMP sites in this study.

Tips on Interpreting Charts, Tables, and Data Overall

Quantitative and Qualitative Data

This evaluation primarily features "quantitative" data that uses statistics. "Qualitative" data, open-ended answers written by the respondents' in their own words were also collected in a limited number of questions.

Survey Questions

Many survey questions asked respondents to think about their experiences "since September 11, 2001". For this and other types of specific information, look at the bottom of survey-related charts and tables for the original survey question

Different Groups' Different Answers to Questions

To get a better sense of what the findings mean, in some cases, the Evaluation Team compared the answers to questions from one group of respondents (such as respondents who indicated that their worksite has a high likelihood of experiencing a catastrophic event caused by a terrorist attack) to other groups (such as respondents who indicated that their worksite has a medium or low likelihood of such an event). These comparisons, sometimes called "cross-tabulations" or "cross-tabs," are used to help see if certain groups of respondents have different perceptions of the issue or experiences than other groups of respondents.

Missing Data

For a variety of reasons, those who fill out surveys do not always answer every question. Respondents were told that they did not have to answer any questions that they did not want to answer. The number of people completing each question is indicated in each chart. Also included in each chart is the percentage of missing responses for this question. This is based on those who did not answer that particular question in relation to the total number of people (125) who completed the survey and are included in the analysis.

Percentages and the Impact of Rounding

When analyzing the data and presenting it in this report, we chose to round numbers to one decimal place. When a value was 5 or above, we rounded up. While this makes it easier to read, it has its drawbacks. You may notice in some findings that percentages do not add up to 100. This is due to rounding.

Averages

All of the charts and tables in this report use percentages to show the proportion of respondents who selected the various response choices. However, in the sections addressing the effectiveness of company actions, and in the Discussion and Conclusions, we also averaged the rating given by respondents about effectiveness. You will notice that there is a gray row in the tables and a gray box in the charts that present the "averages" of the data on a 7-point scale. This number represents how respondents, on average, rated the issue.

To calculate the average, we assigned each choice in the scale a value as follows: *very effective*=7, *moderately effective*=6, *slightly effective*=5, *neutral*=4, *slightly ineffective*=3, *moderately ineffective*=2, and *very ineffective*=1. We calculated the average as follows:

1. Multiplied the number of respondents who indicated a particular response by the value assigned to that response
2. Added up all the products across the different response choices on the 7-point scale
3. Divided the sum of the products by the total number of respondents to get the average

Here is an example:

Response Choices on 7-point scale	Number of respondents who indicated the response choice	Value assigned to each response choice	Number of respondents multiplied by the Value assigned
Very effective	13	7	91
Moderately effective	20	6	120
Slightly effective	22	5	110
Neutral	47	4	188
Slightly ineffective	5	3	15
Moderately ineffective	8	2	16
Very ineffective	9	1	9
Total	**124**		**549**
Average		**4.4**	

Products

Total number of respondents

(549 divided by 124)

Sum of the products

On the 7-point scale, a 4.4 falls between a 4 and a 5. This indicates that on average, respondents rated this issue somewhere between *neutral* and *slightly effective*.

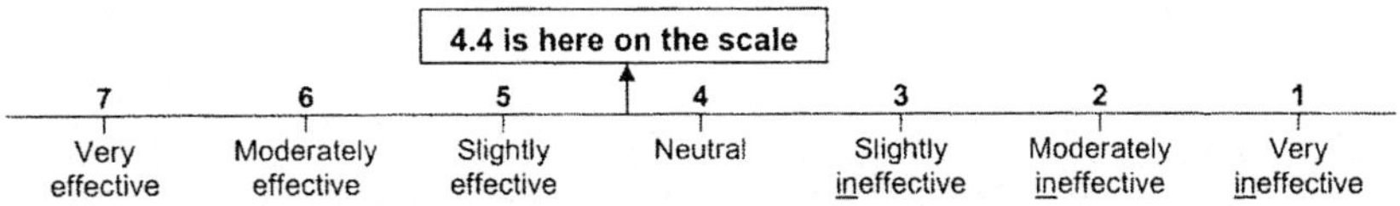

Key Findings

Likelihood of a Catastrophic Event

Over half (54%) of the respondents reported that there was either a *high* or *medium* likelihood of a catastrophic event from a **terrorist attack** at their worksite, while 59% thought there was either a *high* or *medium* likelihood of a catastrophic **unintentional incident**. (See Table 2 below.)

Table 2: Likelihood of a Catastrophic Event

Likelihood	A terrorist attack	An unintentional incident
High	26.2%	21.1%
Medium	27.9%	37.4%
Low	45.9%	41.5%

Q3. What is the likelihood of your worksite experiencing a catastrophic event involving fire, explosion, or a hazardous release caused by the following? (Q3a. Terrorist Attack, Responses = 122, Missing = 2.4%); (Q3b. Unintentional Incident, Responses = 123, Missing = 1.6%)

Preventing a Catastrophic Event

The survey asked respondents about possible preventative actions taken by the company at their worksite since the attacks of 9/11. Two thirds of the sites (66%) reported that the company had assessed vulnerabilities at their sites. (See Table 3 below.) Other most frequently reported preventative actions included:

- 43% improved communication systems
- 38% updated warning systems
- 38% improved training and procedures
- 34% improved containment of potential hazardous releases
- 30% improved quality and availability of personal protective equipment

However, some preventative actions that could directly lessen the likelihood of a catastrophic event were reportedly taken less frequently, such as:

- 17% reduced volumes of hazardous substances
- 17% strengthened plant vessels, tanks, piping or other structures
- 14% improved the siting of hazardous substances or processes

Table 3: Possible Actions to Prevent a Catastrophic Event

Possible actions to prevent a catastrophic event	Was action taken?		
	Yes	No	Don't Know
1. Assessed vulnerabilities	66.4%	12.0%	21.6%
2. Improved communication systems or equipment	42.7%	45.2%	12.1%
3. Updated warning systems	38.4%	48.8%	12.8%
4. Improved training and procedures to prevent possible terrorist attacks	37.6%	54.4%	8.0%
5. Improved containment of potential hazardous releases	33.6%	50.4%	16.0%
6. Improved quality and availability of personal protective equipment	30.4%	57.6%	12.0%
7. Reduced volumes of hazardous substances	16.8%	60.0%	23.2%
8. Strengthened plant vessels, tanks, piping or other structures	16.8%	65.6%	17.6%
9. Improved the siting of hazardous substances or processes to less vulnerable locations	13.6%	68.8%	17.6%

Q4. Since September 11, 2001, has the company at your worksite taken any of the following actions to prevent a catastrophic event caused by a terrorist attack? (1 & 3-9: Responses = 125, Missing = 0.0%); (2: Responses = 124; Missing = 0.8%)

Note: Percents may not add up to 100% due to rounding

Plant Security

Both actions to re-assess worksite security and improve plant security were taken more frequently than the preventative actions previously described. A substantial majority of all the study sites acted in this area.

- 64% re-assessed worksite security in the face of new terrorist threats
- 73% improved systems to guard and secure the plant. (See Chart 5 below.)

Chart 5: Plant Security

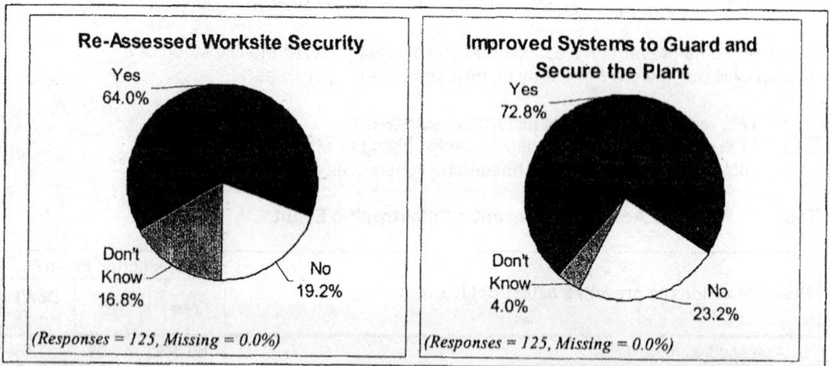

Q5. Since September 11, 2001, has the company at your worksite done any of the following related to plant security in the face of new terrorist threats?

Effectiveness of Prevention Actions

We examined how respondents assessed the effectiveness of actions taken by their company since 9/11 to *lessen the vulnerability* of their worksites to a catastrophic event caused by a **terrorist attack** or an **unintentional incident**. First we consider the effectiveness of company actions to *lessen vulnerability* to an event caused by a **terrorist attack**, and then the effectiveness of company actions to *lessen vulnerability* to an event caused by an **unintentional incident**. After, we report about the effectiveness ratings of all participants regarding company actions to *lessen vulnerabilities* to each of the possible causes, we consider differences among those who judged their sites to be at *high*, *medium*, or *low* likelihood of a catastrophic event.

Efforts to Lessen Vulnerability to a TERRORIST ATTACK. When asked about the overall effectiveness of actions taken by their company since 9/11 to *lessen the vulnerability* of their worksite to a catastrophic event caused by a **terrorist attack**, the respondents' ratings were as follows:

- 44% effective (includes: *very effective*, *moderately effective*, and *slightly effective)*
- 36% *neutral*
- 21% ineffective (includes: *very ineffective*, *moderately ineffective*, and *slightly ineffective)*

A very small contingent of respondents rated their sites actions as *very effective* (3%). Notably, more than one-third reported the effectiveness of the company actions as *neutral*. (See Table 4 and Chart 6, on next page.)

We also analyzed these effectiveness ratings by computing an average of all respondent answers using the 7-point scale with *very effective*=7, *moderately effective*=6, *slightly effective*=5, *neutral*=4, *slightly ineffective*=3, *moderately ineffective*= 2, *and very ineffective*=1. Using this scale, on average respondents rated the actions of the company to reduce vulnerabilities to a **terrorist attack** at their sites a 4.2, only slightly more effective than *neutral*.

Table 4: Effectiveness of Prevention Actions to LESSEN VULNERABILITY to a Catastrophic Event Caused by a TERRORIST ATTACK

Effectiveness of Company Actions	Lessen Vulnerability to a TERRORIST ATTACK
Very effective	3.2%
Moderately effective	14.5%
Slightly effective	25.8%
Neutral	35.5%
Slightly ineffective	4.8%
Moderately ineffective	5.6%
Very ineffective	10.5%
Average on 7 point scale	**4.2**

Q6. Overall, since September 11, 2001, how effective have the actions taken by the company been in lessening the vulnerability of your worksite to a catastrophic event caused by the following? (Q6a. Responses = 124, Missing = 0.8%). **Note: Percents may not add up to 100% due to rounding**

Chart 6: Effectiveness of Prevention Actions to LESSEN VULNERABILITY to a Catastrophic Event Caused by a TERRORIST ATTACK

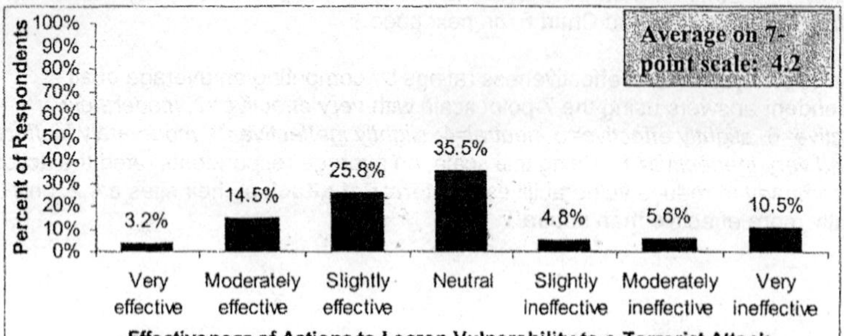

Q6. Overall, since September 11, 2001, how effective have the actions taken by the company been in lessening the vulnerability of your worksite to a catastrophic event caused a ***terrorist attack**?* *(Responses = 124, Missing = 0.8%)*

Note: Percents may not add up to 100% due to rounding.

<u>Effectiveness of Actions to LESSEN VULNERABILITY by *High, Medium*, or Low Likelihood of a Catastrophic Event due to a TERRORIST ATTACK.</u> To further understand the perspectives of respondents, we also examined responses to this question by looking at differences in the effectiveness ratings of those respondents who thought there was a *high* likelihood of a catastrophic event due to a **terrorist attack** at their worksite, and at respondents who thought there was a *medium* or *low* likelihood of an event due to a **terrorist attack**. Noteworthy differences emerged when examining the effectiveness ratings of those who work at sites in which they perceive different vulnerability levels, such as:

- 45% of those who work at sites with a *high* vulnerability, rated company actions as ineffective, more than double either of the other two vulnerability level ratings of ineffective (*medium* vulnerability sites = 18%, *low* vulnerability sites = 11%).
- Respondents from *high* vulnerability worksites average effectiveness rating was 3.5, mid-way between *slightly ineffective* and *neutral*, while those of *medium* and *low* vulnerability sites were almost one point higher (*medium* vulnerability sites = 4.5, *low* vulnerability sites = 4.4). (See Table 5 and Chart 7 below.)

Table 5: Effectiveness of Actions to LESSEN VULNERABILITY by High, Medium, or Low Likelihood of a Catastrophic Event due to a <u>TERRORIST ATTACK</u>

Effectiveness of Actions to Lessen Vulnerability to a TERRORIST ATTACK	Likelihood of a Catastrophic Event at Site Caused by a TERRORIST ATTACK		
	High	Medium	Low
Very effective	3.2%	2.9%	3.6%
Moderately effective	9.7%	17.6%	16.1%
Slightly effective	19.4%	38.2%	21.4%
Neutral	22.6%	23.5%	48.2%
Slightly ineffective	9.7%	5.9%	1.8%
Moderately ineffective	16.1%	5.9%	0.0%
Very ineffective	19.4%	5.9%	8.9%
Average on 7 point scale	3.5	4.5	4.4

Q3. What is the likelihood of your worksite experiencing a catastrophic event involving fire, explosion, or a hazardous release caused by the following? Q6. Overall, since September 11, 2001, how effective have the actions taken by the company been in lessening the vulnerability of your worksite to a catastrophic event caused by the following? **Note: Percents may not add up to 100% due to rounding**

Chart 7: Effectiveness of Actions to LESSEN VULNERABILITY by High, Medium, or Low Likelihood of a Catastrophic Event due to a TERRORIST ATTACK

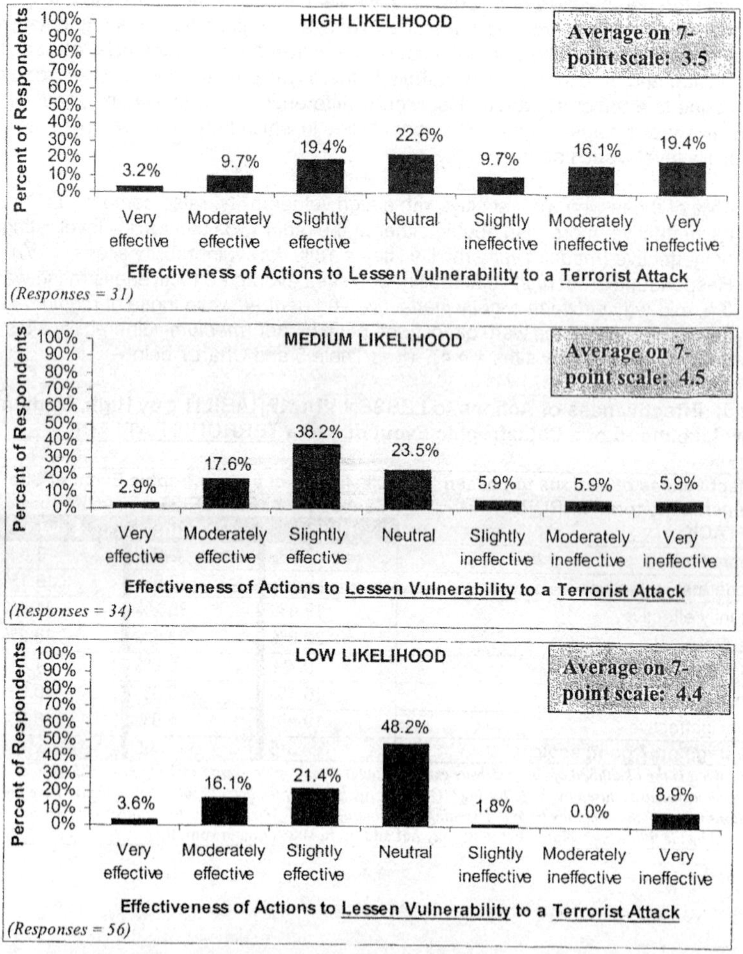

***Questions:* *Q3.* *What is the likelihood of your worksite experiencing a catastrophic event involving fire, explosion, or a hazardous release caused by a* **TERRORIST ATTACK**? **Q6.** *Overall, since September 11, 2001, how effective have the actions taken by the company been in lessening the vulnerability of your worksite to a catastrophic event caused by a* **TERRORIST ATTACK**? **Note: Percents may not add up to 100% due to rounding.**

Efforts to Lessen Vulnerability to an UNINTENTIONAL INCIDENT. Overall, when asked about the effectiveness of actions taken by their company since 9/11 to *lessen the vulnerability* of their worksite to a catastrophic event caused by an **unintentional incident,** respondents rated the effectiveness of the actions of the company at their worksites as follows:

- 33% effective (includes: *very effective*, *moderately effective*, and *slightly effective)*
- 46% *neutral*
- 21% ineffective (includes: *very ineffective*, *moderately ineffective*, and *slightly ineffective)*

Similar to effectiveness ratings if an event were caused by a **terrorist attack**, a very small contingent of respondents rated their sites actions in *lessening vulnerability* to an **unintentional incident** as *very effective* (3%). Notably, nearly half reported the effectiveness of the company actions to *lessen vulnerability* to an **unintentional incident** as *neutral.* (See Table 6 and Chart 8 below.)

We also analyzed these effectiveness ratings by computing an average of all respondent answers using the 7-point scale described above. On average, respondents rated the actions of the company to reduce vulnerabilities to an **unintentional incident** at their sites a 4.1, only slightly more effective than *neutral.* This was about the same as the overall effectiveness average for reducing vulnerabilities to a catastrophic event caused by a **terrorist attack** (4.2).

Table 6: Effectiveness of Prevention Actions to LESSEN VULNERABILITY to a Catastrophic Event Caused by an UNINTENTIONAL INCIDENT

Effectiveness of Company Actions	Lessen Vulnerability to an UNINTENTIONAL INCIDENT
Very effective	3.2%
Moderately effective	16.9%
Slightly effective	12.9%
Neutral	46.0%
Slightly ineffective	6.5%
Moderately ineffective	6.5%
Very ineffective	8.1%
Average on 7 point scale	**4.1**

Q6. Overall, since September 11, 2001, how effective have the actions taken by the company been in lessening the vulnerability of your worksite to a catastrophic event caused by the following? (Q6b. Responses = 124, Missing = 0.8%). **Note: Percents may not add up to 100% due to rounding**

Chart 8: Effectiveness of Company Actions to LESSEN VULNERABILITY to a Catastrophic Event Caused by an UNINTENTIONAL INCIDENT

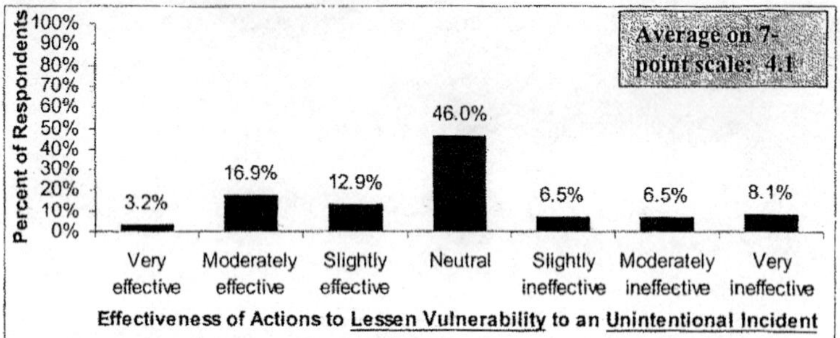

Q6. Overall, since September 11, 2001, how effective have the actions taken by the company been in lessening the vulnerability of your worksite to a catastrophic event caused an unintentional incident? (Responses = 124, Missing = 0.8%)

<u>Effectiveness of Actions to LESSEN VULNERABILITY by *High, Medium*, or *Low* Likelihood of a Catastrophic Event due to an UNINTENTIONAL INCIDENT.</u> To further understand the perspectives of respondents, we also examined responses to this question by looking at differences in the effectiveness ratings of those respondents who thought there was a *high* likelihood of a catastrophic event due to an **unintentional incident** at their worksite, and at respondents who thought there was a *medium* or *low* likelihood of an event due to an **unintentional incident.** Noteworthy differences emerged when examining the effectiveness ratings of those who work at sites with perceived varying vulnerability levels, such as:

- 38% of those who rated their sites as *high* vulnerability, rated company actions as <u>in</u>effective. This was more than double those who either rated the vulnerability of their sites as *medium* vulnerability (17% <u>in</u>effective) or *low* vulnerability (16% <u>in</u>effective).
- Over one-third of respondents who rated their sites as either *high* or *low* vulnerability to catastrophic event caused by an **unintentional event** rated their sites' efforts to *lessen vulnerability* as effective (*high* vulnerability = 38% effective, *low* vulnerability = 36% effective). In contrast, only about one-quarter of respondents who rated their sites as *medium* vulnerability rated their sites' efforts as effective (26%). (See Table 7 and Chart 9 below.)

Table 7: Effectiveness of Actions to LESSEN VULNERABILITY by High, Medium, or Low Likelihood of a Catastrophic Event due to an <u>UNINTENTIONAL INCIDENT</u>

Effectiveness of Actions to Lessen Vulnerability to an UNINTENTIONAL INCIDENT	**Likelihood of a Catastrophic Event at Site Caused by an UNINTENTIONAL INCIDENT**		
	High	**Medium**	**Low**
Very effective	3.8%	0.0%	6.0%
Moderately effective	11.5%	17.4%	20.0%
Slightly effective	23.1%	8.7%	10.0%
Neutral	23.1%	56.5%	48.0%
Slightly <u>in</u>effective	11.5%	8.7%	2.0%
Moderately <u>in</u>effective	11.5%	6.5%	4.0%
Very <u>in</u>effective	15.4%	2.2%	10.0%
Average on 7 point scale	**3.8**	**4.2**	**4.3**

Q3. What is the likelihood of your worksite experiencing a catastrophic event involving fire, explosion, or a hazardous release caused by the following? Q6. Overall, since September 11, 2001, how effective have the actions taken by the company been in lessening the vulnerability of your worksite to a catastrophic event caused by the following? **Note: Percents may not add up to 100% due to rounding**

Chart 9: Effectiveness of Actions to LESSEN VULNERABILITY by High, Medium, or Low Likelihood of a Catastrophic Event due to an UNINTENTIONAL INCIDENT

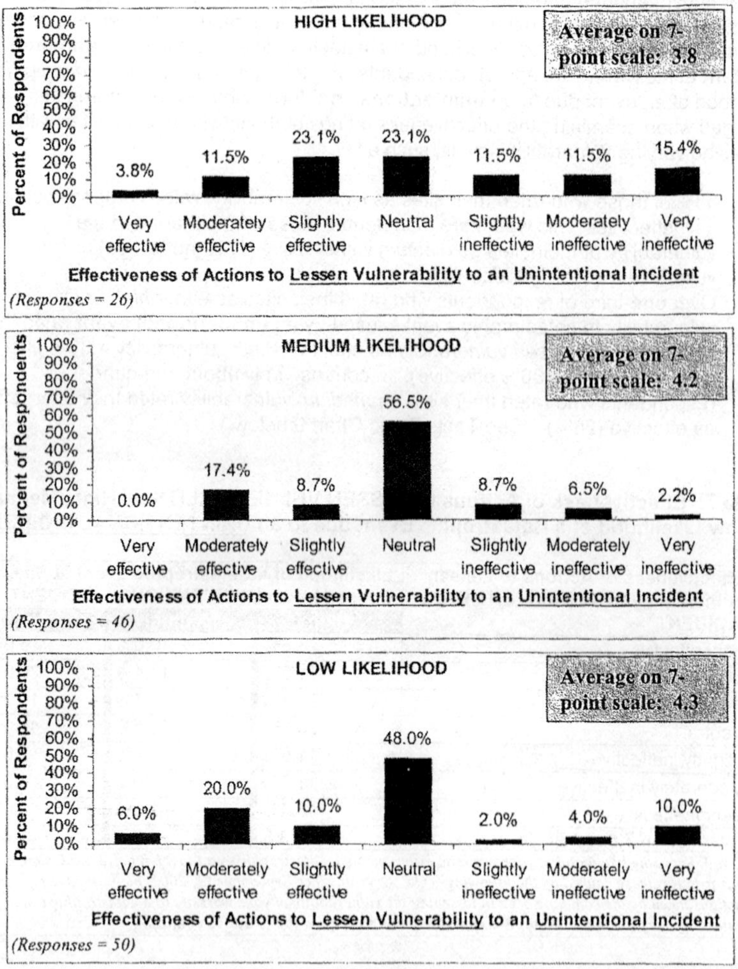

Questions: *Q3. What is the likelihood of your worksite experiencing a catastrophic event involving fire, explosion, or a hazardous release caused by an UNINTENTIONAL INCIDENT? Q6. Overall, since September 11, 2001, how effective have the actions taken by the company been in lessening the vulnerability of your worksite to a catastrophic event caused by an UNINTENTIONAL INCIDENT?* **Note: Percents may not add up to 100% due to rounding.**

Preparing to Respond

Another set of questions asked about actions taken by companies to be better prepared to respond to catastrophic events that might be caused by a **terrorist attack**. The most frequently reported company actions *in preparing to respond* to an event caused by a **terrorist attack** were as follows:

- 68% provided emergency response training to employees in the past 12 months
- 59% conducted emergency response drills for the plant site

Regarding whether the companies at their worksites had updated *facility* emergency response plans since 9/11, respondents reported the following:

- 47% updated emergency response plans
- 33% did not update emergency response plans
- 20% did not know whether the facility updated its emergency response plans

Regarding other company actions to *prepare to respond*, respondents reported the following:

- 46% informed local fire and police departments, HazMat teams, etc. about specific plant hazards
- 15% did not inform fire and police departments, HazMat teams, etc. about specific plant hazards, and 40% said they did not know whether the company at their site had communicated with emergency services
- 42% put additional procedures in place to inform employees of emergencies
- 30% updated shutdown procedures

When considering actions to inform local community services, or nearby residents or update the **community** Emergency Response Plan, respondents reported fewer actions and an increase in *don't know* responses.

- 23% informed local hospitals, health departments and emergency medical personnel about potential health threats from plant-specific exposures (20% did not inform these services, and 57% of respondents did not know)
- 21% updated the Emergency Response Plan for the **community** (34% did not update Emergency Response Plan for the **community**, 45% did not know)
- 15% put in place additional procedures to inform the **community** about an emergency (45% did not put in place additional procedures to inform the **community**, 40% did not know)

(See Table 8 below.)

Table 8: Possible Actions to Be Prepared to Respond to a Catastrophic Event

Possible actions to be prepared to respond to a catastrophic event	Was action taken? Yes	No	Don't Know
1. Provided emergency response training to employees within the past 12 months	67.5%	26.0%	6.5%
2. Conducted emergency response drills for the plant site	58.9%	35.5%	5.6%
3. Updated Emergency Response Plan for **the facility**	46.8%	33.1%	20.2%
4. Informed local fire and police departments, HazMat teams, etc. about potential plant-specific hazards	45.5%	14.6%	39.8%
5. Put in place additional procedures to inform **employees** of an emergency (e.g., alarms, public address system)	41.9%	50.8%	7.3%
6. Updated shutdown procedures for critical equipment in an emergency	29.8%	41.1%	29.0%
7. Informed local hospitals, health departments, emergency medical personnel, etc. about the potential health threats from plant-specific exposures	23.4%	20.2%	56.5%
8. Updated Emergency Response Plan for **the community**	21.0%	33.9%	45.2%
9. Put in place additional procedures to inform the **community** about an emergency (e.g., alarms, public address system)	15.3%	45.2%	39.5%

Q7. Since September 11, 2001, has the company at your worksite taken any of the following actions to be better prepared to respond to a catastrophic event that was caused by a possible terrorist attack? (1&4: Responses = 123, Missing = 1.6%); (2-3 & 5-9: Responses = 124, Missing = 0.8%). **Note: Percents may not add up to 100% due to rounding**

Effectiveness of Actions in Preparing to Respond to a Catastrophic Event

Another set of questions asked respondents about the effectiveness of actions taken by the company *in preparing to respond* to a catastrophic event caused by a **terrorist attack** or an **unintentional incident**. First we consider the effectiveness of company actions *in preparing to respond* to an event caused by a **terrorist attack**, and then the effectiveness of company actions *in preparing to respond* to an event caused by an **unintentional incident**. After, we report about the effectiveness ratings of all participants regarding company actions *in preparing to respond* to each of the possible causes, we consider differences among those who judged their sites to be at *high*, *medium*, or *low* likelihood of a catastrophic event.

Effectiveness of Actions in PREPARING TO RESPOND to an Event Caused by a TERRORIST ATTACK. Overall, when asked about the effectiveness of response preparedness actions taken by their company since 9/11 *in preparing to respond* to an event caused by a **terrorist attack,** respondents rated the effectiveness of the company actions as follows:

- 38% effective (includes: *very effective*, *moderately effective*, and *slightly effective)*
- 38% *neutral*
- 23% ineffective (includes: *very ineffective*, *moderately ineffective*, and *slightly ineffective)*

About 5%, a small contingent, of respondents rated their sites' actions *in preparing to respond* to an event caused by a **terrorist attack** as *very effective*. Notably, more than one third reported the effectiveness of their company's actions as *neutral*. (See Table 9 and Chart 10 below.)

We also analyzed these effectiveness ratings by computing an average of all respondent answers using the 7-point scale described earlier. Using this scale, on average, respondents rated the actions of the company *in preparing to respond* to an event caused by a **terrorist attack** at their sites a 4.1, or nearly *neutral*.

Table 9: Effectiveness of Actions in PREPARING TO RESPOND to a Catastrophic Event Caused by a TERRORIST ATTACK

Effectiveness of Company Actions	Respond to an Event Caused by a TERRORIST ATTACK
Very effective	4.8%
Moderately effective	13.6%
Slightly effective	20.0%
Neutral	38.4%
Slightly ineffective	5.6%
Moderately ineffective	4.8%
Very ineffective	12.8%
Average on 7 point scale	**4.1**

Q8. Overall, since September 11, 2001, how effective have the actions taken by the company been in preparing your worksite to respond to a catastrophic event caused by the following? (Q8a. Responses = 125, Missing = 0.0%). **Note: Percents may not add up to 100% due to rounding**

Chart 10: Effectiveness of Actions in PREPARING TO RESPOND to a Catastrophic Event Caused by a TERRORIST ATTACK

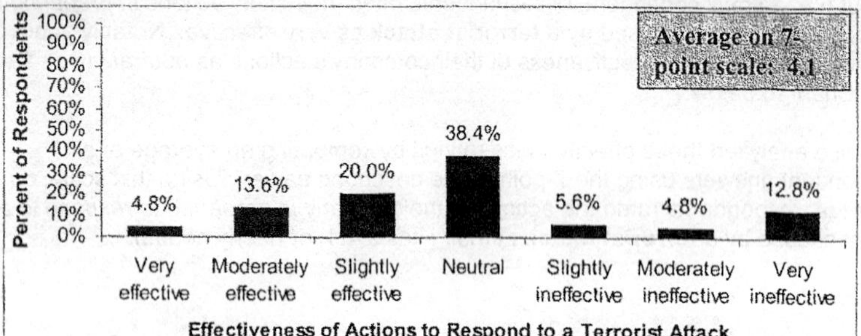

Q8. Overall, since September 11, 2001, how effective have the actions taken by the company been in preparing your worksite to respond to a catastrophic event caused a ***terrorist attack****? (Responses = 125, Missing = 0.0%)*

Note: Percents may not add up to 100% due to rounding

Effectiveness of Actions in PREPARING TO RESPOND to a Catastrophic Event Caused by a TERRORIST ATTACK by *High, Medium,* or *Low* Likelihood of Such an Event. To further understand the perspectives of respondents, we also examined responses to this question by looking at differences in the effectiveness ratings of those respondents who thought there was a *high* likelihood of a catastrophic event caused by a **terrorist attack** at their worksite, and at respondents who thought there was a *medium*, or *low* likelihood of an event caused by a **terrorist attack**. Noteworthy differences emerged when examining the effectiveness ratings of those who work at sites in which they perceive different vulnerability levels, such as:

- 44% of those who work at sites with a *high* vulnerability, rated company actions as ineffective, considerably higher than either of the other two vulnerability level ineffective ratings (*medium* vulnerability = 27% ineffective, *low* vulnerability = 11% ineffective)
- Respondents from *high* vulnerability worksites average effectiveness rating was 3.3 or between *neutral* and *slightly ineffective*. Respondents from *medium* and *low* vulnerability sites were about one point higher or between *neutral* and *slightly effective* (*medium* vulnerability = 4.2, *low* vulnerability = 4.4). (See Table 10 and Chart 11 below.)

Table 10: Effectiveness of Actions in PREPARING TO RESPOND by High, Medium, or Low Likelihood of a Catastrophic Event due to a TERRORIST ATTACK

Effectiveness of Actions in Preparing to Respond to Event Caused by a TERRORIST ATTACK	Likelihood of a Catastrophic Event at Site Caused by a TERRORIST ATTACK		
	High	Medium	Low
Very effective	3.1%	5.9%	5.4%
Moderately effective	9.4%	11.8%	17.9%
Slightly effective	12.5%	32.4%	16.1%
Neutral	31.3%	23.5%	50.0%
Slightly ineffective	9.4%	8.8%	1.8%
Moderately ineffective	6.3%	11.8%	0.0%
Very ineffective	28.1%	5.9%	8.9%
Average on 7 point scale	**3.3**	**4.2**	**4.4**

Q3. What is the likelihood of your worksite experiencing a catastrophic event involving fire, explosion, or a hazardous release caused by the following? Q8. Overall, since September 11, 2001, how effective have the actions taken by the company been in preparing your worksite to respond to a catastrophic event caused by the following? **Note: Percents may not add up to 100% due to rounding**

Chart 11: Effectiveness of Actions in PREPARING TO RESPOND by High, Medium, or Low Likelihood of a Catastrophic Event due to a TERRORIST ATTACK

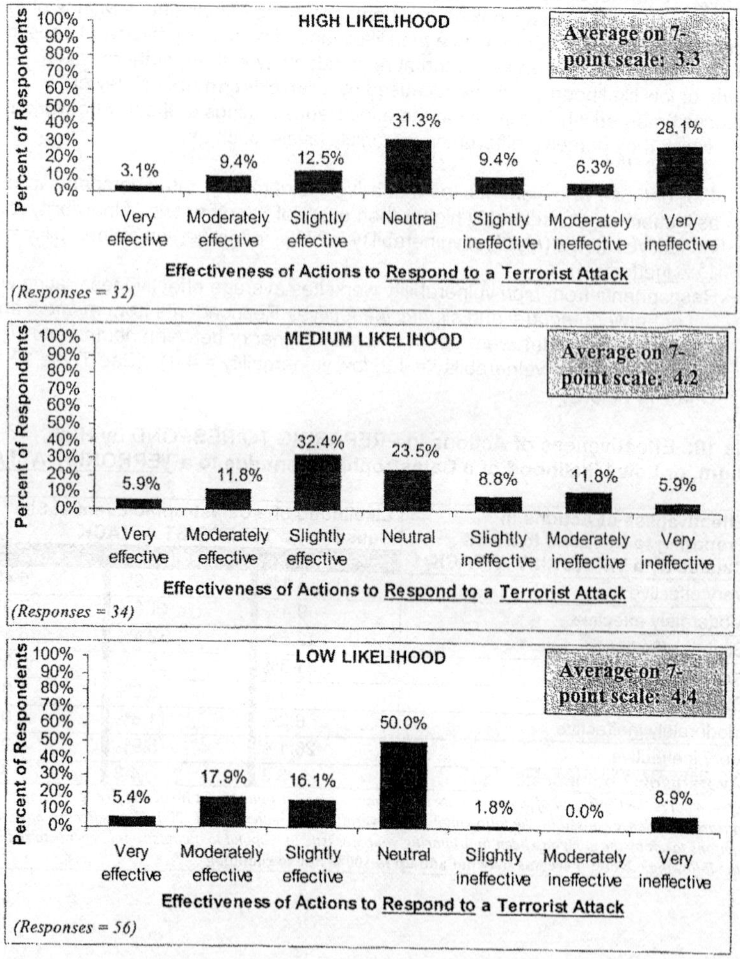

***Questions:** Q3. What is the likelihood of your worksite experiencing a catastrophic event involving fire, explosion, or a hazardous release caused by a **TERRORIST ATTACK**? **Q8.** Overall, since September 11, 2001, how effective have the actions taken by the company been in preparing your worksite to respond to a catastrophic event caused a **TERRORIST ATTACK**?* **Note: Percents may not add up to 100% due to rounding.**

Effectiveness of Actions in PREPARING TO RESPOND to an Event Caused by an UNINTENTIONAL INCIDENT. Overall, when asked about the effectiveness of actions taken by their company since 9/11 *in preparing to respond* to a catastrophic event caused by an **unintentional incident,** respondents rated the effectiveness of the company actions as follows:

- 44% effective (includes: *very effective, moderately effective*, and *slightly effective)*
- 38% *neutral*
- 18% ineffective (includes: *very ineffective, moderately ineffective*, and *slightly ineffective)*

Eleven percent (11%) of respondents rated the actions of the companies at their worksites as *very effective,* more than twice as high as any of the overall *very effective* ratings. Notably, more than one-third (38%) reported the effectiveness of the company actions *in preparing to respond* to an **unintentional incident** as *neutral.* (See Table 11 and Chart 12 below.)

We also analyzed these effectiveness ratings by computing an average of all respondent answers using the 7-point scale described above. Respondents rated the actions of the company *in preparing to respond* to an event caused by an **unintentional incident** at their sites an average of 4.4, slightly more effective than they rated the effectiveness of actions *in preparing to respond* to a catastrophic event caused by a **terrorist attack** (4.1).

Table 11: Effectiveness of Actions in PREPARING TO RESPOND to a Catastrophic Event Caused by an UNINTENTIONAL INCIDENT

Effectiveness of Company Actions	Prepare to Respond to an UNINTENTIONAL INCIDENT
Very effective	10.5%
Moderately effective	16.1%
Slightly effective	17.7%
Neutral	37.9%
Slightly ineffective	4.0%
Moderately ineffective	6.5%
Very ineffective	7.3%
Average on 7 point scale	**4.4**

Q8. Overall, since September 11, 2001, how effective have the actions taken by the company been in preparing your worksite to respond to a catastrophic event caused by the following? (Q8b. Responses = 124, Missing = 0.8%). **Note: Percents may not add up to 100% due to rounding**

Chart 12: Effectiveness of Actions in PREPARING TO RESPOND to a Catastrophic Event Caused by an UNINTENTIONAL INCIDENT

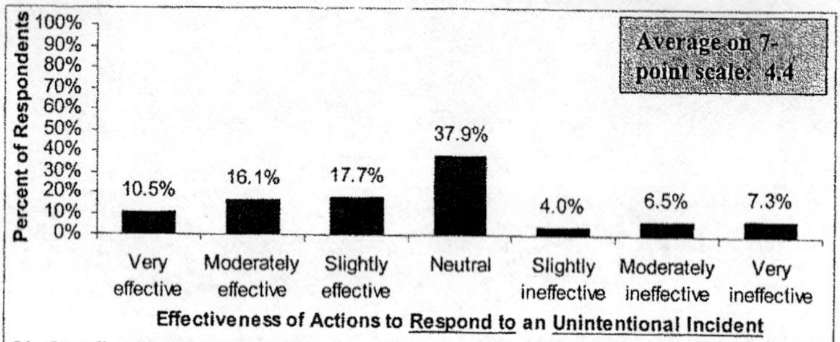

Q8. Overall, since September 11, 2001, how effective have the actions taken by the company been in preparing your worksite to respond to a catastrophic event caused an unintentional incident? (Responses = 124, Missing = 0.8%)

Note: Percents may not add up to 100% due to rounding

Effectiveness of Actions in PREPARING TO RESPOND by *High, Medium, or Low* Likelihood of an Event Caused by an UNINTENTIONAL INCIDENT. To further understand the perspectives of respondents, we also examined responses to this question by looking at differences in the effectiveness ratings of those respondents who thought there was a *high* likelihood of a catastrophic event caused by an **unintentional incident** at their worksite, and those who thought there was a *medium* or *low* likelihood of such an event. The average effectiveness for *preparing to respond* was about the same across the different vulnerability levels (*high* vulnerability = 4.5, *medium* vulnerability = 4.4, *low* vulnerability = 4.4). However, noteworthy differences emerged when examining the levels of effectiveness of those who work at sites perceived to face a *high* likelihood of an **unintentional incident** as compared to those with *medium* or *low* likelihood of **unintentional incidents**, such as:

- 62% of those who work at sites that they rated as *high* vulnerability, rated company actions *in preparing to respond* as effective. This was higher than any other effectiveness rating in this survey. This is notably higher than either of the other two vulnerability level groups' effectiveness ratings regarding their sites' actions *in preparing to respond* to a catastrophic event caused by an **unintentional incident** (*medium* vulnerability = 41% effective; *low* vulnerability = 38% effective).
- Respondents who rated sites as either *medium* or *low* vulnerability were more than twice as likely to rate the effectiveness of their sites' actions *in preparing to respond* to an event caused by an **unintentional incident** as *neutral* (*medium* vulnerability = 41% *neutral* effectiveness, *low* vulnerability = 46% *neutral* effectiveness), when compared to respondents from *high* vulnerability sites (15% *neutral* effectiveness). (See Table 12 and Chart 13 below.)

Table 12: Effectiveness of Actions in PREPARING TO RESPOND by High, Medium, or Low Likelihood of a Catastrophic Event Caused by an UNINTENTIONAL INCIDENT

Effectiveness of Actions in Preparing to Respond to an Event Caused by an UNINTENTIONAL INCIDENT.	Likelihood of a Catastrophic Event at Site Caused by an UNINTENTIONAL INCIDENT		
	High	Medium	Low
Very effective	15.4%	4.3%	14.0%
Moderately effective	11.5%	21.7%	14.0%
Slightly effective	34.6%	15.2%	10.0%
Neutral	15.4%	41.3%	46.0%
Slightly ineffective	3.8%	6.5%	2.0%
Moderately ineffective	7.7%	8.7%	4.0%
Very ineffective	11.5%	2.2%	10.0%
Average on 7 point scale	**4.5**	**4.4**	**4.4**

Q3. What is the likelihood of your worksite experiencing a catastrophic event involving fire, explosion, or a hazardous release caused by the following? Q8. Overall, since September 11, 2001, how effective have the actions taken by the company been in preparing your worksite to respond to a catastrophic event caused the following? **Note: Percents may not add up to 100% due to rounding**

Chart 13: Effectiveness of Actions in PREPARING TO RESPOND by High, Medium, or Low Likelihood of a Catastrophic Event Caused by an UNINTENTIONAL INCIDENT

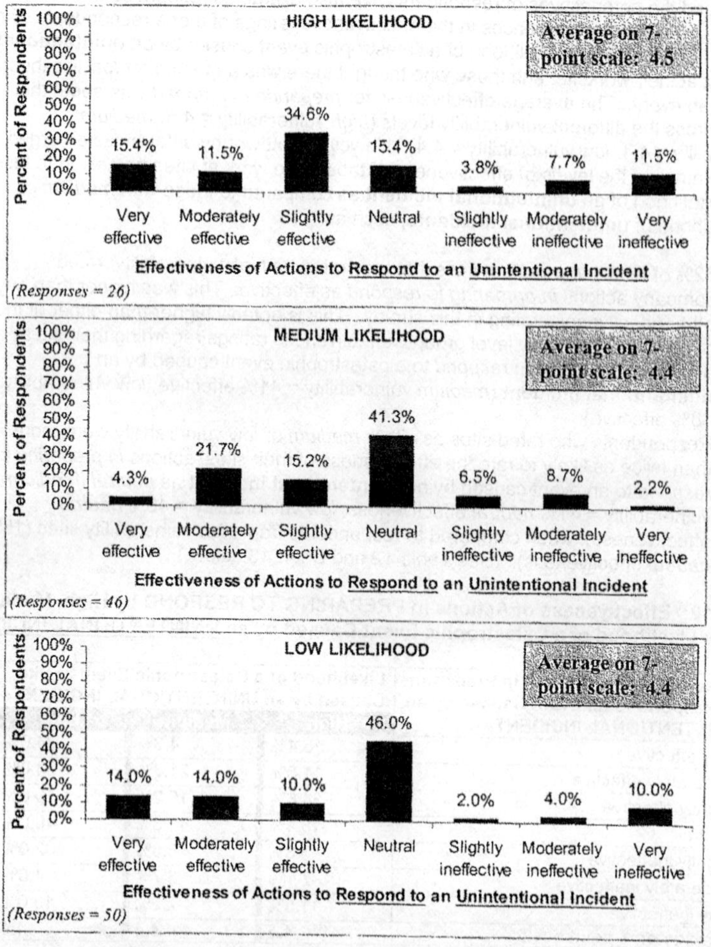

Questions: *Q3. What is the likelihood of your worksite experiencing a catastrophic event involving fire, explosion, or a hazardous release caused by an UNINTENTIONAL INCIDENT? Q8. Overall, since September 11, 2001, how effective have the actions taken by the company been in preparing your worksite to respond to a catastrophic event caused an UNINTENTIONAL INCIDENT?* **Note: Percents may not add up to 100% due to rounding.**

Training: Quality, Scope, and Need

Training employees is a key vehicle in *preventing* or *preparing to respond* to a catastrophic event whether caused by a **terrorist attack** or an **unintentional incident**. Survey questions related to training focused on the following: the extent to which companies provided training to employees since 9/11, whether companies improved training since 9/11, and whether respondents thought members of the bargaining unit at their facilities needed additional training.

Extent of Training. Sixty-eight percent (68%) of employees reported that their employers had provided emergency response training to employees within the *past 12 months.* Regarding *how many* employees at their sites received training *since 9/11*, respondents reported the following:

- About one-third reported that no employees at their sites received training about either *preventing* (34%) or *responding to* (28%) a catastrophic event caused by a terrorist attack.
- 38% reported that half or fewer employees had received *response preparedness* training.
- 27% reported that half or fewer employees had received training in *prevention.*
- 15% or fewer said that more than half to all employees had received *prevention* (15%) or *response preparedness* (13%) training. (See Charts 14 and 15 below.)
- Notably, a sizeable percent of respondents reported that they did not know about training to *prevent* (25%), or *respond* (21%) to catastrophic events at their sites.

Chart 14: Percent of Employees Trained to Prevent a Catastrophic Event Caused by a Terrorist Attack

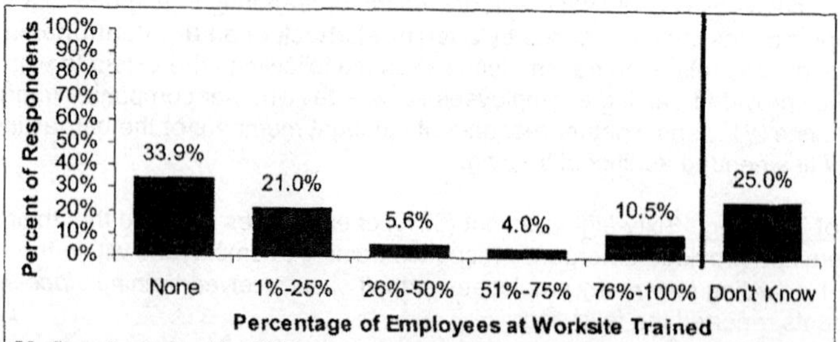

*Q9. Since September 11, 2001, what percentage of employees at your worksite has the company trained about **preventing** or responding to a potential catastrophic event caused by a terrorist attack? (Responses = 124, Missing = 0.8%)*

Chart 15: Percent of Employees Trained to Respond to a Catastrophic Event Caused by a Terrorist Attack

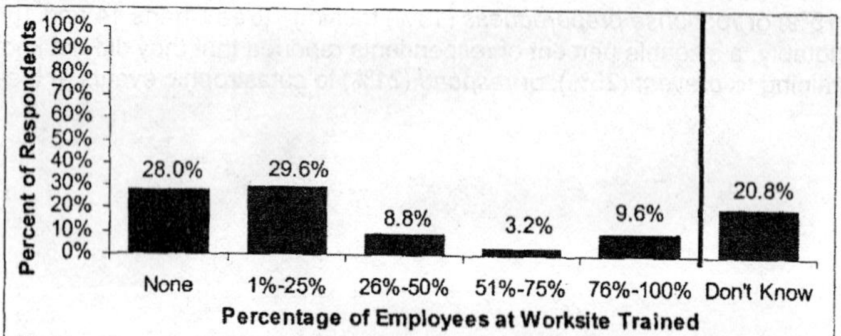

*Q9. Since September 11, 2001, what percentage of employees at your worksite has the company trained about preventing or **responding** to a potential catastrophic event caused by a terrorist attack? (Responses = 125, Missing = 0.0%)*

Need for Additional Training. When asked whether members of the PACE bargaining unit needed additional training related to a potential catastrophic event caused by a **terrorist attack**, respondents reported the following:

- Almost three-quarters (74%) said additional training for members of the bargaining unit was needed.
- 9% said *no* additional training was needed
- 17% responded that they *don't know* whether additional training is needed. (See Chart 16 below.)

Chart 16: Bargaining Unit Needs Additional Training Related to Terrorist Attacks

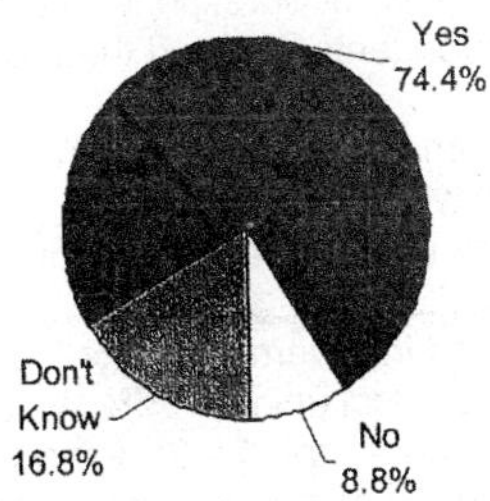

Q10. Do members of the bargaining unit need additional training related to potential catastrophic events caused by a terrorist attack at your worksite? (Responses = 125, Missing = 0.0%)

Involvement in Incident Prevention and Response by Local Unions, Hourly Workers, or Communities

The last parts of the survey questionnaire assessed whether *the company had taken actions* to involve the local union, hourly workers or the community regarding plans or actions related to preventing or responding to potential catastrophic events caused by a **terrorist attack**; and also assessed whether *the local union had taken actions* to improve the company's plans or action in this area. The findings regarding involvement follow.

<u>Company Initiated Action.</u> Overall, respondents reported relatively few actions initiated by the company to involve the local union, hourly workers, or the community regarding its plans or actions to prevent or respond to a catastrophic event caused by a possible **terrorist attack**. Respondents reported the following:

- 28% or fewer reported some type of involvement of any group (local union, hourly workers, community).
- An overwhelming majority report no involvement by local unions or hourly workers.
- About one-quarter reported involvement of the local union and hourly workers in making recommendations (local union = 25%, hourly workers = 22%), and being informed by the company (local union = 21%, hourly workers = 28%)
- The highest area in which the company involved hourly workers and the community was in informing them about plans or possible actions to respond to or prevent a catastrophic event caused by a terrorist attack (hourly workers = 28%, community = 12%).
- Almost two-thirds (63%-66%) reported not knowing if or how the company involved the community. (See Table 13 below.)

When explaining how the company involved others, the most common methods reported included: meetings, committees, letters, and training and drills.

Table 13: Possible Actions Taken by the Company to Involve Others

Possible actions taken by the company to work with the Local Union, Hourly Workers, and the Community	Yes	No	Don't know
LOCAL UNION			
Informed	21.3%	74.6%	4.1%
Involved in Assessment	9.8%	86.9%	3.3%
Involved in Making Recommendations	25.2%	65.0%	9.8%
HOURLY WORKERS			
Informed	27.6%	61.8%	10.6%
Involved in Assessment	12.1%	78.2%	9.7%
Involved in Making Recommendations	21.8%	64.5%	13.7%
COMMUNITY			
Informed	12.2%	24.4%	63.4%
Involved in Assessment	8.1%	26.8%	65.0%
Involved in Making Recommendations	7.3%	26.8%	65.9%

Q11, 13, 14. Since September 11, 2001, has the company worked with {the local union/hourly workers/the community} regarding its plans or actions to prevent or respond to a catastrophic event caused by a possible terrorist attack at your worksite?

Local Union Action. Consistent with findings from above where 25% or fewer of local unions reported being involved by the companies at their sites regarding incident prevention or response, an overwhelming majority of respondents (83%) reported that their local union had taken no action related to improving the company's plans or actions regarding preventing or responding to a catastrophic event caused by a possible **terrorist attack** at their worksite. Ten percent (10%) reported that the local union had taken action, while 7% said they did not know about any action. (See Chart 17 below.) Of those respondents who indicated actions taken by the local union, they reported that the local union had asked the company for additional employee training, and had offered for the local union to work with the company on these issues.

Chart 17: Local Union Taken Action

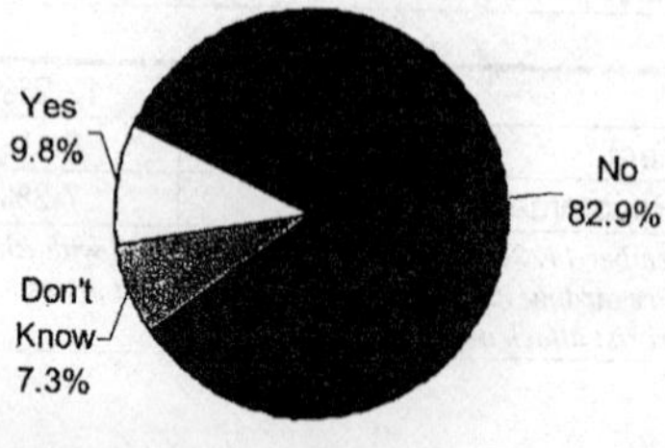

12. Since September 11, 2001, has the local union taken action related to improving the company's plans or actions regarding preventing or responding to a catastrophic event caused by a possible terrorist attack at your worksite? (Responses = 123, Missing = 1.6%)

Study Limitations

This preliminary study and its data are limited and thus these findings cannot be generalized broadly to represent other sites either within or outside of PACE. The key limitations follow:

- The survey looked at employee perceptions and did not include an independent assessment of actual actions taken by companies.
- No baseline or prior data about the perceptions of key people within PACE local unions about their site's vulnerability to a catastrophic event, or of their employer's programs prior to 9/11 are available.
- The study sampling technique may be limited. The survey respondents were selected from a list of Risk Management Program (RMP) sites. However, due to security limitations imposed since 9/11, the most accurate lists of RMP sites are not readily available. Therefore:
    - Some sites which did respond may not be RMP sites any longer
    - Some sites which were not surveyed may be RMP sites at this time
    - Some sites may be high hazard, RMP-like sites but do not have the RMP designation that are not included in this study.
    - Respondents may have underreported whether their sites are RMP sites because the RMP designation has more to do with environmental management than with worker safety. Therefore, respondents may be unfamiliar with this designation.
- The health and safety expertise of respondents and their knowledge of potential catastrophic incidents may have varied from site to site depending on who at the local union responded. While PACE requested that the local union president designate a person knowledgeable about what the company and the local union might be doing to lower the vulnerability of their site, and suggested that appropriate people might include: the local union president, secretary-treasurer, chair or member of the Health and Safety Committee, Health and Safety or TOP Representative, or other health and safety activist. The diversity of respondents' knowledge from site to site may contribute to the data being uneven in some cases, especially regarding those who responded "don't know" to many questions.
- We are unable to conduct any follow-up with the actual respondents because the survey was anonymous so we do not know who completed it.

Readers should be careful not to assume that the findings can be generalized broadly to represent all PACE represented workplaces, all PACE represented sites from a specific industrial sector, or RMP sites in general.

Discussion and Conclusions

The findings contained in this report begin to provide evidence about where sites represented by PACE Union are in preventing and preparing to respond to catastrophic events caused by a terrorist attack or an unintentional incident. We believe it points out areas that may be in need of further examination, discussion, and action to assure that workers at PACE represented workplaces and the communities in which they are located have the best levels of prevention and response possible.

We suggest that PACE staff and leaders at the International level, as well as local union leaders consider how to use the findings and discussion contained in this report. We hope that this report and subsequent dialogue enables you to brainstorm and determine which actions to initiate to advance the opportunities that may have been revealed in these findings.

Reminders about the Data

When you review and deliberate about which actions to take from these findings, it is important to remember all the "Limitations" section statements but especially the following limitations of the report:

- This survey looked at perceptions only. It did not include an independent assessment of, for example, which employees actually received training since September 11, 2001, or which actions companies actually took.

- The survey respondents were selected from a list of Risk Management Program (RMP) sites. However, due to security limitations imposed since 9/11, the most accurate lists of RMP sites are not readily available. Therefore, some sites who did respond may not actually be RMP sites any longer, and some sites who were not surveyed may actually be RMP sites at this time. Readers should be careful not to assume that the findings can be generalized broadly to represent all PACE represented workplaces, all PACE represented sites from a specific industrial sector, or RMP sites in general.

Possibility and Likelihood of A Catastrophic Event

Ninety-five percent (95%) of the respondents reported that their sites have large enough quantities of chemicals to cause a catastrophic event if those materials were involved in a fire, explosion or release. Over half of the sites indicated that they face a *high* or *medium* likelihood of a catastrophic event due to a **terrorist attack** (54%) or **unintentional incident** (59%).

What Companies Are Doing

Company Preventative Actions. In response to these vulnerabilities respondents' reports suggest that most employers assessed their sites vulnerabilities (66%) and worksite security (64%). Company actions appeared to focus more frequently on security, with almost three-quarters (73%) of the respondents reporting improved systems to guard and secure the plant.

All other company actions were reportedly taken at less than half of the sites. These included improved communication systems (43%), improved training and procedures to prevent possible terrorist attacks (38%), updated warning systems (38%), improved containment of potential hazardous releases (34%), and improved quality and availability of personal protective equipment (30%.).

Furthermore, some of the most preventative actions that could directly reduce the likelihood of a catastrophic event were reportedly taken with the least frequency, such as: reduced volumes of hazardous substances (17%); strengthened plant vessels, tanks, piping or other structures (17%); and improved the siting of hazardous substances or processes (14%).

Company Actions To Prepare To Respond. When *preparing to respond* to an event caused by a **terrorist attack**, 68% of the companies provided emergency response training to employees in the past 12 months, and 59% conducted emergency response drills for the plant site. Only about half (47%) of the respondents reported that the companies at their worksites had updated *facility* emergency response plans since 9/11. Other company actions to *prepare for responding* to an event included: 46% informed local fire and police departments, HazMat teams, etc. about specific plant hazards, 42% put additional procedures in place to inform employees of emergencies, and 30% updated shutdown procedures.

However, respondents' use of the *don't know* choice increased considerably in the set of questions about actions to inform local community services, or nearby residents or update the **community** Emergency Response Plan. While 23% knew their employers had informed local hospitals, health departments and emergency medical personnel about potential health threats from plant-specific exposures, 20% said these services were not informed, and 57% reported *don't know*.

It appears that the more distant from rank and file hourly worker experiences the survey queried, the greater the percentage of *don't know* answers. It appears that the gap between hourly workers and community emergency response planning is great.

Effectiveness Of Company Prevention and Response Actions

Effectiveness of Prevention Actions. Less than half (44%) of the respondents indicated that their company's preventative actions, including security efforts, were effective (includes: *very effective, moderately effective*, and *slightly effective)* in *reducing the vulnerabilities* of their site to a catastrophic event caused by a **terrorist attack**. Over one-third (36%) were *neutral* about the effectiveness, and one-fifth (21%) said the actions were ineffective (includes: *very ineffective, moderately ineffective*, and *slightly ineffective)*.

When considering the effectiveness of actions to prevent an event caused by an **unintentional incident**, only one-third (33%) said the company's actions were effective. Forty-six percent (46%) were *neutral* about the effectiveness, and one-fifth (21%) said the actions were ineffective to *reduce their sites' vulnerabilities* to an event caused by an **unintentional incident**.

On average, respondents rated the effectiveness of company actions to *prevent* a catastrophic event only slightly above *neutral* (**terrorist attack** = 4.2 and **unintentional incident** = 4.1) on a 7-point scale.

Respondent assessment of the effectiveness of the company actions to *prevent* a catastrophic event were even more striking when considering perceptions of a site's vulnerability to a catastrophic event (*high, medium, low*). Forty-five percent (45%) of the respondents who rated their sites with a *high* vulnerability level also rated their company's actions to *prevent* an event caused by a **terrorist attack** as ineffective. This ineffective rating is notably higher than ratings given by respondents from *medium* or *low* vulnerability sites who rated their company's actions regarding an event caused by a **terrorist attack** as follows: 18% ineffective and 11% ineffective, respectively.

Furthermore, we noted with interest that respondents rated the effectiveness of company actions to *prevent* an event caused by a **terrorist attack** (44%) higher than one caused by an **unintentional incident** (33%). Is it possible that additional security measures may have reduced some vulnerabilities to terrorist attacks, but that actions to address the inherent dangers of hazardous materials and processes at these industrial workplaces have yet to be taken? (See Table 14, Effectiveness of Prevention and Response Actions)

Effectiveness of Response Actions. Only 38% of the respondents indicated that their company's actions *in preparing to respond* to an event caused by a **terrorist attack** were effective (includes: *very effective, moderately effective*, and *slightly effective)*. As many were *neutral* (38%) about the effectiveness of actions *in preparing to respond* to such an event, while almost one quarter (23%) said the actions were ineffective (includes: *very ineffective, moderately ineffective*, and *slightly ineffective)*. When considering the effectiveness of actions *in preparing to respond* to an event caused by an **unintentional incident**, forty-four percent (44%) said the company's actions were effective. The same percentage (38%) were *neutral* regarding the effectiveness of *preparing to respond* to an **unintentional incident** as they were regarding preparations to respond to an event caused by a **terrorist attack**. Eighteen percent (18%) said the company's actions were ineffective.

On average, respondents rated the effectiveness of company actions to *respond* to a catastrophic event caused by a **terrorist attack** only slightly above *neutral* (4.1) on a 7-point scale. However, respondents' perceptions of the effectiveness of employers' actions *in preparing to respond* to an event caused by an **unintentional incident** was slightly higher at 4.4, midway between *neutral* and *slightly effective*.

When rating the effectiveness of the company actions *in preparing to respond,* respondents from sites rated as having a *high* likelihood of a catastrophic event reported considerable differences from the *medium* or *low* likelihood sites. When considering responding to an event caused by a **terrorist attack**, 44% of respondents who characterized their sites as *high* risk found their company's actions ineffective. This rating is considerably higher than the ineffectiveness ratings given by respondents at sites with a *medium* or *low* likelihood of an event (*medium* likelihood = 27% ineffective, *low* likelihood = 11% ineffective). However, most notable is that when considering the effectiveness of company actions *in preparing to respond* to an **unintentional incident**, the *high*est risk respondents rated their employers' actions with the highest levels of effectiveness in the survey, with 62% indicating that their company's actions were effective.

Is it possible that the sites characterized as *high* risk have developed extensive emergency response programs to respond to **unintentional incidents**, especially when compared to sites that ranked themselves with a *medium* or *low* likelihood of experiencing a catastrophic event caused by an **unintentional incident**? Or could it be that employees from *high* risk sites have confidence in their employer's response plans as a coping/survival strategy for working in workplaces that are intrinsically high hazard?

Table 14: Effectiveness of Prevention and Response Actions

Cause of Event and Report from whom	Average on 7-point scale	Effective*	Neutral	Ineffective**
PREVENTION				
Terrorist Attack				
All Respondents	4.2	43.5%	35.5%	20.9%
High Likelihood Respondents	3.5	32.3%	22.6%	45.2%
Medium Likelihood Respondents	4.5	58.7%	23.5%	17.7%
Low Likelihood Respondents	4.4	41.1%	48.2%	10.7%
Unintentional Incident				
All Respondents	4.1	33.0%	46.0%	21.1%
High Likelihood Respondents	3.8	38.4%	23.1%	38.4%
Medium Likelihood Respondents	4.2	26.1%	56.5%	17.4%
Low Likelihood Respondents	4.3	36.0%	48.0%	16.0%
RESPONSE				
Terrorist Attack				
All Respondents	4.1	38.4%	38.4%	23.2%
High Likelihood Respondents	3.3	25.0%	31.3%	43.8%
Medium Likelihood Respondents	4.2	50.1%	23.5%	26.5%
Low Likelihood Respondents	4.4	39.4%	50.0%	10.7%
Unintentional Incident				
All Respondents	4.4	44.3%	37.9%	17.8%
High Likelihood Respondents	4.5	61.5%	15.4%	23.0%
Medium Likelihood Respondents	4.4	41.2%	41.3%	17.4%
Low Likelihood Respondents	4.4	38.0%	46.0%	16.0%

Q3. What is the likelihood of your worksite experiencing a catastrophic event involving fire, explosion, or a hazardous release caused by the following? Q6. Overall, since September 11, 2001, how effective have the actions taken by the company been in lessening the vulnerability of your worksite to a catastrophic event caused by the following? Q8. Overall, since September 11, 2001, how effective have the actions taken by the company been in preparing your worksite to respond to a catastrophic event caused the following?
Note: Percents may not add up to 100% due to rounding.

***EFFECTIVE:** Includes those who responded *very effective, moderately effective,* and *slightly effective*
****INEFFECTIVE:** Includes those who responded *very ineffective, moderately ineffective,* and *slightly ineffective*

Training

All the survey respondents included in this report's findings indicated that their sites have hazardous materials in quantities large enough to lead to a catastrophic event if involved in fire, explosion or other releases. However, training to *prevent* or *respond* to these risks appears to be lacking. About one-third of respondents reported that no employees at their sites received training about *preventing* (34%) or *responding* (28%) to a catastrophic event caused by a **terrorist attack** since 9/11. At sites where some training occurred, only 38% reported that half or fewer employees received *response preparedness* training, and only 27% reported that half or fewer employees received *prevention* training. Notably, a sizeable percent of respondents reported not knowing about training to *prevent* (25%), or *respond* (21%) to catastrophic events at their sites. Seventy-four percent (74%) reported that additional training was needed for members of their bargaining unit.

The findings indicate that when training did occur, it was more focused on *responding* to emergencies, rather than *preventing* them. The amount of training among the workforce appears very limited, with the majority of the survey sites reporting that fewer than half of the employees have participated in training about *preventing* or *responding* to a potential catastrophic event caused by a terrorist attack. Furthermore, almost three-quarters of the respondents indicated that their members need additional training. These findings suggest a strong need for additional prevention and response training among PACE represented sites.

Involvement Of Hourly Workers, the Local Union Or Community

An overwhelming majority of respondents reported no action had been initiated by the companies at their sites to involve the local union or hourly workers in company plans or actions to prevent or respond to a catastrophic event caused by a possible **terrorist attack**. About one-quarter reported involvement by the local union, and hourly workers in making recommendations (local union = 25%, hourly workers = 22%), and being informed by the company (local union = 21%, hourly workers = 28%). Almost two-thirds (63%-66%) reported *don't know when asked* how the company involved the community. It must be asked, how can company action programs be effectively undertaken and have so many people be unaware of them?

Ten percent (10%) of respondents reported that their local unions had taken action to improve the company's plans or actions regarding prevention of or response to a catastrophic event. However, 83% reported no action had been initiated by their local union. Those respondents who indicated actions taken by the local union, described efforts to ask the company for additional employee training, and offers for the local union to work with the company on these issues.

It appears that companies are working to address prevention and response regarding hazardous materials without meaningfully involving or engaging hourly workers, or local unions. Our findings suggest that this is the same regarding working with communities surrounding the sites. With almost two-thirds of responses in the *don't know* choice regarding community involvement, it suggests that the further away the question focus is from the shop-floor, facility, or direct impact on rank and file workers, the less informed respondents were.

Recommendations for the Future

A number of action-oriented opportunities for PACE Union's Health and Safety Department and local unions emerge from this examination of the survey findings.

The PACE Evaluation Team Incident Prevention and Response Since 9/11 Work Group recommends that local unions examine this report's findings and consider the following questions:

1. What does this data mean for your local and for your site?
2. What actions do you want the company at your site to take regarding the following: preventing catastrophic events; preparing to respond to potential catastrophic events or emergencies; and involving your local union, hourly workers and the communities surrounding your facility?
3. What role should your local union take to initiate or advocate for the highest levels of prevention for your members, the facility, and the communities surrounding your facility?
4. How can your site work more closely in coordination with local emergency responders and health providers who would respond in an emergency?
5. Can your local union organize a training for your members about these issues, using the PACE Health and Safety Department curriculum?

Furthermore, the Evaluation Team Work Group recommends that the PACE Health and Safety Department take the following actions:

A. Educate and train PACE members about more effective actions companies could take to prevent catastrophic events using higher levels of prevention, rather than solely focusing on increased security measures.
B. Develop expanded training opportunities for PACE members about: 1) prevention and response to hazardous materials emergencies, and 2) the variety of roles local unions, hourly workers, and communities can play in prevention and response activities.
C. Increase the level of awareness about these issues within PACE Union.

Preventing and preparing to respond to potential catastrophic events whether caused by terrorist attacks or unintentional incidents are important issues facing PACE's membership. The PACE Evaluation Team hopes this assessment and report contribute to the dialogue and to effective action to meet these serious challenges.

Testimony of

Carol Andress
Economic Development Specialist
Environmental Defense Fund

Before the Senate Committee on
Homeland Security and Government Affairs

July 13, 2005

Good morning, I am Carol Andress, Economic Development Specialist with Environmental Defense.[1] Thank you for the opportunity to speak on the issue of security of America's chemical facilities.

The problem of securing chemical facilities is daunting. Thousands of facilities store and use dangerous chemicals in large quantities that could pose major risks to their neighbors if released. According to EPA records, approximately 2500-2800 facilities would each put any of over 10,000 people at risk of injury or death in the event of a major chemical release.[2] Nearly 5,000 facilities store more than 100,000 pounds of at least one EPA-designated extremely hazardous substance.[3]

The terrorists attacks of 9-11 have focused considerable attention to the threats posed by a deliberate attack on these dangerous stockpiles. However, even before 9-11, environmental and labor organizations highlighted the potential dangers to workers and fenceline communities from these facilities. Accidents, malfunctions, and other workplace incidents can be deadly.

While the increased attention from 9-11 is good, the bad news is that little progress has been made. Progress has been hindered for two reasons: (1) reliance on voluntary measures; and (2) sole focus on physical security. The good news is that many facilities have safer ways of doing business that eliminate or significantly reduce their reliance on dangerous chemicals. The challenge is to spur these changes soon so that security money is well invested. After all, why spend money trying to protect chemicals that don't need to be there?

---

[1] Environmental Defense, a leading national nonprofit organization, represents more than 400,000 members. Since 1967, Environmental Defense has linked science, economics, law and innovative private sector partnerships to create breakthrough solutions to the most serious environmental problems.

[2] Dana Shea, Congressional Research Service. "RMP facilities in the United States as of May 2005," June 27, 2005

[3] Jim Belke, U.S. Environmental Protection Agency. "Chemical accident risks in US industry—a preliminary analysis of accident risks data from US hazardous facilities," September 25, 2000.

Risks are Substantial, Widespread, and Unaddressed

Security experts have repeatedly warned this committee and others of the risks posed by chemical stockpiles and the need for Congressional action to address hazards that vulnerable chemical plants pose to workers, firefighters, police officers, and surrounding communities. Attachment 1 includes a list of 18 federal agencies and organizations that have warned about the dangers of a terrorist attack at a chemical facility, including warnings from--

- Former deputy homeland security advisor Richard Falkenrath who testified to Congress in January, 2005 that "since 9/11 we have essentially done nothing" to reduce the vulnerability of the national chemical sector. Mr. Falkenrath repeated his concerns in testimony to this committee in April.

- Former Department of Homeland Security (DHS) Inspector General Ervin who noted in a February 2005 op-ed, "Complicating the picture further is the fact that 85 percent of America's critical infrastructure is owned by the private sector, which has been reluctant to protect itself (and which the government has been reluctant to prod into protecting itself)."

- The Army Surgeon General's Office, which ranked the potential for attacks on chemical plants second only to bio-terrorism as the top threat confronting America's homeland security.

- A RAND study sponsored by the Air Force reported "Toxic warfare is a threat not just for U.S. forces engaged in military operations but also for civilians within the United States. This risk is increased by the wide availability of toxic materials throughout the United States, together with the proximity of industrial operations to large urban centers."

The Congressional Research Service (CRS) recently complied a report based on companies' estimates of people living within an area around a facility that could be affected by a worst-case chemical accident (see Attachment 2 for CRS report). At each of approximately 110 chemical facilities, more than a million people live in the vulnerable area surrounding the plant; at 550-600 chemical facilities more than 100,000 people live close enough that they could affected by a release. Moreover, the economic impact, which is not reflected in these numbers, could be devastating if neighboring businesses are shuttered following a chemical release.

Nor is this a hypothetical issue. The National Response Center has identified over 3,000 major chemical accidents at industrial facilities over the past 15 years. So far in 2005 alone, I know of three incidents that have resulted in more than 50 deaths and hundreds of injuries, including--

- Ten people killed in Graniteville, South Carolina in January after inhaling pure chlorine gas. Several of the dead or injured were emergency responders.
- BP refinery explosion in Texas in March that killed 15 and injured 170.
- Chlorine release in China in March that resulted in 28 people dead and 350 hospitalized.

In light of all of this, what have facilities been doing to reduce risks? The industry, DHS, Coast Guard and others have made laudatory efforts to boost security, but the fact is that no amount of fenceline security will protect facilities from a deliberate attack.

In fact, investigative reporters have repeatedly found holes in security. At numerous facilities, reporters have been able to document lax or non-existent security. Several of these lapses are at facilities that are members of American Chemistry Council's (ACC) Responsible Care program. In May 2005 a New York Times reporter noted inadequate security at the Kuehne chemical plant, in northern New Jersey, which has 12 million people living within its vulnerability zone.

Carl Prine with the Pittsburgh Tribune Review and CBS news team have made more than 100 visits to facilities in Baltimore, Pittsburgh, Chicago, Houston and elsewhere. Their visits included a facility in Baltimore that is perhaps a poster child for why chemical security focused solely on physical security is insufficient. This facility was subject to three separate, but overlapping programs:

1. American Chemistry Council's guidelines and in fact the facility had already passed the company's mandatory "third party" verification process;
2. Maritime Transportation Security Act (MTSA) because it is located on a navigable waterway; and
3. A Baltimore ordinance on mandatory security plans.

A potential 4th program that also may cover this facility is a chemical security law recently passed by the state of Maryland.

Despite these security requirements, a reporter was still able to enter an unguarded gate and reach two fully-loaded chlorine railcars, then leave without ever being challenged. (See Attachment 3 for copies of Tribune Review and NY Times articles)

So the problem is serious, pervasive, and can't be addressed with only guards, gates and guns.

Safer Options are Available

The good news in protecting against chemical terrorism is that we have options better than increasing physical security and hoping terrorists cannot evade fences and guards. The most cost-effective and sustainable way to achieve security is to design production processes and products in a way that is inherently safer. [4]

Unlike a physical security measure, an inherently safer approach offers many benefits. By reducing the source of the problem—the dangerous chemicals or processes—it cuts the need for security measures and minimizes the likelihood of a major chemical accident. It also reduces regulatory hassles—for example, a facility that cuts its chemical use to below certain thresholds no longer has to submit a Risk Management Plan. Many high-hazard industries could become intrinsically safer and eliminate concerns about terrorist attacks.

---

[4] Kenneth Geiser, "Primary Measures Safer, Cheaper, Better." The Environmental Forum, January/February 2004

Numerous wastewater treatment and drinking water treatment facilities have stopped using deadly chlorine gas in recent years. The added costs are small and are more than made up for by the savings in security expenses and the peace of mind that comes from knowing that residents and workers are no longer at risk.

For example, in 1999, after 85 years of using chlorine gas to disinfect drinking water, the Cleveland Water Division started to systematically eliminate chlorine gas at three local drinking water treatment plants. By early 2001, railcars of chlorine gas were gone. The costs of switching to safer chemicals were manageable (about $700,000 for one of the plants) and easily absorbed by an agency that spends several million dollars on capital improvements annually.[5] Several years earlier, the local sewage utility, Northeast Ohio Regional Sewage District (NEORSD) , converted its three wastewater treatment facilities to a liquid bleaching system with good results.

The impact of these changes means that over a million Cleveland residents, who otherwise could have been in harm's way in the case of a terrorist act or accident, no longer have to fear a chlorine gas release from local water utilities. Tim Tigue, Director of Operations for NEORSD, said" We'll never go back to chlorine gas. We owe it to our ratepayers and our workers."

Sample of wastewater and water facilities that have eliminated chemical hazards

Facility	City	State	New disinfection method	Population previously in vulnerability zone
Middlesex County Utilities Authority	Sayreville	NJ	hypochlorite	10,740,000
Northeast Water Pollution Control Plant	Philadelphia	PA	hypochlorite	1,575,971
Back River Wastewater Treatment	Baltimore	MD	hypochlorite	1,470,000
Baldwin Water Treatment Facility	Cleveland	OH	hypochlorite	1,400,000
R. M. Clayton WRC	Atlanta	GA	ultraviolet light	1,151,993
Wyandotte Wastewater Treatment Facility	Wyandotte	MI	ultraviolet light	1,100,000
Niagara Falls	Niagara Falls	NY	hypochlorite	1,100,000
Nottingham Water Treatment Facility	Cleveland	OH	hypochlorite	1,100,000
Mill Creek WWTP	Cincinnati	OH	hypochlorite	860,000
Jefferson Parish East Bank WWTP	Harahan	LA	hypochlorite	790,000
East Section Reclamation Plant	Renton	WA	hypochlorite	650,000
Little Falls Water Treatment Plant	Totowa	NJ	hypochlorite	430,000
Buckman Water Reclamation Facility	Jacksonville	FL	ultraviolet light	360,000
Portland	Portland	OR	hypochlorite	157,000
South Valley Water Reclamation Facility	West Jordan	UT	ultraviolet light	131,968

In addition, two thirds of the nation's oil refineries use safer processes that do not rely on highly toxic hydrofluoric acid in their processing.[6] Power plants too could eliminate their

[5] Cleveland Water Division's website notes that in 2003 Capital Improvement Program expenditures totaled $81.9 million (www.clevelandwater.com/annual_report).
[6] "Needless Risk: Oil Refineries and Hazard Reduction," US Public Interest Research Group Education Fund, October 2003.

reliance on extremely hazardous chemicals. For example, the 69 power plants using a aqueous ammonia pose a substantially smaller danger than the 166 power plants using ammonia gas.[7]

Other similarly situated plants have yet to implement common sense solutions to reduce hazards. To spur more widespread progress, Congress should enact a program with the several key elements:

1. Mandatory Safety and Security

Federal legislation to address the problem of terrorism at chemical facilities must put a priority on cutting the presence of extremely dangerous chemicals in populated areas. This cannot be done through voluntary programs; market mechanisms are simply inadequate to achieve the important goal of reducing potential catastrophic hazards. From my research of hundreds of wastewater and drinking water facilities, I found that facilities that are facing daily questions about operational efficiency, water quality standards, and financial performance, have little interest in dealing with catastrophic hazards that appear remote.

Facilities that did improve often did so as a result of external pressure, such as pressure from local officials or neighbors. The mayor of Cincinnati, Ohio, for example, set up a task force after 9-11 to examine local risks. They found that the Mill Creek Wastewater Treatment Plant was the single greatest risk to residents and then worked to eliminate that risk in a matter of weeks. In Cleveland OH, it took an aggressive Local Emergency Planning Committee to persuade utility officials to eliminate chlorine gas. Blue Plains Wastewater Treatment in Washington DC accelerated their plans to change following 9-11, but it helped that they already had plans to change in response to complaints from neighbors, including Bolling AFB and the Anacostia Naval Research Center.

Federal legislation must supply uniform outside pressure by establishing mandatory policies. Specifically, Congress should mandate that facilities using large quantities of dangerous chemicals must evaluate ways to:

(1) switch to safer chemicals or processes,
(2) reduce the amount of dangerous chemical used, or
(3) reduce the amount stored on site.

When those options are practical, the facility should be required to implement them. Facilities, especially high-risk facilities, should be expected to make significant investments in reducing the quantity and nature of the hazardous chemicals on site.

Even so, not every facility will be able to eliminate or significantly reduce hazards. When facilities find that—

(1) there is no safer process that is technologically feasible;
(2) all identifiable safer processes are prohibitively expensive in comparison to the potential damages of an accidental release; or

---

[7] "Unnecessary Dangers: Emergency Chemical Release Hazards at Power Plants," Working Group on Community Right-to-Know, July 2004.

(3) available alternatives would create an equal or greater hazard to public health and the environment;

they should provide a justification for why an alternative approach is not practicable.

Three states—New Jersey, Massachusetts, and California--have laws aimed at inspiring facilities to cut their use of certain toxic chemicals. New Jersey's Toxic Catastrophe Prevention Act (TCPA) is the only one that focuses specifically on accident prevention. Under that program facilities that have an "extraordinarily hazardous substance" at or above a certain threshold must submit a risk management plan for state approval, pay a fee based on the amount and type of chemical on site, and assess ways to prevent accidents. As a result of this program, numerous facilities have reduced or eliminated their use of dangerous chemicals, including 290 wastewater treatment facilities that have eliminated their use of chlorine gas.

2. Trust but Verify

Most facilities will make a good faith effort to implement safer approaches. However, this is far too important to rely solely on good intentions; facility owners and operators must be accountable to federal authorities and the public for reducing hazards. Accountability measures should include:

Government oversight, including federal review and approve of security/safety plans.

Public disclosure of the certification, signed by the CEO, that the company was unable to implement alternative approaches. The certificate should be public, in the same controlled manner as the Risk Management Plans, so that communities and agencies can know if facilities are doing everything reasonable to reduce catastrophic hazards.

Government intervention: DHS should have the authority to intervene in cases where the agency finds that a facility has acted in bad faith and not done a credible job of implementing cost-effective safer approaches. For example, when a large portion of an industry sector has reduced or eliminated risks and yet a similarly situated facility that could endanger thousands of people has refused to act.

Linking public funds with safer operations: Taxpayer money should not be spent at facilities that pose an unnecessary risk to the American public. This is especially applicable to sewage treatment and water treatment facilities that historically have received large amounts of federal and state funding as part of the State Revolving Loan Fund. There is simply no excuse anymore for chlorine gas to be used at urban water and sewer utilities—the risks are substantial and affordable alternatives are readily available. Congress should allocate future funding only to facilities that eliminate or significantly reduce chemical hazards to workers and nearby residents.

3. No Loopholes for Voluntary Programs

Safety and security requirements established by federal legislation must apply to all facilities that pose a significant risk and avoid creating loopholes for specific sectors simply because they are part of a voluntary program.

For example, the chemical industry has long argued that it should be allowed to implement its own voluntary programs rather than comply with federal standards. The industry supports having federal standards, so long as their companies get special exemptions.

We commend these early efforts to prevent a terrorist incident and under a federal program these facilities should get credit for their prior work. However, allowing facilities to follow their own standards has not been deemed acceptable for airports or nuclear plants, and should not be acceptable for chemical plants.

The best-known voluntary program established so far, the American Chemistry Council's (ACC) security program called "Responsible Care," is wholly inadequate. The code includes vague guidelines, not prescriptive standards and focuses on physical site security (i.e., guards and gates). ACC has touted third-party verification requirements, but has not established suitable qualifications for who can certify a plant's security plan as adequate. Finally, neither ACC or anyone else collects information sufficient to verify compliance at member companies.

Sal DePasquale, formerly with Georgia Pacific, was involved in developing the ACC program. He recently testified to the House of Representatives Homeland Security Committee about the program:

> The result of guidelines and nice sounding best practices is to create a smoke and mirrors exercise that makes it appear that something serious is being accomplished, when it, indeed, is not."... "In response to September 11, the ACC required its members to conduct a vulnerability analysis. This is a noteworthy exercise, but it does not require the companies to actually do anything in response to the analysis nor does it establish any minimum standards for defense against the most obvious exposures. Indeed, it is another exercise in smoke and mirrors and makes it seem like something substantive is occurring, when it is not.[8]

Congress should allow companies to submit work performed to comply with other laws (for instance, a vulnerability assessment done under the Bioterrorism Act), as part of meeting their obligations under chemical security legislation. Companies would only need to supplement those submissions with any additional information required by the chemical security bill.

It is particularly important that work done as part of a voluntary industry program be strictly scrutinized. DHS should review this material on a facility-by-facility basis to ensure compliance with each element of the law.[9] It is one thing to recognize the security efforts performed under other federal statutes if those efforts meet the requirements of

[8] Sal DePasquale, testimony to the House of Representatives Homeland Security Committee, June 15, 2005.
[9] Qualified third party verification will be an important supplement to government oversight for facilities that do not pose a high risk. To ensure that the audits are objective, DHS must establish strict criteria for third party qualifications and independence. A bill must require such third parties to have expertise in alternative approaches, prevent conflicts-of-interest, and ensure timely DHS audits of some third party certifications.

this proposed law – it is completely unacceptable to rubber stamp voluntary measures that have not been evaluated or enforced by any federal agency.

In addition to the principles described above, an effective chemical security program also should include the following provisions:

- **Worker participation and training.** Workers and first responders are most immediately affected by a major chemical release and so have a direct, personal interest as well as the expertise to ensure that vulnerability assessments and security plans are adequate. Facilities should be required to consult with workers and first responders in the development of their plans. It should also provide for training on inherent safety for state and local officials as well as owners and operators and employees.

- **Additional population protection and emergency preparedness.** If a facility is unable to implement a safer process, technology or chemical to reduce the consequences of a successful terrorist attack, it should be required to meet a higher standard of protection, including the use of buffer zones around the perimeter of the facility that reduce the number of people who might be injured in the event of a chemical incident. In addition, facilities should be required to develop emergency plans and conduct evacuation drills of employees and simulate community evacuations coordinated by local first responders and volunteers.

- **Technical support and coordination with government experts on hazardous chemicals.** The Homeland Security Department's primary expertise is in assessing and addressing security, while EPA has primary expertise and years of regulatory experience with the various industries that use large volumes of hazardous materials. Both agencies should play critical roles in a chemical security program and should be directed to coordinate.

  In addition, EPA should be directed to establish a national clearinghouse whose function is to encourage the "use of inherently safer technology" through exchange of information.

- **Restrictions on siting of new facilities in populated areas.** In this day and age it seems foolhardy to allow new facilities that use large amounts of dangerous chemicals to be located in heavily populated areas. Congress should direct DHS to develop rules to avoid creating new catastrophic risks.

Efforts to protect Americans from terrorist attacks are often costly and complicated. Instances when protection of the public can be achieved in a cost-effective manner should be aggressively pursued. That some of these options have side benefits, such as eliminating the potential for chemical accidents makes them all the more appealing. Congress should insist that facilities take all reasonable steps to reduce risks of catastrophic chemical release.

Attachment 1: Who Has Warned About Terrorism at Chemical Plants?

Many experts have cautioned that terrorists can target industrial facilities that use extremely hazardous substances. Government agencies, research institutes, trade associations, labor unions, and public interest groups have warned of the dangers posed by hazardous chemicals in communities. These published warnings include reports by:

- Department of Homeland Security;[i]
- Department of Justice;[ii]
- Environmental Protection Agency;[iii]
- General Accounting Office;[iv]
- Congressional Research Service;[v]
- Congressional Budget Office;[vi]
- Agency for Toxic Substances and Disease Registry;[vii]
- Naval Research Laboratory;[viii]
- Army Surgeon General;[ix]
- American Chemistry Council;[x]
- PACE International Union;[xi]
- Brookings Institution;[xii]
- Rand Corporation;[xiii]
- Center for Strategic and International Studies;[xiv]
- Environmental Defense;[xv]
- Safe Hometowns Initiative;[xvi]
- U.S. Public Interest Research Group;[xvii]
- Working Group on Community Right-to-Know.[xviii]

Compiled by the Working Group on Community Right-to-Know, www.crtk.org, March 2005

---

[i] Press Release: Statement by the Department of Homeland Security on Continued Al-Quada Threats, Department of Homeland Security, November 21, 2003.

[ii] Assessment of the Increased Risk of Terrorist or Other Criminal Activity Associated With Posting Off-site Consequence Analysis Information on the Internet, U.S. Department of Justice, April 18, 2000; and, A Method to Assess the Vulnerability of U.S. Chemical Facilities, National Institute of Justice, U.S. Department of Justice, November 2002.

[iii] Strategic Plan for Homeland Security, U.S. Environmental Protection Agency, September 2002.

[iv] Homeland Security: Voluntary Initiatives Are Under Way at Chemical Facilities, but the Extent of Security Preparedness is Unknown, U.S. General Accounting Office, GAO-03-439, March 14, 2003.

[v] CRS Report to Congress: Chemical Plant Security, Congressional Research Service, January 2003.

[vi] Homeland Security and the Private Sector, Congressional Budget Office, December 2004.

[vii] Industrial Chemicals and Terrorism: Human Health Threat Analysis, Mitigation and Prevention, Agency for Toxic Substances and Disease Registry, 1999; and, Terrorist Use of Expedient Chemical Agents: Health Risk Assessment and Las Vegas Case Study, Agency for Toxic Substances and Disease Registry, undated.

[viii] Testimony of Dr. Jay Boris of the Naval Research Laboratory before the Committee on Public Works and the Environment of the Council of the District of Columbia, January 23, 2004.

[ix] Study Assesses Risk of Attack on Chemical Plant, *Washington Post*, March 12,2002.

[x] The Terrorist Threat in America, Chemical Manufacturers Association (American Chemistry Council), April 1998.

[xi] PACE International Union Survey: Workplace Incident Prevention and Response Since 9/11, Paper, Allied-Industrial, Chemical and Energy Workers International Union (PACE), October 27, 2004.

[xii] Protecting the American Homeland, Brookings Institution, March 2002.

[xiii] Toxic Warfare, RAND Corporation, 2002.

[xiv] News Release: Chemical Facilities Vulnerable, Center for Strategic and International Studies, December 23, 2003.

[xv] Eliminating Hometown Hazards: Cutting Chemical Risks at Wastewater Treatment Facilities, Environmental Defense, December 2003.

[xvi] The Safe Hometowns Guide, The Safe Hometowns Initiative, 2002.

[xvii] Needless Risk: Oil Refineries and Hazard Reduction, U.S. Public Interest Research Group, October 2003.

[xviii] Unnecessary Dangers: Emergency Chemical Release Hazards at Power Plants, Working Group on Community Right-to-Know, July 21, 2004.

Congressional Research Service

Memorandum

June 27, 2005

TO: Honorable Edward Markey
Attention: Michal Freedhoff

FROM: Dana A. Shea
Analyst in Science and Technology Policy
Resources, Science, and Industry Division

SUBJECT: RMP Facilities in the United States as of May 2005

This memorandum responds to your request regarding facilities submitting Risk Management Plans (RMPs) to the U.S. Environmental Protection Agency (EPA). You requested an analysis of RMP facilities within the United States by potentially affected population.

Under the Clean Air Act, Section 112(r), the EPA established a program requiring risk management plans to be provided to the EPA by facilities possessing greater than certain threshold quantities of 140 chemicals.[1] As part of this reporting requirement, facilities are required to determine the worst-case scenario release from a single chemical process, using EPA criteria and guidelines.[2] Facilities are also required to estimate the population potentially at risk from this worst-case scenario release by calculating the population that resides within a circle surrounding the facility, with the radius of the circle determined by the distance the worst-case scenario release might travel.[3]

Since the population potentially affected under an EPA worst-case scenario release is calculated in a circle around the facility, it is unlikely that this entire population would be affected by any single chemical release, even if it is a result of a worst-case accident. In the event of an actual catastrophic chemical release, meteorologic effects will determine the direction of the release, and therefore those potentially affected, and effects on the health of those individuals affected would vary, depending on many factors. In addition, worst-case

[1] The list of 140 chemicals, 77 toxic chemicals and 63 flammable chemicals, and their threshold quantities are found at 40 CFR 68.130.

[2] The criteria and guidelines for determining the worst-case scenario release are found at 40 CFR 68.25.

[3] This requirement is found at 40 CFR 68.30. The criteria for determining the distance a worst-case scenario release might travel are found at 40 CFR 68.22.

scenarios do not take into account emergency response measures that might be taken by operators of the facilities or others to mitigate harm.

Facilities may register and deregister from the RMP program as their chemical processes and the amounts of chemicals they store and use change. Facilities are required to review and update the RMP plan filed with the EPA at least once every five years.[4] Possible reasons that facilities might not review and update the filed RMP plan include: the facility is out of compliance; the facility is no longer in business; the facility has reduced the amount of reportable chemical to below threshold levels, but neglected to inform the EPA; or the facility fell under the Chemical Safety Information, Site Security and Fuels Regulatory Relief Act (CSISSFRA) and is no longer covered by the RMP requirement.

In 1999, Congress passed the Chemical Safety Information, Site Security and Fuels Regulatory Relief Act.[5] This act removes from coverage by the RMP program any flammable fuel when used as fuel or held for sale as fuel by a retail facility. In implementing this Act, the EPA allowed facilities that had previously filed under the RMP program the options of withdrawing from the program, which would delete the information from the EPA database, or taking no further action, which would leave the information in the EPA database as a voluntary submission.[6] As a result, some entries in the EPA database which have not been updated within the five year requirement are likely to be facilities falling under CSISSFRA that opted to take no action.

At your request, I searched the May 2005 update of the EPA RMP*National Database (with off-site consequence analysis (OCA) data) for facilities that have registered under the RMP program. Facilities that have deregistered from the RMP program were excluded. You requested that these facilities be classified by state according to the population potentially affected by a worst-case release, according to the EPA worst-case scenario criteria, using thresholds of 1,000 people, 10,000 people, 100,000 people, and 1,000,000 people. Additionally, you requested that facilities with out-dated RMP filings be identified and subtracted from each population category. Facilities required to update their RMP filing by April 1, 2005 that had not done so were considered out of date for the purposes of this analysis and were excluded. Therefore, each category is described by a range of values, with the lower value being current, compliant RMP facilities and the upper value being all registered RMP facilities.

Facilities may register and deregister from the RMP program as chemical processes and amounts of chemicals stored and used change. Therefore, the number of facilities listed above should be considered as illustrative of the current industry profile, rather than absolute.

If you have any further questions regarding this topic or questions regarding the information in this memorandum, please contact me at 7-6844.

---

[4] This requirement is found at 40 CFR 68.36. Facilities not excluded by CSISSFRA that do not review and update the RMP plan are not in compliance with the RMP regulation. They may be subject to enforcement actions by EPA under the Clean Air Act, Section 113.

[5] P.L. 106-40.

[6] See 65 *Fed. Reg.* March 13, 2000, p. 13,247.

Table 1. Compliant and Total RMP Facilities in Each State, by Potential Affected Population (Parameters Designated by Requester)

	Compliant and Total Number of Facilities with a Worst-Case Release Potentially Affecting a Population of:				
State	0 - 999	1,000 - 9,999	10,000 - 99,999	100,000 - 999,999	1,000,000+
AK	14 - 18	10 - 11	0	0	0
AL	78 - 103	65 - 86	35 - 42	12 - 13	0
AR	49 - 59	66 - 80	44 - 51	3	0
AS	0	0	0 - 1	0	0
AZ	26 - 42	40 - 46	28 - 37	4 - 5	2
CA	274 - 339	230 - 298	258 - 294	52 - 58	11 - 13
CO	119 - 128	63 - 67	24	1	1
CT	8 - 11	19 - 24	7 - 12	1	0
DC	0	1	1	0	0
DE	11	15	4	3	2
FL	81 - 90	156 - 176	112 - 125	21 - 22	7
GA	119 - 132	134 - 143	48 - 48	7	1
GU	2 - 4	0	0	0	0
HI	5 - 6	8 - 9	2	0	0
IA	476 - 527	380 - 395	55 - 60	3	0
ID	24 - 29	23 - 25	14 - 16	0	0
IL	530 - 630	290 - 317	60 - 70	20 - 25	12 - 13
IN	213 - 265	140 - 160	50 - 62	13 - 14	3 - 4
KS	493 - 540	199 - 217	31 - 35	4 - 5	0
KY	78 - 86	74 - 81	32 - 36	16	0
LA	121 - 138	88 - 106	50 - 57	47 - 50	2
MA	22 - 27	24 - 34	22 - 27	1	1
MD	37 - 38	21 - 26	42 - 73	7	3
ME	10 - 13	12 - 14	4 - 5	1 - 2	0
MI	79 - 92	78 - 91	38 - 47	11 - 12	5
MN	193 - 281	154 - 196	45 - 54	8	3
MO	164 - 214	126 - 151	37 - 40	6 - 8	0

MS	49 - 54	60 - 69	42 - 45	2	0
MT	45 - 56	20 - 22	7	3	0
NC	106 - 138	90 - 108	42 - 46	7 - 8	1
ND	232 - 266	71 - 78	11	0	0
NE	303 - 339	192 - 207	35 - 36	2 - 3	0
NH	5 - 7	5 - 8	1	1	1
NJ	44 - 46	20	19 - 20	6 - 7	7
NM	40 - 46	12	6 - 7	2	0
NV	23 - 29	6 - 7	4 - 5	3 - 4	1
NY	53 - 60	66 - 70	32 - 35	15 - 16	3
OH	158 - 167	151 - 169	88 - 95	16 - 17	8
OK	158 - 214	79 - 103	23 - 25	7	0
OR	50 - 55	39 - 40	25	4	0
PA	101 - 111	144 - 159	80 - 82	16 - 18	2
PR	9 - 16	38 - 58	38 - 53	1	0
RI	1 - 5	4 - 6	6 - 7	4	0
SC	66 - 73	107 - 109	20 - 21	9	0
SD	44 - 46	29 - 32	5	0	0
TN	62 - 69	92 - 101	31 - 34	19 - 20	0
TX	466 - 598	321 - 423	260 - 311	59 - 67	28 - 29
UT	41 - 43	18 - 20	11	5	1
VA	56 - 64	67 - 70	21 - 21	9	0
VI	0	0	1	0	0
VT	2 - 4	4 - 6	0	0	0
WA	125 - 135	79 - 82	30 - 33	8	1
WI	89 - 124	94 - 116	50 - 54	6	0
WV	24 - 27	27	18 - 20	8	0
WY	53 - 57	9	3	0	0

Source: CRS analysis of the EPA RMP*National Database (with off-site consequence analysis (OCA) data), updated May 2005.

Note: Facilities required to update their RMP filing by April 1, 2005 that had not done so were considered out of compliance and excluded when considering the compliant facility universe. In cases where facilities report multiple worst-case scenario releases, the worst-case scenario potentially affecting the most people has been considered. When all facilities in a given category are compliant, only a single value is reported.

Attachment 3: Reporters Find Gaps In Voluntary Industry Programs

As Congress debates protecting America's chemical plants against terrorists, investigative news reporters are finding open gates, holes in fences, no guards, and other lax security at facilities that store extremely hazardous chemicals. In response, the chemical industry's lobbying arm, the American Chemistry Council (ACC), suggests that these are *not* facilities covered by the industry's voluntary "Responsible Care" safety and security code. However, reporters have entered or found lax security at more than 20 ACC member or partner company facilities, listed below. Plainly, even major chemical facilities are vulnerable. This shows why eliminating unnecessary chemical dangers is the most certain way to deter terrorists and protect public safety.

Facilities of current ACC members (June 2005):

- NALCO (Chicago)[1]
- Rhodia (Chicago)[1]
- Flexsys America (Akzo Nobel subsidiary; Monongahela, Pennsylvania)[1]
- Ashland Specialty Chemical (Pittsburgh, Pennsylvania)[1]
- Calgon Carbon (Pittsburgh, Pennsylvania)[1]
- Sunoco (Neville Island, Pennsylvania)[1]
- PVS Technologies (Houston)[1]
- W.R. Grace (Chicago)[1]
- Dow Chemical (Houston)[1,2]
- Unspecified facility near Los Angeles, Calif.[3]

Facilities of former ACC member companies (members at time of investigation):

- Neville Chemical (Pennsylvania)[1]
- Three LaRoche Industries facilities (Chicago, Baltimore and Pennsylvania)[1]
- Millennium Chemicals (Baltimore)[1]
- BP Chemical[4]
- BP (Chicago)[1]
- Noveon (Louisville)[5,6]

Facilities of "Responsible Care" partner companies:

- Conrail
- CSX Transportation
- Union Pacific Railroad

---

1 Investigated by Pittsburgh-Tribune reporter Carl Prine.
2 Incidental entry only – not near tanks.
3 Investigated by CBS 60 Minutes.
4 Investigated by CBS News.
5 Reported by Louisville Courier-Journal.
6 Participation in Responsible Care is a condition of membership in the Synthetic Organic Chemical Manufacturers Association (SOCMA).

Compiled by: Working Group on Community Right-to-Know, a project of OMB Watch, June 2005.

PITTSBURGH
Tribune-Review

Chemical sites still vulnerable

By Carl Prine
TRIBUNE-REVIEW
Sunday, November 16, 2003

Two years after 9/11, terrorists still have unfettered access to potentially catastrophic amounts of toxins and explosives nationwide, the Pittsburgh Tribune-Review and "60 Minutes" have found.

The news organizations' odyssey through facilities making, storing or shipping deadly chemicals follows Trib investigations last year that uncovered shoddy security at more than 60 plants in the Pittsburgh area and in Baltimore, Chicago and Houston.

Beginning in August, the Trib and the CBS newsmagazine jointly scouted security at 15 facilities around Pittsburgh and Baltimore. CBS continued on to California, Illinois, New Jersey and Texas.

The Trib and "60 Minutes" have combined to inspect more than 50 plants over the last four months, finding:

- Lax security. A Trib reporter, "60 Minutes" correspondent Steve Kroft and a CBS cameraman strolled to the tanks of lethal boron trifluoride at Neville Chemical Co. on Neville Island. Crossing through open or unlocked gates, they spent more than 30 minutes at the unguarded works during two undetected visits. Plant officials called the police only after the journalists confronted Neville's security chief with their findings. Neville Township police then cited the men for defiant trespass. According to Neville's filings with the Environmental Protection Agency, a catastrophic release of the corrosive vapors would threaten the lives of nearly 38,000 within three miles.
- Open rail lines. The easiest entrance to Neville Chemical and five other plants was through unguarded rail corridors. Because of just-in-time delivery and a lack of space at older yards, companies such as the James Austin Co. in Butler County and Univar in Forward store their chlorine gas on the tracks. Industrial chlorine is corrosive enough to eat through human teeth. A lone tanker at Univar's warehouse endangers 1.2 million people, according to the EPA.
- Unlocked gates and broken fences. At the Wilkinsburg Penn Joint Water Treatment Facility in Verona, a broken fence and an unlocked door allowed a Trib reporter to reach 20 tons of chlorine gas and millions of gallons of drinking water. If the chlorine tank ruptured, the gas could lap neighborhoods up to three miles away, threatening more than 100,000 people. Nearly every Pennsylvania facility examined suffered from dilapidated wire or open gates.

□ See-nothing guards and workers. Inattentive guards allowed easy access to five facilities, including Giant Eagle's Chartiers warehouse in the West End. A reporter popped through a fence hole to get to the grocer's warehouse and its 20,000 pounds of anhydrous ammonia, a coolant for refrigeration. In a break room at the warehouse, workers sipping sodas chatted with him about the Steelers. Giant Eagle's ammonia tank puts nearly 43,000 people -- including children in 24 schools -- at risk of death, burns or blindness, according to company filings with local emergency planners.

Federal officials were most concerned about the easy penetration of security at the nation's potentially deadliest plants. At the mammoth Sony Technology Center in Westmoreland County, an unsecured gate, distracted guards and unconcerned employees let a reporter reach 200,000 pounds of chlorine gas. No one stopped him as he touched train derailing levers, waved to security cameras, and photographed chlorine tankers and a nitric acid vat. If ruptured, one Sony railcar could spew gas 13 miles, endangering 190,000 people. Two other plants penetrated by the Trib and "60 Minutes" -- Univar and Millennium Chemical in Baltimore -- each put more than 1 million neighbors at risk of chlorine poisoning.

In February, Homeland Security Secretary Tom Ridge issued a bulletin warning that "al-Qaida operatives may attempt to launch conventional attacks against U.S. nuclear/chemical-industrial infrastructure to cause contamination, disruption and terror." When told how the Trib and "60 Minutes" easily punctured plant security in several states, he was concerned but expressed optimism that long-term federal reforms will protect Americans from toxic catastrophes.

"I think what we need to understand is that this enormously complex and diverse economy, worth trillions of dollars, has many potential targets," Ridge said. "And we have to begin to understand that we can't eliminate the risk. We have to manage the risk. And the way we manage the risk is by starting to take a look at those that are most vulnerable, whose use or destruction could result in a catastrophic loss of life or economic damage."

Ridge said federal teams recently began scrutinizing security deficiencies at "nearly two dozen" facilities the agency considers most tantalizing to terrorists. On Friday, Homeland Security announced that National Guard troops had visited about 150 sensitive sites, of which "more than half" were chemical facilities. Details of the visits were not disclosed.

But the plants' neighbors want tighter security and more openness about potential dangers sooner, not later.

"They've never told us anything about the chlorine there. I've never even heard they had all of that there," said Nancie Bluebaugh, of East Huntingdon, who lives a few blocks from Sony. "I have a child here. We see the trains coming and going, but we had no clue what was in them.

"I'll do a lot of praying now."

Yvette Leto, who lives a few blocks downwind from Neville Chemical's boron trifluoride, believes federal agencies should outlaw catastrophic chemical storage near cities. According to Neville's filings with emergency planners, the plant also could unleash deadly hydrogen fluoride, anhydrous ammonia, benzene, styrene, phosphoric acid and 10 other toxins that burn flesh, blind eyes, flood lungs with blood or cause cancer.

"The big shots who run the corporations aren't worried about us," Leto said. "They're fine because they don't live here. Are they willing to come down and live next to these plants, like we do? I bet they wouldn't do it. But they'll put the chemicals here."

Neville Chemical officials would not comment.

Frank Leto, Yvette's father-in-law and next-door neighbor, believes federal regulations should balance the risk of disaster with the need for well-paying manufacturing jobs. A retired Aristech and Pittsburgh Coke and Chemical employee, he said the chemical industry keeps the Neville Island economy afloat.

"I worked there for 50 years, so I know how dangerous chemicals can be," he said. "But you can't have it both ways. People complain about the dangers and the smells and all that, but they'd complain even more if the companies packed up and left town."

Reforms fail

When told of the latest incursions by the Trib and "60 Minutes," most plant officials immediately pledged for the second year in a row to investigate security snafus.

AK Steel authorities said they always work to improve security, citing a $25 million upgrade that recently reduced use of nitric acid at two Butler plants. AK also installed dikes to significantly reduce hydrofluoric acid dangers.

Giant Eagle immediately repaired a broken fence and assured the Trib no one else would reach its chemicals.

After a reporter spent more than 20 minutes probing sensitive purification rooms, Oakmont Water Authority officials vowed to add gates to block access to their Hulton Water Treatment Plant. In Beaver, a township supervisor and the Chippewa Township Sanitary Authority have discussed placing the water-treatment plant under tighter vigilance.

Several Pennsylvania facilities failed the Trib's latest test even after major security upgrades since 9/11.

After last year's incursion by a Trib reporter, the Wilkinsburg Penn authority spent more than $100,000 scripting a security plan and adding electric gates, camera detectors and worker identification badges at both the Tyler Road treatment plant and a pump station along the Allegheny River.

But a Trib reporter twice scooted through a fence hole and an unlocked door that led to 20 tons of chlorine gas. The treatment plant and its Nadine Road pump station suffer from century-old layouts that are perfect for saving money on utility operations but difficult to secure from intruders.

"I've been here 23 years," said Wilkinsburg Penn director Mark Lerch. "Back then, security was never an issue. The water treatment was out of sight, out of mind. But 9/11 showed our vulnerability to terrorists. They can hit our natural gas or our electricity and we will survive. But you can't go without water."

Lerch lectured workers on lax security and tightened plant perimeters.

After a Trib reporter penetrated Univar's security last year, the company erected high fences at its Bunola yard, instituted round-the-clock guards, installed cameras and even fortified its river dock, making the works impregnable from nearly every direction -- except the railroad.

Managers asked the rail line to let them fence off track where a chlorine tanker parks daily. But federal safety laws wouldn't allow it. So the Trib and "60 Minutes" were able to make four undetected trips up the rails to 90 tons of chlorine gas.

"We really have done everything we can to make our facility secure," said Univar manager Cliff Moll. "I really think we went the extra mile and did everything anyone could do. But we can't do anything about the railroad."

The extra mile?

At other plants, workers and neighbors questioned whether management had done anything to stiffen security since the Trib's visits last year.

A reporter easily canvassed the sprawling Allegheny Ludlum mill in Brackenridge three days in a row, following a path down a bluff, across the railroad, behind a guard shack and up to 100,000 pounds of hydrogen fluoride, a lethal toxin used to "pickle" stainless steel.

Longtime Brackenridge employees blamed lax security on recent guard cutbacks and indifference. If released, the mill's acid could waft nearly a mile and threaten more than 16,000 residents with blindness, severe burns and death. A spill also would jeopardize water supplies drawn from the Allegheny and Ohio rivers.

Allegheny Ludlum officials declined to comment.

"I know they put in surveillance cameras, but we don't know if anyone is really watching," said Gerard Magoc, a Brackenridge steelworker for 31 years. "They put on a big show about searching cars, though. They're big on theft. ... They care more about protecting their toilet paper than they do about their hazardous materials."

James Austin Co. managers also didn't discuss breaches.

On Oct. 28, EPA officials asked Austin to resubmit disaster plans, citing inaccurate estimates of the population endangered by its railcars. The bleach manufacturer claimed a chlorine plume could reach 12 miles, affecting only 5,500 people in the North Hills and Butler County. Trib research of U.S. Census figures, however, shows that the gas endangers 260,000 neighbors, making Austin one of the 700 potentially deadliest plants nationwide.

Sony officials said they would have "locked down" the East Huntingdon yard had the FBI warned them a terrorist or reporter was coming. Because Sony is in a rural area, corporate authorities believe it isn't a likely sabotage target.

"We respond to a threat if it's reported to us," said Sony security director Tim Pratt. "We can close the place down if something happens."

Since the Trib's surprise visit last year, Sony officials have added a concrete bulwark, metal fences and a video camera to aid security at their chlorine railcars, where a rupture would endanger 190,000 people.

Counterterrorism experts say that's not good enough. They increasingly advocate the use of barbed wire, heavily armed guards or technologies that reduce or eliminate the threat of toxic releases -- security standards common to the nuclear industry because of federal regulations.

The New York Times

Row of Loosely Guarded Targets Lies Just Outside New York City

May 9, 2005 | DAVID KOCIENIEWSKI

KEARNY, N.J., May 7 - It is the deadliest target in a swath of industrial northern New Jersey that terrorism experts call the most dangerous two miles in America: a chemical plant that processes chlorine gas, so close to Manhattan that the Empire State Building seems to rise up behind its storage tanks.

According to federal Environmental Protection Agency records, the plant poses a potentially lethal threat to 12 million people who live within a 14-mile radius.

Yet on a recent Friday afternoon, it remained loosely guarded and accessible. Dozens of trucks and cars drove by within 100 feet of the tanks. A reporter and photographer drove back and forth for five minutes, snapping photos with a camera the size of a large sidearm, then left without being approached.

That chemical plant is just one of dozens of vulnerable sites between Newark Liberty International Airport and Port Elizabeth, which extends two miles to the east. A Congressional study in 2000 by a former Coast Guard commander deemed it the nation's most enticing environment for terrorists, providing a convenient way to cripple the economy by disrupting major portions of the country's rail lines, oil storage tanks and refineries, pipelines, air traffic, communications networks and highway system.

Since 9/11, those concerns have only been magnified. Law enforcement officials have warned of the need to prepare for an assault on one of the four major chemical plants in the area or an attempt to ship nuclear or biological weapons through its two port complexes.

Trying to safeguard more than 100 potential terrorist targets in two miles surrounded by residential communities, industrial areas and commuter corridors has proved a daunting challenge. Federal, state and local officials have spent hundreds of millions of dollars to install gates, roadblocks and security cameras and to provide additional patrols, surveillance and intelligence operations.

But even those in charge of the effort say the job is incomplete, bogged down by obstacles that are a microcosm of the nation's struggle against potential terrorist threats.

After distributing tens of billions to state and local governments since 9/11, the federal Department of Homeland Security cut New Jersey's financing this year to about $60 million from $99 million last year. Many security experts have complained that the formula - which provides Montana with three times as much money per capita as New Jersey - is guided more by politics than by the likelihood of an attack.

Meanwhile, security at Newark Airport, while more rigorous and time-consuming for passengers, has been marred by embarrassing breakdowns, as screeners have repeatedly failed to prevent federal officials from sneaking weapons and fake bombs onto planes.

The time and expense of screening shipping containers has slowed attempts to tighten security at Port Newark and Port Elizabeth, where customs officials say their radiation screening devices are ineffective and need replacement.

The private companies that own 80 percent of the most dangerous targets have given varying degrees of cooperation, officials said, and the chemical industry has effectively blocked attempts in Washington to mandate stricter regulations.

As a result, many of the most crucial security tasks are left to local police departments, some of which say they are too understaffed and poorly equipped to mount a proper counterterrorism effort.

"They tell us to patrol, do this, do that, but don't give us the money or equipment," said Sgt. Michael Cinardo of the Kearny Police Department, one of several law enforcement agencies responsible for patrolling around the chlorine plant.

He said the department requires patrol officers to stop by the plant at least five times each shift.

Security against terrorism is a particularly sensitive issue in New Jersey. More than 700 people killed on 9/11 lived there. And, in October 2001, the first major bioterrorism attack on United States soil was launched from a New Jersey post office when a series of anthrax-laced letters were mailed to members of Congress and the news media. The State Health Department's muddled response came to symbolize the nation's need to prepare itself to face new threats.

Since then, New Jersey officials have spent more than $350 million in state tax money on counterterrorism, building an apparatus that is run by seasoned law enforcement experts and is generally well regarded.

New Jersey's Homeland Security Department, established in 2002, has helped to train, coordinate and increase staffing at local law enforcement and emergency medical agencies; assembled a 1,000-person task force to focus on urban areas; and purchased boats, decontamination suits, radio systems and a computerized intelligence network so federal agents and the New Jersey State Police can share information with all 566 municipalities.

In the most dangerous two miles, they have erected concrete barriers outside hospitals and office buildings and put fences along elevated highways that pass chemical plants. The State Police patrol the skies, highways and coastal waters, and federal officials have used various surveillance techniques. On the New Jersey Turnpike, troopers try to check any vehicle that stops for as little as five minutes.

But given the sheer number of vulnerable sites - three major oil and natural gas pipelines, heavily traveled rail lines and more than a dozen chemical plants - many security experts acknowledge that the response is inadequate.

In the months after 9/11, government officials routinely refused to discuss the most mundane aspects of security, saying that they did not want to offer inside information to potential enemies. Now, said Sidney J. Caspersen, the director of the state's Office of Counterterrorism, there is more risk in remaining silent.

"The terrorists already know what's out here," Mr. Caspersen said. "They have been found with blueprints of our buildings, and a lot of the information is available over the Internet or at a public library. The only question is whether we will find a way to protect these targets before they find a way to attack them."

The answer to that question will depend largely on the ability to operate with limited money and a tangle of bureaucracies.

In several instances, counterterrorism money sent to the state has been used for questionable purposes: the city of Newark spent $300,000 on two air-conditioned garbage trucks, and New Jersey Transit has proposed using $36 million in security money to overhaul the Hoboken Ferry terminal. Even groups like Taxpayers for Common Sense say that places like New Jersey, Houston and Long Beach, Calif., deserve more federal dollars.

As for the ports, the federal Homeland Security Department's inspector general's office recently criticized the agency for directing much of its $517 million in port security money to relatively low-risk sites in places like Kentucky and Tennessee, and not giving enough to busy, vulnerable facilities like Port Newark. Although the Port of New York and New Jersey recently received an additional $42 million for counterterrorism efforts, Port Newark lacks the up-to-date equipment now used to search cargo at ports like Hong Kong.

"We put more resources into securing the average large bank in Manhattan than we do for the entire security of Port Newark," said Stephen Flynn, a former Coast Guard commander who is now a security analyst for the Council on Foreign Relations and who conducted the study that first identified this part of North Jersey as the nation's most terror-prone two miles. "That's just irresponsible."

Some New Jersey officials have hoped that the newly appointed secretary of homeland security, Michael Chertoff, will be sympathetic to the state's situation because he is a native of Elizabeth. But when he visited New Jersey during a terror drill last month, Mr. Chertoff was noncommital about restoring cuts.

"Frankly, it's not a matter of spending a great lot of money," he said. "It's a matter of taking resources we have and having a plan in place so we use them effectively."

New Jersey officials say that the cuts will force them to reduce surveillance of possible targets, cancel training sessions for first responders and counterterrorism experts, and forestall the purchase of equipment to detect chemical, nuclear or biological agents. The state has said it will also have to scale back plans to fortify storage facilities and rail lines near the Pulaski Skyway, an area known as "chemical alley."

Even if New Jersey were to receive more money, however, its counterterrorism effort would still face other difficulties.

At Newark Airport, which handles 32 million passengers a year, the federal government and the Port Authority of New York and New Jersey have spent tens of millions of dollars on high-tech baggage screening equipment, more guards and other security improvements. But Transportation Security Administration employees failed to detect weapons or fake bombs in about a quarter of the 81 tests conducted between last June and September. In December, when a machine detected a simulated explosive, baggage screeners lost track of it and it was loaded onto a flight to Holland.

Meanwhile, even less has been done to secure the nation's greatest vulnerability to terror attacks, its 15,000 chemical plants, 123 of which pose a threat to at least 1 million people, according to the Environmental Protection Agency. A spokeswoman for the Chemistry Council, an industry group representing 150 of the nation's largest chemical plants, said its members had already invested $2 billion in improved security and were working with Congress to establish federal safety guidelines.

"We want to work with the Department of Homeland Security and Congress to make these plants safer in a way that works for everyone," Kate McGloon, the spokeswoman, said.

Michelle Petrovich, a Department of Homeland Security spokeswoman, said agency officials had visited more than half the nation's 300 most dangerous plants and urged the companies to enhance perimeter security and switch to less hazardous chemicals and processes. As a result, Ms. Petrovich said, she believes North Jersey is "one of the safer areas because it has received the most attention in terms of protective measures."

But Richard A. Falkenrath, a former deputy homeland security adviser to the White House, said that effort has done little to make the public safer. "Saying that you're doing something doesn't mean you're actually making a difference," said Mr. Falkenrath, who recently testified before Congress, urging tighter regulation of the chemical industry.

Since 2001, at least two major efforts to bolster chemical plant security have been stalled, in part by industry lobbyists.

The latest proposal to tighten security at chemical plants, which appears to be gaining support in Congress, would establish safety guidelines. But Senator Jon S. Corzine said that it is only a half measure because it would not mandate that plants in densely populated areas stop using highly dangerous chemicals like chlorine gas and switch to more benign alternatives, like sodium hypochlorite. The plants use such chemicals to make antiseptics for water purification plants.

For those who live in the shadow of these plants, there is little expectation that the federal government will mount a more vigorous security response.

Carolyn M. Chapluske of Kearny, who has lived in North Jersey all her life, said, "People pay taxes and deserve to be protected. But they probably won't. It's just the way things work."

U. S. Department of
Homeland Security

United States
Coast Guard

Commandant
United States Coast Guard

2100 Second Street, S.W.
Washington, DC 20593-0001
Staff Symbol: G-ICA
Phone: (202) 366-4280
FAX: (202) 366-7124

DEPARTMENT OF HOMELAND SECURITY

U. S. COAST GUARD

STATEMENT OF

REAR ADMIRAL CRAIG E. BONE

ON

CHEMICAL FACILITY SECURITY

BEFORE THE

HOMELAND SECURITY AND GOVERNMENTAL AFFAIRS COMMITTEE

U. S. SENATE

JULY 27, 2005

Introduction

Good morning Madam Chairperson and distinguished members of the Committee. It is a pleasure to be here today to discuss the U.S. Coast Guard's role in securing the chemical facilities on the navigational waterways of the United States.

The men and women of the U.S. Coast Guard and the Department of Homeland Security remain committed to improving maritime homeland security each and every day through continued interagency cooperation and assistance from our partners at the local, state and international levels, as well as maritime industry stakeholders. This is true of the maritime facilities for which we have approved security plans, including chemical facilities.

Protecting the Marine Transportation System

Considering the vast economic utility of our ports, waterways and coastal approaches, it is clear that a terrorist incident against a facility in our marine transportation system could have a disastrous impact on public safety, the environment, our nation's economy, and international trade. Such an incident, if it were to occur in a strategic port, could also threaten our military mobilization capabilities. An incident at one of the 350 chemical facilities that operate next to our navigable waterways would seriously threaten and could have a profound impact on the health and well being of the local community if it led to a release of dangerous chemicals. Clearly, the security of the chemical sector is vitally important to the U.S. Coast Guard (USCG), the Department of Homeland Security and the environment and economic well being of the nation.

Of more than 3,000 port facility security plans (FSPs) that the Coast Guard has reviewed and approved under the Maritime Transportation Security Act (MTSA), we have approved 300 for chemical facilities.

Each security plan, including those for chemical facilities located in or near a port, explains how the facility will meet its security obligations under the regulations found in Title 33, Code of Federal Regulations, Part 105, including: access control, designation of restricted areas, handling of cargo, delivery of vessel stores and bunkers, monitoring of the facility and security incident procedures. There are additional requirements for certain dangerous cargo facilities as well. There are also requirements placed on the owner or operator of the facility, the facility security officer, and all facility personnel with security duties. These regulations detail requirements for everyone on the facility for specific security training, drills and exercises, records to be maintained, communications (particularly with vessels), and dedicated security systems. Each plan must also explain how facility personnel will respond to changes in Maritime Security threat levels as directed by the Commandant of the Coast Guard in consultation with the Secretary of Homeland Security.

The Coast Guard also approved an Alternative Security Program (ASP) for the American Chemical Council (ACC). An ASP is an option afforded to facility operators under MTSA. Instead of creating their own facility plan, operators of facilities required to meet Title 33, Code of Federal Regulations, Parts 101 through 106, may choose to utilize an existing plan which has previously been approved by the Coast Guard. One example of such a program is the Responsible Care Security Code developed by the American Chemical Council (ACC). The ACC represents about 150 companies that handle approximately 80-90% of the chemical production by capacity. ACC estimates that their member companies have spent over $3 billion nationwide toward security since September 11, 2001. Approximately 50 chemical facility operators have chosen to use an ASP rather than create their own, individual plans. Members of an organization with an approved ASP must implement that ASP in its entirety to be deemed in compliance. Also, the owners and operators of facilities under an ASP must conduct a specific assessment and verification of implementation for their facility.

The Coast Guard has completed compliance inspections of all facilities that currently have FSPs or ASPs to verify that they are operating within their respective plans. The compliance by the maritime industry has been commendable, but there have been some isolated instances where control actions against the facility were necessary. Since the July 1st, 2004, implementation date for MTSA, the Coast Guard has taken control actions (restrictions to, or suspension of, operations) against 45 facilities. Two of those facilities were from the chemical industry.

The Coast Guard's work in implementing MTSA for waterfront facilities has been a collaborative effort with other federal, state and local agencies, as well as private industry partners. We have worked in conjunction with the Information Analysis and Infrastructure Protection Directorate (IAIP) within the Department of Homeland Security (DHS) to ensure that all approved MTSA plans are consistent with IAIP's approach to non-waterfront chemical facilities.

This collaborative approach extends to the local port level. Each Area Maritime Security Committee (AMSCs), operating in support of the Coast Guard Captain of the Port (COTP), has identified their port's specific vulnerabilities, and created a plan to address those vulnerabilities. The AMSCs, which normally include representatives from the oil and chemical sector, developed their Area Maritime Security Plans (AMSPs) to address the risks specific to their ports. These area plans focus on critical port infrastructure which include those regulated under MTSA, as well as those facilities merely located in close proximity to navigable waterways, but do not engage in marine transfer operations. Such facilities would not be regulated under MTSA. These plans address how local, state and Federal resources will be deployed to prevent terrorist attacks and protect critical infrastructure in our ports, waterways and coastal areas

As part of the Coast Guard's port security assessments of the 55 militarily and economically strategic ports, certain facilities were identified as key assets by the AMSCs and we conducted limited scope assessments using plausible threat scenarios. This program is designed to assess the vulnerability of facilities to terrorist attack and identify countermeasures and mitigation strategies that can be communicated to the facility owner/operator and the local COTP. Over the last year, the Port Security Assessment Team has visited 10 chemical plants located on waterfront property and identified by the AMSCs. Generally, we have recommended better intrusion detection and prevention systems, waterfront surveillance, and maritime domain awareness for these facilities.

The Coast Guard has also contracted for a special assessment of inland barges which carry Certain Dangerous Cargo (CDC), their vulnerabilities and the blast consequence analysis of various types of Improvised Explosive Devices (IED) and stand off weapon attacks. The modeling for this study has been accomplished, and we are awaiting the final report from the contractor. These reports help us to understand vulnerabilities and consequences of intentional attacks and identify if additional protective security measures are necessary.

In addition to working to prevent a maritime transportation security incident, the Coast Guard has also focused on preparing to respond to an incident. We have been involved extensively in the development and implementation of the National Response Plan, an all-contingency response plan which includes hazardous materials releases as a scenario, and the National Incident Management System (NIMS). Through the Coast Guard's vice-chairmanship of the 17-agency National Response Team, and the multi-agency support through the Oil Pollution Act of 1990, and the Comprehensive Environmental Response, Compensation, and Liability Act (CERCLA), we work cooperatively with Federal, state and local agencies as well as industry partners to mitigate the effects of all types of chemical incidents.

Exercises and drills to test plans are a vital component of any security program and are required to be conducted by regulated MTSA facilities and vessels, testing prevention and response preparedness/capabilities. Each plan holder conducts a drill every three months and is required to conduct an exercise every year with no more than eighteen months between exercises. Additionally, national level exercises have recently had a component that focused on chemical releases. Exercise DETERMINED PROMISE 04, sponsored by U.S. Northern Command, simulated a Sarin gas release, and a chlorine gas release from a tank truck. The Department of Homeland Security sponsored the recent TOPOFF 3 exercise which involved a simulated mustard gas release near another port area. In both exercises, the Coast Guard's Atlantic Area Commander was the primary USCG planner, with planning and exercise participation from personnel at USCG Headquarters; the First Coast Guard District in Boston; the Fifth Coast Guard District in Portsmouth, VA; Marine Safety Offices in Long Island Sound and Hampton Roads; and Station New London, CT. In 2007, the Spill of National Significance exercise will include a number of hazardous material facilities as it will involve a large scale earthquake in the Mississippi River Valley.

The Coast Guard has been an active contributing agency in the Government Coordinating Council (GCC) sponsored by IAIP, which was formed under the National Infrastructure Protection Plan as a single point of contact to facilitate organization and coordination of sector policy and planning. Of the thirteen critical sectors of GCC, the Chemical Sector GCC is to provide effective coordination of chemical security strategies and activities, policy and communication across government, and between the government and the private sector to support the nation's homeland security mission. It acts as the counterpart and partner to the private industry-led Chemical Sector GCC to plan, implement and execute sufficient and necessary sector-wide security programs for the nation's Chemical Critical Infrastructure. This includes sector-wide planning, development of sector best practices and cross sector coordination.

The USCG is working with DHS on the Interagency Security Plan initiatives. The Coast Guard has developed a security matrix under Operation NEPTUNE SHIELD, which is our internal plan to identify, prevent and protect facility and vessel operations with the potential for material consequences from a terrorist attack. This matrix establishes a protocol of risk awareness, surveillance, vessel tracking, air patrols, cutter presence, security zones, security boardings of vessels and positive control measures to mitigate the vulnerabilities inherent in the ports, waterways and maritime domain.

Closing Gaps

We are working with DHS and appropriate local jurisdictions and companies to develop new tools for assessing risk for the chemical sector, such as the Risk Analysis and Management for Critical Asset Protection (RAMCAP) tool. RAMCAP provides a common framework for homeland security risk analysis decision-making through common terminology, common metrics for comparing risks across industry sectors, common basis for reporting results and a common basis for informing resource allocation decisions for countermeasures and consequence mitigation actions.

The Coast Guard, in concert with other Federal agencies, state and local authorities, and partners in industry, will continue to refine tools and analysis to aid senior leaders and first responders alike in their ability to protect, prevent and rapidly respond to a maritime transportation security incident, minimize damage in such an incident, and aid in recovery operations.

The Coast Guard will continue to perform FSP compliance examinations and spot checks on waterfront facilities regulated under Title 33, Code of Federal Regulations, Part 105, including facilities identified as chemical production and storage operations. Those facilities will continue to be held to a standard commensurate with the vulnerabilities at the facility, the threat to the facility and the consequences of a successful attack. Execution of FSPs, in concert with the other efforts of the Coast Guard and the Department of Homeland Security, are designed to mitigate the exposure to attack or illicit use by terrorist interests through effective and realistic countermeasures tailored to meet the specific vulnerabilities of the regulated operations.

Conclusion

Since 9/11, the USCG has worked closely with Federal, state and local agencies, and members of the chemical industry, to enhance the security of the chemical sector in the maritime region and the marine transportation system. We have established a robust strategy to enhance public safety from potential threats to these chemical facilities located in the maritime region. We have conducted vulnerability assessments, implemented comprehensive security plans and exercised these plans against realistic scenarios. We have researched, developed and tested tools to address high risk cargos and operations. We have ensured compliance and continue to work with partners to close gaps. Through plans, compliance inspections, sector coordinating councils and area committees, we will help bolster the chemical sector's security posture. As future needs change, we will continue to use risk-based strategies to close remaining high risk security gaps and protect the public from harm.

Thank you for the opportunity to testify today. I will be happy to answer any questions you may have.

Testimony of

Beth Turner

Director – Global Operations Security
E.I. duPont de Nemours and Company, Inc.
1007 Market Street
Wilmington, DE 19898
302-774-1000

Before the
Senate Homeland Security
and Governmental Affairs Committee

"Chemical Facility Security: What is the Appropriate Federal Role? (Part II)"

July 27, 2005

Madam Chair, Senator Lieberman, and distinguished Members of the Committee, my name is Beth Turner. I am the Director of Global Operations Security for DuPont. In this role, I am responsible for the security of our operating assets around the world. I also chair the Chemical Sector Coordinating Council, a mechanism that enables our critical sector of the nation's infrastructure to interact with the Department of Homeland Security (DHS). I co-led the American Chemistry Council (ACC) team that developed the original Responsible Care® Security Code in 2002 and the team that reassessed the Code in 2004. I have been a member of the ACC Security Team since its formation. Thank you for this opportunity to speak today. My testimony will first address the actions DuPont has taken to protect our employees, the communities in which we operate, and our facilities; second, our work on industry programs; and third, our views regarding critical legislation in this area.

DuPont is a global corporation founded 203 years ago. The founders of our company established an uncompromising commitment to safety that continues today. DuPont began as a manufacturer of black powder for the U.S. government in 1802, operations that, by their very nature, required a focus on safety and technology. The company's founder, E. I. duPont, built safety into the very fabric of DuPont culture by requiring his managers to live on the first manufacturing sites. That culture and clear personal accountability remain just as strong today. These form the foundation for every system and process in DuPont. In fact, it has been the underpinning for many DuPont products through the years. Our discovery of nylon, for example, made safer parachutes for D-Day, and our development of Neoprene®, a synthetic rubber, made military transportation easier and safer. In today's war on terrorism, products such as Kevlar® high-performance fiber, for example, are used in applications such as body and vehicle armoring; and our Sentry-glass® technology helps to protect the occupants of the Pentagon, embassies and civilians around the world.

Today, our core businesses often involve producing valuable products from a wide range of chemical feedstocks in high temperature, high-pressure reactions. This requires an intense focus on safety and security. DuPont is recognized as one of the safest companies in the world. In fact, the DuPont workplace safety record in the 1920s was actually better than the U.S. industry average in the year 2004. Our focus on safety and security is driven by what we at DuPont know as our core value commitment to our employees and our communities.

DuPont Voluntary Implementation of Security Measures

While security has long been part of our site operations, the world-changing events of September 11, 2001, compelled us to view security in a different light. Our reaction was both tactical and strategic. We immediately hardened security at our facilities and then began focusing on a longer-term strategic program that we are implementing in phases and continually strengthening. We have also truly integrated security into our DuPont culture so that we safeguard our employees and our neighbors, as well as protect the company's assets. I'd like to review some of the steps we took to make this happen. I will discuss physical security actions, as well as process safety measures, administrative procedures, and emergency response preparedness, all of which have a vital role in an integrated security plan.

Quickly after the 2001 attacks, senior corporate leadership made security an even higher priority by integrating security into the company's longstanding safety core value. This move sent a very clear and powerful message throughout the corporation. It immediately resulted in organizational changes and revisions to corporate policy, as well as extensive communications with employees throughout the company. Since the fall of 2001, security has continually been reinforced as an integral part of our long-standing safety culture. This corporate leadership decision to manage security as we have managed safety for over 200 years allowed DuPont to aggressively step up to the new realities we face as a nation and a world.

In early 2002, DuPont launched a global security survey to better understand specific security measures in place at over 500 locations around the world. We assessed operating facilities and prioritized them using a risk-based approach. In conjunction with the criteria established by the American Chemistry Council, we placed approximately 1/3 of our U.S. sites in the highest priority grouping named DuPont Category 1 sites. Of these, about 2/3 are covered by the U.S. EPA Risk Management Program. The remaining DuPont sites had no potential for off-site release or theft of materials and were placed in Category 2.

A site security leader at each location was designated as the focal point for security. Networks of site security leaders were formed around the world and function today to exchange best practices and to enhance security skills. I want to give special recognition to these site security leaders who have worked tirelessly since the 2001 attacks. It is their outstanding work that I review with you today.

To develop our vulnerability assessment (VA) methodology for high priority sites, we collaborated with the Sandia National Laboratories. DuPont trained U.S. Category 1 site security and process safety leaders in July of 2002, and those individuals led the site security assessment teams, with oversight and support from corporate headquarters.

The assessments identified potential vulnerabilities and appropriate upgrades to procedures, equipment, staffing, and manufacturing processes. Local law enforcement, emergency planning and response organizations, and others conducted third-party verifications of the upgrades at U.S. sites. DuPont accelerated the timing for the overall vulnerability assessment process and completed upgrades and verification at all DuPont U.S. Category 1 facilities between 9 and 12 months sooner than the American Chemistry Council deadline.

Vulnerability assessments were also conducted at DuPont U.S. Category 2 facilities. Security enhancements were identified and implemented at these sites as well.

Category 1 sites in Asia, Europe, and Latin America were identified using this same risk-based approach and assessment methodology. Security upgrades at these facilities are currently being implemented.

While I cannot speak publicly about specific actions at specific sites, I can describe in general terms the types of upgrades that have been implemented at our U.S. Category 1 sites. Upgrades have been, and continue to be, made to equipment, staffing, processes, procedures, and preparedness to secure the sites from a range of reasonably predictable and defensible threats.

First, I will review security equipment upgrades. These varied from site to site but included fencing, motorized gates, turnstiles, signage, access control systems, video surveillance, lighting, electronic intrusion detection and alarm monitoring, crash gates, and barricades. We have, subsequently, implemented a special maintenance program to ensure the new equipment remains functional and reliable.

In addition to new capital equipment, security practices were enhanced as appropriate at every site. Perimeter patrols were increased, and strict site access control measures were implemented. These include significant reductions in on-site vehicular traffic and increased inspections of rail cars, trucks, and other vehicles, as well as stricter policies for checking the identification of those seeking access to the sites. Suspicious activities are quickly identified with the new equipment and immediately reported to law enforcement for immediate investigation.

Security officer staffing has been significantly increased, and we have added officers with prior experience in security, the military, or law enforcement. Security officers received additional training and are continually retrained. Security supervisors receive an incentive to take additional training and become professionally certified as Certified Protection Officers.

Second, strong process safety management is one of the most important means of protecting our employees and our communities. A typical large chemical facility includes miles of piping and thousands of pressure vessels, tanks, pumps, valves, instruments, and other components. Each process unit is uniquely sized and designed to handle a range of chemicals and operate under a variety of temperatures and pressures.

The long-standing DuPont process safety management system is designed and applied to address these complexities. It includes engineered safety systems such as interlocks, excess flow and automatic shutoff devices, relief valves, spill containment structures, and emergency shutdown devices. Our strong process safety program also includes extensive operating procedures, testing and inspection of equipment, and thorough safety reviews of any process and equipment changes. Process safety also includes formal, detailed, and regular analyses of operations. These analyses are one of the most critical components of process safety. Such analyses are complex and must be conducted by engineers and other experts skilled in hazard evaluation methodologies and chemical processes. Process safety analyses for both new and existing facilities consistently include evaluation of opportunities to use inherently safer approaches and lower risk chemicals. I will expand upon the topic of inherent safety in a moment.

Another key component of process safety is the DuPont requirement to conduct detailed audits on all individual process units. The audits are conducted by highly experienced auditors who use a formal protocol of nearly 300 items. Each audit takes a team of two to three auditors approximately five days to conduct. Corporate headquarters tracks audit results and suggested improvements until all items are completed An external third-party auditor conducts an annual evaluation of the audit program effectiveness. Results are reviewed with senior leadership, and findings are addressed.

DuPont has long had a robust cyber security program aimed at ensuring business continuity. In 2002, we implemented a program to secure all high and medium risk process control systems at sites around the world. Sites conducted cyber security vulnerability assessments on critical manufacturing and control systems. As a result of the assessments, measures were taken to install special firewalls to protect critical process control systems from remote access.

Key programs in place prior to the 2001 attacks have been further strengthened. As one means to provide a safe and secure work environment, DuPont has a robust workplace violence prevention program. We aggressively respond to all threats made against employees or contractors. Formal training is conducted periodically to train supervisors on how to recognize, respond to, and investigate all threats made in the workplace.

Screening workers' backgrounds is a critical, longstanding part of our safety and security program. In the U.S., and as laws allow in other countries, DuPont requires criminal background checks of all employees upon hiring and all contractors seeking access to a DuPont site. The U.S. background check looks for misdemeanors and felonies for the seven-year period prior to the proposed date of access to the site. The checks must be of court records. In addition, DuPont does not accept checks performed under other programs unless they are conducted in strict conformance with our own internal standards. To provide perspective, one of our large facilities has a normal work force of 2600 employees and contractors. This swells by an additional 600-700 contractors during major maintenance overhauls. Contract firms that work at this site conduct about 2500 criminal background checks per year.

Long-standing relationships with local law enforcement and emergency planners have also been reinforced, and new relationships have been formed at the local level with federal groups such as the FBI and the Joint Terrorism Task Forces. DuPont sites and these local and federal groups work together to train, identify and investigate suspicious activities, patrol areas around our sites, and conduct tabletop and full-scale exercises. In addition, strong relationships have been developed with the Coast Guard at DuPont facilities regulated under the Maritime Transportation Security Act of 2002. Each of these sites has been successfully inspected by the Coast Guard. Perhaps a few brief examples of interactions with local law enforcement and first responders would be helpful:

- One site, for example, sets aside a month each year where all local community response units (i.e, HAZMAT, rescue, divers, river boat, fire, etc.) tour the facility to understand the layout, chemicals, fire system, and capabilities of the DuPont emergency response team. This allows for a much more effective, integrated response.

- In 2004, one site's annual emergency response drill with local volunteer fire departments was a simulated railcar accident releasing a flammable material. The drill resulted in the local departments gaining an appreciation for DuPont's capability to supply firewater, and an agreement was reached to work on improving the unified command and control structure.

- Another site drilled with the local SWAT team on a scenario of a hostile employee taking a hostage inside the plant. After the drill, the site met with the SWAT team, other local law enforcement agents, emergency responders, and the DuPont emergency preparedness team to discuss what had been learned.

- At one of our sites, the Coast Guard has visited several times to witness our response to drills. The local area conducted a drill, using $250,000 of funding from the federal Department of Homeland Security. It simulated two terrorist attacks, one at the airport and the other on the highway. As part of the drill, the Coast Guard escalated security to Maritime Security Level 3, allowing us to test the site's preparedness.

We work with a range of trade associations and federal government agencies, including the Department of Homeland Security, to develop effective national programs to secure key operations. As illustrated by the examples provided above, we have frequent interaction with many local, state, and federal entities and have found government agencies to be willing and helpful partners in furthering our security efforts.

Emergency response planning has long been a DuPont priority. Our sites have formal emergency response plans and, as noted above, conduct drills and exercises with both our own response personnel and with local first responders. We actively participate in Local Emergency Planning Commissions (LEPC) and believe that these organizations are very important in creating strong, integrated emergency management systems. The most effective LEPCs are those in which the members share a strong commitment to protect the local community, have strong leadership and participation, and are adequately funded. DuPont site participation in LEPCs across the country includes, for example:

- Regular meetings and collaboration on security preparedness.
- Security drills involving police, SWAT forces, emergency services and fire departments.
- Committee leadership.
- Joint review of and learning from community incidents ranging from gasoline spills to a small chemical release in a local high school chemistry class.
- Shelter-in-place awareness campaigns.
- Joint support and guidance for local emergency warning systems.

DuPont also works with the American Chemistry Council CHEMTREC® program to provide technical assistance for chemical emergencies. In addition, we are active in the TRANSCAER® program to provide training, conferences, and other educational programs to local communities, law enforcement agencies, and first responders. DuPont is actively involved in many local mutual aid groups. These groups frequently discuss procedures and lessons from incidents, participate in drills, and share unique expertise. It allows members to have access to wide-ranging emergency response capability and equipment.

In addition, DuPont maintains its own regional transportation emergency response teams for incidents involving our own materials and as mutual assistance to other companies. In 2004, we provided assistance to approximately 35 communities throughout the US. In addition to incident mutual aid, DuPont has rail tank cars and other transportation containers that are used to train public sector emergency responders in managing chemical releases during rail, ocean or highway accidents. In 2004, 5000 first responders were trained in approximately 150 communities in the United States, Canada, and Mexico. In Asia last year, DuPont also trained various government agencies and 1200 first responders.

In addition to site security, there is a significant focus on transportation security. DuPont actively works with the railroads to better coordinate rail service to our sites and to store cars within the secure site perimeters. Rail carriers work with us to secure key shipments and have implemented significant security plans. For truck transportation, DuPont has partnered with truck carriers on measures such as anti-theft devices and electronic surveillance, formal driver security training, planning safe routes, and improved identification of drivers.

My company became one of the first members of the voluntary Customs-Trade Partnership Against Terrorism (C-TPAT) program in November 2002 and is actively working to tighten security of goods entering the U.S. Product security and other aspects of securing the value chain have been, and continue to be, integrated into processes to screen new customers, audit contractors, and conduct product reviews and risk assessments.

Each DuPont U.S. Category 1 site has carefully planned for special security actions that might be required in extreme circumstances – e.g., routing all deliveries and people to remote locations for special screening. Operations security is a critical component of the global crisis management process. A special automated crisis notification system has been implemented so that the security leaders at all DuPont U.S. Category 1 sites can be contacted within 10 minutes or less.

When the national threat level is escalated, security measures are immediately assessed by both headquarters and sites. This assessment occurs even if there is no direct connection to the chemical industry or DuPont. Additional measures are determined based on the specific threat environment at the time. Measures may be implemented broadly or in a more focused manner. An example is DuPont's internal response to the recent London bombings. Although the DHS escalation to orange pertained only to mass transit, we directed all DuPont U.S. Category 1 sites to inspect 100% of inbound rail cars, along with trucks and all other vehicles, if they were not already doing so.

Perhaps the most powerful security preparedness initiative activated since 9/11 involves the ongoing vigilance of our individual DuPont employees and contractors. A cornerstone of DuPont safety has always been the personal accountability each of us has as a DuPont employee for our own safety and that of fellow employees and neighbors. The same is now true for security. Our employees and contractors have readily accepted this new responsibility. We have delivered extensive security awareness training to all employees and contractors and routinely communicate information to maintain the awareness.

We often find that individuals who do not have direct security responsibility are some of the best security officers. I'd like to share a few examples of actions by employees eager to contribute to our security effort:

- Employees question unknown cars in the parking lot or aircraft flying near the site. They ask questions or bring things to security's attention if something is different. They notice if a door isn't locked when it should be, and they report unusual activities on and around the plant.

- At one of our sites, an employee observed a contractor taking pictures at the plant. This is a violation of site policy. The incident was immediately reported, investigated, and communicated to everyone on site. At the plant manager's weekly meeting, employees discussed the incident, the proliferation of photographic devices, and the site policy.
- Security is fully integrated into site safety in the minds of our employees. When a site leader asks at a safety meeting, 'How many people are responsible for security at our site?', the reply from employees is always, 'All of us.'

I've touched on a wide range of security measures DuPont has implemented. It is important to note that the American Chemistry Council's Responsible Care® Security Code has certainly raised the bar by setting high standards for its members. Those companies that have implemented the Code have been recognized by agencies such as the Department of Homeland Security, the United States Coast Guard, U.S. Customs and others as having strong security practices in place across the country. On a global basis, the U.S. Responsible Care® Security Code has been a foundation on which site-specific, country-specific, and company-specific elements have been added to further strengthen our efforts.

As I have described, DuPont has very strong, thorough security systems and processes. However, we recognize that an effective security program is a journey. Hence, our security enhancements are ongoing. It requires constant vigilance and continual improvement. As with safety, there will **never** be a time when DuPont says, "We have done as much as we can on security."

Federal Chemical Security Legislation

Let me turn now to the issue of federal legislation. We appreciate the thoughtful approach this Committee has taken to this issue, and we look forward to working with you as you develop legislation.

While we believe DuPont has taken appropriate actions and that the Responsible Care® Security Code has helped other members of the American Chemistry Council do the same, we recognize that government has an important role in protecting this critical sector and ensuring all chemical sites are taking appropriate actions. Accordingly, we support meaningful and effective federal chemical security legislation. We also support the core principles for chemical security that were outlined on June 15 by Department of Homeland Security Assistant Secretary Robert Stephan in his testimony to this Committee.

I would like to briefly share our thoughts on important elements that should be addressed in the legislation and resulting regulations.

First, and foremost, it is important that the legislation have a clear security focus. We think it is critical to keep facilities focused on core security activities to ensure that they can get the job done in a timely and effective manner. To do this, legislation should not direct facilities to pursue actions that are fundamentally not security matters. Doing so would potentially delay and dilute essential security enhancement work.

Second, it is important that the legislation be risk-based so that government and private sector resources are allocated where they are most needed and can provide the greatest benefits. The prioritization of DuPont sites, as I discussed earlier, has been invaluable in channeling resources commensurate with the potential risk. Department of Homeland Security Secretary Chertoff and his predecessor Secretary Ridge have consistently emphasized a risk-based approach for the agency's work. A risk-based approach, which allows security measures to be escalated when the threat environment changes, is key to helping our country and the chemical industry be secure and competitive.

Third, DuPont believes that regulatory authority for chemical security should reside with the Department of Homeland Security. The main reason DHS was created was to provide central federal oversight of security. The agency has established a close relationship with the chemical sector through mechanisms such as the Chemical Sector Coordinating Council that I currently chair and the DHS Chemical Sector Specialist role that serves as a direct liaison with the individual sites and companies in the chemical sector. While DHS faces challenges as a new agency, it is making good progress in addressing them.

DHS and the chemical sector are already working together on programs such as the Risk Analysis and Management for Critical Asset Protection (RAMCAP), the Homeland Security Information Network (HSIN), and the Buffer Zone Protection Plan (BZPP). DHS is also a critical conduit for intelligence and threat information and is connected into other government agency threat streams. In addition, DHS is the means by which the chemical sector connects with all other infrastructure sectors, many of which are vital to the business continuity and security of our sector. Finally, DHS already administers the Maritime Transportation Security Act of 2002, which is the only existing set of security regulations impacting chemical facilities. In providing DHS with regulatory authority over chemical security, it is critically important that the program DHS is asked to implement be appropriately resourced and staffed.

With lead regulatory authority vested in DHS, we believe other agencies should continue to play a role, in consultation with DHS. These include the Federal Bureau of Investigation, the Department of Transportation, and the Environmental Protection Agency that administers the Risk Management Program (RMP) of the Clean Air Act Amendments of 1990. Regulatory jurisdiction for facilities already covered by the Maritime Transportation Security Act of 2002 should remain with the Coast Guard in DHS. We believe chemical plant security should be guided by a clear federal program rather than a patchwork of state and local programs.

Fourth, it is important to recognize the different, yet complimentary roles for government and the private sector in security matters. The private sector can, and should, take reasonable steps to secure its facilities against reasonable threats. However, there are clear boundaries where industry's capability ends and the government responsibility for defense of the nation's critical infrastructure begins. We cannot be expected to fully defend against military-style assaults, when it is clearly a lead role for government. Requirements for chemical plant security must be developed in the context of public security measures.

Fifth, flexibility is important. Prior witnesses have discussed the significant diversity of facilities producing and managing chemicals. DuPont operates thousands of different processes employing a wide variety of raw materials. These range from facilities producing soy protein to plants managing flammable materials and from rural to urban locations. Chemical security legislation should establish the main principles that are fundamental to a risk-based approach. Then, the legislation should allow flexibility in the rule-making process and implementation to accommodate the significant diversity of the chemical sector. Legislation needs to permit DHS to tailor its regulations to implement a timely and efficient program, focus its oversight, and deploy its resources on those matters that will best enhance security.

Sixth, the Maritime Transportation Security Act of 2002 (MTSA) has proven to be an effective security regulation for DuPont facilities covered under Part 105 of the Coast Guard rules. I respectfully recommend that it be used as a model for chemical security regulation of the highest priority facilities. MTSA Part 105 addresses the essential elements of a strong security program including staffing, training, security vulnerability assessments, facility security plans, exercises, incident reporting, and audits. It provides for protection of sensitive security information, as well as civil penalties for non-conformances. Parts of the MTSA regulations would need to be broadened beyond maritime transportation and considered for other minor modifications; but, fundamentally, MTSA is strong and effectively secures covered facilities. A case in point--one of our DuPont facilities, initially covered by MTSA Part 105, no longer receives regulated materials at its wharf. However, since the facility has already invested in meeting the MTSA requirements and believes that these requirements are enhancing site security, the facility has decided to voluntarily continue using the MTSA Part 105 requirements as the basis for their site security program.

Next, for both efficiency and fairness, new legislation and regulation should take into account the work already done under programs such as Responsible Care® and the Maritime Transportation Security Act. Both of these have materially enhanced security at impacted facilities, and the prior efforts should be given credit as a regulatory framework is developed.

Another important issue is protection of information that could present a security risk if made public. We believe strongly in transparency and accountability. However, in the security arena, information disclosure presents unique risks. Two federal rules have proven to be effective in protecting sensitive information from disclosure. These are the "Sensitive Security Information" (SSI) rules of the Transportation Security Administration and Department of Transportation, and the "Protected Critical Infrastructure Information" (PCII) rules of the Department of Homeland Security. However, limitations to both rules could leave some sensitive information requested or obtained by the government unprotected from disclosure. The SSI rules apply only to transportation security and, for the most part, only to the aviation and maritime industries. The PCII rules apply only to information voluntarily submitted to the PCII Program of DHS. PCII does not protect information that is required to be submitted to the government or information that is submitted to other branches of DHS or to other federal or state agencies. Finally, PCII does not support the ability of DHS to share information with state and local governments unless formal, written agreements have been reached with those governments. Any new chemical security legislation must have comprehensive provisions to ensure security-sensitive information is protected and to facilitate an effective exchange of information to and between governmental agencies.

My final comments pertain to inherent safety, commonly referred to as inherently safer technology (IST). At DuPont, inherent safety has been an integral part of our plant design and process safety systems since the 1960s. Over the decades, we have assessed and implemented many inherently safer solutions. The following are examples of inherently safer process modifications that DuPont has recently implemented after extensive analysis and re-engineering:

- Converted from shipment and temporary storage of propylene (a flammable) in river barges to on-site storage of a much smaller quantity, resulting in inventory reduction by a factor of 5.
- Redesigned the process chemistry and manufacturing technology to eliminate sulfur dioxide (an inhalation hazard) and convert to sodium bisulfide (very low toxicity).

These solutions cover all facets of risk reduction, not just the "chemical replacement" option that is frequently cited at the exclusion of other options. DuPont does not support that narrow definition. Instead, we must also include reducing the quantities of hazardous materials, moderation of operating conditions such as pressures or temperatures, simplification of process equipment, and provision of containment structures.

While inherently safer analysis is fundamentally risk-based, it consistently includes an evaluation of opportunities to reduce the inherent hazard of our processes and raw materials. Inherently safer solutions must be applied at the local chemical process and technology level and not through a "cookie cutter" approach. Each process segment must have the flexibility to evaluate inherently safer options on a case-by-case basis. This is necessary because each option for change has unique implications that must be considered.

Evaluations must assess technical feasibility, benefits of risk reduction, introduction of new risks or trade-offs, impact on proprietary technologies, effects on product quality, and implementation and ongoing operating cost. It is important to ensure that risk is not reduced in one place and increased in another and that the site can continue to operate the process reliably and safely. It is also important to ensure the ability of our suppliers to provide timely delivery of any new raw materials. In addition, changes to our product formulations must not create unintended consequences in our customers' operations. Finally, we must not disrupt the supply chain as we produce high quality products at competitive prices that customers demand.

Given the breadth and depth of an inherent safety evaluation, it typically takes several months or years to complete. This needs to be considered relative to the kind of short timeframes appropriate in a security context. Another factor to consider is technical capability. The expertise for inherently safer analyses resides in the business enterprises themselves with the technical and operations personnel who are intimately familiar with the technology, hazards, and overall risk of each process - and not in government.

For years, the Responsible Care® Process Safety Code has required consideration of inherent safety approaches. The ACC Responsible Care® Security Code additionally required members to consider inherent safety approaches to process design, engineering and administrative controls, and prevention and mitigation measures to address risks identified in the security vulnerability assessment. The security vulnerability assessment methodologies developed by the Sandia Laboratories and the Center for Chemical Process Safety, and used by ACC members, focus on intentional acts. These methodologies expressly require consideration of alternatives for safer technologies and chemicals. DuPont security vulnerability assessment teams include both security and process safety experts to ensure that both perspectives are considered at every site.

DuPont believes that inherently safer technology and chemicals are mainstream components of process safety and have a role to play as companies evaluate security. However, DuPont does not believe that inherent safety can, or should, be mandated by regulation. Previous testimony has discussed material substitutions at water and sewage treatment plants. This is a very straightforward and basic inherently safer application because it involves a simple mixing operation and no chemical reactions. These basic water and sewage treatment processes are very different from the complex processes at most chemical manufacturing plants. This complexity, along with the unique nature of each process, means that companies need the flexibility to assess and decide options. Prescriptive inherently safer technology requirements are unworkable. As Chairman Collins stated in her closing comments at the last chemical security hearing, a risk-based approach will provide incentive to companies to consider inherently safer options as a means to move to a lower risk tier.

Chemical Sector Coordinating Council

I was also asked to comment on the Chemical Sector Coordinating Council, in my role as chair of this group.

Many across the chemical sector share the commitment to security. That is why, in June of last year, a group of 16 national chemical trade associations, with DHS' guidance, formed the Chemical Sector Coordinating Council (CSCC). President Bush, in Homeland Security Presidential Directive-7, encouraged formation of such sector-specific bodies to "(a) identify, prioritize, and coordinate the protection of critical infrastructure and key resources, and (b) facilitate sharing of information about physical and cyber threats, vulnerabilities, incidents, potential protective measures, and best practices." The sector councils fulfill many of the operational roles that the Homeland Security Act establishes for private sector members of critical infrastructure.

While membership in the Chemical Sector Coordinating Council is composed of trade associations, DHS asked that an owner/operator representative serve as leader, or chair, of the Council to provide specific, front-line perspective and guidance. I presently serve in that capacity.

Members of the Chemical Sector Coordinating Council define the sector as "entities engaged in the production of chemicals, as well as those engaged in the storage, transportation, delivery and use of chemicals not adequately addressed by other critical infrastructure sectors." For example, it does not include water treatment facilities or transportation modes, both of which have separate and robust sector coordination mechanisms.

It is also important to note for the Committee what the Chemical Sector Coordinating Council is not. It is not a trade association, and there are no dues. The Council does not receive any federal funding – it is a sweat equity activity. Further, because it is an operational rather than an advisory body, the Council does not take policy positions. Each member association is responsible for developing and communicating its own policy positions and is free to develop independent relationships with DHS. So, for example, the Council will not take a position on any proposed legislation affecting security in the chemical sector.

In addition to serving as a routine information-sharing mechanism, the Council helped develop and ensure broad participation in a DHS-sponsored "tabletop exercise" earlier this year; provided critical input to the development of a Terrorist Threat Reporting Guide for use by companies in the sector; participated in the recent TOPOFF-3 exercise; and is working closely with DHS to develop, refine and disseminate its "Risk Analysis and Management for Critical Asset Protection" (RAMCAP) methodology. RAMCAP will allow DHS to compare the vulnerabilities of disparate assets and resources against a series of benchmark threat scenarios, thereby enabling DHS to allocate protective resources rationally, on the basis of risk, across all critical infrastructures and key resources.

I'm pleased to report the Council enjoys a high level of participation and engagement from its member organizations and, after only one year of existence, has tackled a number of substantive subjects.

In closing my remarks, Madam Chair, I want to thank you and the members of the committee for allowing me to share what DuPont has done to build strong security systems and processes. We have very successfully integrated security responsibility and engagement into our culture. We also know there is more to do. We take very seriously the confidence and trust placed in us by the public and government. DuPont leadership is committed to continually strengthening security at our sites, in transportation, and along the value chain. We recognize that the security and safety of our operations are critical to our employees, our neighbors and, in fact, essential to the future of our company. Thank you for the opportunity to testify before the Committee on the important issue of chemical plant security. We appreciate the important work of this Committee and stand ready to work with you as you move forward.

The Fertilizer Institute

Nourish, Replenish, Grow

Testimony of

Jim Schellhorn

of

Terra Industries Inc.

on behalf of The Fertilizer Institute

Before the

Senate Homeland Security and Governmental Affairs Committee

Regarding

Chemical Facility Security

DESCRIPTION OF TESTIMONY

A description of the many homeland security efforts American fertilizer and agricultural producers have recently undertaken and steps Congress could take to assist the agricultural community.

July 27, 2005

Union Center Plaza
820 First Street, NE Suite 430
Washington, DC 20002

202 962 0490
202 962 0577 fax
www.tfi.org

Madam Chairman and members of the committee, I am Jim Schellhorn. I am the director of environmental, health and safety for Terra Industries and am responsible for security for Terra's North American operations. I am here today to testify on behalf of The Fertilizer Institute, or TFI. TFI is the leading voice of the nation's fertilizer industry, representing the public policy, communication and statistical needs of manufacturers, producers, retailers and transporters of fertilizer. I very much appreciate the opportunity to appear here today.

Terra is headquartered in Sioux City, Iowa. We are a leading international producer of nitrogen fertilizers. We have production facilities and terminals in Arkansas, Iowa, Louisiana, Mississippi, Oklahoma, Texas, Ontario, and the United Kingdom, and 50 percent interest in a facility on the island of Trinidad. Worldwide we produce 5.8 million short tons of anhydrous ammonia and ammonium nitrate fertilizers annually, which are considered hazardous materials by the Department of Transportation (DOT) and U.S. Coast Guard. We also produce 4.6 million short tons of urea and urea ammonium nitrate solution—or UAN—annually. These products are not considered hazardous and are not regulated as hazardous materials by the DOT. Anhydrous ammonia is produced not only for sale as a finished product, but also as a feedstock for our other nitrogen products. Our facilities operate 24 hours a day, 7 days a week and Terra employs approximately 1,200 people in North America and the United Kingdom. Although Terra Industries operates a small trucking company, we largely depend on outside trucking companies in addition to rail and barge companies to ship much of our fertilizer and related products.

Fertilizers Role in Food Production

Fertilizer is essential to food production. Without fertilizers contribution to crop production roughly one-third of the world's population would be without food. Because food production depletes soil nutrient supplies, farmers rely on fertilizers to keep the soil productive. With the help of commercial fertilizer, North American farmers are able to produce the most abundant and affordable food in the world.

The air we breathe is about 78 percent nitrogen, but there are very few plants that can make direct use of nitrogen in the air. To make this nitrogen available to support life, fertilizer producers take nitrogen out of the atmosphere and convert it into a form plants can easily use. Nitrogen fertilizer manufacturing captures naturally occurring atmospheric nitrogen, and combines it with hydrogen from natural gas form anhydrous ammonia. Ammonia is then used to make other nitrogen fertilizer products.

Fertilizer and Security

Shortly after the events of Sept. 11, 2001, TFI formed a security task force, of which Terra is a member. In September 2002, TFI's Security Task Force developed and TFI's Board of Directors adopted an industry security code of management practices designed to help the fertilizer industry secure the manufacture and transport of our products using a risk-based approach. The voluntary code calls on the industry to use methodologies developed by the Center for Chemical Process Safety (CCPS) or the Synthetic Organic Chemical Manufacturers Association (SOCMA) – or an equivalent methodology – when conducting security vulnerability assessments (SVA) and making security-related improvements **(Exhibit A)**. The code establishes benchmarks for conducting security vulnerability assessments, implementing security measures, conducting employee training and drills, communicating with law enforcement, conducting periodic audits and verifying physical site security measures through a third party. The code details timelines for these activities by ranking facilities at high, medium and low risk levels. TFI's Security Task Force monitors code implementation.

The fertilizer industry is very diverse. Companies such as Terra produce and sell fertilizer into the retail distribution system, which in turn sells it to farmer customers. Most of Terra's production and storage facilities, like many others in our industry, are located in rural communities. For instance, Terra's Verdigris plant near Claremore, Oklahoma, where I am located, is approximately six miles from Claremore and our nearest neighbor is more than one-quarter mile away. Because we produce and store anhydrous ammonia and other fertilizers at the Verdigris plant, we are subject to many federal safety, security and environmental regulations. For example, we must adhere to the Occupational Safety and Health Administration's, or OSHA's, process safety management regulations, the U.S. Coast Guard's Maritime Transportation Security Act, or MTSA, regulations and the EPA's Risk Management Program requirements. Terra has five facilities along the inland waterway system that are subject to the MTSA regulations and nine facilities that are subject to DOT security regulations for the transportation of hazardous materials. We have nine facilities subject to the Clean Air Act Risk Management Program, or RMP, requirements. Ammonia is the principle product we produce and store that is subject to the RMP regulations.

I would like to take a moment and discuss the specific measures Terra has taken and continues to undertake to secure our facilities and the products we produce.

After TFI developed the industry security code, Terra immediately began to conduct security assessments and audits at all our facilities. We used both outside law enforcement experts and internal resources to identify vulnerabilities, implement countermeasures and develop security plans. At each stage of the process, we ranked both our facilities and our vulnerabilities based upon risk. Using those rankings, we began to address the highest risks first. All of our facilities now have active security plans and our waterfront facilities are in compliance with the MTSA regulations.

Terra has installed additional lighting, fences, physical barricades and video monitors at strategic locations. In addition, all gates are locked when unattended and facility access is tightly controlled by security personnel or employees 24 hours a day, 7 days a week. Specifically, all product trucking companies and drivers are pre-approved; all deliveries to our facilities are checked at the gate prior to entering the facility; and criminal background checks are now required for contractors as well as Terra employees. We have also recently implemented a system to ensure delivery receipts for all truck shipments of ammonium nitrate from Terra-owned facilities.

Terra Industries and other members of TFI have undertaken tremendous efforts to ensure that criminals intent on harming our country could not purchase and misuse fertilizer products that are vital to feeding America and the world.

For example, immediately after the tragedy in Oklahoma City, the fertilizer industry partnered with the Bureau of Alcohol, Tobacco, Firearms and Explosives in an outreach program called *Be Aware for America,* which was aimed at protecting our products in our places of business. A few years later, again in partnership with ATF, we developed *Be Secure for America*, which provided additional information about keeping our products secure. Both programs were widely distributed to law enforcement and within our industry. After the terrorist attack on Sept. 11, 2001, the fertilizer industry announced our most stringent program yet, called *America's Security Begins with You*. This program has been endorsed by ATF, the Department of Homeland Security and the Association of American Plant Food Control Officials, who regulate fertilizer at the state level. The campaign urges that security plans be developed and implemented, records of sales be maintained, and that law enforcement be alerted to any suspicious activity.

These voluntary programs have primarily focused on ammonium nitrate, the fertilizer used in the Oklahoma City bombing. Recognizing the changing nature of the nation's security, Sens. Thad Cochran (R-Miss.), Mark Pryor (D-Ark.), Pat Roberts (R-Kansas) and Saxby Chambliss (R-Ga.) recently introduced the "Secure Handling of Ammonium Nitrate Act of 2005," (S. 1141). The bill directs the Department of Homeland Security to promulgate regulations requiring all facilities that handle ammonium nitrate fertilizer to register at the state level and maintain records for all purchases of ammonium nitrate fertilizer. The fertilizer industry's support of the Senate legislation – and parallel legislation introduced in the House – takes the industry's voluntary programs to the next level through the creation of a uniform federal set of rules for sellers and purchasers of ammonium nitrate.

What More Needs to Be Done?

We believe that chemical facilities will most effectively address security when given the flexibility to use measures that will address the risks specific to each facility. Quite simply, we at Terra and others in the industry have not employed a "one size fits all" approach at our facilities, and believe that any legislation requiring us to do so would be counterproductive.

Equally important, Congress must recognize the security measures already taken and facilities covered under other federal regulations, such as the MTSA, to avoid duplicate regulations.

There has also been considerable debate over whether Congress should mandate the use of inherently safer technologies (IST). IST is not a security measure – it is a safety concept that has been misapplied by some groups in a way that we fear could lead to the ban or restricted use of basic nitrogen fertilizers. For instance, if anhydrous ammonia manufacture was banned in the United States as a result of an IST mandate there would be no nitrogen fertilizer manufacturing in the United States because ammonia is the basic feedstock for all other nitrogen fertilizers. U.S. farmers would have to rely on imported fertilizer to grow their crops, and indirectly, the American public would have to rely on foreign fertilizer for their food supply. What's more, without additional security restrictions on imports, foreign ammonia brought in by ship from overseas would pose additional security risks.

Terra and the fertilizer industry are not opposed to evaluating the chemical process safety of their operations and considering ways that process safety can be improved. On the contrary, the process hazard analyses and risk assessments we have conducted as part of our PSM and RMP programs, and the security vulnerability assessments we have performed include consideration of ways to minimize the hazards that are identified. However, this type of hazard assessment can only work when applied by a site owner's engineers who truly understand the facility's operations.

Conclusion

Madam Chairman and members of the committee, American farmers, fertilizer producers and retailers are committed to security. We have demonstrated that commitment through the significant number of voluntary security steps we've taken and will continue to take. Without question, we very much want to help Congress in its endeavors to shield this country from acts of terrorism. We support Department of Homeland Security (DHS) Secretary Chertoff's efforts to evaluate all of the nation's vulnerabilities and then prioritize the federal government's response based on sound risk assessments.

As the federal government proposes its suggestions for chemical facility security legislation, we recommend such proposals are based on reasonable, clear and equitable performance standards. TFI and its members—Terra among them – believe that to be effective, fair, realistic or feasible to implement, the legislation must:

1. Provide for the varying levels of risk posed by different kinds of chemical facilities.
2. Recognize the security measures our industry has already taken and complement the federal regulations with which we already comply.
3. Reject attempts to mandate IST.

Furthermore, we urge that the federal regulations preempt any such action by state or local governments. Layering federal regulation upon a patchwork of state regulations is at best inefficient and at its worst an impediment to efficient compliance.

I thank you for the opportunity to testify today and look forward to answering any questions you might have.

Contacts:

Jim Schellhorn
Director, Environmental, Health and Safety
Terra Industries Inc.
(918) 266-9653
jschellhorn@terraindustries.com

Kathy Mathers
Vice President, Public Affairs
The Fertilizer Institute
(202) 515-2703
kmathers@tfi.org

Pam Guffain
Director of Government Relations
The Fertilizer Institute
(202) 515-2704
pguffain@tfi.org

The Fertilizer Institute
Nourish, Replenish, Grow

Exhibit A:

The Fertilizer Institute's
Security Code of Management Practices for the Fertilizer Industry

September 23, 2002

Purpose and Scope

The purpose of The Fertilizer Institute's (TFI's) **"Security Code of Management Practices for the Fertilizer Industry"** is to help the fertilizer industry protect people, property, products, processes, information and information systems by enhancing security, including security against a potential terrorist attack. The fertilizer industry encompasses manufacturers, retailers and distributors.

This code is designed to help the fertilizer industry achieve continuous security performance using a risk-based approach to identify, assess and address vulnerabilities, prevent or mitigate incidents, enhance training and response capabilities, and maintain and improve relationships with key state, local and federal government partners. The code is implemented with the understanding that security is a shared responsibility requiring actions by all stakeholders including carriers, customers, suppliers, service providers, government officials and agencies.

Relationship to Other Industry Commitments

The fertilizer industry's commitment to protecting its employees and the public is demonstrated by the implementation of this security code and other good management practices. The fertilizer industry should regularly reassess these security-related practices in an effort to continually improve performance and identify potential vulnerabilities.

Management Practices

A risk-based security management system for people, property, products, processes, information and information systems throughout the fertilizer industry should be implemented. The fertilizer industry encompasses manufacturers, retailers and distributors.

The security management system must include the following management practices:

1. Leadership Commitment

Senior management commits to continuous improvement through accountability, published policies, and provision of sufficient and qualified resources.

2. Analysis of Threats, Vulnerabilities and Consequences

Use available security vulnerability assessment (SVA) methodologies, prioritize and periodically analyze potential security threats, vulnerabilities and consequences. The writers of this code encourage manufacturing facilities to conduct vulnerability assessments using methods developed by the Center for Chemical Process Safety (CCPS), Synthetic Organic Chemical Manufacturers Association (SOCMA), or other equivalent methods.

Writers of this code encourage retailers and distributors to conduct vulnerability assessments using methods developed by the Agribusiness Security Working Group (whose members include the Agricultural Retailers Association (ARA), CropLife America and The Fertilizer Institute (TFI)), or methods developed by CCPS, SOCMA, or other equivalent methods.

3. Implementation of Security Measures

Develop and implement security measures commensurate with identified risks.

4. Information and Cyber-Security

Protect information and information systems as a critical component of a sound security management system.

5. Documentation

Document key elements in security management programs, processes and procedures.

6. Training, Drills and Guidance

Train, drill, and provide guidance for employees, contractors, service providers, and others, as appropriate, to enhance awareness and capability.

7. Communications, Dialogue and Information Exchange

Communicate, foster dialogue and exchange information on appropriate security issues with employees, contractors, communities, customers, suppliers, service providers and government officials, agencies and law enforcement officials. This dialogue and information exchange should be balanced with safeguards for sensitive information.

8. Response to Security Threats

Evaluate, respond, report and communicate security threats as appropriate. Fertilizer facilities will promptly evaluate the real and credible threats and will report and communicate to the fertilizer industry and law enforcement personnel as appropriate.

9. Response to Security Incidents

Evaluate, respond, investigate, report, communicate and take corrective action for security incidents. If an incident should occur, the fertilizer facility will promptly respond and involve government agencies as appropriate. After investigating the incident, the fertilizer facility will incorporate lessons learned and will, as appropriate, share those lessons with others in the fertilizer industry and government agencies and implement corrective actions.

10. Audits

Conduct periodic audits of fertilizer facilities to assess security programs and processes, and implementation of corrective actions.

11. Third-Party Verification

Verification by a third-party, that facilities with potential off-site impacts have implemented the physical site security measures to which they have committed.

12. Management of Change

Evaluate and manage security issues associated with changes involving people, property, products, processes and information or information systems.

13. Continuous Improvement

Utilize continuous performance improvement processes entailing planning, establishment of goals and objectives, monitoring of progress and performance, analysis of trends and development and implementation of corrective actions.

The fertilizer industry will share information on effective security practices within the fertilizer industry and with external, qualified security professionals. The fertilizer industry will continue to expand the awareness of and commitment to enhanced security practices throughout the fertilizer industry. TFI will continue to provide guidance, including sharing examples of effective member security practices, to assist the fertilizer industry in implementation of this code. It will periodically review and, as appropriate, revise the guidance, and will continue to serve as the industry clearinghouse for the exchange of information on security through the secure members only Web site: http://www.npknet.org.

Due to the rapidly evolving nature of security issues and related expertise, TFI will reassess this security code, its management practices and implementation timetable two years after code adoption or earlier as appropriate. Security code implementation guidance will be updated as necessary in the interim.

Time Schedule:

One of the first SVA activities is to perform an initial prioritization of potential security hazards at all facilities operated by the enterprise. This initial prioritization assessment, or enterprise level screening process, will establish the "timeframe tier" for the facility. The enterprise level screening process separates facilities into different tiers based on potential severity of attack, difficulty of attack and attractiveness of the target(s). Based on this screening, the company can then focus energies to complete site security vulnerability assessments and implement specific steps to improve security where it is most needed.

The fertilizer industry should implement all security code practices using the initial prioritization timetable below commencing on the date this code is approved. Timelines for completion of site security vulnerability assessments, implementation of site security measures and verification are found below in Table 1.

For example, a Tier I facility would fall into the highest risk level, Tier II medium risk level, and Tier III low risk level.

Table 1: Schedule for Implementation of Security Assessment

Security Process	Timeframe Tier I	Timeframe Tier II	Timeframe Tier III
Complete Site Security Vulnerability	6 months	12 months	18 months

Assessment			
Complete Implementation of Site Security Measures	18 months	24 months	30 months
Verification of Physical Site Security	21 months	27 months	33 months

The Fertilizer Institute (TFI) represents by voluntary membership the nation's fertilizer producers, manufacturers, retailers, trading firms and equipment manufacturers. This security code of manufacturing practices was developed in keeping with TFI's efforts to protect and promote the nation's fertilizer industry. For more information, please contact TFI at (202) 962-0490 or visit TFI's Web site at http://www.tfi.org.

WRITTEN STATEMENT OF

JOHN CHAMBERLAIN

Security Manager Asset Protection Services
Corporate Security, Shell Oil Company

On behalf of Shell Oil Company and the American Petroleum Institute

Before the

SENATE COMMITTEE ON HOMELAND SECURITY
AND GOVERNMENTAL OPERATIONS

Hearing on "Chemical Facility Security: What Is the Appropriate Federal Role?"

July 27, 2005

Senate Dirksen Office Building, Room 562, 10:00 AM

Chairman Collins, Ranking Member Lieberman and Members of the Committee: My name is John Chamberlain and I am a manager within Corporate Security for the Shell Oil Company. I also serve as Vice Chairman of the Security Committee for the American Petroleum Institute. I have many years of experience working with Shell's refineries, chemical plants, and distribution terminals. I also have 30 years of law enforcement experience.

I am pleased to appear before you today to testify on the issue of chemical security, representing Shell Oil and the American Petroleum Institute (API). Shell Oil Company is an affiliate of the Shell Group, a global group of energy and petrochemical companies, operating in more than 140 countries and territories, employing more than 112,000 people. Approximately 24,000 Shell employees are based in the U.S. Shell Oil Company, including its consolidated companies and its share in equity companies, is one of America's leading oil and natural gas producers, natural gas marketers, gasoline marketers and petrochemical manufacturers. API is the national trade association for the U.S. oil and natural gas industry, represents all sectors of the industry, including exploration, transportation, refining, storage, distribution and marketing.

The U.S. oil and natural gas industry is committed to protecting the reliable supply network of fuels and products to keep our economy growing. The industry has long operated globally, often in unstable regions overseas where security is an integral part of providing for the world's energy needs. Although we are in the energy business, some proposals to address the security of chemical sites could have affected the energy industry, as well as agriculture, water treatment, food and dairy processing, pharmaceuticals, bulk storage terminals, and small businesses. These U.S. industries are essential for our national security and economic vitality, but not traditionally thought of as "chemical industry" facilities.

My testimony today will first outline the three stages of addressing industrial security: finding vulnerabilities, enacting countermeasures and sharing intelligence. While it is rarely reported on, the industry in partnership with the government has taken many actions to improve industrial security. We have operated under federal security law, federal security partnerships, industrial security practices and intelligence sharing and support these ongoing efforts.

Oil, natural gas and chemical plant operations are now safer and more secure as a result of the public-private partnerships and numerous new federal security requirements. Little has been communicated about the actions that Congress, industry, government agencies, and state and local first responders have taken, but these public-private partnerships and new security laws have improved security to ensure the reliable flow of energy to consumers.

Since we support continued security enhancements, my testimony will then address specific proposals that we believe would be disruptive to our industrial security operations. Congress has been wise to avoid passing environmental mandates and public release of security information proposed in the name of protecting the industry from terrorist attacks. These would be disruptive to ongoing security operations.

We hope that you would avoid provisions that would be counterproductive to the gains we have made in security since 9-11. Whether or not new security legislation is passed, we will continue, in partnership with the government, to re-evaluate and improve security in U.S. oil and gas operations. Our oil and gas infrastructure is the most reliable in the world and our aim is to continue our coordinated efforts to enhance our infrastructure security.

The U.S. oil and natural gas industry has long operated globally, often in unstable regions overseas where security is an integral part of providing for the world's energy needs. After September 11th, 2001, the industry partnered with federal and local authorities to reevaluate and strengthen our domestic security. The world of corporate security changed forever on 9/11, as we had to more seriously address the possibility of intentional acts to harm our facilities and employees instead of just accidental events.

Within months of the attack, the industry developed security measures for all segments of the oil and gas network – including pipelines, refineries, terminals, and others. One reason the industry was able to move so quickly is that we have high caliber security professionals working for companies who are experts in physical security and protecting our assets. At Shell, our corporate security staff has extensive background in federal, state or local law enforcement as well as experience in military security.

Nationwide, oil and gas companies made major investments in hardening facility protection, training and communications all the way from wellheads and offshore platforms to tankers, ports, pipelines, refineries, storage tanks, and most importantly, employees, our contractors and their communities.

All of these steps were carried out in close partnerships with federal, state and local law enforcement and security officials. The partnership forged between the oil and natural gas industry and government at all levels is now working to protect hundreds of facilities across the country from the potential of terrorist attacks. With this history in mind, we ask that you recognize these efforts and avoid disrupting them as you consider if new proposals to improve the security of U.S. refineries and chemical plants are needed. A terrorist, unlike a pollutant or physical workplace, is clever and has the ability to adapt against a checklist of rules. This is one reason we value our close, professional partnerships of the government, industry and local communities in addition to security requirements.

Some legislators may be tempted to treat security as a concern to be addressed with inflexible regulations. We ask that you recognize that a terrorist, unlike a pollutant or physical workplace environment, is clever and has the ability to adapt against a checklist of rules. This is one reason we value our close, professional partnerships of the government, industry and local communities.

Three steps: Find Vulnerabilities, Enact Countermeasures and Communicate Threats

At every stage, we have been mindful that we are protecting not only our facilities and our petroleum products, but also the people who work for Shell Oil Co., our on-site contractors and the neighboring communities. This reassessment of facility security since 9/11 has been a three-stage process: (1) reevaluate our threats and vulnerabilities, (2) carefully put security standards and countermeasures into place and (3) improve systems for communication with federal, state and local law enforcement about terrorist threats quickly.

API and the National Petrochemical and Refiners Association produced an industry-wide method for managers to identify security vulnerabilities in their operations. The SVA methodology is a sophisticated, risk-based tool used to identify the security hazards, threats and vulnerabilities of a facility, and to evaluate the best measures to provide safe facility operations to protect employees and surrounding communities. In other words, it provides the framework for a complete security analysis of the facility and its operations.

The SVA covers both physical and cyber security, process safety, facility and process design and operations, emergency response, management and law enforcement.

In 2004, the oil and natural gas industry expanded the SVA methodology to include to pipeline, truck, rail and liquefied natural gas (LNG) operations. DHS has recognized the SVA methodology and even uses it to train its own employees.

API and federal security personnel also completed the "Security Guidelines for the Petroleum Industry," to help managers protect facilities and respond to changes in the threat level. This guidance is now in routine use as a roadmap for companies in deciding how best to protect all sectors of the industry against the threat of attack. These are the working methods and countermeasures the oil sector uses to protect all segments of the industry.

The guidelines are important because they allow companies to effectively manage security risks and provide a reference to federal security laws and regulations that have an impact on petroleum operations. The Secretary of Energy and later the Undersecretary for the Department of Homeland Security have endorsed the industry guidelines. These security protocols are constantly being updated and improved. A third edition was published in April 2005.

I am submitting the API Security Guidelines and the API/NPRA Security Vulnerability Assessment to be made part of the hearing record.

To sort out and streamline communications from law enforcement agencies to the industry, API and DOE founded the "Energy ISAC", or Energy Information Sharing and Analysis Center. The Energy-ISAC is an Internet-based, secure, early warning system for making sure that threats and suspicious behavior are relayed between oil and gas operators and homeland security agencies.

The Department of Homeland Security is updating this intelligence sharing system with the Homeland Security Information Network. Along with other companies, Shell is participating in the HSIN, which links owners and operators with each other and with DHS and the intelligence community to permit secure conversations about physical and cyber threats, incidents, vulnerabilities and best practices.

Industry-led actions

Over the last four years, the Department of Homeland Security has assumed primary responsibility for the security of domestic infrastructure. Both the federal Energy and Transportation Departments also have key roles for guaranteeing energy assurance and transport of hazardous materials.

In that time, government inspectors have examined refineries and other key energy production assets and conducted cyber-attack vulnerability tests on critical oil and gas facilities. On the West Coast, DOE and DHS conducted an oil sector system-wide assessment on counter terrorism measures.

Shell has also participated in dozens of industry workshops and training to establish common practices within the company so every sector adopts the strongest possible plan for self-protection.

Like other integrated oil companies, we have joined with DHS in developing a common system for comparing security risks across the nation's varied critical infrastructure. The system, called Risk Assessment Methodology for Critical Asset Protection, (RAMCAP) will give Congress and the Executive Branch the tools they need to make decisions and allocate money on security.

Companies like Shell are also working with DHS, state and local governments to protect and secure the areas surrounding our facilities by establishing clearly identified buffer zones. Shell is also a participant in several security information sharing programs. They include the Energy Information Sharing & Analysis Center and the Oil & Natural Gas Sector Homeland Security Coordinating Council.

In addition, Shell has implemented voluntary actions as part of the American Chemistry Council's Responsible Care Security Code, which further enhances security of our chemicals facilities, our communities and our products. The Security Code addresses facility, cyber and transportation security and has been widely recognized by local, state and federal governments as a model for other U.S. industries.

Complying with Post 9-11 Security Laws

The industry has worked hard to meet and exceed new security requirements enacted by the Congress since September 11th. Under the Maritime Transportation Security Act, the U.S. Coast Guard inspects oil tankers, barges, many refineries, chemical plants, and numerous other storage and shipping facilities. API prepared for a new era in the regulation of oil tankers by meeting on multiple occasions with the Coast Guard and DOE. The Coast Guard's Director of Port Security praised API and company efforts that led to no major interruptions in energy supplies to the U.S. as the MTSA regulations were implemented.

Under the new law, refineries and other waterfront facilities in the United States submitted security plans that have been reviewed and approved by the Coast Guard. MTSA also requires companies to designate facility security officers (FSO) who oversee the implementation of their security plans and conduct quarterly drills and an annual exercise to test how well the facility's security plan has been carried out. We believe that overall the new law is working as intended, a view that is shared by the Coast Guard.

Under the Patriot Act, carriers of hazardous materials are subject to background checks and must prepare security plans to protect themselves and their employees. API adopted recommended practices for managers of offshore platforms to prepare for possible terrorist attacks. This practice is to be used as a reference standard for the federal government.

To provide an early warning against potential cyber terrorist attacks against pipeline computer systems, API published new standards for monitoring the movement of oil through pipelines. The standard is called Supervisory Control and Data Acquisition. In the summer of 2004, 19 oil and gas associations created the Oil and Natural Gas Homeland Security Coordination Council to give the government a single point of reference for the industry when it is needed.

Views on new security proposals

You invited me here today to discuss the possibility of new federal legislation to enhance security at refineries and the rest of the energy delivery system. As I mentioned earlier, we support the continued security enhancements that Congress, federal agencies, state and local first responders and industry has put into operation. We would caution that whatever direction the committee decides to take, that you be careful not to disrupt security practices and partnerships that are already in place. Perhaps, the old caution to medical students, "First do no harm," should apply here as well.

We ask that you keep in mind the principle that security requirements should be risk-based and site-specific. In other words, a one-size-fits-all approach will not work and would only provide a roadmap for terrorists to use to determine industry's security countermeasures.

Some legislators may be tempted to treat security as a concern to be addressed with inflexible regulations. We ask that you recognize that a terrorist, unlike a pollutant or physical workplace environment, is clever and has the ability to adapt against a checklist of rules. This is one reason we value our close, professional partnerships of the government, industry and local communities.

MTSA

Should the committee conclude that legislation is needed, we suggest that it not apply to facilities under existing coverage of the MTSA. Anther option for avoiding disruption is to require all facilities in compliance with this law shall be deemed to satisfy the requirements of new security law.

We also suggest that sites that contain areas only partially-covered by MTSA have the option for the entire facility to be covered by MTSA instead of the new law, thus avoiding duplication in regulations at a single facility.

Examining the MTSA security law, I would like to highlight a few characteristics for your consideration. In implementing a broad new security law last year, the U.S. Coast Guard has, overall, done a successful job in implementing security countermeasures without impeding the commerce it protects. This is in large credit to the U.S. Coast Guard's centuries-long experience in protecting onshore and offshore commerce of the U.S., as well as the existing relationships of local stakeholders and respective Captains of the Port. Without this security expertise and these existing relationships of private sector operations, the MTSA would not have been successful.

The MTSA, like all other federal security laws, protected and strengthened our infrastructure, instead of having a federal bureaucracy attempt to redraw or micromanage how private operators function. In other words, it has a risk-based philosophy; the required security protections must meet the risk under which we operate. The security regulations were not an attempt to change the modern environmental and safety regulations with which we comply.

Complying with other security standards

Facilities should be deemed to satisfy the requirements of new security requirements if they operate under a security program in partnership with federal agencies, such as the Pipeline Security Program in the Department of Transportation and the API Security Guidance Program. Facilities should also be deemed to satisfy the requirements of this new law if they comply with other security requirements that are substantially equivalent. DHS should have the discretion to recognize state security law for designating early compliance status.

IST

During discussions about security at refineries, the subject of "inherent safety' has arisen. Inherent safety is a safety strategy whereby substitute chemicals or processes are used to reduce or eliminate hazards. We strongly oppose any mandates for the inherently safer technology because it would be counterproductive to protecting our infrastructure.

Unfortunately, while the concept of inherent safety is well understood, the application of inherent safety in facilities is less understood. For example, there is no agreed upon methodology to measure the effectiveness of inherent safety or inherently safer options. If you can't measure or evaluate inherent safety, it is unreasonable to impose regulations to mandate it. In addition, the complexity of refineries prevents a prescriptive approach to inherent safety - judgments about process safety hazards are best made by process safety experts at each site. We contend that it is more important to manage the overall risk of a facility using risk assessment methodologies.

Infrastructure security laws already passed by Congress, such as last year's Maritime Transportation Security Act and BioTerrorism Act, authorize enforcement of vulnerability assessments and security plans for private facilities, but do not create a new requirement for "IST". No other security law requires IST for good reasons.

First, creating an "Inherently Safer Technology" requirement for American farms and businesses in the name of national security may actually increase risks. For example, reducing the volume of a hazardous chemical stored at a facility may reduce on-site risk but will increase truck, rail, or barge traffic to maintain raw material, thereby potentially increasing overall risk.

Security law covering companies should be risk-based, not seek out the elimination of all risk, which is impossible. Private farms and company facilities that need to use substances will necessarily intensify their security plans - based on the risk level of these substances. This obviates the need for IST.

Under new IST authority, a Government order for changes to materials or processes might very well create new liability should those orders result in accidental or intentional harm.

Inherently safer technology is already incorporated under existing federal requirements for health and safety -- the Occupational Safety and Health Administration's Process Safety Management Program and the Environmental Protection Agency's Risk Management Program. American farms and company facilities will continue to comply with federal, state and local requirements.

Farms and company facilities, through self-interest, consider the safest, most innovative, and cost-effective technologies. However, new government mandates for IST would require bureaucrats and courts to determine the best technologies for businesses. Creating a new "security IST" authority will allow government micromanagement in mandating substitutions of all processes and substances. This would limit operational flexibility and innovation.

Penalties for non-compliance

Enforcement penalties should reflect the view that farmers, owners and operators covered under a new law are meant to be protected from criminals, not treated as criminals for good faith efforts. In other words, DHS should be required to make several attempts to correct the non-compliance before assessing penalties. Penalties for non-compliance should not include prison sentences for owners or operators. Culpability for terrorist acts is addressed in other law. If the enforcement action is to shut down the facility, DHS will consider the effect to national energy reliability before enforcement.

Protection of Information

DHS should withhold and protect from disclosure under the Freedom of Information Act any record related to vulnerability of and threats to critical infrastructure in their possession, or any information derived there from, as long as (1) the provider would not customarily make the record available to the public; and (2) the record is designated and certified by the provider, in the manner specified by DHS, as confidential and not customarily made available to the public. Any employee of DHS that releases such information should be subject to criminal penalties, and where appropriate, disbarment from government employment. In addition from FOIA exemption, additional protections should be made to prevent the leak of vulnerability information, which would provide a "roadmap" for terrorists and other criminals. Such information should be protected from civil discovery except in a lawsuit brought pertaining to the operator's compliance with the security related provisions of these requirements. In the case of an enforcement proceeding, the information in the case should remain classified.

Security clearances for classified information for both government and non-government personnel should be updated to reflect the unified offices within the DHS. Classifications should correspond to these clearances, ensuring a "read what you write" ability by appropriate company employees.

Conclusion

Oil and natural gas operations are now safer and more secure as a result of the public-private partnerships and numerous new federal security requirements. These public-private partnerships and new security laws have strengthened the reliable flow of energy to consumers.

Congress has been wise to avoid passing environmental mandates and public release of security information proposed in the name of protecting the industry from terrorist attacks. These would be disruptive to ongoing security operations. We urge the Committee to carefully consider the effect any new federal law would have upon existing successful laws, programs and practices.

The oil and gas industry is committed to protecting the reliable supply network of fuels and products to keep our economy growing. Our oil and gas infrastructure is the most reliable in the world and our aim is to continue our coordinated efforts to enhance our infrastructure security.

United States Senate
Committee on Homeland Security and Governmental Affairs

Testimony

Chief Robert A. Full
Allegheny County, Pennsylvania
July 27, 2005

Good Morning Chairman Collins and Senators. It is distinct honor and privilege to be invited here to testify on behalf of "Chemical Facility Security" and it's impacts at the Local and County level of Government. My County Chief Executive Dan Onorato extends his appreciation for this opportunity as well.

I speak this morning as a 32 year First Responder as a Firefighter, Paramedic, and Hazardous Materials Technician as both a career professional and volunteer from Allegheny County in Southwestern Pennsylvania. I serve as my County's Emergency Management Coordinator, Local Emergency Planning Committee (LEPC) Chairman, and Chairman of one of our State's 9 Regional Counter Terrorism Taskforces in Southwestern Pennsylvania commonly referred to as PA Region 13 which is an organization formed in November 1998 by 13 Counties and Pittsburgh under an intergovernmental agreement for mutual aid during terrorism events and all hazard emergencies.

Allegheny County with the City of Pittsburgh as the county seat is famous for Three Rivers, Steel Making, Research Centers, World Class Medical Systems, Education Institutions such as the University of Pittsburgh, Duquesne and Carnegie Mellon, Major Transportation Systems, and the Pittsburgh Steelers and Pirates covers 730 square miles with a population of 1.3 residents and 130 local municipalities.

This morning as I awoke early to fly here I took a shower and made my coffee with crystal clean and safe water. My clothes have synthetics in them. The breakfast fruits I enjoyed were free from bacteria and hearty from the vine. The fuel in my car and the plane I flew performed as manufactured thank goodness, and as I look around here I see so much of the positives and the need for a strong and safe chemical industry. It has been said and reinforced that one of the main reasons that the United States enjoys the highest standard of living is thru it use of chemicals in all aspects of our daily lives. On behalf of those I represent and myself we couldn't agree more in the need to support and protect the Chemical industry.

As a First Responder paramount to the success of doing your job is to be able to protect, save lives and property during emergencies. An individual comes into Public Safety as a first responder and he/she is primarily trained to deal with the aftermath of an event which was caused by accident, act of god, or intentionally. Every day in this country the men and women of all our Public Safety Departments (Police, Fire, EMS, 9-1-1) demonstrate great courage and conviction to be the best that they can be. These folks plan, train, exercise, and respond to any emergency no matter what the case.

No matter how good a Public Safety organization is there will be times that their training, skills, knowledge and capabilities will be overwhelmed or they may not have the expertise to deal effectively with the situation. To minimize this scenario having a strong emergency plan and relationships with pertinent persons in advance pays dividends each day at the local level across America. It's cliché, but it's not the time or place to exchange business cards during the emergency.

I would like to focus now on Chemical Safety.

In 1986 the Federal Government enacted the SARA Title III, "Emergency Planning and Community Right to Know Act. The overall success of this law can not be over stated, and can be measured in my County and throughout the Country by the reduction in chemical spill emergencies, better informed employees and responders during emergencies, Federal, State, and Local Government input and Coordination, and so much more.

In my 32-year career I have had an opportunity to specialize in Hazardous Material emergencies. I was the first Chief of the City of Pittsburgh Hazardous Materials Team and held that position for 13 years. Today, I oversee 5 Hazardous Materials Teams in my County. I have logged over 2,000 hazardous material calls. I come to see first hand the potential life threatening situations that are involved when chemicals are accidentally or intentionally released from their containers and processes. The chemicals and materials are found in fixed facilities during production, transfer, and storage along with the transportation to and from market via highway, railroad, water, air, and pipeline. Responding to chemical spills requires quick informed decision making along with specialized tools and equipment. Incidence of vapor clouds, running liquid spills, unidentified products and fires severely complicate local response actions. Many times to the point that a community may not be able to react fast enough to save it's residents. Transportation accidents involving chemicals provide even a greater challenge as they move in and out of our neighborhoods, by our schools, homes and places of business.

SARA Title III targeting fixed chemical facilities followed by similar legislation in our case enacted by the Commonwealth of Pennsylvania has directly contributed to saving lives, property, and environment. The SARA Law has allowed us to be "proactive" through planning, training, and networking versus "reactive" in always responding to the unknown and not knowing the players when you get there. The Federal Government has served us all well with this law, but we need to update some of the provisions to meet the needs of today.

I believe we all knew it would come some day. Never did any of us expect it to come in a manner so coordinated with such devastating results. It did, and we should of learned from it and not forget. I was always told by my Father that mistakes and accidents can and will happen. Most importantly you learn and work to make sure you don't make the same mistake twice. We may of missed it the first time to a degree, but let's do everything to prevent it from happening a second time. The next time when it comes we are told by the Top Security minds in our Governments it may be greater in magnitude with even more loss of life and property. Utilizing Weapons of Mass Destruction (WMD) involving Chemicals, Biological, Nuclear, Radioactive, and Explosive devices. We need to get and be ready, now.

At the Local Government and First Responder levels we are concerned that we that our residents believe that we can protect them effectively against the threat of WMD, which could easily be one of our Chemical facilities. Our men and women on the front lines in our communities have been working hard in getting some of the special training and have begun to reap the benefit of some of the generous Homeland Funding made available by this Congress and President by putting new specialized equipment in the hands of the First Responders and Local Governments. The sharing of intelligence between levels of Governments hasn't been better. However, we are not were we need to be yet and have a long way to go, but we are better off today than yesterday.

Terrorism Threat Assessments and a Uniform Strategy to deal with them are a common requirement and need at all levels of Government. In looking at all the existing and potential hazards and threats to our Communities Chemical Facilities and there Transportation concerns rise with a few to the top. It's not that we don't know what's in the Plant or what being transported in most cases we do, through the impact of Federal and State Laws, but it's we don't know for sure what safety and security measures are in place to keep someone or thing from getting to them. Can the "Bad Guy's" use them against us? The fact is that there are some chemicals and materials if released from their containers for whatever reason or by a terrorist that can cause great injury and death to an unprotected public. We have to make sure that we do everything in our collective powers to make sure that we understand and make Chemical Facilities and there Transportation safe from attack through enhance security assessment and coordination possibly using the frame work of the SARA Title III Law and the Local Emergency Planning Committee's (LEPC), Joint Terrorism Task Forces (JTTF), and U.S. Department of Homeland Security Specialist's. In that regard we believe that new Legislation or a revision to SARA Title III is necessary to bring security standardization to Chemical site plans.

Chairman and Members of this Committee today you are hearing from some of the most notable and responsible Chemical Companies in our Country. I have had the opportunity work with a number of their response personnel over the years. The companies are top notch, well trained, have excellent plans, in very good financial state, and have in most cases good security systems. Unfortunately, that is not the case around the Country for many of the other Companies. Many smaller Chemical companies often suffer from lack of finances, resources, and do not have good security systems and accountability. The American Chemistry Council has done a good job in stepping up to the plate with providing a voluntary program with materials and training on Chemical Plant Security. A problem exists that first it is "Voluntary" and second, not all Companies belong to the Council.

Today, we have an opportunity to be proactive versus reactive. Chemical Plant and their Transportation is an issue that needs and should be addressed on a National level to ensure uniformity. I don't have the political or legislative expertise on whether a new law or tweaking an old one is the best way to go. I leave that you. I was around in the 80's when there was great out cry from the Chemical Industry about how the SARA Title III Law was unnecessary and that the industry voluntary program for planning, and response etc. was more than adequate. The Law almost was not enacted. It took a real wake up call. Several thousand lost their lives and tens of thousands were injured in Bhopal, India in 1984 and another incident in West Virginia to raise enough concern to move on the Law.

Today we may hear some of the same. Security, trade secrets, plans, products we have heard it all before. What if it gets out? Etc. Together we can work out the details. I don't know of many trade secrets or critical information given out to competitors or the general public. If so there should be appropriate accountability and even sanctions for those who do. LEPC's have been a great tool to ensure effective planning and community safety. We can have experienced security people look over plans as necessary. I do not advocate LEPC's as a policing agency for security, but I do advocate that you cannot appreciate or effectively plan for incidents within your jurisdiction without the full benefit of all aspects of hazards and vulnerabilities.

The Public is counting on us. "Shame on all of us if we wait until it's to late"

October 2004

Security Vulnerability Assessment Methodology for the Petroleum and Petrochemical Industries, Second Edition

American Petroleum Institute

NPRA

October 2004

Security Vulnerability Assessment Methodology for the Petroleum and Petrochemical Industries, Second Edition

American Petroleum Institute
1220 L Street, NW
Washington, DC
20005-4070

National Petrochemical &
Refiners Association
1899 L Street, NW
Suite 1000
Washington, DC
20036-3896

PREFACE

The American Petroleum Institute (API) and the National Petrochemical & Refiners Association (NPRA) are pleased to make this Second Edition of this Security Vulnerability Assessment Methodology available to members of petroleum and petrochemical industries. The information contained herein has been developed in cooperation with government and industry, and is intended to provide a tool to help maintain and strengthen the security of personnel, facilities, and industry operations; thereby enhancing the security of our nation's energy infrastructure.

API and NPRA wish to express sincere appreciation to the member companies who have made personnel available to work on this document. We especially thank the Department of Homeland Security and its Directorate of Information Analysis & Infrastructure Protection and the Department of Energy's Argonne National Laboratory for their invaluable contributions. The lead consultant in developing this methodology has been David Moore of the AcuTech Consulting Group, whose help and experience was instrumental in developing this document. Lastly, we want to acknowledge the contributions of the Centers for Chemical Process Safety for their initial work on assessing security vulnerability in the chemical industry.

This methodology constitutes but one approach for assessing security vulnerabilities at petroleum and petrochemical industry facilities. However, there are several other vulnerability assessment techniques and methods available to industry, all of which share common risk assessment elements. Many companies, moreover, have already assessed their own security needs and have implemented security measures they deem appropriate. This document is not intended to supplant measures previously implemented or to offer commentary regarding the effectiveness of any individual company efforts.

The focus of this second edition was to expand the successful first edition by including additional examples of how the methodology can be applied to a wide range of assets and operations. This includes petroleum refining and petrochemical manufacturing operations, pipelines, and transportation including truck and rail. The methodology was originally field tested at two refinery complexes, including an interconnected tank farm, marine terminal and lube plant before the publication of the first edition. Since then, it has been used extensively at a wide variety of facilities involving all aspects of the petroleum and petrochemical industries.

API and NPRA are not undertaking to meet the duties of employers, manufacturers, or suppliers to train and equip their employees, nor to warn any who might potentially be exposed, concerning security risks and precautions. Ultimately, it is the responsibility of the owner or operator to select and implement the security vulnerability assessment method and depth of analysis that best meet the needs of a specific location.

CONTENTS

CHAPTER 1 INTRODUCTION 1
1.1 INTRODUCTION TO SECURITY VULNERABILITY ASSESSMENT 1
1.2 OBJECTIVES, INTENDED AUDIENCE AND SCOPE OF THE GUIDANCE 1
1.3 SECURITY VULNERABILITY ASSESSMENT AND SECURITY MANAGEMENT PRINCIPLES 2
CHAPTER 2 SECURITY VULNERABILITY ASSESSMENT CONCEPTS 3
2.1 INTRODUCTION TO SVA TERMS 3
2.2 RISK DEFINITION FOR SVA 3
2.3 CONSEQUENCES 4
2.4 ASSET ATTRACTIVENESS 4
2.5 THREAT 5
2.6 VULNERABILITY 5
2.7 SVA APPROACH 5
2.8 CHARACTERISTICS OF A SOUND SVA APPROACH 7
2.9 SVA STRENGTHS AND LIMITATIONS 8
2.10 RECOMMENDED TIMES FOR CONDUCTING AND REVIEWING THE SVA 8
2.11 VALIDATION AND PRIORITIZATION OF RISKS 8
2.12 RISK SCREENING 9
CHAPTER 3 SECURITY VULNERABILITY ASSESSMENT METHODOLOGY 9
3.1 OVERVIEW OF THE SVA METHODOLOGY 9
3.2 SVA METHODOLOGY 15
3.3 STEP 1: ASSETS CHARACTERIZATION 18
3.4 STEP 2: THREAT ASSESSMENT 23
3.5 SVA STEP 3: VULNERABILITY ANALYSIS 25
3.6 STEP 4: RISK ANALYSIS/RANKING 28
3.7 STEP 5: IDENTIFY COUNTERMEASURES: 28
3.8 FOLLOW-UP TO THE SVA 29
ATTACHMENT 1 – EXAMPLE SVA METHODOLOGY FORMS 31
ABBREVIATIONS AND ACRONYMS 41

APPENDIX A—SVA SUPPORTING DATA REQUIREMENTS 43
APPENDIX B—SVA COUNTERMEASURES CHECKLIST 45
APPENDIX C—SVA INTERDEPENDENCIES AND INFRASTRUCTURE CHECKLIST 67
APPENDIX C1—REFINERY SVA EXAMPLE 115
APPENDIX C2—PIPELINE SVA EXAMPLE 123
APPENDIX C3—TRUCK TRANSPORTATION SVA EXAMPLE 135
APPENDIX C4—RAIL TRANSPORTATION SVA EXAMPLE 145
References 155

Figures
2.1 Risk Definition 3
2.2 SVA Risk Variables 3
2.3 Asset Attractiveness Factors 4
2.4 Overall Asset Screening Approach 6
2.5 Recommended Times for Conducting and Reviewing the SVA 9

3.1 Security Vulnerability Assessment Methodology Steps 11
3.1a Security Vulnerability Assessment Methodology—Step 1 12
3.1b Security Vulnerability Assessment Methodology—Step 2 13
3.1c Security Vulnerability Assessment Methodology—Steps 3 – 5 14
3.2 SVA Methodology Timeline 15
3.3 SVA Team Members 16
3.4 Sample Objectives Statement 16
3.5 Security Events of Concern 17
3.6 Description of Step 1 and Substeps 19
3.7 Example Candidate Critical Assets 20
3.8 Possible Consequences of Security Events 21
3.9 Example Definitions of Consequences of the Event 22
3.10 Description of Step 2 and Substeps 23
3.11 Threat Rating Criteria 25
3.12 Target Attractiveness Factors (for Terrorism) 25
3.13 Attractiveness Factors Ranking Definitions (A) 26
3.14 Description of Step 3 and Substeps 26
3.15 Vulnerability Rating Criteria 27
3.16 Description of Step 4 and Substeps 28
3.17 Risk Ranking Matrix 29
3.18 Description of Step 5 and Substeps 29
A SVA Methodology Flow Diagram 124

Security Vulnerability Assessment Methodology for the Petroleum and Petrochemical Industries

Chapter 1 Introduction

1.1 INTRODUCTION TO SECURITY VULNERABILITY ASSESSMENT

The first step in the process of managing security risks is to identify and analyze the threats and the vulnerabilities facing a facility by conducting a Security Vulnerability Assessment (SVA). The SVA is a systematic process that evaluates the likelihood that a threat against a facility will be successful. It considers the potential severity of consequences to the facility itself, to the surrounding community and on the energy supply chain.

The SVA process is a team-based approach that combines the multiple skills and knowledge of the various participants to provide a complete security analysis of the facility and its operations. Depending on the type and size of the facility, the SVA team may include individuals with knowledge of physical and cyber security, process safety, facility and process design and operations, emergency response, management and other disciplines as necessary.

The objective of conducting a SVA is to identify security hazards, threats, and vulnerabilities facing a facility, and to evaluate the countermeasures to provide for the protection of the public, workers, national interests, the environment, and the company. With this information security risks can be assessed and strategies can be formed to reduce vulnerabilities as required. SVA is a tool to assist management in making decisions on the need for countermeasures to address the threats and vulnerabilities.

1.2 OBJECTIVES, INTENDED AUDIENCE AND SCOPE OF THE GUIDANCE

This document was prepared by the American Petroleum Institute (API) and the National Petrochemical & Refiners Association (NPRA) Security Committees to assist the petroleum and petrochemical industries in understanding security vulnerability assessment and in conducting SVAs. The guidelines describe an approach for assessing security vulnerabilities that is widely applicable to the types of facilities operated by the industry and the security issues they face. During the development process it was field tested at two refineries, two tank farms, and a lube plant, which included typical process equipment, storage tanks, marine operations, infrastructure, pipelines, and distribution terminals for truck and rail. Since then, it has been used extensively at a wide variety of facilities involving all aspects of the petroleum and petrochemical industry.

This methodology constitutes one approach for assessing security vulnerabilities at petroleum and petrochemical industry facilities. However, there are several other vulnerability assessment techniques and methods available to industry, all of which share common risk assessment elements. Many companies, moreover, have already assessed their own security needs and have implemented security measures they deem appropriate. This document is not intended to supplant measures previously implemented or to offer commentary regarding the effectiveness of any individual company efforts.

Ultimately, it is the responsibility of the owner/operator to choose the SVA method and depth of analysis that best meets the needs of the specific location. Differences in geographic location, type of operations, and on-site quantities of hazardous substances all play a role in determining the level of SVA and the approach taken. Independent of the SVA method used, all techniques include the following activities:

- Characterize the facility to understand what critical assets need to be secured, their importance and their interdependencies and supporting infrastructure;
- Identify and characterize threats against those assets and evaluate the assets in terms of attractiveness of the targets to each adversary and the consequences if they are damaged or stolen;
- Identify potential security vulnerabilities that threaten the asset's service or integrity;
- Determine the risk represented by these events or conditions by determining the likelihood of a successful event and the consequences of an event if it were to occur;
- Rank the risk of the event occurring and, if high risk, make recommendations for lowering the risk;
- Identify and evaluate risk mitigation options (both net risk reduction and benefit/cost analyses) and re-assess risk to ensure adequate countermeasures are being applied.

This guidance was developed for the industry as an adjunct to other available references which includes:

- American Petroleum Institute, "Security Guidelines for the Petroleum Industry", May, 2003;
- API RP 70, "Security for Offshore Oil and Natural Gas Operations", First Edition, April, 2003;

- "Guidelines for Analyzing and Managing the Security Vulnerabilities of Fixed Chemical Sites", American Institute of Chemical Engineers (AIChE) Center for Chemical Process Safety (CCPS"), August, 2002;
- "Vulnerability Analysis Methodology for Chemical Facilities (VAM-CF)", Sandia National Laboratories, 2002.

API and NPRA would like to acknowledge the contribution of the Center for Chemical Process Safety (CCPS) compiled in their "Guidelines for Analyzing and Managing the Security of Fixed Chemical Sites." It was this initial body of work that was used as a basis for developing the first edition of the API NPRA SVA methodology. Although similar in nature, the SVA Method was developed for the petroleum and petrochemical industry, at both fixed and mobile systems. Examples have been added that demonstrate applicability at various operating segments of the industry. Owner/Operators may want to use any of the methods above, or another equivalent and appropriate methodology in conducting their SVAs. These guidelines should also be considered in light of any applicable federal, state and local laws and regulations.

The guidance is intended for site managers, security managers, process safety managers, and others responsible for conducting security vulnerability analyses and managing security at petroleum and petrochemical facilities.

The method described in this guidance may be widely applicable to a full spectrum of security issues, but the key hazards of concern are malevolent acts, such as terrorism, that have the potential for widespread casualties or damage.

These guidelines provide additional industry segment specific guidance to the overall security plan and SVA method presented in Part I of the API Security Guidelines for the Petroleum Industry.

1.3 SECURITY VULNERABILITY ASSESSMENT AND SECURITY MANAGEMENT PRINCIPLES

Owner/Operators should ensure the security of facilities and the protection of the public, the environment, workers, and the continuity of the business through the management of security risks. The premise of the guidelines is that security risks should be managed in a risk-based, performance-oriented management process.

The foundation of the security management approach is the need to identify and analyze security threats and vulnerabilities, and to evaluate the adequacy of the countermeasures provided to mitigate the threats. Security Vulnerability Assessment is a management tool that can be used to assist in accomplishing this task, and to help the owner/operator in making decisions on the need for and value of enhancements.

The need for security enhancements will be determined partly by factors such as the degree of the threat, the degree of vulnerability, the possible consequences of an incident, and the attractiveness of the asset to adversaries. In the case of terrorist threats, higher risk sites are those that have critical importance, are attractive targets to the adversary, have a high level of consequences, and where the level of vulnerability and threat is high.

SVAs are not necessarily a quantitative risk assessment, but are usually performed qualitatively using the best judgment of the SVA Team. The expected outcome is a qualitative determination of risk to provide a sound basis for rank ordering of the security-related risks and thus establishing priorities for the application of countermeasures.

A basic premise is that all security risks cannot be completely prevented. The security objectives are to employ four basic strategies to help minimize the risk:

1. Deter
2. Detect
3. Delay
4. Respond

Appropriate strategies for managing security can vary widely depending on the individual circumstances of the facility, including the type of facility and the threats facing the facility. As a result, this guideline does not prescribe security measures but instead suggests means of identifying, analyzing, and reducing vulnerabilities. The specific situations must be evaluated individually by local management using best judgment of applicable practices. Appropriate security risk management decisions must be made commensurate with the risks. This flexible approach recognizes that there isn't a uniform approach to security in the petroleum industry, and that resources are best applied to mitigate high-risk situations primarily.

All Owner/Operators are encouraged to seek out assistance and coordinate efforts with federal, state, and local law enforcement agencies, and with the local emergency services and Local Emergency Planning Committee. Owner/Operators can also obtain and share intelligence, coordinate training, and tap other resources to help deter attacks and to manage emergencies.

Chapter 2 Security Vulnerability Assessment Concepts

2.1 INTRODUCTION TO SVA TERMS

A Security Vulnerability Assessment (SVA) is the process that includes determining the likelihood of an adversary successfully exploiting vulnerability and estimating the resulting degree of damage or impact. Based on this assessment, judgments can be made on degree of risk and the need for additional countermeasures. To conduct a SVA, key terms and concepts must be understood as explained in this chapter.

2.2 RISK DEFINITION FOR SVA

For the purposes of a SVA, the definition of risk is shown in Figure 2.1. The risk that is being analyzed for the SVA is defined as an expression of the likelihood that a defined threat will target and successfully attack a specific security vulnerability of a particular target or combination of targets to cause a given set of consequences. The complete SVA may evaluate one or more issues or sum the risk of the entire set of security issues. The risk variables are defined as shown in Figure 2.2.

A high-risk event, for example, is one which is represented by a high likelihood of a successful attack against a given critical target asset. Likelihood is determined by considering several factors including its attractiveness to the adversary, the degree of threat, and the degree of vulnerability. Criticality is determined by the asset's importance or value, and the potential consequences if attacked. If the likelihood of a successful attack against an important asset is high, then the risk is considered high and appropriate countermeasures would be required for a critical asset at high risk.

For the SVA, the risk of the security event is normally estimated qualitatively. It is based on the consensus judgment of a team of knowledgeable people as to how the likelihood and consequences of an undesired event scenario compares to other scenarios. The assessment is based on best available information, using experience and expertise of the team to make sound risk management decisions. The team may use a risk matrix, which is a graphical representation of the risk factors, as a tool for risk assessment decisions.

The API NPRA SVA Methodology has a two step screening process to focus attention on higher risk events. The key variables considered in the first screening are Consequences and Target Attractiveness. If either of those are either not sufficiently significant, the asset is screened out from further specific consideration. Later, the complete set of risk variables shown in Figure 2.1 are used in the second screen to determine the need for additional specific countermeasures.

Figure 2.1—Risk Definition

Security Risk is a function of: • **Consequences** of a successful attack against an asset and • **Likelihood** of a successful attack against an asset.
Likelihood is a function of: • the **Attractiveness** to the adversary of the asset, • the degree of **Threat** posed by the adversary, and • the degree of **Vulnerability** of the asset.

Figure 2.2—SVA Risk Variables[4]

Consequences	*Consequences* are the potential adverse impacts to a facility, the local community and/or the nation as a result of a successful attack.
Likelihood	*Likelihood* is a function of the chance of being targeted for attack, and the conditional chance of mounting a successful attack (both planning and executing) given the threat and existing security measures. This is a function of Threat, Vulnerability, and Target Attractiveness (see Figure 2.1).
Attractiveness	*Attractiveness* is a surrogate measure for likelihood of attack. This factor is a composite estimate of the perceived value of a target to a specific adversary.
Threat	*Threat* is a function of an adversary's intent, motivation, capabilities, and known patterns of operation. Different adversaries may pose different threats to various assets within a given facility or to different facilities.
Vulnerability	*Vulnerability* is any weakness that can be exploited by an adversary to gain access and damage or steal an asset or disrupt a critical function. This is a variable that indicates the likelihood of a successful attack given the intent to attack an asset.

[4]Ibid, AIChE.

2.3 CONSEQUENCES

The severity of the consequences of a security event at a facility is generally expressed in terms of the degree of injury or damage that would result if there were a successful attack. Malevolent acts may involve effects that are more severe than expected with accidental risk. Some examples of relevant consequences in a SVA include:

- Injuries to the public or to workers.
- Environmental damage.
- Direct and indirect financial losses to the company and to suppliers and associated businesses.
- Disruption to the national economy, regional, or local operations and economy.
- Loss of reputation or business viability.
- Need to evacuate people living or working near the facility.
- Excessive media exposure and related public concern affecting people that may be far removed from the actual event location.

The estimate of consequences may be different in magnitude or scope than is normally anticipated for accidental releases. In the case of security events, adversaries are determined to cause maximize damage, so a worse credible security event should be defined. Critical infrastructure especially may have dependencies and interdependencies that need careful consideration.

In addition, theft of hazardous materials should be included in SVAs as applicable. Adversaries may be interested in theft of hazardous materials to either cause direct harm at a later date, use them for other illicit purposes such as illegal drug manufacturing, or possibly to make chemical weapons using the stolen materials as constituents.

Consequences are used as one of the key factors in determining the criticality of the asset and the degree of security countermeasures required. During the facility characterization step, consequences are used to screen low value assets from further consideration. For example, terrorists are assumed to be uninterested in low consequence assets (those that do not meet their criteria for valuable impacts).

2.4 ASSET ATTRACTIVENESS

Not all assets are of equal value to adversaries. A basic assumption of the SVA process is that this perception of value from an adversary's perspective is a factor that influences the likelihood of a security event. Asset attractiveness is an estimate of the real or perceived value of a target to an adversary based on such factors as shown in Figure 2.3.

During the SVA, the attractiveness of each asset should be evaluated based on the adversary's intentions or anticipated level of interest in the target. Security strategies can be developed around the estimated targets and potential threats. This factor, along with consequences, are used to screen facilities from more specific scenario analysis and from further specific countermeasures considerations during the first screening of the methodology.

Figure 2.3—Asset Attractiveness Factors

Type of effect:
• Potential for causing maximum casualties
• Potential for causing maximum damage and economic loss to the facility and company
• Potential for causing maximum damage and economic loss to the geographic region
• Potential for causing maximum damage and economic loss to the national infrastructure
Type of target:
• Usefulness of the process material as a weapon or to cause collateral damage
• Proximity to a national asset or landmark
• Difficulty of attack including ease of access and degree of existing security measures (soft target)
• High company reputation and brand exposure
• Iconic or symbolic target
• Chemical or biological weapons precursor chemical
• Recognition of the target

2.5 THREAT

Threat can be defined as any indication, circumstance, or event with the potential to cause loss of, or damage, to an asset. It can also be defined as the intention and capability of an adversary to undertake actions that would be detrimental to valued assets. Sources of threats may be categorized as:

- Terrorists (international or domestic);
- Activists, pressure groups, single-issue zealots;
- Disgruntled employees or contractors;
- Criminals (e.g., white collar, cyber hacker, organized, opportunists).

Threat information is important reference data to allow the Owner/Operator to understand the adversaries interested in the assets of the facility, their operating history, their methods and capabilities, their possible plans, and why they are motivated. This information should then be used to develop a design basis threat or threats.

Adversaries may be categorized as occurring from three general types:

- Insider threats
- External threats
- Insiders working as colluders with external threats

Each applicable adversary type should be evaluated against each asset as appropriate to understand vulnerabilities.

2.6 VULNERABILITY

Vulnerability is any weakness that can be exploited by an adversary to gain unauthorized access and subsequent destruction or theft of an asset. Vulnerabilities can result from, but are not limited to, weaknesses in current management practices, physical security, or operational security practices. In a SVA, vulnerabilities are evaluated either by broadly considering the threat and hazards of the assets they could attack or affect, or analyzed by considering multiple potential specific sequences of events (a scenario-based approach). For this SVA methodology, each critical asset is analyzed from at least an asset-based approach at first by considering consequences and attractiveness. If it is a specific high value target, then it is recommended to analyze the asset further using scenarios.

2.7 SVA APPROACH

The general approach is to apply risk assessment resources and, ultimately, special security resources primarily where justified based on the SVA results. The SVA process involves consideration of each facility from both the general viewpoint and specific asset viewpoint. Consideration at the general level is useful for determination of overall impacts of loss, infrastructure and interdependencies at the facility level, and outer perimeter analysis including access control and general physical security. For example, all facilities will maintain a minimum level of security with general countermeasures such as the plant access control strategy and administrative controls. Certain assets will justify a more specific level of security, such as additional surveillance or barriers, based on their value and expected level of interest to adversaries. The benefit of evaluating specific assets is that individual risks can be evaluated and specific countermeasures applied where justified in addition to more general countermeasures.

This SVA methodology uses this philosophy in several ways. The method is intended to be comprehensive and systematic in order to be thorough. First, it begins with the SVA team gaining an understanding of the entire facility, the assets that comprise the facility, the critical functions of the facility, and the hazards and impacts if these assets or critical functions are compromised. This results in an understanding of which assets and functions are 'critical' to the business operation. This is illustrated in Figure 2.4.

Criticality is defined both in terms of the potential impact to the workers, community, the environment and the company, as well as to the business importance of the asset. For example, a storage tank of a hazardous material may not be the most critical part of the operation of a process, but if attacked, it has the greatest combined impact so it may be given a high priority for further analysis and special security countermeasures.

Based on this first level of screening from all assets to critical assets, a critical asset list is produced. Next, the critical assets are reviewed in light of the threats. Adversaries may have different objectives, so the critical asset list is reviewed from each adversary's perspective and an asset attractiveness ranking is given. This factor is a quick measure of whether the adversary would value damaging, compromising, or stealing the asset, which serves as an indicator of the likelihood that an adversary would want to attack this asset and why.

If an asset is both critical (based on value and consequences) and attractive, then it is considered a "target" for purposes of the SVA. A target may optionally receive further specific analysis, including the development of scenarios to determine and test perceived vulnerabilities.

As shown in Figure 2.4, all assets receive at least a general security review. This is accomplished by the SVA team's initial consideration of assets, along with a baseline security survey. General security considerations may be found in security references such as the countermeasures checklist provided in Appendix B.

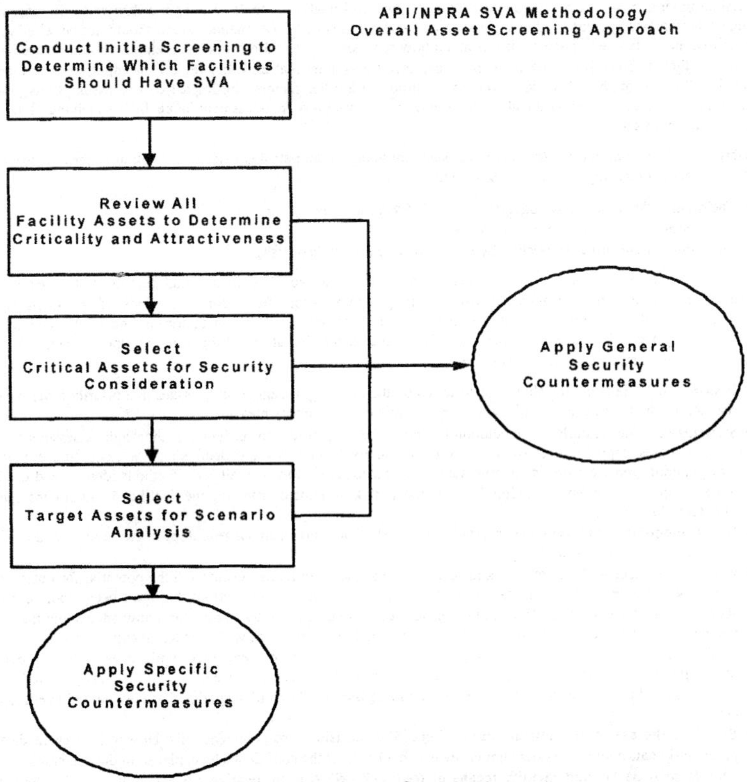

Figure 2.4—Overall Asset Screening Approach

All facilities should establish a security strategy. The general strategy is to protect against unauthorized access at the facility perimeter, and to control the access of authorized persons on the facility. Certain assets will be protected with added layers of protection, due to their attractiveness and consequences of loss. The specific security countermeasures provided to those assets would be to deter, detect, delay, and respond to credible threats against the assets to limit the risk to a certain level.

2.8 CHARACTERISTICS OF A SOUND SVA APPROACH

It is important to distinguish between a security risk management process and any given SVA methodology. Security risk management is the management framework that includes the SVA, development and implementation of a security plan, and the application of needed countermeasures to enhance security. SVA is the estimation of risk for the purposes of decision-making. SVA methodologies can be very powerful analytical tools to integrate data and information, and help understand the nature and locations of risks of a system. However, SVA methods alone should not be relied upon to establish risk, nor solely determine decisions about how risks should be addressed. SVA methods should be used as part of a process that involves knowledgeable and experienced personnel that critically review the input, assumptions, and results. The SVA team should integrate the SVA output with other factors, the impact of key assumptions, and the impact of uncertainties created by the absence of data or the variability in assessment inputs before arriving at decisions about risk and actions to reduce risk.

A variety of different approaches to SVA have been employed in the petroleum sector as well as other industries. The major differences among approaches are associated with:

- The relative "mix" of knowledge, data, or logic SVA methods;
- The complexity and detail of the SVA method; and
- The nature of the output (probabilistic versus relative measures of risk).

Ultimately, it is the responsibility of the owner/operator to choose the SVA method that best meets the needs of the company, the facilities and the agencies tasked with providing additional security in times of imminent danger. Therefore, it is in the best interest of the owner/operator to develop a thorough understanding of the various SVA methods in use and available, as well as the respective strengths and limitations of the different types of methods, before selecting a long-term strategy. A SVA should be:

- **Risk-based**—The approach should be to focus on the most significant security issues in a priority order based on risk. Risk can also be used to judge the adequacy of existing security measures.
- **Structured**—The underlying methodology must be structured to provide a thorough assessment. Some methodologies employ a more rigid structure than others. More flexible structures may be easier to use; however, they generally require more input from subject matter experts. However, all SVA methods identify and use logic to determine how the data considered contributes to risk in terms of affecting the likelihood and/or consequences of potential incidents.
- **Given adequate resources**—Appropriate personnel, time, and financial resources must be allocated to fit the detail level of the assessment.
- **Experience-based**—The frequency and severity of past security related events and the potential for future events should be considered. It is important to understand and account for any actions that have been made to prevent security related events. The SVA should consider the system-specific data and other knowledge about the system that has been acquired by field, operations, and engineering personnel as well as external expertise.
- **Predictive**—A SVA should be investigative in nature, seeking to identify recognized as well as previously unrecognized threats to the facility service and integrity. It should make use of previous security related events, but focus on the potential for future events, including the likelihood of scenarios that may never have happened before.
- **Based on the use of appropriate data**—Some SVA decisions are judgment calls. However, relevant data and particularly data about the system under review should affect the confidence level placed in the decisions.
- **Able to provide for and identify means of feedback**—SVA is an iterative process. Actual field drills, audits, and data collection efforts from both internal and external sources should be used to validate (or invalidate) assumptions made.

2.9 SVA STRENGTHS AND LIMITATIONS

Each of the SVA methods commonly used has its strengths and limitations. Some approaches are well suited to particular applications and decisions, but may not be as helpful in other situations. In selecting or applying SVA methods, there are a number of questions that should be considered. Some of the more significant ones are summarized below.

- Does the scope of the SVA method encompass and identify significant security related events and risks of the facility or along the system? If not, how can the risks that are not included in the SVA method be assessed and integrated in the future?
- Will all data be assessed, as it really exists along the system? Data should be location specific so that additive effects of the various risk variables can be determined. Can the assessment resolution be altered, e.g. station-by-station or mile-by-mile, dependent on the evaluation needs?
- What is the logical structure of variables that are evaluated to provide the qualitative and quantitative results of the SVA? Does this provide for straightforward data assimilation and assessment?
- Does the SVA method use numerical weights and other empirical factors to derive the risk measures and priorities? Are these weights based on the experience of the system, operator, industry, or external sources?
- Do the basic input variables of the SVA method require data that are available to the operator? Do operator data systems and industry data updating procedures provide sufficient support to apply the SVA method effectively? What is the process for updating the SVA data to reflect changes in the system, the infrastructure, and new security related data? How is the input data validated to ensure that the most accurate, up-to-date depiction of the system is reflected in the SVA?
- Does the SVA output provide adequate support for the justification of risk-based decisions? Are the SVA results and output documented adequately to support justification of the decisions made using this output?
- Does the SVA method allow for analysis of the effects of uncertainties in the data, structure, and parameter values on the method output and decisions being supported? What sensitivity or uncertainty analysis is supported by the SVA method?
- Does the SVA method focus exclusively on RMP-based "worst case" events or is it structured to determine "most probable worst case" events that may at times be less severe than postulated in an RMP or include additive effects of adjacent assets to yield consequences more severe than postulated in the RMP?

2.10 RECOMMENDED TIMES FOR CONDUCTING AND REVIEWING THE SVA

The SVA process or SVA methods can be applied at different stages of the overall security assessment and evaluation process. For example, it can be applied to help select, prioritize, and schedule the locations for security assessments. It can also be performed after the security assessment is completed to conduct a more comprehensive SVA that incorporates more accurate information about the facility or pipeline segment.

There are six occasions when the SVA may be required, as illustrated in Figure 2.5.

2.11 VALIDATION AND PRIORITIZATION OF RISKS

Independent of the process used to perform a SVA, the owner/operator must perform a quality control review of the output to ensure that the methodology has produced results consistent with the objectives of the assessment. This can be achieved through a review of the SVA data and results by a knowledgeable and experienced individual or, preferably, by a cross-functional team consisting of a mixture of personnel with skill sets and experience-based knowledge of the systems or segments being reviewed. This validation of the SVA method should be performed to ensure that the method has produced results that make sense to the operator. If the results are not consistent with the operator's understanding and expectations of system operation and risks, the operator should explore the reasons why and make appropriate adjustments to the method, assumptions, or data. Some additional criteria to evaluate the quality of a SVA are:

- Are the data and analyses handled competently and consistently throughout the system? (Can the logic be readily followed?)
- Is the assessment presented in an organized and useful manner?
- Are all assumptions identified and explained?
- Are major uncertainties identified, e.g., due to missing data?
- Do evidence, analysis, and argument adequately support conclusions and recommendations?

Once the SVA method and process has been validated, the operator has the necessary information to prioritize risks. To determine what risk mitigation actions to take, the operator considers which systems (or segments of systems) have the highest risks and then looks at the reasons the risks are higher for these assets. These risk factors are known as risk drivers since they drive the risk to a higher level for some assets than others do.

2.12 RISK SCREENING

Security issues exist at every facility managed by the petroleum and petrochemical industry, but the threat of intentional acts is likely to be different across the industry. This is captured by the factor known as 'asset attractiveness', whereby certain assets are considered to be more attractive to adversaries than others. Based on many reported threat assessments, intelligence reports, and actual events around the world, these factors can be used to evaluate target attractiveness.

It is likely that most facilities have no specific threat history for terrorism. As a result, the assumption must be made that potential malevolent acts are generally credible at each facility and this is then tempered by the site-specific factors. A screening process may contain the following factors:

1. Target attractiveness or target value;
2. Degree of threat;
3. Vulnerability;
4. Potential consequences (casualties, environmental, infrastructure and economic).

These are the same factors as are used for evaluating an individual asset risk, but the difference is that this is done at a generalized facility level for the risk screening instead of at a target asset level. Note that target attractiveness itself includes the factors of consequences and vulnerability. Target attractiveness is an aggregate of factors, which shows the complexity of the process of targeting. Consequences are listed again separately since they have such importance in targeting.

Consequence and target attractiveness are the dominant factors in determining terrorist risk. This is particularly true in the target-rich environment of the United States, where the rare nature of any particular terrorist act vs. the potential number of targets poses a major risk dilemma. Priority should first be given to the consequence ranking, but then consideration should be given to the attractiveness ranking when making assessments. In this way resources can be appropriately applied to assets where they are most likely to be important. This philosophy may be adopted by a company at an enterprise level to help determine both the need to conduct detailed (vs. simpler checklist analyses or audits), and the priority order for the analysis.

Figure 2.5—Recommended Times for Conducting and Reviewing the SVA

1	An initial review of all relevant facilities and assets per a schedule set during the initial planning process
2	When an existing process or operation is proposed to be substantially changed and prior to implementation (revision or rework)
3	When a new process or operation is proposed and prior to implementation (revision or rework)
4	When the threat substantially changes, at the discretion of the manager of the facility (revision or rework)
5	After a significant security incident, at the discretion of the manager of the facility (revision or rework)
6	Periodically to revalidate the SVA (revision or rework)

Chapter 3 Conducting the Security Vulnerability Assessment Methodology

3.1 OVERVIEW OF THE SVA METHODOLOGY

The SVA process is a risk-based and performance-based methodology. The user can choose different means of accomplishing the general SVA method so long as the end result meets the same performance criteria. The overall 5-step approach of the SVA methodology is described as follows:

Step 1: Asset Characterization

The asset characterization includes analyzing information that describes the technical details of facility assets as required to support the analysis, identifying the potential critical assets, identifying the hazards and consequences of concern for the facility and its surroundings and supporting infrastructure, and identifying existing layers of protection.

Step 2: Threat Assessment

The consideration of possible threats should include internal threats, external threats, and internally assisted threats (i.e., collusion between insiders and outside agents). The selection of the threats should include reasonable local, regional, or national intelligence information, where available. This step includes determining the target attractiveness of each asset from each adversary's perspective.

Step 3: Vulnerability Analysis

The vulnerability analysis includes the relative pairing of each target asset and threat to identify potential vulnerabilities related to process security events. This involves the identification of existing countermeasures and their level of effectiveness in reducing those vulnerabilities.

The degree of vulnerability of each valued asset and threat pairing is evaluated by the formulation of security-related scenarios or by an asset protection basis. If certain criteria are met, such as higher consequence and attractiveness ranking values, then it may be useful to apply a scenario-based approach to conduct the Vulnerability Analysis. It includes the assignment of risk rankings to the security-related scenarios developed. If the asset-based approach is used, the determination of the asset's consequences and attractiveness may be enough to assign a target ranking value and protect via a standard protection set for that target level. In this case, scenarios may not be developed further than the general thought that an adversary is interested in damaging or stealing an asset.

Step 4: Risk Assessment

The risk assessment determines the relative degree of risk to the facility in terms of the expected effect on each critical asset as a function of consequence and probability of occurrence. Using the assets identified during Step 1 (Asset Characterization), the risks are prioritized based on the likelihood of a successful attack. Likelihood is determined by the team after considering the attractiveness of the targeted assets assessed under Step 2, the degree of threats assessed under Step 2, and the degree of vulnerability identified under Step 3.

Step 5: Countermeasures Analysis

Based on the vulnerabilities identified and the risk that the layers of security are breached, appropriate enhancements to the security countermeasures may be recommended. Countermeasure options will be identified to further reduce vulnerability at the facility. These include improved countermeasures that follow the process security doctrines of deter, detect, delay, respond, mitigate and possibly prevent. Some of the factors to be considered are:

- Reduced probability of successful attack
- Degree of risk reduction by the options
- Reliability and maintainability of the options
- Capabilities and effectiveness of mitigation options
- Costs of mitigation options
- Feasibility of the options

The countermeasure options should be re-ranked to evaluate effectiveness, and prioritized to assist management decision making for implementing security program enhancements. The recommendations should be included in a SVA report that can be used to communicate the results of the SVA to management for appropriate action.

Once the SVA is completed, there is a need to follow-up on the recommended enhancements to the security countermeasures so they are properly reviewed, tracked, and managed until they are resolved. Resolution may include adoption of the SVA team's recommendations, substitution of other improvements that achieve the same level of risk abatement, or rejection. Rejection of a SVA recommendation and related acceptance of residual risk should be based on valid reasons that are well documented.

This SVA process is summarized in Figure 3.1 and illustrated further in the flowcharts that follow in Figures 3.1a through 3.1c. Section 3.2 of this chapter describes the preparation activities, such as data gathering and forming the SVA team. Sections 3.3 through 3.8 provide details for each step in the SVA methodology. These steps and associated tasks are also summarized in Figure 3.5.

Figure 3.1—Security Vulnerability Assessment Methodology Steps

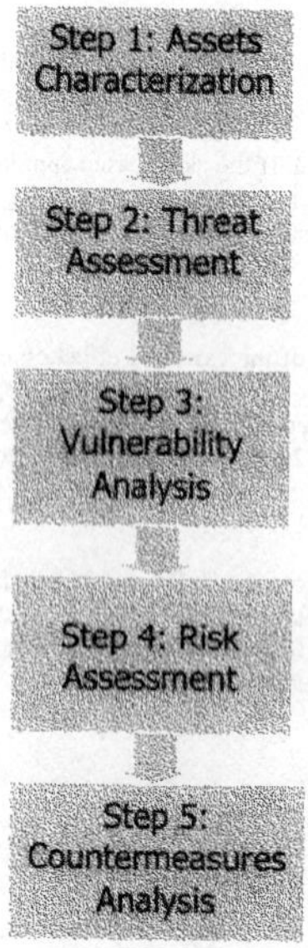

1.1 Identify critical assets and infrastructure
1.2 Evaluate existing countermeasures
1.3 Evaluate severity of impacts

2.1 Adversary identification
2.2 Adversary characterization
2.3 Target attractiveness
2.4 Select targets for further analysis

3.1 Define scenarios and evaluate specific consequences
3.2 Evaluate effectiveness of existing security measures
3.3 Identify vulnerabilities and estimate degree of vulnerability

4.1 Estimate likelihood of attack by vulnerability, threat, and attractiveness
4.2 Evaluate risk and need for additional countermeasures

5.1 Identify and evaluate countermeasures options
5.2 Prioritize potential enhancements by cost, effectiveness, and other factors

Figure 3.1a—Security Vulnerability Assessment Methodology—Step 1

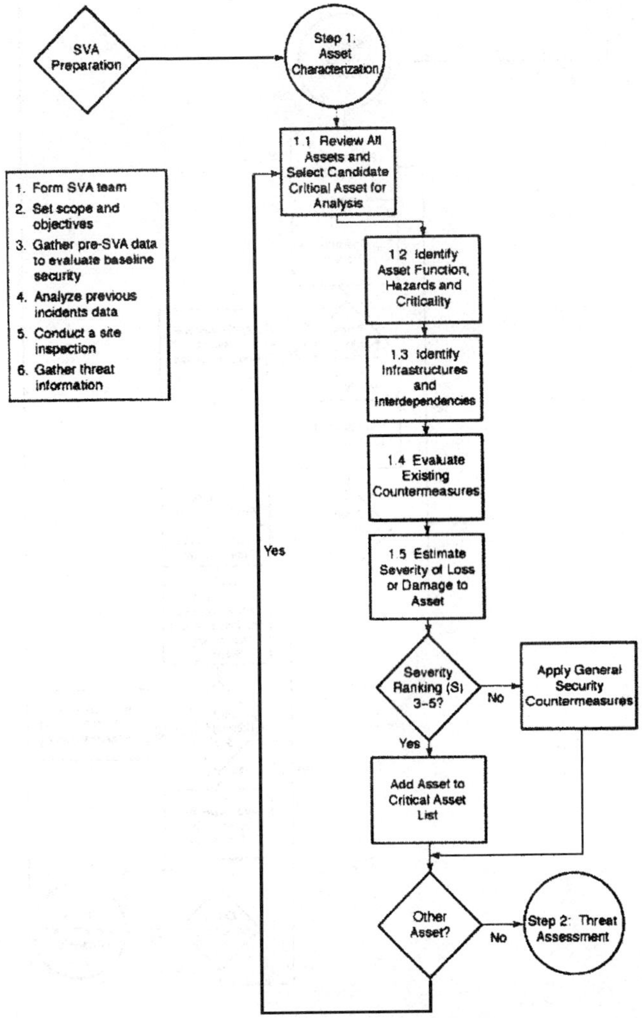

Figure 3.1b—Security Vulnerability Assessment Methodology—Step 2

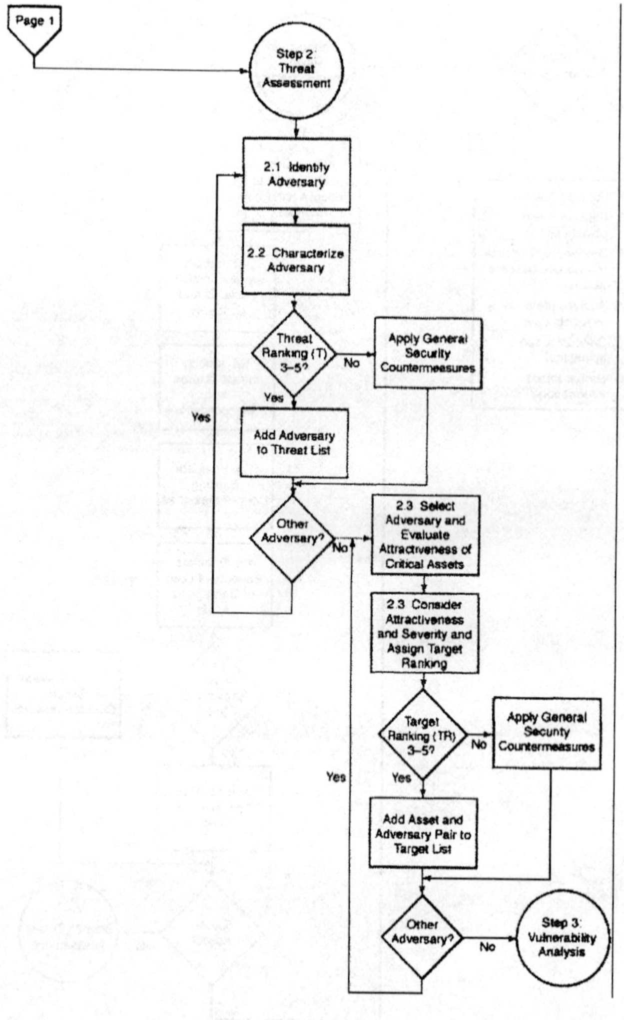

Figure 3.1c—Security Vulnerability Assessment Methodology—Steps 3 – 5

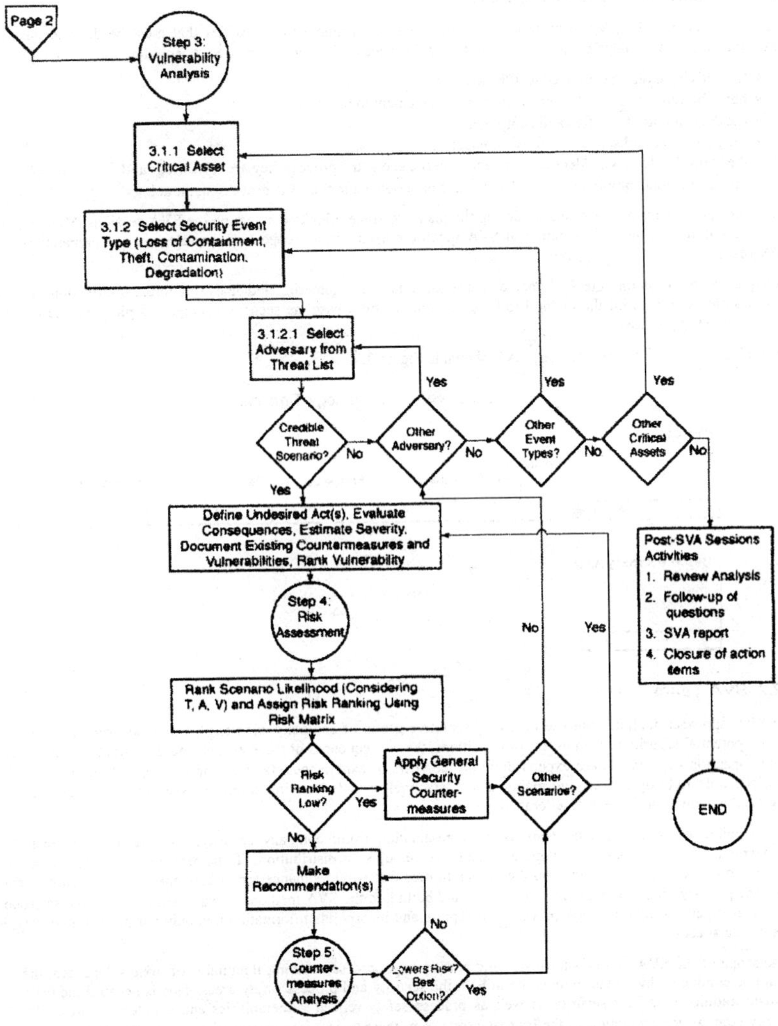

3.2 SVA PREPARATION

3.2.1 Planning for Conducting a SVA

Prior to conducting the SVA team-based sessions, there are a number of activities that must be done to ensure an efficient and accurate analysis. There are many factors in successfully completing a SVA including the following:

- the activity should be planned well in advance;
- have the full support and authorization by management to proceed;
- the data should be verified and complete;
- the objectives and scope should be concise;
- the team should be knowledgeable of and experienced at the process they are reviewing; and,
- the team leader should be knowledgeable and experienced in the SVA process methodology.

All of the above items are controllable during the planning stage prior to conducting the SVA sessions. Most important for these activities is the determination of SVA specific objectives and scope, and the selection and preparation of the SVA Team.

Prerequisites to conducting the SVA include gathering study data, gathering and analyzing threat information, forming a team, training the team on the method to be used, conducting a baseline security survey, and planning the means of documenting the process.

The typical timeline for conducting a SVA is shown in Figure 3.2.

Figure 3.2—SVA Methodology Timeline

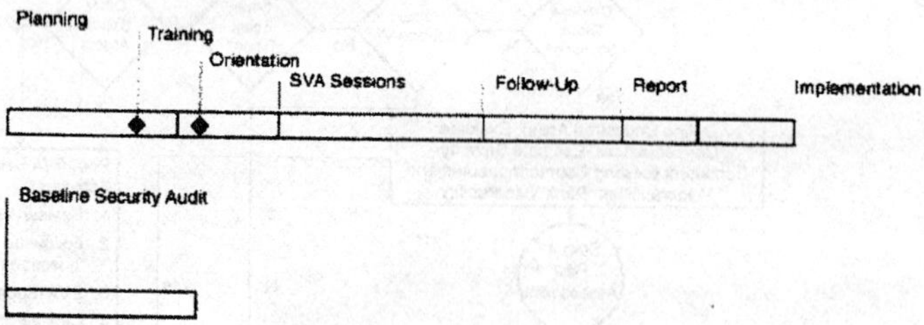

3.2.2 SVA Team

The SVA approach includes the use of a representative group of company experts plus outside experts if needed to identify potential security related events or conditions, the consequences of these events, and the risk reduction activities for the operator's system. These experts draw on the years of experience, practical knowledge, and observations from knowledgeable field operations and maintenance personnel in understanding where the security risks may reside and what can be done to mitigate or ameliorate them.

Such a company group typically consists of representation from: company security, risk management, operations, engineering, safety, environmental, regulatory compliance, logistics/distribution, IT and other team members as required. This group of experts should focus on the vulnerabilities that would enhance the effectiveness of the facility security plan. The primary goal of this group is to capture and build into the SVA method the experience of this diverse group of individual experts so that the SVA process will capture and incorporate information that may not be available in typical operator databases.

If the scope of the SVA includes terrorism and attacks on a process handling flammable or toxic substances, the SVA should be conducted by a team with skills in both the security and process safety areas. This is because the team must evaluate traditional facility security as well as process-safety related vulnerabilities and countermeasures. The final security strategy for protection of the process assets from these events is a combination of security and process safety strategies.

It is expected that a full time 'core' team is primarily responsible, and that they are led by a Team Leader. Other part-time team members, interviewees and guests are used as required for efficiency and completeness. At a minimum, SVA teams should possess the knowledge and/or skills listed in Figure 3.3. Other skills that should be considered and included, as appropriate, are included as optional or part-time team membership or as guests and persons interviewed.

The SVA Core Team is typically made up of three to five persons, but this is dependent on the number and type of issues to be evaluated and the expertise required to make those judgments. The Team Leader should be knowledgeable and experienced in the SVA approach.

3.2.3 SVA Objectives and Scope

The SVA Team leader should develop an objectives and scope statement for the SVA. This helps to focus the SVA and ensure completeness. An example SVA objectives statement is shown in Figure 3.4.

A work plan should then be developed to conduct the SVA with a goal of achieving the objectives. The work plan needs to include the scope of the effort, which includes which physical or cyber facilities and issues will be addressed.

Given the current focus on the need to evaluate terrorist threats, the key concerns are the intentional (malevolent) misuse of petroleum and hazardous to cause catastrophic consequences. Given this focus, the key events and consequences of interest include the four listed in Figure 3.5. Other events may be included in the scope as determined by the SVA Team, but it is recommended that these four primary security events be addressed first since these are the events that make the petroleum and petrochemical industry unique from other industries.

Figure 3.3—SVA Team Members

The SVA Core Team members should have the following skill sets and experience:

- Team leader—knowledge of and experience with the SVA methodology;
- Security representative—knowledge of facility security procedures, methods and systems;
- Safety representative—knowledge of potential process hazards, process safety procedures, methods, and systems of the facility;
- Facility representative—knowledge of the design of the facility under study including asset value, function, criticality, and facility procedures;
- Operations representative—knowledge of the facility process and equipment operation;
- Information systems/Automation representative (for cyber security assessment) —knowledge of information systems technologies and cyber security provisions; knowledge of process control systems.

The SVA Optional/Part-time Team members may include the following skill sets and experience:

- Security specialist—knowledge of threat assessment, terrorism, weapons, targeting and insurgency/guerilla warfare, or specialized knowledge of detection technologies or other countermeasures available;
- Cyber security specialist—knowledge of cyber security practices and technologies;
- Subject matter experts on various process or operations details such as process technologies, rotating equipment, distributed control systems, electrical systems, access control systems, etc.;
- Process specialist—knowledge of the process design and operations
- Management—knowledge of business management practices, goals, budgets, plans, and other management systems.

Figure 3.4—Sample Objectives Statement[8]

To conduct an analysis to identify security hazards, threats, and vulnerabilities facing a fixed facility handling hazardous materials, and to evaluate the countermeasures to provide for the protection of the public, workers, national interests, the environment, and the company.

[7]Ibid, AIChE.
[8]Ibid, AIChE.

Figure 3.5—Security Events of Concern

Security Event Type	Candidate Critical Assets
Loss of Containment, Damage, or Injury	Loss of containment of process hydrocarbons or hazardous chemicals on the plant site from intentional damage of equipment or the malicious release of process materials, which may cause multiple casualties, severe damage, and public or environmental impact. Also included is injury to personnel and the public directly or indirectly
Theft	Hydrocarbon, chemical, or information theft or misuse with the intent to cause severe harm at the facility or offsite
Contamination	Contamination or spoilage of plant products or information to cause worker or public harm on or offsite
Degradation of Assets	Degradation of assets or infrastructure or the business function or value of the facility or the entire company through destructive acts of terrorism.

3.2.4 Data Gathering, Review, and Integration

The objective of this step is to provide a systematic methodology for Owner/Operators to obtain the data needed to manage the security of their facility. Most Owner/Operators will find that many of the data elements suggested here are already being collected. This section provides a systematic review of potentially useful data to support a security plan. However, it should be recognized that all of the data elements in this section are not necessarily applicable to all systems.

The types of data required depend on the types of risks and undesired acts that are anticipated. The operator should consider not only the risks and acts currently suspected in the system, but also consider whether the potential exists for other risks and acts not previously experienced in the system, e.g., bomb blast damage. This section includes lists of many types of data elements. The following discussion is separated into four subsections that address sources of data, identification of data, location of data, and data collection and review.

Appendix A includes a list of potentially useful data that may be needed to conduct a SVA. Appendix B is a checklist of countermeasures that may be used as a data collection form prior to conducting a SVA. Similarly, Appendix C is a checklist for infrastructure and interdependencies that can be used both before and after a SVA for ensuring completeness.

3.2.4.1 Data Sources

The first step in gathering data is to identify the sources of data needed for facility security management. These sources can be divided into four different classes.

1. **Facility and Right of Way Records.** Facility and right of way records or experienced personnel are used to identify the location of the facilities. This information is essential for determining areas and other facilities that either may impact or be impacted by the facility being analyzed and for developing the plans for protecting the facility from security risks. This information is also used to develop the potential impact zones and the relationship of such impact zones to various potentially exposed areas surrounding the facility i.e., population centers, and industrial and government facilities.
2. **System Information.** This information identifies the specific function of the various parts of the process and their importance from a perspective of identifying the security risks and mitigations as well as understanding the alternatives to maintaining the ability of the system to continue operations when a security threat is identified. This information is also important from a perspective of determining those assets and resources available in-house in developing and completing a security plan. Information is also needed on those systems in place, which could support a security plan such as an integrity management program and IT security functions.
3. **Operation Records.** Operating data are used to identify the products transported and the operations as they may pertain to security issues to facilities and pipeline segments which may be impacted by security risks. This information is also needed to prioritize facilities and pipeline segments for security measures to protect the system, e.g., type of product, facility type and location, and volumes transported. Included in operation records data gathering is the need to obtain incident data to capture historical security events.
4. **Outside Support and Regulatory Issues.** This information is needed for each facility or pipeline segment to determine the level of outside support that may be needed and can be expected for the security measures to be employed at each facility or pipeline segment. Data are also needed to understand the expectation for security preparedness and coordination from the regulatory bodies at the government, state, and local levels. Data should also be developed on communication and other infrastructure issues as well as sources of information regarding security threats, e.g., ISACs (Information Sharing and Analysis Centers).

3.2.4.2 Identifying Data Needs

The type and quantity of data to be gathered will depend on the individual facility or pipeline system, the SVA methodology selected, and the decisions that are to be made. The data collection approach will follow the SVA path determined by the initial expert team assembled to identify the data needed for the first pass at SVA. The size of the facility or pipeline system to be evaluated and the resources available may prompt the SVA team to begin their work with an overview or screening assessment of the most critical issues that impact the facility or pipeline system with the intent of highlighting the highest risks. Therefore, the initial data collection effort will only include the limited information necessary to support this SVA. As the SVA process evolves, the scope of the data collection will be expanded to support more detailed assessment of perceived areas of vulnerability.

3.2.4.3 Locating Required Data

Operator data and information are available in different forms and format. They may not all be physically stored and updated at one location based on the current use or need for the information. The first step is to make a list of all data required for security vulnerability assessment and locate the data. The data and information sources may include:

- Facility plot plans, equipment layouts and area maps
- Process and Instrument Drawings (P&IDs)
- Pipeline alignment drawings
- Existing company standards and security best practices
- Product throughput and product parameters
- Emergency response procedures
- Company personnel interviews
- LEPC (Local Emergency Planning Commission) response plans
- Police agency response plans
- Historical security incident reviews
- Support infrastructure reviews

3.2.4.4 Data Collection and Review

Every effort should be made to collect good quality data. When data of suspect quality or consistency are encountered, such data should be flagged so that during the assessment process, appropriate confidence interval weightings can be developed to account for these concerns.

In the event that the SVA approach needs input data that are not readily available, the operator should flag the absence of information. The SVA team can then discuss the necessity and urgency of collecting the missing information

3.2.5 Analyzing Previous Incidents Data

Any previous security incidents relevant to the security vulnerability assessment may provide valuable insights to potential vulnerabilities and trends. These events from the site and, as available, from other historical records and references, should be considered in the analysis. This may include crime statistics, case histories, or intelligence relevant to facility.

3.2.6 Conducting a Site Inspection

Prior to conducting the SVA sessions, it is necessary for the team to conduct a site inspection to visualize the facility and to gain valuable insights to the layout, lighting, neighboring area conditions, and other facts that may help understand the facility and identify vulnerabilities. The list of data requirements in Appendix A and the checklist in Appendix B may be referenced for this purpose.

3.2.7 Gathering Threat Information

The team should gather and analyze relevant company and industry or government-provided threat information, such as that available from the Energy ISAC, DHS, FBI, or other local law enforcement agency.

3.3 STEP 1: ASSETS CHARACTERIZATION

Characterization of the facility is a step whereby the facility assets and hazards are identified, and the potential consequences of damage or theft to those assets is analyzed. The focus is on processes which may contain petroleum or hazardous chemicals and key assets, with an emphasis on possible public impacts. The Asset Attractiveness, based on these and other factors, is included in the facility characterization. These two factors (severity of the consequences and asset attractiveness) are used to screen the facility assets into those that require only general vs. those that require more specific security countermeasures.

The team produces a list of candidate critical assets that need to be considered in the analysis. Attachment 1—Step 1: Critical Assets/Criticality Form is helpful in developing and documenting the list of critical assets. The assets may be processes, operations, personnel, or any other asset as described in Chapter 3.

Figure 3.6 below summarizes the key steps and tasks required for Step 1.

Step 1.1—Identify Critical Assets

The SVA Team should identify critical assets for the site being studied. The focus is on petroleum or chemical process assets, but any asset may be considered. For example, the process control system may be designated as critical, since protection of it from physical and cyber attack may be important to prevent a catastrophic release or other security event of concern. Figure 3.7 is an example list of specific assets that may be designated as critical at any given site. Assets include the full range of both material and non-material aspects that enable a facility to operate.

Figure 3.6—Description of Step 1 and Substeps

Step	Task
Step 1: Assets Characterization	
1.1 Identify critical assets	Identify critical assets of the facility including people, equipment, systems, chemicals, products, and information.
1.2 Identify critical functions	Identify the critical functions of the facility and determine which assets perform or support the critical functions.
1.3 Identify critical infrastructures and interdependencies	Identify the critical internal and external infrastructures and their interdependencies (e.g., electric power, petroleum fuels, natural gas, telecommunications, transportation, water, emergency services, computer systems, air handling systems, fire systems, and SCADA systems) that support the critical operations of each asset.
1.4 Evaluate existing countermeasures	Identify what protects and supports the critical functions and assets. Identify the relevant layers of existing security systems including physical, cyber, operational, administrative, and business continuity planning, and the process safety systems that protect each asset.
1.5 Evaluate impacts	Evaluate the hazards and consequences or impacts to the assets and the critical functions of the facility from the disruption, damage, or loss of each of the critical assets or functions.
1.6 Select targets for further analysis	Develop a target list of critical functions and assets for further study.

Figure 3.7—Example Candidate Critical Assets

Security Event Type	Candidate Critical Assets
Loss of Containment, Damage, or Injury	• Process equipment handling petroleum and hazardous materials including processes, pipelines, storage tanks • Marine vessels and facilities, pipelines, other transportation systems • Employees, contractors, visitors in high concentrations
Theft	• Hydrocarbons or chemicals processed, stored, manufactured, or transported • Metering stations, process control and inventory management systems • Critical business information from telecommunications and information management systems including Internet accessible assets
Contamination	• Raw material, intermediates, catalysts, products, in processes, storage tanks, pipelines • Critical business or process data
Degradation of Assets	• Processes containing petroleum or hazardous chemicals • Business image and community reputation • Utilities (electric power, steam, water, natural gas, specialty gases) • Telecommunications Systems • Business systems

The following information should be reviewed by the SVA Team as appropriate for determination of applicability as critical assets:

- Any applicable regulatory lists of highly hazardous chemicals, such as the Clean Air Act 112(r) list of flammable and toxic substances for the EPA Risk Management Program (RMP) 40 *CFR* Part 68 or the OSHA Process Safety Management (PSM) 29 *CFR* 1910.119 list of highly hazardous chemicals;
- Inhalation poisons or other chemicals that may be of interest to adversaries;
- Large and small scale chemical weapons precursors as based on the following lists:
  - Chemical Weapons Convention list;
  - FBI Community Outreach Program (FBI List) for Weapons of Mass Destruction materials and precursors;
  - The Australia Group list of chemical and biological weapons
- Material destined for the food, nutrition, cosmetic or pharmaceutical chains;
- Chemicals which are susceptible to reactive chemistry.

Owner/Operators may wish to consider other categories of chemicals that may cause losses or injuries that meet the objectives and scope of the analysis. These may include other flammables, critically important substances to the process, explosives, radioactive materials, or other chemicals of concern.

In addition, the following personnel, equipment and information may be determined to be critical:

- Process equipment
- Critical data
- Process control systems
- Personnel
- Critical infrastructure and support utilities

Step 1.2—Identify Critical Functions

The SVA Team should identify the critical functions of the facility and determine which assets perform or support the critical functions. For example, the steam power plant of a refinery may be critical since it is the sole source of steam supply to the refinery.

Step 1.3—Identify Critical Infrastructures and Interdependencies

The SVA team should identify the critical internal and external infrastructures and their interdependencies (e.g., electric power, petroleum fuels, natural gas, telecommunications, transportation, water, emergency services, computer systems, air handling systems, fire systems, and SCADA systems) that support the critical operations of each asset. For example, the electrical substation may be the sole electrical supply to the plant, or a supplier delivers raw material to the facility via a single pipeline. Appendix C, Interdependencies and Infrastructure Checklist, can be used to identify and analyze these issues. Note that some of these issues may be beyond the control of the owner/operator, but it is necessary to understand the dependencies and interdependencies of the facility, and the result of loss of these systems on the process.

Step 1.4—Evaluate Existing Countermeasures

The SVA team identifies and documents the existing security and process safety layers of protection. This may include physical security, cyber security, administrative controls, and other safeguards. During this step the objective is to gather information on the types of strategies used, their design basis, and their completeness and general effectiveness. A pre-SVA survey is helpful to gather this information. The data will be made available to the SVA team for them to form their opinions on the adequacy of the existing security safeguards during Step 3: Vulnerability Analysis and Step 5: Countermeasures Analysis.

Appendix B—Countermeasures Survey Form can be used to gather information on the presence and status of existing safeguards or another form may be more suitable. Existing records and documentation on security and process safety systems, as well as on the critical assets themselves, can be referenced rather than repeated in another form of documentation.

The objective of the physical security portion of the survey is to identify measures that protect the entire facility and/or each critical asset of the facility, and to determine the effectiveness of the protection. Appendix B contains checklists that may be used to conduct the physical security portion of the survey.
Note that the infrastructure interdependencies portion of the survey will identify infrastructures that support the facility and/or its critical assets (e.g., electric power, water, and telecommunications). A physical security review of these vital infrastructures should also be conducted.

Step 1.5—Evaluate Impacts

The Impacts Analysis step includes both the determination of the hazards of the asset being compromised as well as the specific consequences of a loss. The SVA team should consider relevant chemical use and hazard information, as well as information about the facility. The intent is to develop a list of target assets that require further analysis partly based on the degree of hazard and consequences. Particular consideration should be given to the hazards of fire, explosion, toxic release, radioactive exposure, and environmental contamination.

The consequences are analyzed to understand their possible significance. The Appendix A—Attachment 1—Step 1: Critical Assets/Criticality Form is useful to document the general consequences for each asset. The consequences may be generally described but consideration should be given to those listed in Figure 3.8.

Figure 3.8—Possible Consequences of Security Events

Public fatalities or injuries
Site personnel fatalities or injuries
Large-scale disruption to the national economy, public or private operations
Large-scale disruption to company operations
Large-scale environmental damage
Large-scale financial loss
Loss of critical data
Loss of reputation or business viability

The consequence analysis is done in a general manner. If the security event involves a toxic or flammable release to the atmosphere, the EPA RMP offsite consequence analysis guidance can be used as a starting point. If it is credible to involve more than the largest single vessel containing the hazardous material in a single incident, the security event may be larger than the typical EPA RMP worst-case analysis.

A risk ranking scale can be used to rank the degree of severity. Figure 3.9 illustrates a set of consequence definitions based on four categories of events—A. Fatalities and injuries; B. Environmental impacts; C. Property damage; and D. Business interruption.

Figure 3.9—Example Definitions of Consequences of the Event

DESCRIPTION	RANKING
A. Possible for any offsite fatalities from large-scale toxic or flammable release; possible for multiple onsite fatalities B. Major environmental impact onsite and/or offsite (e.g., large-scale toxic contamination of public waterway) C. Over $X property damage D. Very long term (> X years) business interruption/expense; Large-scale disruption to the national economy, public or private operations; Loss of critical data; Loss of reputation or business viability	**S5 – Very High**
A. Possible for onsite fatalities; possible offsite injuries B. Very large environmental impact onsite and/or large offsite impact C. Over $ X – $ Y property damage D. Long term (X months – Y years) business interruption/expense	**S4 – High**
A. No fatalities or injuries anticipated offsite; possible widespread onsite serious injuries B. Environmental impact onsite and/or minor offsite impact C. Over $ X -$ Y property damage D. Medium term (X months – Y months) business interruption/expense	**S3 – Medium**
A. Onsite injuries that are not widespread but only in the vicinity of the incident location; No fatalities or injuries anticipated offsite B. Minor environmental impacts to immediate incident site area only C. $ X – $ Y loss property damage D. Short term (up to X months) business interruption/expense	**S2 – Low**
A. Possible minor injury onsite; No fatalities or injuries anticipated offsite B. No environmental impacts C. Up to $X Property Damage D. Very short term (up to X weeks) business interruption/expense	**S1 – Very Low**

The consequences of a security event at a facility are generally expressed in terms of the degree of acute health effects (e.g., fatality, injury), property damage, environmental effects, etc. This definition of consequences is the same as that used for accidental releases, and is appropriate for security-related events. The key difference is that they may involve effects that are more severe than expected with accidental risk. This difference has been considered in the steps of the SVA.

The SVA Team should evaluate the potential consequences of an attack using the judgment of the SVA team. If scenarios are done, the specific consequences may be described in scenario worksheets.

Team members skilled and knowledgeable in the process technology should review any off-site consequence analysis data previously developed for safety analysis purposes or prepared for adversarial attack analysis. The consequence analysis data may include a wide range of release scenarios if appropriate.

Proximity to off-site population is a key factor since it is both a major influence on the person(s) selecting a target, and on the person(s) seeking to defend that target. In terms of attractiveness to a terrorist, if the target could expose a large number of persons, this type of target is likely to be a high-value, high-payoff target.

Step 1.6—Select Targets for Further Analysis

For each asset identified, the criticality of each asset must be understood. This is a function of the value of the asset, the hazards of the asset, and the consequences if the asset was damaged, stolen, or misused. For hazardous chemicals, consideration may include toxic exposure to workers or the community, or potential for the misuse of the chemical to produce a weapon or the physical properties of the chemical to contaminate a public resource.

The SVA Team develops a Target Asset List which is a list of the assets associated with the site being studied that are more likely to be attractive targets, based on the complete list of assets and the identified consequences and targeting issues identified in the previous steps. During Step 3: Vulnerability Analysis, the Target Asset List will be generally paired with specific threats and evaluated against the potential types of attack that could occur.

The SVA methodology uses ranking systems that are based on a scale of 1 –s 5 where 1 is the lowest value and 5 is the highest value. Based on the consequence ranking and criticality of the asset, the asset is tentatively designated a candidate critical target asset. The attractiveness of the asset will later be used for further screening of important assets.

3.4 STEP 2: THREAT ASSESSMENT

The threat assessment step involves the substeps shown in Figure 3.10.

Step 2.1—Adversary Identification

The next step is to identify specific classes of adversaries that may perpetrate the security-related events. The adversary characterization sub-step involves developing as complete an understanding as is possible of the adversary's history, capabilities and intent. A threat matrix is developed to generally pair the assets with each adversary class as shown in Attachment 1—Step 2: Threat Assessment Form.

Figure 3.10—Description of Step 2 and Substeps

Step	Task
Step 2: Threat Assessment	
2.1 Adversary identification	Evaluate threat information and identify threat categories and potential adversaries. Identify general threat categories. Consider threats posed by insiders, external agents (outsiders), and collusion between insiders and outsiders.
2.2 Adversary characterization	Evaluate each adversary and provide an overall threat assessment/ ranking for each adversary using known or available information. Consider such factors as the general nature/history of threat; specific threat experience/history to the facility/operation; known capabilities/methods/weapons; potential actions, intent/ motivation of adversary.
2.3 Analyze target attractiveness	Conduct an evaluation of target (from assets identified in Step 1) attractiveness from the adversary perspective.

Depending on the threat, the analyst can determine the types of potential attacks and, if specific information is available (intelligence) on potential targets and the likelihood of an attack, specific countermeasures may be taken. Information may be too vague to be useful, but SVA Teams should seek available information from Federal, State, and Local law enforcement officials in analyzing threats. Absent specific threat information, the SVA can still be applied based on assuming general capabilities and characteristics of typical hypothetical adversaries.

Threat assessment is an important part of a security management system, especially in light of the emergence of international terrorism in the United States. There is a need for understanding the threats facing the industry and any given facility or operation to properly respond to those threats. This section describes a threat assessment approach as part of the security management process. Later in Section 3.0 the use of the threat assessment in the SVA process will be more fully explained.

A threat assessment is used to evaluate the likelihood of adversary activity against a given asset or group of assets. It is a decision support tool that helps to establish and prioritize security-program requirements, planning, and resource allocations. A threat assessment identifies and evaluates each threat on the basis of various factors, including capability, intention, and impact of an attack.

Threat assessment is a process that must be systematically done and kept current to be useful. The determination of these threats posed by different adversaries leads to the recognition of vulnerabilities and to the evaluation of required countermeasures to manage the threats. Without a design basis threat or situation specific threat in mind, a company cannot effectively develop a cost-effective security management system.

In characterizing the threat to a facility or a particular asset for a facility, a company should examine the historical record of security events and obtains available general and location-specific threat information from government organizations and other sources. It should then evaluate these threats in terms of company assets that represent likely targets.

Some threats are assumed continuous, whereas others are assumed to be variable. As such, this guidance follows the Department of Homeland Security's Homeland Security Advisory System (HSAS) and the U.S.C.G. Maritime Security (MARSEC) security levels for management of varying threat levels to the industry. The threat assessment determines the estimated general threat level, which varies as situations develop. Depending on the threat level, different security measures over baseline measures will likely be necessary.

While threat assessments are key decision support tools, it should be recognized that, even if updated often, threat assessments might not adequately capture emerging threats posed by some adversary groups. No matter how much we know about potential threats, we will never know that we have identified every threat or that we have complete information even about the threats of which we are aware. Consequently, a threat assessment must be accompanied by a vulnerability assessment to provide better assurance of preparedness for a terrorist or other adversary attack.

Intelligence and law enforcement agencies assess the foreign and domestic terrorist threats to the United States. The U.S. intelligence community—which includes the Central Intelligence Agency, the Defense Intelligence Agency, and the State Department's Bureau of Intelligence and Research, among others—monitors the foreign-origin terrorist threat to the United States. The FBI gathers information and assesses the threat posed by domestic sources of terrorism.

Threat information gathered by both the intelligence and law enforcement communities can be used to develop a company-specific threat assessment. A company attempts to identify threats in order to decide how to manage risk in a cost-effective manner. All companies are exposed to a multitude of threats, including terrorism or other forms of threat.

A threat assessment can take different forms, but the key components are:

1. Identification of known and potential adversaries;
2. Recognition and analysis of their intentions, motivation, operating history, methods, weapons, strengths, weaknesses, and intelligence capabilities;
3. Assessment of the threat posed by the adversary factors mentioned above against each asset, and the assignment of an overall criticality ranking for each adversary.

Threats need to be considered from both insiders and outsiders, or a combination of those adversaries working in collusion. Insiders are defined as those individuals who normally have authorized access to the asset. They pose a particularly difficult threat, due to the possibility for deceit, deception, training, knowledge of the facilities, and unsupervised access to critical information and assets.

The threat categories to be considered are those that include intent and capability of causing major catastrophic harm to the facilities and to the public or environment. Typical adversaries that may be included in a SVA are: international terrorists, domestic terrorists (including disgruntled individuals/"lone wolf" sympathizers), disgruntled employees, or extreme activists.

All companies are encouraged to discuss threats with local and Federal law enforcement officials, and to maintain networking with fellow industrial groups to improve the quality of applicable threat information.

The threat assessment is not necessarily based on perfect information. In fact, for most facilities, the best available information is vague or nonspecific to the facility. A particularly frustrating part of the analysis can be the absence of site-specific information on threats. A suggested approach is to make an assumption that international terrorism is possible at every facility that has adequate attractiveness to that threat. Site-specific information adjusts the generic average rankings accordingly.

To be effective, threat assessment must be considered a dynamic process, whereby the threats are continuously evaluated for change. During any given SVA exercise, the threat assessment is referred to for guidance on general or specific threats facing the assets. At that time the company's threat assessment should be referred to and possibly updated as required given additional information and analysis of vulnerabilities.

Figure 3.11 includes a five level ranking system for defining threats against an asset.

Step 2.2—Adversary Characterization

Insiders, outsiders or a combination of the two may perpetrate an attack. Insiders are personnel that have routine, unescorted access within the facility. Outsiders do not. Collusion between the two may be the result of monetary gain (criminal insider/terrorist outsider), ideological sympathy, or coercion.

The adversary characterization will assist in evaluating the attack issues associated with insider, outsider, and colluding adversary threats. The SVA team should consider each type of adversary identified as credible, and generally define their level of capabilities, motivation, and likelihood of threat.

Step 2.3—Analyze Target Attractiveness

The team assigns the target attractiveness ranking. To facilitate this use Attachment 1—Threat Assessment: Target Attractiveness Form can be used.

The attractiveness of the target to the adversary is a key factor in determining the likelihood of an attack. Examples of issues that may be addressed here include:

- Proximity to a symbolic or iconic target, such as a national landmark
- Unusually high corporate profile among possible terrorists, such as a major defense contractor
- Any other variable not addressed elsewhere, when the SVA Team agrees it has an impact on the site's value as a target or on the potential consequences of an attack.

The SVA Team should use the best judgment of its subject matter experts to assess attractiveness. This is a subjective process as are all vulnerability assessments whether qualitative or quantitative in nature.

Each asset is analyzed to determine the factors that might make it a more or less attractive target to the adversary. Attractiveness is used to assess likelihood of the asset being involved in an incident.

Target Attractiveness is an assessment of the target's value from the adversary's perspective, which is one factor used as a surrogate measure for likelihood of attack. Note that target attractiveness itself includes the other factors of consequences and difficulty of attack/vulnerability. Target attractiveness is an aggregate of factors, which shows the complexity of the process of targeting and anti-terrorism efforts. Arguably target attractiveness is the dominant factor in determining terrorist risk. This is particularly true in the target-rich environment of the United States, where the rare nature of any particular terrorist act vs. the potential number of targets poses a major risk assessment dilemma.

The attractiveness of assets varies with the adversary threat including their motivation, intent, and capabilities. For example, the threat posed by an international terrorist and the assets they might be interested in could greatly vary from the threat and assets of interest to a violent activist or environmental extremist.

Figure 3.12 shows the factors that should be evaluated when evaluating target attractiveness for terrorism. The team can use these factors and rank each asset against each adversary by the scale shown in Figure 3.13. Other adversaries may be interested in other factors, and the user of the SVA is encouraged to understand the relevant factors and substitute them for those in Figure 3.12 as applicable.

3.5 SVA STEP 3: VULNERABILITY ANALYSIS

The Vulnerability Analysis step involves three steps, as shown in Figure 3.14. Once the SVA Team has determined how an event can be induced, it should determine how an adversary could make it occur. There are two schools of thought on methodology: the scenario-based approach and the asset-based approach. Both approaches are identical in the beginning, but differ in the degree of detailed analysis of threat scenarios and specific countermeasures applied to a given scenario. The assets are identified, and the consequences and target attractiveness are analyzed as per Step 2, for both approaches. Both approaches result in a set of annotated potential targets, and both approaches may be equally successful at evaluating security vulnerabilities and determining required protection.

Figure 3.11—Threat Rating Criteria

Threat Level	Description
5 – Very High	Indicates that a credible threat exists against the asset and that the adversary demonstrates the capability and intent to launch an attack, and that the subject or similar assets are targeted on a frequently recurring basis.
4 – High	Indicates that a credible threat exists against the asset based on knowledge of the adversary's capability and intent to attack the asset or similar assets.
3 – Medium	Indicates that there is a possible threat to the asset based on the adversary's desire to compromise similar assets.
2 – Low	Indicates that there is a low threat against the asset or similar assets and that few known adversaries would pose a threat to the assets.
1 – Very Low	Indicates no credible evidence of capability or intent and no history of actual or planned threats against the asset or similar assets.

Figure 3.12—Target Attractiveness Factors (for Terrorism)

Type of effect:
• Potential for causing maximum casualties
• Potential for causing maximum damage and economic loss to the facility and company
• Potential for causing maximum damage and economic loss to the geographic region
• Potential for causing maximum damage and economic loss to the national infrastructure
Type of target:
• Usefulness of the process material as a weapon or to cause collateral damage
• Proximity to national asset or landmark
• Difficulty of attack including ease of access and degree of existing security measures (soft target)
• High company reputation and brand exposure
• Iconic or symbolic target
• Chemical or biological weapons precursor chemical
• Recognition of the target

Figure 3.13—Attractiveness Factors Ranking Definitions (A)

Ranking Levels	Adversary Ranking (1 – 5)
1 – Very Low	Adversary would have no level of interest in the asset
2 – Low	Adversary would have some degree of interest in the asset
3 – Medium	Adversary would have a moderate degree of interest in attacking the asset
4 – High	Adversary would have a high degree of interest in the asset
5 – Very High	Adversary would have a very high degree of interest in the asset

Figure 3.14—Description of Step 3 and Substeps

Step	Task
Step 3: Vulnerability Analysis	
3.1 Define scenarios and evaluate specific consequences	Use scenario-analysis and/or use asset-based analysis to document the adversary's potential actions against an asset.
3.2 Evaluate effectiveness of existing security measures	Identify the existing measures intended to protect the critical assets and estimate their levels of effectiveness in reducing the vulnerabilities of each asset to each threat or adversary.
3.3 Identify vulnerabilities and estimate degree of vulnerability	Identify the potential vulnerabilities of each critical asset to applicable threats or adversaries. Estimate the degree of vulnerability of each critical asset for each threat-related undesirable event or incident and thus each applicable threat or adversary.

Step 3.1—Define Scenarios and Evaluate Specific Consequences

Each asset in the list of critical target assets from Step 2 is reviewed in light of the threat assessment, and the relevant threats and assets are paired in a matrix or other form of analysis, as shown in Attachment 1—Steps 3 – 5—Scenario Based Vulnerability Worksheet/Risk Ranking/Countermeasures Form. The importance of this step is to develop a design basis threat statement for each facility.

Once the SVA Team has determined how a malevolent event can be induced, it should determine how an adversary could execute the act.

The action in the Scenario-based approach follow the SVA method as outlined in Chapter 3. To establish an understanding of risk, scenarios can be assessed in terms of the severity of consequences and the likelihood of occurrence of security events. These are qualitative analyses based on the judgment and deliberation of knowledgeable team members.

Step 3.2—Evaluate Effectiveness of Existing Security Measures

The SVA Team will identify the existing measures intended to protect the critical assets and estimate their levels of effectiveness in reducing the vulnerabilities of each asset to each threat or adversary.

Step 3.3—Identify Vulnerabilities and Estimate Degree of Vulnerability

Vulnerability is any weakness that can be exploited by an adversary to gain unauthorized access and the subsequent destruction or theft of an asset. Vulnerabilities can result from, but are not limited to, weaknesses in current management practices, physical security, or operational security practices.

For each asset, the vulnerability or difficulty of attack is considered using the definitions shown in Figure 3.15.

The Scenario-based approach is identical to the Asset-based approach in the beginning, but differs in the degree of detailed analysis of threat scenarios. The scenario-based approach uses a more detailed analysis strategy and brainstorms a list of scenarios to understand how the undesired event might be accomplished. The scenario-based approach begins with an onsite inspection and interviews to gather specific information for the SVA Team to consider.

The following is a description of the approach and an explanation of the contents of each column of the worksheet in Attachment 1—Steps 3 – 5 Scenario Based Vulnerability Worksheet/Risk Ranking/Countermeasures Form.

Figure 3.15—Vulnerability Rating Criteria

Vulnerability Level	Description
5 – Very High	Indicates that there are no effective protective measures currently in place to Deter, Detect, Delay, and Respond to the threat and so an adversary would easily be capable of exploiting the critical asset.
4 – High	Indicates there are some protective measures to Deter, Detect, Delay, or Respond to the asset but not a complete or effective application of these security strategies and so it would be relatively easy for the adversary to successfully attack the asset.
3 – Medium	Indicates that although there are some effective protective measures in place to Deter, Detect, Delay, and Respond, there isn't a complete and effective application of these security strategies and so the asset or the existing countermeasures could likely be compromised.
2 – Low	Indicates that there are effective protective measures in place to Deter, Detect, Delay, and Respond, however, at least one weakness exists that an adversary would be capable of exploiting with some effort to evade or defeat the countermeasure given substantial resources.
1 – Very Low	Indicates that multiple layers of effective protective measures to Deter, Detect, Delay, and Respond to the threat exist and the chance that the adversary would be able to exploit the asset is very low.

The SVA Team devises a scenario based on their perspective of the consequences that may result from undesired security events given a postulated threat for a given asset. This is described as an event sequence including the specific malicious act or cause and the potential consequences, while considering the challenge to the existing countermeasures. It is conservatively assumed that the existing countermeasures are exceeded or fail in order to achieve the most serious consequences, in order to understand the hazard. When considering the risk, the existing countermeasures need to be assessed as to their integrity, reliability, and ability to deter, detect, and delay.

In this column the type of malicious act is recorded. As described in Chapter 2, the four types of security events included in the objectives of a SVA at a minimum include:

1. Theft/Diversion of material for subsequent use as a weapon or a component of a weapon
2. Causing the deliberate loss of containment of a chemical present at the facility
3. Contamination of a chemical, tampering with a product, or sabotage of a system
4. An act causing degradation of assets, infrastructure, business and/or value of a company or an industry.

Given the information collected in Steps 1 – 3 regarding the site's key target assets, the attractiveness of these targets, and the existing layers and rings of protection, a description of the initiating event of a malicious act scenario may be entered into the Undesired Event column. The SVA team brainstorms the vulnerabilities based on the information collected in Steps 1 – 3. The SVA team should brainstorm vulnerabilities for all of the malicious act types that are applicable at a minimum. Other scenarios may be developed as appropriate.

Completing the Worksheet

The next step is for the team to evaluate scenarios concerning each asset/threat pairing as appropriate. The fields in the worksheet are completed as follows:

1. **Asset:** The asset under consideration is documented. The team selects from the targeted list of assets and considers the scenarios for each asset in turn based on priority.
2. **Security Event Type:** This column is used to describe the general type of malicious act under consideration. At a minimum, the four types of acts previously mentioned should be considered as applicable.
3. **Threat Category:** The category of adversary including terrorist, activist, disgruntled employee, etc.
4. **Type:** The type of adversary category whether (I) – Insider, (E) – External, or (C) – Colluded threat.
5. **Undesired Act:** A description of the sequence of events that would have to occur to breach the existing security measures is described in this column.
6. **Consequences:** Consequences of the event are analyzed and entered into the Consequence column of the worksheet. The consequences should be conservatively estimated given the intent of the adversary is to maximize their gain.

   It is recognized that the severity of an individual event may vary considerably, so SVA teams are encouraged to understand the expected consequence of a successful attack or security breach.
7. **Consequences Ranking:** Severity of the Consequences on a scale of 1 – 5 as shown in Figure 3.8. The severity rankings are assigned based on a conservative assumption of a successful attack.

8. **Existing Countermeasures:** The existing security countermeasures that relate to detecting, delaying, or deterring the adversaries from exploiting the vulnerabilities may be listed in this column. The countermeasures have to be functional (i.e., not bypassed or removed) and sufficiently maintained as prescribed (i.e., their ongoing integrity can be assumed to be as designed) for credit as a countermeasure.
9. **Vulnerability:** The specific countermeasures that would need to be circumvented or failed should be identified.
10. **Vulnerability Ranking:** The degree of vulnerability to the scenario rated on a scale of 1 – 5 as shown in Figure 3.15.
11. **L(ikelihood):** The likelihood of the security event is assigned a qualitative ranking in the likelihood column. The likelihood rankings are generally assigned based on the likelihood associated with the entire scenario, assuming that all countermeasures are functioning as designed/intended. Likelihood is a team decision and is assigned from the Likelihood scale based on the factors of Vulnerability, Attractiveness, and Threat for the particular scenario considered.
12. **R(isk):** The severity and likelihood rankings are combined in a relational manner to yield a risk ranking. The development of a risk-ranking scheme, including the risk ranking values is described in Step 4.
13. **New Countermeasures:** The recommendations for improved countermeasures that are developed are recorded in the New Countermeasures column.

3.6 STEP 4: RISK ANALYSIS/RANKING

In either the Asset-based or the Scenario-based approach to Vulnerability Analysis, the next step is to determine the level of risk of the adversary exploiting the asset given the existing security countermeasures. Figure 3.16 lists the substeps.

The scenarios are risk-ranked by the SVA Team based on a simple scale of 1 – 5. The risk matrix shown in Figure 3.17 could be used to plot each scenario based on its likelihood and consequences. The intent is to categorize the assets into discrete levels of risk so that appropriate countermeasures can be applied to each situation.

Note: For this matrix, a Risk Ranking of "5 x 5" represents the highest severity and highest likelihood possible.

3.7 STEP 5: IDENTIFY COUNTERMEASURES:

A Countermeasures Analysis identifies shortfalls between the existing security and the desirable security where additional recommendations may be justified to reduce risk. In assessing the need for additional countermeasures, the team should ensure each scenario has the following countermeasures strategies employed:

- **DETER** an attack if possible
- **DETECT** an attack if it occurs
- **DELAY** the attacker until appropriate authorities can intervene
- **RESPOND** to neutralize the adversary, to evacuate, shelter in place, call local authorities, control a release, or other actions.

The SVA Team evaluates the merits of possible additional countermeasures by listing them and estimating their net effect on the lowering of the likelihood or severity of the attack. The team attempts to lower the risk to the corporate standard.

Figure 3.16—Description of Step 4 and Substeps

Step	Task
Step 4: Risk Assessment	
4.1 Estimate risk of successful attack	As a function of consequence and probability of occurrence, determine the relative degree of risk to the facility in terms of the expected effect on each critical asset (a function of the consequences or impacts to the critical functions of the facility from the disruption or loss of the critical asset, as evaluated in Step 1) and the likelihood of a successful attack (a function of the threat or adversary, as evaluated in Step 2, and the degree of vulnerability of the asset, as evaluated in Step 3).
4.2 Prioritize risks	Prioritize the risks based on the relative degrees of risk and the likelihoods of successful attacks.

Figure 3.17—Risk Ranking Matrix

		SEVERITY				
LIKELIHOOD		5	4	3	2	1
	5	[illegible]	[illegible]	[illegible]	Med	Med
	4	[illegible]	[illegible]	Med	Med	Low
	3	[illegible]	Med	Med	Low	Low
	2	Med	Med	Low	Low	Low
	1	Med	Low	Low	Low	Low

Figure 3.18—Description of Step 5 and Substeps

Step	Task
Step 5: Countermeasures Analysis	
5.1 Identify and evaluate enhanced countermeasures options	Identify countermeasures options to further reduce the vulnerabilities and thus the risks while considering such factors as: • Reduced probability of successful attack • The degree of risk reduction provided by the options • The reliability and maintainability of the options • The capabilities and effectiveness of these mitigation options • The costs of the mitigation options • The feasibility of the options Rerank to evaluate effectiveness.
5.2 Prioritize potential enhancements	Prioritize the alternatives for implementing the various options and prepare recommendations for decision makers

3.8 FOLLOW-UP TO THE SVA

The outcome of the SVA is:

- the identification of security vulnerabilities;
- a set of recommendations (if necessary) to reduce risk to an acceptable level.

The SVA results should include a written report that documents:

- The date of the study;
- The study team members, their roles and expertise and experience;
- A description of the scope and objectives of the study;
- A description of or reference to the SVA methodology used for the study;
- The critical assets identified and their hazards and consequences;
- The security vulnerabilities of the facility;
- The existing countermeasures;
- A set of prioritized recommendations to reduce risk.

Once the report is released, it is necessary for a resolution management system to resolve issues in a timely manner and to document the actual resolution of each recommended action.

Attachment 1—Example SVA Methodology Forms

The following four forms can be used to document the SVA results. Blank forms are provided, along with a sample of how each form is to be completed. Other forms of documentation that meet the intent of the SVA guidance can be used.

Step 1: Critical Assets/Criticality Form		
Facility Name:		
Critical Assets Form		
Critical Assets	**Criticality/Hazards**	**Asset Severity Ranking**
1.		
2.		
3.		
4.		
5.		
6.		
7.		
8.		
9.		

Step 2: Threat Assessment Form

General Adversary Type	Source	General Threat History	Site Specific Threat History	Potential Actions	Adversary Capability	Adversary Motivation/ Intent	Overall Assessment	Threat Ranking
1. International Terrorists								
2. Domestic Terrorists								
3. Domestic Disgruntled Employee or Contractor								
4. Domestic Activists								

Step 2: Attractiveness/Target Ranking Form

Facility Name:

Critical Assets	Function/Hazards/ Criticality	Asset Severity Ranking	Asset Attractiveness						
			Foreign/Domestic Attractiveness Rationale	A1	Employee/ Contractor Attractiveness Rationale	A2	Activist Attractiveness Rationale	A3	TR
1.									
2.									
3.									
4.									
5.									
6.									

Step 3 – 5: Vulnerability Analysis/Risk Ranking/Countermeasures Form

Facility Name:

Critical Assets:

Attractiveness:

Security Event Type	Threat Category	Type	Undesired Act	Consequences	S	Existing Countermeasures	Vulnerability	V	L	R	New Countermeasures

Glossary of Terms[12]

Adversary: Any individual, group, organization, or government that conducts activities, or has the intention and capability to conduct activities detrimental to critical assets. An adversary could include intelligence services of host nations, or third party nations, political and terrorist groups, criminals, rogue employees, and private interests. Adversaries can include site insiders, site outsiders, or the two acting in collusion.

Alert levels: Describes a progressive, qualitative measure of the likelihood of terrorist actions, from negligible to imminent, based on government or company intelligence information. Different security measures may be implemented at each alert level based on the level of threat to the facility.

Asset: An asset is any person, environment, facility, material, information, business reputation, or activity that has a positive value to an owner. The asset may have value to an adversary, as well as an owner, although the nature and magnitude of those values may differ. Assets in the SVA include the community and the environment surrounding the site.

Asset category: Assets may be categorized in many ways. Among these are:

- People
- Hazardous materials (used or produced)
- Information
- Environment
- Equipment
- Facilities
- Activities/Operations
- Company reputation

Benefit: Amount of expected risk reduction based on the overall effectiveness of countermeasures with respect to the assessed vulnerabilities.

Capability: When assessing the capability of an adversary, two distinct categories need to be considered. The first is the capability to obtain, damage, or destroy the asset. The second is the adversary's capability to use the asset to achieve their objectives once the asset is obtained, damaged, or destroyed.

Checklist: A list of items developed on the basis of past experience that is intended as a guide to assist in applying a standard level of care for the subject activity and to assist in completing the activity in as thorough a manner.

Consequences: The amount of loss or damage that can be expected, or may be expected from a successful attack against an asset. Loss may be monetary but may also include political, morale, operational effectiveness, or other impacts. The impacts of security events, which should involve those that are extremely severe. Some examples of relevant consequences in a SVA include fatality to member(s) of the public, fatality to company personnel, injuries to member(s) of the public, injuries to company personnel, large-scale disruption to public or private operations, large-scale disruption to company operations, large-scale environmental damage, large-scale financial loss, loss of critical data, and loss of reputation.

Cost: Includes tangible items such as money and equipment as well as the operational costs associated with the implementation of countermeasures. There are also intangible costs such as lost productivity, morale considerations, political embarrassment, and a variety of others. Costs may be borne by the individuals who are affected, the corporations they work for, or they may involve macroeconomic costs to society.

Cost-Benefit analysis: Part of the management decision-making process in which the costs and benefits of each countermeasure alternative are compared and the most appropriate alternative is selected. Costs include the cost of the tangible materials, and also the on-going operational costs associated with the countermeasure implementation.

Countermeasures: An action taken or a physical capability provided whose principal purpose is to reduce or eliminate one or more vulnerabilities. The countermeasure may also affect the threat(s) (intent and/or capability) as well as the asset's value. The cost of a countermeasure may be monetary, but may also include non-monetary costs such as reduced operational effectiveness, adverse publicity, unfavorable working conditions, and political consequences.

Countermeasures analysis: A comparison of the expected effectiveness of the existing countermeasures for a given threat against the level of effectiveness judged to be required in order to determine the need for enhanced security measures.

Cyber security: Protection of critical information systems including hardware, software, infrastructure, and data from loss, corruption, theft, or damage.

Delay: A countermeasures strategy that is intended to provide various barriers to slow the progress of an adversary in penetrating a site to prevent an attack or theft, or in leaving a restricted area to assist in apprehension and prevention of theft.

Detection: A countermeasures strategy to that is intended to identify an adversary attempting to commit a security event or other criminal activity in order to provide real-time observation as well as post-incident analysis of the activities and identity of the adversary.

Deterrence: A countermeasures strategy that is intended to prevent or discourage the occurrence of a breach of security by means of fear or doubt. Physical security systems such as warning signs, lights, uniformed guards, cameras, bars are examples of countermeasures that provide deterrence.

Hazard: A situation with the potential for harm.

Intelligence: Information to characterize specific or general threats including the motivation, capabilities, and activities of adversaries.

Intent: A course of action that an adversary intends to follow.

Layers of protection: A concept whereby several independent devices, systems, or actions are provided to reduce the likelihood and severity of an undesirable event.

Likelihood of adversary success: The potential for causing a catastrophic event by defeating the countermeasures. LAS is an estimate that the security countermeasures will thwart or withstand the attempted attack, or if the attack will circumvent or exceed the existing security measures. This measure represents a surrogate for the conditional probability of success of the event.

Mitigation: The act of causing a consequence to be less severe.

Physical security: Security systems and architectural features that are intended to improve protection. Examples include fencing, doors, gates, walls, turnstiles, locks, motion detectors, vehicle barriers, and hardened glass.

Process Hazard Analysis (PHA): A hazard evaluation of broad scope that identifies and analyzes the significance of hazardous situations associated with a process or activity.

Response: The act of reacting to detected or actual criminal activity either immediately following detection or post-incident.

Risk: The potential for damage to or loss of an asset. Risk, in the context of process security, is the potential for a catastrophic outcome to be realized. Examples of the catastrophic outcomes that are typically of interest include an intentional release of hazardous materials to the atmosphere, or the theft of hazardous materials that could later be used as weapons, or the contamination of hazardous materials that may later harm the public, or the economic costs of the damage or disruption of a process.

Risk assessment: Risk (R) assessment is the process of determining the likelihood of an adversary (T) successfully exploiting vulnerability (V) and the resulting degree of consequences (C) on an asset. A risk assessment provides the basis for rank ordering of risks and thus establishing priorities for the application of countermeasures.

Safeguard: Any device, system or action that either would likely interrupt the chain of events following an initiating event or that would mitigate the consequences.4

Security layers of protection: Also known as concentric 'rings of protection', a concept of providing multiple independent and overlapping layers of protection in depth. For security purposes, this may include various layers of protection such as counter-surveillance, counterintelligence, physical security, and cyber security.

Security management system checklist: A checklist of desired features used by a facility to protect its assets.

Security plan: A document that describes an owner/operator's plan to address security issues and related events, including security assessment and mitigation options. This includes security alert levels and response measures to security threats.

Security Vulnerability Assessment (SVA): A SVA is the process of determining the likelihood of an adversary successfully exploiting vulnerability, and the resulting degree of damage or impact. SVAs are not a quantitative risk analysis, but are performed qualitatively using the best judgment of security and safety professionals. The determination of risk (qualitatively) is the desired outcome of the SVA, so that it provides the basis for rank ordering of the security-related risks and thus establishing priorities for the application of countermeasures.

Target attractiveness: An estimate of the value of a target to an adversary based on the factors shown below. Experience has shown that, particularly for terrorist attacks, certain targets better accomplish the objectives of the adversaries than do others. Since the SVA is a risk-based analytical approach, consideration must be given to these factors in defining the threat and in determining the need for any enhanced countermeasures.

- Potential for mass casualties/fatalities
- Extensive property damage
- Proximity to national assets or landmarks
- Possible disruption or damage to critical infrastructure
- Disruption of the national, regional or local economy
- Ease of access to target
- Media attention or possible interest of the media
- Company reputation and brand exposure

Technical security: Electronic systems for increased protection or for other security purposes including access control systems, card readers, keypads, electric locks, remote control openers, alarm systems, intrusion detection equipment, annunciating and reporting systems, central stations monitoring, video surveillance equipment, voice communications systems, listening devices, computer security, encryption, data auditing, and scanners.

Terrorism: The FBI defines terrorism as, "the unlawful use of force or violence against persons or property to intimidate or coerce a Government, the civilian population, or any segment thereof, in furtherance of political or social objectives."

Threat: Any indication, circumstance, or event with the potential to cause the loss of, or damage to an asset. Threat can also be defined as the intention and capability of an adversary to undertake actions that would be detrimental to critical assets.

Threat categories: Adversaries may be categorized as occurring from three general areas:

- Insiders
- Outsiders
- Insiders working in collusion with outsiders

Undesirable events: An event that results in a loss of an asset, whether it is a loss of capability, life, property, or equipment.

Vulnerabilities: Any weakness that can be exploited by an adversary to gain access to an asset. Vulnerabilities can include but are not limited to building characteristics, equipment properties, personnel behavior, locations of people, equipment and buildings, or operational and personnel practices.

Abbreviations and Acronyms

A	Attractiveness
ACC	American Chemistry Council
AT	Target attractiveness
AIChE	American Institute of Chemical Engineers
API	American Petroleum Institute
AWCS	Accidental Worst-Case Scenario
C	Consequence
CCPS"	Center for Chemical Process Safety of the American Institute of Chemical Engineers (AIChE)
CCTV	Closed Circuit Television
CEPPO	Chemical Emergency Preparedness and Prevention Office (USEPA)
CMP	Crisis Management Plan
CSMS	Chemical Security Management System
CW	Chemical Weapons
CWC	Chemical Weapons Convention
D	Difficulty of Attack
DCS	Distributed Control Systems
DHS	Department of Homeland Security
DOE	Department of Energy
DOT	U. S. Department of Transportation
EHS	Environmental, Health, and Safety
EPA	U. S. Environmental Protection Agency
ERP	Emergency Response Process
EHS	Environmental, Health, and Safety
FBI	U. S. Federal Bureau of Investigation
FC	Facility Characterization
HI	Hazard Identification
HSAS	Homeland Security Advisory System
IPL	Independent Protection Layer
IT	Information Technology
LA	Likelihood of Adversary Attack
LAS	Likelihood of Adversary Success
LOPA	Layer of Protection Analysis
MARSEC	Maritime Security Levels
MOC	Management of Change
NPRA	National Petrochemical and Refiners Association
OSHA	Occupational Safety and Health Administration
PHA	Process Hazard Analysis
PLC	Programmable Logic Controller
PSI	Process Safety Information
PSM	Process Safety Management (Also refers to requirements of 29 *CFR* 1910.119)
R	Risk
RMP	Risk Management Process (Also refers to requirements of EPA 40 *CFR* Part 68)
S	Severity of the Consequences
SOCMA	Synthetic Organic Chemical Manufacturers Association
SOP	Standard Operating Procedure
SVA	Security Vulnerability Assessment
T	Threat
TSA	Transportation Security Agency
V	Vulnerability
WMD	Weapons of Mass Destruction

APPENDIX A—SVA Supporting Data Requirements

SVA Methodology Supporting Data	
Category*	**Description**
A	Scaled drawings of the overall facility and the surrounding community (e.g., plot plan of facility, area map of community up to worst case scenario radius minimum)
A	Aerial photography of the facility and surrounding community (if available)
A	Information such as general process description, process flow diagrams, or block flow diagrams that describes basic operations of the process including raw materials, feedstocks, intermediates, products, utilities, and waste streams
A	Information (e.g., drawings that identify physical locations and routing) that describes the infrastructures upon which the facility relies (e.g., electric power, natural gas, petroleum fuels, telecommunications, transportation [road, rail, water, air], water/wastewater)
A	Previous security incident information
A	Description of guard force, physical security measures, electronic security measures, security policies
A	Threat information specific to the company (if available)
B	Specifications and descriptions for security related equipment and systems. Plot plan showing existing security countermeasures
B	RMP information including registration and offsite consequence analysis (if applicable, or similar information)
B	Most up-to-date PHA reports for processes considered targets
B	Emergency response plans and procedures (site, community response, and corporate contingency plans)
B	Information on material physical and hazard properties (MSDS)
B	Crisis management plans and procedures (site and corporate)
B	Complete a SVA chemicals checklist to determine whether the site handles any chemicals on the following lists:
C	• EPA Risk Management Program (RMP) 40 *CFR* Part 68;
C	• OSHA Process Safety Management (PSM) 29 *CFR* 1910.119;
C	• Chemical Weapons Convention, Schedule 2 and specifically listed Schedule 3 chemicals;
C	• FBI Community Outreach Program (FBI List) for WMD precursors;
C	• The Australia Group list of chemical and biological weapons.
C	Design basis for the processes (as required)
C	Unit plot plans of the processes
C	Process flow diagrams (PFDs) and piping and instrument diagrams (P&IDs) for process streams with hazardous materials
C	Safety systems including fire protection, detection, spill suppression systems
C	Process safety systems including safety instrumented systems (SIS), PLC's, process control systems
C	Operating procedures for start-up, shutdown, and emergency (operators may provide general overview of this information, with written information available as required)
C	Mechanical equipment drawings for critical equipment containing highly hazardous chemicals
C	Electrical one-line diagrams
C	Control system logic diagrams
C	Equipment data information
C	Information on materials of construction and their properties
C	Information on utilities used in the process
C	Test and maintenance procedures for security related equipment and systems

*Categories: A = Documentation to be provided to SVA team as much in advance as possible before arrival for familiarization;
B = Documentation to be gathered for use in SVA team meetings on site;
C = Documentation that should be readily available on an as-needed basis.

APPENDIX B—SVA Countermeasures Checklist

Appendix B Table of Contents

Page

SVA Countermeasures Survey ... 47
IDENTIFICATION OF PHYSICAL SECURITY SYSTEMS ... 47
IDENTIFICATION OF PROCESS SAFETY SYSTEMS ... 47
SECURITY PROGRAM MANAGEMENT ... 48
(a) Security Organization ... 48
(b) Security Plans and Policies ... 48
(c) Security Resources ... 48
(d) Senior Management Security ... 48
(e) Security Audits ... 48
(f) Handling of Sensitive Information ... 49
(g) Internal Communications ... 49
THREAT DETECTION AND EVALUATION CAPABILITIES ... 50
(a) Threat Analysis Working Group ... 50
(b) Organization's Response to Threat Updates ... 51
PERIMETER BARRIERS – FENCES, GATES ... 52
(a) Fences ... 52
(b) Gates ... 52
(c) Vehicle Barriers ... 52
BUILDING BARRIERS – WALLS, ROOF/CEILING, WINDOWS, DOORS ... 53
(a) Walls ... 53
(b) Roof/Ceiling ... 53
(c) Windows ... 53
(d) Doors ... 54
INTRUSION DETECTION ... 55
(a) Intrusion Sensors (If Applicable) ... 55
(b) Intrusion Alarm Deployment (If Applicable) ... 55
(c) Intrusion Alarm Assessment ... 55
CLOSED CIRCUIT TELEVISION ... 55
(a) CCTV ... 55
ACCESS CONTROL ... 56
(a) Personnel Access ... 56
(b) Vehicle Access ... 56
(c) Contraband Detection ... 56
(d) Access Point Illumination ... 56
SECURITY FORCE ... 57
(a) Protective Force ... 57
(b) Local Law Enforcement Agencies ... 57
INFORMATION, COMPUTER, NETWORK, AND INTELLECTUAL PROPERTY SECURITY ... 58
(a) Information, Computer, Network, and Intellectual Property Security ... 58
PREVENTING AND CONTROLLING RELEASES OF HAZARDOUS MATERIALS ... 60
(a) Hardening Processes ... 60
(b) Reducing the Quantity and Hazard of a Release from a Malicious Act ... 61
(c) Mitigating a Release from a Malicious Act ... 63
(d) Emergency Response, Crisis Management, and Community Coordination ... 64

SVA Countermeasures Survey

The objective of the physical security portion of the survey is to identify measures that protect the entire facility and/or each critical asset of the facility, and to determine the effectiveness of the protection. This attachment contains checklists that are used to conduct the physical security portion of the survey. The Security Program Management Checklist is used to identify physical security measures that may be present to protect the entire facility or a critical asset at the facility. The remaining checklists are used to specifically evaluate the individual elements of the physical security system that are present. The conclusion of whether a particular element provides adequate protection is to be reported as part of the findings in the body of the SVA. A "set" of checklists should be completed for the facility as a whole and if appropriate, for each of the critical assets within the facility.

Note that the infrastructure interdependencies portion of the survey will identify infrastructures that support the facility and/or its critical assets (e.g., electric power, water, and telecommunications). A physical security review of these vital infrastructures should also be conducted.

IDENTIFICATION OF PHYSICAL SECURITY SYSTEMS			
Date: [MONTH XX, 2002]			Facility: [FACILITY]
This checklist applies to [the entire facility/ASSET]			
Instructions: This checklist identifies the physical security elements that may be used to protect the entire facility and/or a critical asset. Identify which elements are present for the facility or the critical asset listed above. Once physical security elements are identified, they can be reviewed by using the applicable checklists. At the completion of the reviews, the effectiveness of the elements is to be documented in the body of the survey report.			
Physical Security System Element	**Element Present**		**COMMENTS**
	Yes	**No**	
Perimeter Barriers			
Building Barriers			
Intrusion Detection			
Access Controls			
Security Force			

IDENTIFICATION OF PROCESS SAFETY SYSTEMS			
Date: [MONTH XX, 2002]			Facility: [FACILITY]
This checklist applies to [the entire facility/ASSET]			
Instructions: This checklist identifies the process safety elements that may be used to protect the entire facility and/or a critical asset. Identify which elements are present for the facility or the critical asset listed above. Once physical security elements are identified, they can be reviewed by using the applicable checklists. At the completion of the reviews, the effectiveness of the elements is to be documented in the body of the survey report.			
Process Safety System Element	**Element Present**		**COMMENTS**
	Yes	**No**	
Hardening Processes			
Emergency Response			
Chemical Detection			
Fire Detection			
Fire Suppression			

SECURITY PROGRAM MANAGEMENT	
Date: [MONTH XX, 2002]	Facility: [FACILITY]
	COMMENTS
(a) Security Organization	
1. Is there a senior level security working group with representatives from each major office or department to establish security policies (including physical security, operations security, and infrastructure interdependencies security) and integrate them across all elements of the organization? • If there is a senior level security working group, describe the membership, the lines of communication, and any scheduled periodic meetings to resolve security issues. • If there is not such a group, how are security policies established?	
2. Is there a security office that is responsible for implementing security policies and procedures (including physical security, operations security, and infrastructure interdependencies security)? • If there is a security office, where does it report in the organization, how many people are in the office, and are resources adequate? Also describe any training received. • If there is not such an office, how are security policies implemented?	
(b) Security Plans and Policies	
3. Is there a mission statement describing the physical security, operations security, and infrastructure security programs?	
4. Is there a formal security plan and statement of security policies? If there is, describe it including how it is communicated to employees.	
5. Is there a formal threat definition and assessment statement? If there is, describe it including how it is communicated to employees.	
(c) Security Resources	
1. Are the resources (budget and staffing) applied to security (including physical security, operations security, and infrastructure interdependencies security) considered adequate?	
2. Do security personnel feel that they have adequate training to accomplish their functions?	
(d) Senior Management Security	
1. Is there an executive protection program for senior executives/managers? If there is such a program, describe it.	
2. Is public information on senior executives/managers controlled? If it is, describe how it is controlled.	
(e) Security Audits	
1. Is there a regular security assessment or audit? If there is, describe how it is done, by whom, and how frequently.	

2. Has the most recent audit indicated any weaknesses? Summarize the results of the audit, particularly any weaknesses identified.	
3. Have any corrective measures been implemented recently? Describe them.	
(f) Handling of Sensitive Information	
1. How is sensitive information identified and marked?	
2. Who has access to sensitive security information?	
3. How is sensitive information protected, stored, accessed, transmitted, and destroyed?	
4. How do senior executives/managers protect sensitive security information?	
(g) Internal Communications	
1. How does management provide security information to employees at the site?	
2. Describe the process for obtaining feedback from employees on security related issues.	

THREAT DETECTION AND EVALUATION CAPABILITIES	
Date: [MONTH XX, 2002]	Facility: [FACILITY]
This checklist applies to the entire facility	
	COMMENTS
(a) Threat Analysis Working Group	
1. Is the organization a member of a local threat analysis working group? Describe the group	
2. If the organization is a member of such a group, list the organizations that participate in the working group (e.g., local, county, state, and federal agencies, the military).	
3. Are there other industry partners participating in the working group? Describe them.	
4. Are active efforts being made to recruit other meaningful participants into the working group? Describe the efforts.	
5. Do the participants in the working group have management support, requirements, and funding to participate? Describe the situation.	
6. Are the members of the working group willing participants and do they work against bureaucratic obstacles that may prevent the success of the group? Describe the situation.	
7. Do the members of the working group have the authority to share information with other members of the group? Describe the situation.	
8. Have the members of the working group been given appropriate U.S. government clearances to share in threat information? Describe the situation.	
9. Do the members of the working group have access to the National Infrastructure Protection Center (NIPC), Analytical Services, Inc., (ANSER), FBI-sponsored InfraGuard, Carnegie Mellon University's CERT", and other information system security warning notices? List the threat information systems they use.	
10. Indicate the frequency and regularity of the working group meetings.	
11. Do the members of the working group have processes in place to obtain real-time information from the field (e.g., on-duty offices, civilian neighborhood watch programs, local businesses, other working groups in the area)? Describe these processes.	
12. Do members of the working group have the ability to initiate information-gathering requests back into the field environment? Describe the capability.	
13. Are the threat statements developed by the working group specific to the organization or the industry, versus general nationwide warnings? Describe the process for gathering these statements.	

14. Do some members of the working group conduct scheduled meetings with the public to discuss concerns and observations? Describe these interactions.	
15. Do the members of the working group know what the critical assets of the organization are? Describe the extent of their knowledge.	
16. Do the members of the working group understand industry interdependencies and work with other industry members to address these potential concerns? Describe the extent of these interactions.	
17. What are the roles and responsibilities of the working group members during response and recovery activities?	
(b) Organization's Response to Threat Updates	
1. Does senior management support and/or participate in the threat analysis working group? Describe the extent of the support/participation.	
2. Does the organization receive as-needed threat briefings from local, state, and federal agencies? Describe the nature and extent of the briefings.	
3. Does the organization have the ability to distribute organization-specific threat warnings in real time? Describe the process.	
4. Does the organization have the ability to augment security programs based on threat updates? Describe the process.	
5. Does the organization conduct historical trending analysis for security events (both planned and actual) and implement security activates to mitigate them? Describe the analysis.	
6. Does the organization create possible threat scenarios based on input from the threat analysis working group and conduct related security exercises? Describe the exercises.	

PERIMETER BARRIERS—FENCES, GATES	
Date: [MONTH XX, 2002]	Facility: [FACILITY]
This checklist applies to [the entire facility/ASSET]	
	COMMENTS
(a) Fences	
1. Characterize fence construction and rate the level of security it provides as low, moderate to high, or other (specify). • Low: no fence or only a 6-foot chain-link fence. • Moderate to high: 8-foot chain-link fence with outriggers, 10 to 12-foot chain-link fence, or over 12-foot chain-link fence with outriggers.	
2. Characterize fence signage as no signs, posted "No Trespassing," or other (specify).	
3. Characterize the fence alarm system as no alarms, fence sensors (taut wire, vibration, strain, electric field, or multiple sensors), or other (specify).	
4. Fence area: • Is the fence within 2 inches of firm hard ground? • Is the fence line clear of vegetation, trash, equipment, and other objects that could impede observation? • Is the area free of objects that would aid in traversing the fence? • Is physical protection installed for all points where utilities (e.g., electric power lines, natural gas pipelines, telecommunication lines, water supply, storm sewers, drainage swells) intersect the fence perimeter?	
5. How is the fence protected from vehicles (aircraft cable, concrete barriers or median, guard rails, steel posts, a ditch, crash I-beams, train barrier, or other [specify])?	
6. Fence illumination: • Is there security lighting for the fences? Describe the security lighting system. • Do alarms or infrared detectors trigger the lighting? Describe the triggering process.	
(b) Gates	
1. Characterize the gates as no gate closure, vehicle bar, chain-link fence, or other (specify).	
2. Characterize the gate locks as no lock, lock not used, gate unlocked, gate attended by personnel when unlocked, ID actuated lock, padlock, or other (specify).	
3. How is access to gate keys controlled?	
4. Gate lighting: • Describe the security lighting for the gates. • Do alarms or infrared detectors trigger the lighting? Describe the triggering process.	
(c) Vehicle Barriers	
1. Characterize vehicle barriers as none, a vehicle bar, blocked by vehicle when gate open, hydraulic wedge, or other (specify).	

BUILDING BARRIERS—WALLS, ROOF/CEILING, WINDOWS, DOORS	
Date: [MONTH XX, 2002]	Facility: [FACILITY]
This checklist applies to [the entire facility/ASSET]	
	COMMENTS
(a) Walls	
2. Characterize wall construction and rate the level of security wall provide as low, moderate, or high. • Low: chain-link mesh, 16-gauge metal, wood studs and dry wall, wood studs and plywood, or other (specify). • Moderate: clay block, 8-inch hollow block, 8-inch filled block, or other (specify). • High: 8-inch filled rebar block, 12-inch filled rebar block, 2-inch precast concrete tees, 4-inch reinforced concrete, 8-inch reinforced concrete, 12-inch reinforced concrete, 24-inch reinforced concrete, or other (specify).	
3. Do the walls extend from the floor to the structural ceiling?	
(b) Roof/Ceiling	
1. Characterize the roof material and rate the level of security it provides as low, moderate, or high. • Low: 20-gauge metal with insulation, ½-inch wood, or other (specify). • Moderate: 20-gauge metal built-up roof, concrete built-up roof with T-beams, or other (specify). • High: 5-½-inch concrete roof, 8-inch concrete roof, 3-foot earth cover, 3-foot soil/cement/earth cover, or other (specify).	
2. Does the interior drop ceiling extend beyond the structural walls?	
(c) Windows	
1. Characterize the window materials and rate the level of security they provide as low or moderate. • Low: standard windows or other (specify). • Moderate: 9-gauge expanded mesh, ½-inch diameter x 1-½-inch quarry screen, ½-inch diameter bars with 6-inch spacing, $^{3}/_{16}$-inch x 2-½-inch grating, or other (specify).	
2. Characterize the window alarms (for windows that would be accessible by foot or ladder) as none, vibration sensor, glass breakage sensor, conducting tape, grid mesh, multiple sensors, or other (specify).	

(d) Doors	
1. Characterize door materials and rate the level of security they provide as low, moderate, or high. • Low: wood, 9-gauge wire mesh, hollow-core metal, no lock/hinge, or other (specify). • Moderate: hollow-core metal, tempered-glass panel, security-glass panel, half-height turnstile, or other (specify). • High security: ½-inch steel plate, turnstile – aluminum, Class V or VI vault, or other (specify).	
2. Characterize the door locks and rate the level of security they provide as low, moderate, or high. • Low: none, lock not used, or other (specify). • Moderate: door unlocked, attended by personnel when unlocked, ID actuated lock, padlock, keyed cylinder lock, combination lock, mechanically coded lock, or other (specify). • High: electronically coded lock, two-person rule lock system, lock inaccessible from the door exterior, or other (specify).	
3. How is access to the keys for the door locks controlled?	
4. Door Alarms: • Is door position monitored? • Indicate the type of door penetration sensor (vibration, glass breakage, conducting tape, grid mesh, or other [specify]).	

INTRUSION DETECTION	
Date: [MONTH XX, 2002]	Facility: [FACILITY]
This checklist applies to [the entire facility/ASSET]	
	COMMENTS
(a) Intrusion Sensors (If Applicable)	
1. Characterize the exterior intrusion sensors as seismic buried cable, electric field, infrared, microwave, video motion, or other (specify).	
2. Characterize the interior intrusion sensors as sonic, capacitance, video motion, infrared, ultrasonic, microwave, or other (specify).	
(b) Intrusion Alarm Deployment (If Applicable)	
1. Characterize intrusions alarm deployment in terms such as: • continuously monitored, • positioned to prevent gaps in coverage, • detection zone kept clear of obstructions (e.g., dips, equipment, snow, ice, grass, debris), • tamper and system problem indicators provided, • compensatory measures employed when alarms are not operating, • backup power provided, and • other (specify).	
(c) Intrusion Alarm Assessment	
1. Characterize the assessment of intrusion alarms as not assessed, closed circuit TV, automatic deployment of protective force, or other (specify).	

CLOSED CIRCUIT TELEVISION	
Date: [MONTH XX, 2002]	Facility: [FACILITY]
This checklist applies to [the entire facility/ASSET]	
Note: Different access points to the facility and/or to critical assets may have different access controls. The comments should clearly distinguish whether the evaluation applies to all access points or to specific access points.	
	COMMENTS
(a) CCTV	
1. Describe the current CCTV system in use at the site.	
2. Characterize cameras in use and what asset(s) the cameras cover (PTZ, Autodome type, Fixed, Day/Night)	
3. Who monitors the CCTV cameras (Operations and/or Security) and what are the protocols for camera operation?	
4. Describe the policy for review of information recorded on CCTV system.	
5. Describe the preventive maintenance program for the CCTV system.	

ACCESS CONTROL	
Date: [MONTH XX, 2002]	Facility: [FACILITY]
This checklist applies to [the entire facility/ASSET]	
Note: Different access points to the facility and/or to critical assets may have different access controls. The comments should clearly distinguish whether the evaluation applies to all access points or to specific access points.	
	COMMENTS
(a) Personnel Access	
1. Characterize access point control as unmanned, unarmed guard, armed guard, or other (specify).	
2. Characterize the identification check process as none in place, casual recognition, credential check (e.g., drivers license, passport, state ID), picture badge, PIN, exchange badge, retinal scan, hand geometry, speech pattern, signature dynamics, fingerprint, or other (specify).	
3. Characterize the organization's badging policy in terms such as no badging policy, visitor badges required, badge issuance and control procedures in place (describe), and badges show permission to access specific areas (describe).	
(b) Vehicle Access	
1. Characterize vehicle access point controls as unmanned, unarmed guard, armed guard, or other (specify).	
2. Characterize the vehicle access identification process as none in place, vehicle stickers, vehicle stickers with personnel identification, automated system (describe), or other (specify).	
3. Describe the vetting process for incoming/outgoing bulk shipments of items by vehicle. Are deliveries scheduled or are is a list of drivers provided prior to delivery.	
(c) Contraband Detection	
1. Characterize item and vehicle search procedures as none, cursory, or detailed	
2. Is there a policy for incoming/outgoing drivers that report the possession of weapons? If so describe the policy/procedure.	
(d) Access Point Illumination	
1. Access Point Illumination: • Is there security lighting for the access points? Describe the security lighting system. • Do alarms or infrared detectors trigger the lighting? Describe the triggering process.	

SECURITY FORCE	
Date: [MONTH XX, 2002]	Facility: [FACILITY]
This checklist applies to [the entire facility/ASSET]	
	COMMENTS
(a) Protective Force	
1. Specify the size of the protective force in terms or total number and the number on duty during working hours, non-working hours, and weekends/holidays.	
2. Specify the equipment available to the protective force such as uniforms; vehicles (specify number); weapons (describe); communications devices (describe); and other equipment (describe).	
3. Describe the training of the protective force. Specifically, describe the initial training, any continuing training (e.g., on-the-job), and drills and exercises.	
4. Describe the organization of the protective force. Specifically, describe the command structure, their mission as defined, any established policies and procedures, and established emergency response plans.	
5. Are there provisions for a back-up force (e.g., recalling off-duty personnel)? Describe the provisions in place.	
6. Protective Force Command Center: • Is there a protective force command and control center? Describe it. • Is there a backup center? Describe it.	
7. Are protective force operations disguised to prevent intelligence about the facility from being inadvertently released? Describe how this is done.	
8. Describe protective force procedures for responding to alarms.	
9. Does the protective force provide security escort for visitors? Describe the nature of the escort.	
(b) Local Law Enforcement Agencies	
1. Describe the interaction of the protective force with local law enforcement agencies in terms of memoranda of agreement or other agreements in place (describe), protection responsibilities defined (describe), communication procedures developed (describe), and participation in drills and exercises.	
2. What is the approximate response time for local law enforcement personnel?	

INFORMATION, COMPUTER, NETWORK, and INTELLECTUAL PROPERTY SECURITY	
Date: [MONTH XX, 2002]	Facility: [FACILITY]
This checklist applies to [the entire facility/ASSET]	
	COMMENTS
(a) Information, Computer, Network, and Intellectual Property Security	
1. Have steps been taken to protect technical and business information that could be of use to potential adversaries (sometimes referred to as operational security or OPSEC)?	
2. Have the documentation/computer files that need to be protected for confidentiality been systematically identified and regularly backed-up?	
3. Is sensitive information in research and development and laboratory areas safeguarded against inadvertent disclosure?	
4. Is sensitive information in maintenance areas safeguarded against inadvertent disclosure?	
5. Are computers as well as disks, tapes, and other media adequately secured physically from theft?	
6. Are procedures followed to reduce the likelihood that spoken information (in face-to-face conversations, phone calls, and radio communications) could be picked up by adversaries?	
7. If the content of radio communications cannot be restricted for operational reasons, have they been voice-encrypted?	
8. Are user authorizations granted on the basis of "need to know," "least access," and "separation of functions" rather than position or precedent (note: this has to be balanced against the process safety concepts of employee access to process safety information and employee participation)?	
9. Are appropriate procedures followed for protecting and destroying sensitive documents that could provide key information on critical process operation or vulnerabilities?	
10. Is the computer/server room secured?	
11. Is the computer/server room on the second floor (to protect it from flooding and to reduce the likelihood of theft), and away from outside walls?	
12. Is the computer/server room equipped with adequate communications capability?	
13. Is access to the computer/server room limited to only authorized personnel who need entry?	
14. Are appropriate hardware, software, and procedural techniques used for protecting computers and networks, such as:	
a. Firewalls?	
b. User ID?	

c. Password controls, including the regular changing of passwords?	
d. Encryption?	
15. Virus protection?	
16. Are computer transaction histories periodically analyzed to look for irregularities that might indicate security breaches?	
17. Is Internet access disabled in all application software or operating systems that are pre-packaged?	
18. Are measures in place to control access to or otherwise secure process-specific operating information (e.g., including diagrams, procedures, control loop/DCS information), both electronic and hardcopy versions?	
19. Are process control systems protected from external manipulation (e.g., hacking into control system to operate equipment or delete or alter software codes)?	
20. Is access to process control systems via the Internet or Intranet been restricted? If access is allowed, is the access allowed only to the absolute minimum number of personnel necessary, using user ID, password, separate authentication, and encryption controls as appropriate?	
21. Are temporary passwords restricted from use except for new employees, or when a password is forgotten or is inactive?	
22. Are vendor-supplied passwords changed immediately after installation?	
23. Do users have screen saver with password available and in use when leaving computers on and unattended?	

PREVENTING AND CONTROLLING RELEASES OF HAZARDOUS MATERIALS	
Date: [MONTH XX, 2002]	Facility: [FACILITY]
This checklist applies to [the entire facility/ASSET]	
	COMMENTS
(a) Hardening Processes	
1. Have existing security countermeasures been designed using the concept of rings of protection? Are the critical assets that may qualify as attractive targets at the center of concentric rings of layered protective features?	
2. Have process and systems been designed using the concept of layers of protection? Are there adequate independent protective layers that would detect, prevent, or mitigate a release of hazardous materials?	
3. Are critical process areas and equipment protected with traffic barriers, bollards, dikes, or other measures (e.g., diversionary structures that prevent vehicles from accelerating along a clear path to the process/equipment) to prevent ramming with vehicles?	
4. Are process "unit roads or streets" (i.e., roadways that provide access into specific process areas) provided with gates and, if so, are they securely closed when not in use (these gates may help limit direct vehicular access to critical equipment)?	
5. Are vehicles (except necessary material transport vehicles and/or authorized plant vehicles) prohibited from parking near critical process equipment (300 feet is considered a minimum distance)?	
6. Are full tank trailers or rail cars containing highly hazardous materials (i.e., those materials that could be targeted by terrorists) stored away from fence lines or perimeter areas to reduce their vulnerability to attack?	
7. Are full tank trailers or rail cars containing flammable or explosive materials stored away from critical process areas and equipment to prevent propagation of effects to critical processes?	
8. Are critical processes or equipment, such as tanks storing highly hazardous materials, protected from explosion or fire at adjacent processes (e.g., blast walls)?	
9. Is good housekeeping practiced in critical process areas and are trash dumpsters or receptacles located away from critical processes and equipment (trash dumpsters and poor housekeeping may make it easier to hide a bomb)?	
10. Are doors to interior buildings (e.g., process buildings) and control rooms locked or otherwise secured, where appropriate?	

11. Are hinge pins on doors to critical process areas on the inside of the door? (Note: May not be possible and still maintain easy egress in fire/emergency situations—doors must open out.)	
12. Are critical process areas surrounded with locked and secure fencing (in addition to site perimeter fencing) or located within locked buildings? (Note: Locked and secured fencing or buildings may create confined space issues.)	
13. If critical process areas are not surrounded by fencing or within buildings or if infeasible to do so, are the processes patrolled or monitored continuously by security personnel?	
14. Are highly reactive materials (e.g., water-reactive chemicals) stored in a location and manner that minimizes the potential for intentional contamination (e.g., stored in locked building away from water hose connections, situated away from pipelines/connections with potential incompatible chemicals)?	
15. Are key valves, pumps, metering stations, and open-ended lines on critical processes, especially those in remote or uncontrolled/unrestricted areas, locked closed, located in locked secure structures (e.g., pump house), surrounded by locked secure fencing, and/or constructed of heavy-duty, tamper-resistant materials?	
16. Are ingredients for products potentially targeted for contamination unloaded, stored, transferred, and added to the process in a manner that is monitored and checked?	
17. Can exposed/remote equipment on critical processes feasibly be re-located to more secure/less vulnerable locations?	
18. Can critical process equipment that is highly recognizable from the ground and/or site perimeter be made less recognizable? (Note: This must be balanced against emergency responders need to readily identify equipment)	
19. Can critical processes or equipment be recognized readily from the air (consult aerial photos, if available) and, if so, can they be made less recognizable? (Note: This must be balanced against safety and code issues, such as painting of certain storage tanks in light colors.)	
(b) Reducing the Quantity and Hazard of a Release from a Malicious Act	
1. Has a review of site utility systems been conducted to identify and assess vulnerability of utilities that are essential to safe operation and shutdown of critical processes? Examples of possible critical utilities are:	

a. Electrical power	
b. Cooling water	
c. Compressed air	
d. Natural gas or other fuels	
e. Steam	
f. Nitrogen or other inert gases	
g. Secondary containment (drainage and sewer systems)	
h. Communications systems	
2. Are utility areas that can affect critical processes appropriately secured and monitored? (e.g., cooling water systems and agitation systems on reactive chemical processes that may be particularly important)	
3. Where appropriate, has safe and rapid manual shutdown capability been provided for critical processes and utilities?	
4. Where loss or reduction of utilities can potentially lead to uncontrolled reactions on critical processes, is the operating status of the utilities monitored and/or to alert personnel (e.g., an alarm sounds when cooling water flow is lost or reduced below critical levels)?	
5. Where loss or reduction of utilities can potentially lead to uncontrolled reactions on critical processes, are feed systems interlocked to agitation, cooling systems, and other appropriate utilities in the event of loss of those utilities or systems?	
6. Are appropriate back-up power supplies available for critical processes to allow a safe shutdown? (Note: UPS can be compromised by adversaries.)	
7. In the event of loss of power or pneumatics, do valves and other equipment fail to a safe position in critical processes?	
8. Are container storage areas secured or otherwise monitored, especially those outside of process buildings or in remote areas? (Note: A fire or explosion involving multiple containers can lead to smoke/combustion by-products that present offsite hazards and can serve as a diversion or a "statement.")	
9. Have storage and process inventories of hazardous chemicals been reduced to the extent practicable?	
10. Where appropriate, are critical processes containing highly hazardous chemicals "segmented" (either automatically or via manual action) to prevent release of the majority of process contents (i.e., only the quantity in the compromised "segment" would be released)?	
11. Are pipelines containing highly hazardous materials equipped with low-pressure interlock systems that shut valves or take other action to minimize the release quantity?	

12. Are open-ended lines or other lines or vessel drain systems on critical processes equipped with excess flow valves?	
13. Where appropriate, are hazardous materials being procured in smaller containers instead of maintaining large inventories in a single vessel?	
14. Has a review been conducted to determine if hazardous materials can be purchased and used in a less hazardous form? (Note: This may be particularly applicable to solvents/carriers and waste or water treatment chemicals.)	
15. If materials can be purchased and used in less hazardous forms, is this approach being addressed in an expedited manner?	
16. Has the feasibility and merit of storing large inventories of highly hazardous materials in underground tanks or other systems (e.g., aboveground vaults) that would limit the release rate been evaluated? (Note: This must be balanced against environmental concerns and other liabilities.) If found to be of merit, are plans in place to pursue this approach?	
17. Where appropriate and feasible, are tanks, vessels, and tank trailers/rail cars disconnected from delivery or transfer piping when not in use? (Note: The piping may be more vulnerable than the vessel.)	
(c) Mitigating a Release from a Malicious Act	
1. Are appropriate passive mitigation systems in-place for addressing large volume releases from critical processes?	
2. Have passive mitigation systems been assessed for integrity (i.e., are they being tested and/or maintained as required periodically) and vulnerability to be compromised?	
3. Has passive leak-limiting technology been used where possible (e.g., gaskets resistant to blowout, excess flow valves, etc.)?	
4. Are appropriate active mitigation systems in-place for addressing large volume releases at critical processes?	
5. Have active mitigation systems been assessed for integrity (i.e., are they being tested and/or maintained as required periodically) and vulnerability to be compromised?	
6. Are key control valves, pumps, and other equipment associated with active mitigation systems been locked or secured in operational/ready positions or located within secure structures?	
7. Has expanding the areas of the site where potential ignition sources are limited or eliminated (e.g., expanding the area of site subject to Class I/Div 1 or 2 electrical classification) been evaluated?	

(d) Emergency Response, Crisis Management, and Community Coordination	
1. Is the site's emergency response plan updated for current personnel and organizational functions?	
2. Do emergency plans address security worst case events, or events that are equivalent to security worst case events?	
3. Do emergency plans address malicious acts, especially responder actions in the event of a suspected terrorist/saboteur attack?	
4. Do emergency shutdown procedures address actions to take in the event of catastrophic releases or other terrorist-type event to safely shutdown the process and limit the release? If not, are shutdown procedures being reviewed and updated accordingly?	
5. Does the crisis management plan account for events such as:	
a. Bomb threats?	
b. Elevated homeland security warning status?	
c. Civil disturbance?	
6. Are operating personnel trained in the above-referenced emergency shutdown procedures, especially where they have been updated to address catastrophic or terrorist events?	
7. Has emergency equipment stationed near critical processes (e.g., hose connections) been assessed for vulnerability to compromise and, where appropriate, secured, monitored, or otherwise protected?	
8. If responding to a malicious act, are emergency responders aware that secondary "sucker-punch" devices (i.e., additional incendiary/explosive devices) or effects may be present if flammables are released or explosions are involved?	
9. Are procedures in-place (and responders trained accordingly) to address preservation of evidence due to the area being considered a crime scene?	
10. Where other nearby targets may exist (especially those that may present a greater risk than processes at our site), are plans in place to coordinate with local responders to ensure that those targets are monitored or otherwise protected in the event of a potential "diversionary" attack on our site?	
11. Have plans been developed with adjacent or nearby industry and local officials to facilitate timely communication of suspicious activity between potentially concerned parties?	
12. Have evacuation and shelter-in-place plans been fully developed and coordinated with local offsite emergency responders?	
13. Have local residents and business been instructed on how to shelter-in-place?	

14. Are local police, fire departments, health care providers, and other emergency responders aware of the hazardous materials at the site?	
15. Are plans in place to communicate information to local offsite emergency responders and officials in the event of a release?	
16. Do periodic emergency drills address malicious acts or other security-related emergencies?	
17. Is there a drill/exercise critique system in place to assure that experience from drills and actual emergencies are incorporated into the emergency response plan?	

APPENDIX C— SVA Interdependencies and Infrastructure Checklist

Appendix C Table of Contents

Page

INTERDEPENDENCIES TABLES 71
INFRASTRUCTURE OVERSIGHT AND PROCEDURES 72
(a) Infrastructure Oversight 72
(b) Infrastructure Procedures 72
INTERNAL ELECTRIC POWER SUPPLY AND DISTRIBUTION SYSTEM 73
(a) Primary Source of Electric Power 73
(b) Electric Distribution System 73
(c) Backup Electric Power Systems 73
INTERNAL HVAC SYSTEM 74
(a) Primary HVAC System 74
(b) Supporting Infrastructures 74
(c) Backup HVAC Systems 73
INTERNAL TELEPHONE SYSTEMS 75
(a) Primary Telephone System 75
(b) Data Transfer 75
(c) Cellular/Wireless/Satellite Systems 75
INTERNAL MICROWAVE/RADIO COMMUNICATIONS SYSTEM 76
(a) On-site Fixed Components 76
(b) Mobile and Remote Components 76
INTERNAL INTRANET AND E-MAIL SYSTEM 77
(a) Contained within a Larger System 77
(b) Separate System 77
(c) Access 78
INTERNAL COMPUTERS AND SERVERS 79
(a) Electric Power Sources 79
(b) Environmental Control 79
(c) Protection 79
INTERNAL FIRE SUPPRESSION AND FIRE FIGHTING SYSTEM 80
(a) Alarms 80
(b) Fire Suppression 80
(c) Fire Fighting 80
(d) Other Systems 80
INTERNAL SCADA SYSTEM 81
(a) Type of System 81
(b) Control Centers 81
(c) Electric Power Sources 81
(d) Communications Pathways 82
(e) Remote Components 82
(f) Dedicated SCADA Computers and Servers 83
INTERNAL DOMESTIC WATER SYSTEM 84
(a) Primary System 84
(b) External Water Supply System 84
(c) Internal Water Supply System 84
(d) Backup System 85
INTERNAL INDUSTRIAL WATER/WASTEWATER SYSTEM 86
(a) Primary Water System 86
(b) External Water Supply System 86
(c) Internal Water Supply System 86
(d) Backup Water System 87
(e) Primary Industrial Wastewater System 87
f) Backup Wastewater System 88

INTERNAL PHYSICAL SECURITY SYSTEM 89
(a) Electric Power Sources 89
(b) Communications Pathways 89
(c) Computer Support 90
INTERNAL HUMAN RESOURCES SUPPORT 92
(a) Electric Power Sources 92
(b) Communications Pathways 92
(c) Computer Support 93
INTERNAL FINANCIAL SYSTEM 94
(a) Electric Power Sources 94
(b) Communications Pathways 94
(c) Computer Support 95
EXTERNAL ELECTRIC POWER INFRASTRUCTURE 96
(a) Electric Power Sources 96
(b) Electric Power Pathways 96
(c) Electric Power Contracts 96
(d) Historical Reliability 96
EXTERNAL NATURAL GAS INFRASTRUCTURE 97
(a) Sources of Natural Gas 97
(b) Pathways of Natural Gas 97
(c) Natural Gas Contracts 97
(d) Historical Reliability 98
EXTERNAL PETROLEUM FUELS INFRASTRUCTURE 99
(a) Uses of Petroleum Fuels 99
(b) Reception Facilities 99
(c) Supply Contracts 99
EXTERNAL TELECOMMUNICATIONS INFRASTRUCTURE 100
(a) Telecommunications Carriers 100
(b) Pathways of Telecommunications Cables 100
(c) Historical Reliability 100
(d) Backup Communications Systems 101
EXTERNAL WATER AND WASTEWATER INFRASTRUCTURE 102
(a) Water Supply Reliability 102
(b) Wastewater System Reliability 102
EXTERNAL ROAD TRANSPORTATION INFRASTRUCTURE 103
(a) Road Access 103
(b) Road Access Control 103
EXTERNAL RAIL TRANSPORTATION INFRASTRUCTURE 104
(a) Rail Access 104
(b) Rail Access Control 104
EXTERNAL AIR TRANSPORTATION INFRASTRUCTURE 105
(a) Airports and Air Routes 105
EXTERNAL WATER TRANSPORTATION INFRASTRUCTURE 106
(a) Waterway Access 106
(b) Waterway Access Control 106
EXTERNAL PIPELINE TRANSPORTATION INFRASTRUCTURE 107
(a) Pipeline Access 107
(b) Pipeline Access Control 107
OPSEC TABLES 108
HUMAN RESOURCES SECURITY PROCEDURES 108
(a) Responsibilities 108
(b) Background Checks 108
(c) Insider Threats 108
(d) Disciplinary Procedures 108
(e) Security Training 108
(f) Travel 108

FACILITY ENGINEERING 109
(a) Responsibilities 109
(b) Facility Engineering Information 109
(c) Public Access to Facility 109
FACILITY OPERATIONS 110
(a) Responsibilities 110
(b) Facility Operations Control 110
(c) Facility Construction, Repair, and Maintenance 110
ADMINISTRATIVE SUPPORT ORGANIZATIONS 111
(a) Procurement 111
(b) Legal 111
(c) Budget and Finance 111
(d) Marketing 111
(e) Internal Information 111
TELECOMMUNICATIONS AND INFORMATION TECHNOLOGY 112
(a) Telecommunications 112
(b) Information Technology 112
PUBLICLY RELEASED INFORMATION 113
(a) Responsibilities 113
(b) General Procedures 113
(c) Report Release 113
(d) Press Contacts 113
(e) Briefing and Presentations 113
(f) Public Testimony 113
(g) Internet Information 113
TRASH AND WASTE HANDLING 114
(a) Responsibilities 114
(b) Trash Handling 114
(c) Paper Waste Handling 114
(d) Salvage Material Handling 114
(e) Dumpster Control 114

INTERDEPENDENCIES TABLES

INTERNAL AND EXTERNAL INFRASTRUCTURES TO BE INCLUDED

Date: ______________ Facility: ______________________________

This checklist applies to:

Entire Facility

Critical Asset ______________________________

(Note: Not all infrastructures supporting each asset/facility need to be included in this survey. Only those infrastructures that are important to the asset's/facility's ability to continue to carry out its critical functions and activities need be considered in detail. In addition, the time and resources allotted for the survey may limit the infrastructures that can be examined.)

INFRASTRUCTURE	YES	NO	RATIONALE FOR EXCLUSION/INCLUSION
Internal			
Electric Power Supply and Distribution System			
HVAC System			
Telephone System			
Microwave/Radio Communications System			
Intranet and E-mail System			
Computers and Servers			
Fire Suppression/ Fire Fighting System			
SCADA System			
Domestic Water System			
Industrial Water System			
Physical Security System			
Human Resources Support			
Financial System			
External			
Electric Power			
Natural Gas			
Petroleum Fuels			
Telecommunications			
Water and Wastewater			
Road Transportation			
Rail Transportation			
Air Transportation			
Water Transportation			

INFRASTRUCTURE OVERSIGHT AND PROCEDURES	
Date: [MONTH XX, 2002]	Facility: [FACILITY]
	COMMENTS
(a) Infrastructure Oversight	
Does the facility have a central office or department (such as building management, plant services, facility management) that is responsible for overseeing all or most the infrastructures? Indicate the office/department and list the infrastructures for which they have responsibility and the extent of their responsibilities.	
What coordination or oversight role does the physical security office have in regards to the infrastructures that support critical functions or activities?	
(b) Infrastructure Procedures	
In general, are operating procedures in place for the systems that make up the internal infrastructures and for the physical connections and contracts with the external infrastructures that support them? Describe the extent of these procedures, their format, their availability to relevant staff, and the extent to which they are regularly followed. (Note: details about procedures for specific individual infrastructures are addressed in the relevant checklists.)	
Are contingency procedures in place for the systems that make up the internal infrastructures and for the physical connections and contracts with the external infrastructures that support them? Describe the extent of these procedures, their format, and their availability to relevant staff. (Note: Contingencies refer to situations brought about by a failure or disruption within an infrastructure or the infrastructures that support it.)	
If they exist, have the contingency procedures been tested and are they exercised regularly either as a part of normal operations as through specially designed drills? Describe the drills and their results.	
Are emergency procedures in place for the systems that make up the internal infrastructures and for the physical connections and contracts with the external infrastructures that support them? Describe the extent of these procedures, their format, and their availability to relevant staff. (Note: Emergencies refer to situations brought about external stress on the facility such as high demands.)	
If they exist, have the emergency procedures been tested and are they exercised regularly through specially designed drills? Describe the drills and their results.	

INTERNAL ELECTRIC POWER SUPPLY AND DISTRIBUTION SYSTEM

Date: ______________ Facility: ______________________________

This checklist applies to:

Entire Facility

Critical Asset ______________________________

	COMMENTS
(a) Primary Source of Electric Power	
If the primary source of electric power is a commercial source, are there multiple independent feeds? If so, describe the feeds and their locations.	
If the primary source of electric power is a system operated by the facility or asset, what type of system is it?	
If a facility operated primary electric generation system is used, what is the fuel or fuels used?	
If petroleum fuel is used, what quantity of fuel is stored on site for the primary electric generation system and how long it will last under different operating conditions?	
If the fuel is stored on site, are arrangements and contracts in place for resupply and management of the fuel?	
(b) Electric Distribution System	
Are the components of the electric system that are located outside of buildings (such as generators, fuel storage facilities, transformers, transfer switches) protected from vandalism or accidental damage by fences or barriers? If so, describe the type of protection and level of security it provides.	
Are the various sources of electric power and the components of the internal electric distribution systems - such that they may be isolated for maintenance or replacement without affecting the critical functions of the asset/facility? If not, describe the limitations.	
Have any single points of failure been identified for the electrical power supply and distribution system? If so, list them and describe.	
(c) Backup Electric Power Systems	
Are there additional emergency sources of electric supply beyond the primary system (such as multiple independent commercial feeds, backup generators, uninterruptible power supply [UPSs])? If there are, describe them.	
If there is a central UPS, does it support all the critical functions of the asset/facility in terms of capacity and connectivity? Specify for how long it can operate on battery power and list any potentially critical functions that are not supported.	
If there is a backup generator system, does it support all the critical functions of the facility in terms of capacity and connectivity? Specify the fuel and list any potentially critical functions that are not supported.	
Is the fuel for the backup generator system a petroleum fuel? If yes, specify the quantity stored on site and how long it will last.	
If the fuel is stored on site, are arrangements and contracts in place for resupply and management of the fuel?	

INTERNAL HVAC SYSTEM
(Including Heating Plants and Cooling Towers)

Date: ______________ Facility: ____

This checklist applies to:

Entire Facility

Critical Asset ____

	COMMENTS
(a) Primary HVAC System	
Can critical functions and activities dependent on environmental conditions continue without the HVAC system? If yes, specify which functions and for how long they can continue under various external weather conditions.	
Is the HVAC system that supplies the areas of the asset/facility where critical functions dependent on environmental conditions are carried out separate from or separable fromthe general asset/facility-wide HVAC system?	
(b) Supporting Infrastructures	
Does the HVAC system (or critical portion thereof) depend on the primary electric power supply and distribution system to supply electric power? Specify under what conditions and for how long.	
Besides or in addition to electric power, on what fuel or fuels does the HVAC system (or critical portion thereof) depend?	
If the HVAC system (or critical portion thereof) depends on natural gas, are there provisions for alternative fuels during a natural gas outage? Specify the fuel and how long the HVAC system can operate on it.	
If the HVAC system (or critical portion thereof) depends on petroleum fuels for adequate operation, specify the type of fuel and how long the HVAC system can operate on the fuel available on site.	
If the HVAC system (or critical portion thereof) depends on petroleum fuels, are arrangements and contracts in place for resupply and management of the fuel?	
Does the HVAC system (or critical portion thereof) depend on water? If it does, specify if the water need is continuous or for make-up purposes only and the quantities/rates involved.	
If the HVAC system (or critical portion thereof) depends on water, is a backup supply in place such as well and pump, storage tank, or tank trucks? Specify how long the HVAC can operate on the backup water supply system.	
(c) Backup HVAC Systems	
Is there a separate backup to the HVAC system? If yes, describe the system and the energy and water supply systems it requires.	
Are there contingency procedures in place to continue with the critical functions and activities that take place at the asset/facility during an HVAC outage? If yes, briefly describe them.	
How long can the critical functions and activities at the asset/facility continue using the backup HVAC system or under the contingency procedures?	

INTERNAL TELEPHONE SYSTEMS
(Including Voice, FAX, and Data Transfer)

Date: ____________ Facility: ________________________________

This checklist applies to:

Entire Facility

Critical Asset ________________________________

	COMMENTS
(a) Primary Telephone System	
What types of telephone systems are used within the asset/facility? Are there multiple independent telephone systems? Specify the types of systems, their uses, and whether they are copper-wire or fiber-optic based.	
If there are there multiple independent telephone systems within the asset/facility, is each one adequate to support the critical functions and activities? Indicate any limitations.	
If there are multiple (from independent systems) or redundant (from built-in backups) switches and cables, are they physically separated and isolated to avoid common causes of failure?	
Are the telephone switches located in limited-access or secured areas away from potential damage due to weather or water leaks? Specify types of protection provided.	
(b) Data Transfer	
For large volume and high-speed data transfer within the asset/facility, is there a separate system of switches and cables within the asset/facility? Specify the type of system and whether it is copper-wire or fiber-optic based.	
If there is a separate system for large-volume and high-speed data transfer, are there redundant switches and cables? If yes, describe the situation.	
If there are redundant switches and cables, are they physically separated and isolated to avoid common causes of failure?	
Are the data-transfer switches located in limited-access or secured areas away from potential damage due to weather or water leaks? Specify the types of protection provided.	
(c) Cellular/Wireless/Satellite Systems	
Are cellular/wireless telephones and pagers in widespread use within the asset/facility? If yes, briefly describe their uses.	
If cellular/wireless telephones and pagers are in widespread use, are they adequate to support the critical functions and activities? Specify any limitations.	
Are satellite telephones or data links in widespread use within the asset/facility? If yes, briefly describe their uses.	
If satellite telephones or data links are in widespread use, are they adequate to support the critical functions and activities? Specify any limitations.	

INTERNAL MICROWAVE/RADIO COMMUNICATIONS SYSTEM

Date: ____________ Facility: ________________________________

This checklist applies to:

Entire Facility

Critical Asset ________________________________

	COMMENTS
(a) On-site Fixed Components	
Are there multiple or redundant radio communications systems in place within the asset/facility? If yes, specify the types of systems and their uses.	
If there are multiple radio communications systems, is more than one system adequate to support all the critical functions and activities of the asset/facility? Specify any limitations.	
Are there provisions within the asset's/facility's primary electric power supply and distribution system to supply power for the radio communications systems? If yes, indicate under what conditions and for how long.	
Do the radio communications systems have their own backup electric power supply? If yes, specify the type and how long it can operate.	
Are the components of the system located outside of buildings (such as antennae, on-site towers) protected from vandalism or accidental damage by fences or barriers? If protected, specify the types of protection and level of security they provide.	
(b) Mobile and Remote Components	
Are there mobile components to the radio communications system (such as on vehicles or vessels)? If yes, describe the mobile components.	
Are the mobile components of the radio communications system protected from vandalism or accidental damage by locked boxes or lockable vehicle cabs? Specify the types of protection and level of security they provide.	
Are there remote components to the radio communications system (such as relay towers)? If yes, describe them and their uses.	
Are there backup sources of electric power for these remote components? If yes, indicate the type of backup, the fuels used, and the expected length of operations.	
Are there environmental controls required for the remote components (such as heating, cooling)? If yes, describe them.	
Are there backup environmental controls for these remote components? If yes, indicate the type of backup, the fuels used, and the expected length of operations.	
Is physical security provided for the remote components of the radio communications system? If yes, specify the types of security and the level of protection provided.	
Are there alarms at the remote components of the radio communications system for such things as intrusion, loss of electric power, loss of environmental control, and fuel reserves? If yes, specify the types of alarms, how they are monitored, and the response procedure.	

INTERNAL INTRANET AND E-MAIL SYSTEM

Date: ____________ Facility: ____

This checklist applies to:

Entire Facility

Critical Asset ____

	COMMENTS
(a) Contained within a Larger System	
Is the asset's/facility's intranet and e-mail system dependent on the asset's/facility's computers and servers? If yes, describe the dependence.	
Is the asset's/facility's intranet and e-mail system dependent on the asset's/facility's telephone system? If yes, describe the dependence.	
(b) Separate System	
If the asset's/facility's intranet and e-mail system is a separate system, are there provisions within the asset's/facility's primary electric power supply and distribution system to supply power for the intranet and e-mail system? If yes, specify under what conditions and for how long.	
If the asset's/facility's intranet and e-mail system is a separate system, does it have its own backup electric power supply, such as local UPSs? If yes, specify the type and how long it can operate.	
If the asset's/facility's intranet and e-mail system is a separate system, does the asset's/facility's central HVAC system provide environmental control for important components or does it have its own independent environmental control system? If it has its own, specify the type.	
If the asset's/facility's intranet and e-mail system is a separate system, can it operate with a loss of all environmental control? If yes, specify for how long under various conditions.	
If the asset's/facility's intranet and e-mail system is a separate system, are there any backup environmental controls explicitly for the system? If yes, indicate the type of backup and the expected maximum duration of operation.	
If the asset's/facility's intranet and e-mail system is a separate system, is there special physical security provided for the important components? If yes, specify the type of security and the level of protection provided.	
If the asset's/facility's intranet and e-mail system is a separate system, is there special fire suppression equipment for the important components such as Halon, Inergen, inert gases, or carbon dioxide? If yes, specify the type of system.	
If the asset's/facility's intranet and e-mail system is a separate system, are there special features or equipment in the area of the important components to limit flooding or water intrusion? If yes, indicate the precautions taken.	

If the asset's/facility's intranet and e-mail system is a separate s ystem, are there alarms for the area of the important components for such things as unauthorized intrusion, loss of electric power, loss of environmental control, fire, and flooding or water intrusion? If yes, specify the types of alarms, how they are monitored, and the response procedure.	
(c) Access	
Does the asset/facility have a backup or redundant intranet and e-mail system? If yes, describe the system and the amount of backup it provides.	
Do areas where critical functions and activities take place have multiple or redundant access to the intranet and e-mail system?	
If there are multiple access routes, is each one adequate to support the critical functions and activities? If not, specify any limitations.	

INTERNAL COMPUTERS AND SERVERS
(Including Mainframes, Firewalls, and Router Equipment)

Date: ____________ Facility: ______________________________

This checklist applies to:

Entire Facility

Critical Asset ______________________________

	COMMENTS
(a) Electric Power Sources	
Are there provisions within the asset's/facility's primary electric power supply and distribution system to supply power for the computers and servers? If yes, indicate under what conditions and for how long.	
Do the computers and servers have their own backup electric power supply (such as local UPSs or generators)? If yes, specify the types of backup and how long they can operate.	
(b) Environmental Control	
Does the asset's/facility's central HVAC system provide environment control to the computer and server areas or do the computer and server areas have their own independent environmental control system? If they have their own system, specify the type.	
Can the computers and servers operate with a loss of all environmental control? If yes, specify for how long under various conditions.	
Are there any backup environmental controls explicitly for the computer and server areas? If yes, indicate the type of backup and the expected maximum duration of operation.	
(c) Protection	
Is there special physical security provided for the computer and server areas? If yes, specify the type of security and the level of protection provided.	
Is there special fire suppression equipment in the computer and server areas such as Halon, Inergen, inert gases, or carbon dioxide? If yes, specify the type.	
Are there special features or equipment in the computer and server areas to limit flooding or water intrusion? If yes, describe them.	
Are there alarms for the computer and server areas for such things as unauthorized intrusion, loss of electric power, loss of environmental control, fire, and flooding or water intrusion? If yes, specify the types of alarms, how they are monitored, and the response procedure.	

INTERNAL FIRE SUPPRESSION AND FIRE FIGHTING SYSTEM

Date: ____________ Facility: ______________________________

This checklist applies to:

Entire Facility

Critical Asset ______________________________

	COMMENTS
(a) Alarms	
Does the entire asset/facility (or at least most of it) have a fire and/or smoke detection and alarm system? If yes, specify the type of system, how it is monitored, and the response procedure.	
(b) Fire Suppression	
Does the entire asset/facility (or at least most of it) have a fire suppression system such as an overhead sprinkler system? If yes, specify the medium (usually water) and whether it is of the flooded-pipe or pre-armed type.	
Does the water supply for the fire suppression system come from city water mains or an on-site system, such as wells, rivers, or reservoir?	
If the water supply for the fire suppression system comes from city water mains, specify whether there are separate city fire mains and if the pipe from the main to the asset/facility is separate from the domestic water supply.	
If the water supply for the fire suppression system comes from an on-site system, specify the source, indicate the adequacy of the supply's capacity, and indicate if it is gravity feed or requires active pumps (generally electric).	
(c) Fire Fighting	
Does the asset/facility have its own fire-fighting department? If yes, describe it in terms of adequacy to protect the asset/facility.	
Are city or community fire-fighting services available to the facility? If yes, indicate the type of service and the estimated response time.	
Does the water supply for the fire-fighting hydrants come from city water mains? If yes, specify the number of hydrants and indicate their coverage and accessibility.	
If the water supply for the fire fighting hydrants comes from an on-site system (such as wells, rivers, or reservoir), specify the source, indicate the adequacy of the supply's capacity, and indicate if it is gravity feed or requires active pumps (generally electric). Also, specify the number of hydrants and indicate their coverage and accessibility.	
(d) Other Systems	
Is there special fire suppression equipment, such as Halon, Inergen, inert gases, or carbon dioxide in certain areas such as computer or telecommunications areas? If yes, indicate the types and adequacies of these special systems.	

INTERNAL SCADA SYSTEM

Date: ____________ Facility: ______________________________

This checklist applies to:

Entire Facility

Critical Asset ______________________________

	COMMENTS
(a) Type of System	
Does the asset/facility make use of a substantial SCADA system (i.e., one that covers a large area or a large number of components and functions)? If yes, indicate what functions are monitored and/or controlled, the type of system, and the extent of the system.	
Is the SCADA system independent of the asset's/facility's primary electric power supply and distribution system?	
Is the SCADA system independent of the asset's/facility's telephone system?	
Is the SCADA system independent of the asset's/facility's microwave or radio communications system?	
Is the SCADA system independent of the asset's/facility's computers and servers?	
(b) Control Centers	
Where is the primary control center for the SCADA system located?	
Is there a backup control center? If yes, where is it located? Is it sufficiently remote from the primary control center to avoid common causes of failure such as fires, explosions, or other large threats?	
Are there backups to the SCADA computers and servers at the backup control center or at some other location? If yes, indicate the location of the backup computers and servers, whether they are completely redundant or cover only the most critical functions, and whether they are active "hot" standbys or have to be activated and initialized when needed.	
Note: *The following sets of questions on electric power sources and communications pathways apply to the control centers as well as the other components of the SCADA system.*	
(c) Electric Power Sources	
Are there multiple sources of electric supply (such as multiple independent commercial feeds, backup generators, UPSs) explicitly for the SCADA system? If yes, indicate the types.	
If there is a special UPS, does it support all the functions of the SCADA system in terms of capacity? Specify for how long it can operate on battery power.	
If there is a special backup generator system, does it support all the functions of the SCADA system in terms of capacity?	
What is the fuel or fuels used by the special SCADA backup generator system? If stored on site, specify the quantity stored and how long it will last.	

If the SCADA backup generator fuel is stored on site, are arrangements and contracts in place for resupply and management of the fuel?	
(d) Communications Pathways	
Are there dedicated multiple independent telephone systems or dedicated switches and cables supporting the SCADA system? If yes, specify whether copper-wire or fiber-optic based.	
If there are dedicated multiple independent telephone systems or dedicated switches and cables supporting the SCADA system, is each one individually adequate to support the entire system? Specify any limitations.	
Are the redundant telephone systems or switches and cables physically separated and isolated to avoid common causes of failure? If not, indicate any potential points of common failure.	
Are the dedicated SCADA telephone switches and data-transfer switches located in a limited access or secured area away from potential damage due to weather or water leaks? If so, specify type of protection.	
Are there dedicated multiple or redundant radio communications systems in place to support the SCADA system? If yes, indicate the types.	
If there are multiple radio communications systems, is each one individually adequate to support the entire SCADA system? If not, specify any limitations.	
Are there provisions within the asset's/facility's primary electric power supply and distribution system to supply power for the special SCADA radio communications systems? If yes, specify under what conditions and for how long.	
Do the special SCADA radio communications systems have their own backup electric power supply? If yes, specify the type and how long it can operate.	
Are the components of the special SCADA radio communications system located outside of buildings (such as antennae, on-site towers) protected from vandalism or accidental damage by fences or barriers? If protected, specify the types of protection and level of security provided.	
(e) Remote Components	
Are there remote components to the special SCADA radio communications system (such as relay towers)? If yes, identify the components and there locations.	
Are there backup sources of electric power for these remote components? If yes, indicate the type of backup, the fuels used, and the expected length of operations.	
Are there environmental controls required for the remote components of the special SCADA radio communications system (such as heating, cooling)? If yes, describe them.	

Are there backup environmental controls for these remote components? If yes, indicate the type of backup, the fuels used, and the expected length of operations.	
Is physical security provided for the remote components of the special SCADA radio communications system? If yes, specify the types of security and the level of protection provided.	
Are there alarms at the remote components of the special SCADA radio communications system for such things as intrusion, loss of electric power, loss of environmental control, and fuel reserves? If yes, specify the types of alarms, how they are monitored, and to the response procedure.	
(f) Dedicated SCADA Computers and Servers	
Are there provisions within the asset's/ facility's primary electric power supply and distribution system to supply power for the special dedicated SCADA computers and servers? If yes, specify under what conditions and for how long.	
Do the special dedicated SCADA computers and servers have their own backup electric power supply, such as local UPSs? If yes, specify the types and how long they can operate.	
Does the asset's/facility's central HVAC system provide environment control for the separate special SCADA computer and server areas?	
How long can the separate dedicated SCADA computers and servers operate with a loss of all environmental control? Indicate the conditions that could affect the length of time.	
Do the separate dedicated SCADA computer and server areas have their own independent environmental control system? If yes, specify the type.	
Are there any backup environmental controls explicitly for the dedicated SCADA computer and server areas? If yes, indicate the type of backup and the expected maximum duration of operation.	
Is there special physical security provided for the separate SCADA computer and server areas? If yes, specify the type of security and the level of protection provided.	
Is there special fire suppression equipment in the separate dedicated SCADA computer and server areas such as Halon, Inergen, inert gases, or carbon dioxide? If yes, specify the type of system.	
Are there special features or equipment in the separate SCADA computer and server areas to limit flooding or water intrusion? If yes, indicate the precautions taken.	
Are there alarms for the separate SCADA computer and server areas for such things as unauthorized intrusion, loss of electric power, loss of environmental control, fire, and flooding or water intrusion? If yes, specify the types of alarms, how they are monitored, and the response procedure.	

INTERNAL DOMESTIC WATER SYSTEM

Date: ______________ Facility: ____

This checklist applies to:

Entire Facility

Critical Asset ____

	COMMENTS
(a) Primary System	
Does the asset/facility have a domestic water system? If yes, specify the uses of the water (such as restrooms, locker rooms, kitchens, HVAC makeup water).	
Does the water supply for the domestic water system come from an external source (such as community, city, or regional water mains) or from an internal system (such as wells, river, or reservoir)? If internal, describe the system.	
(b) External Water Supply System	
What type of external water supply system provides the domestic water? Indicate whether it is public or private and its general size (such as community, city, or regional).	
Are on-site pumps and/or storage tanks used to boost the pressure or provide for periods of peak usage? If yes, briefly describe them and their purpose.	
Are the on-site booster water pumps normally dependent upon the asset's/facility's primary electric power supply and distribution system?	
Are there multiple sources of electric supply (such as multiple independent commercial feeds, backup generators, UPSs) explicitly for the on-site booster water pumps? If yes, specify them.	
If there is a special UPS, can it support the on-site booster pumps at required levels? Specify for how long it can operate on battery power.	
If there is a special backup generator system, can it support the on-site booster pumps at required levels? Also indicate the type of fuel or fuels used.	
If the fuel for the dedicated backup generator system for the booster pumps is a petroleum fuel, indicate the quantity stored on site and how long it will last.	
If the fuel for the dedicated backup generator for the booster pumps is stored on site, are arrangements and contracts in place for resupply and management of the fuel?	
(c) Internal Water Supply System	
Indicate the source of the water (such as wells, river, or reservoir), the adequacy of the supply's capacity, and whether it is gravity feed or requires active pumps (generally electric).	
Are the on-site domestic water system pumps independent of the asset's/facility's primary electric power supply and distribution system?	

Are there multiple sources of electric supply (such as multiple independent commercial feeds, backup generators, UPSs) explicitly for the on-site domestic water system pumps? If yes, specify them.	
If there is a special UPS, can it support the on-site domestic water system pumps at required levels? Specify for how long it can operate on battery power.	
If there is a special backup generator system, can it support the on-site domestic water system pumps at the required levels? Also indicate the type of fuel or fuels used.	
If the fuel for the dedicated backup generator system for the on-site domestic water system pumps is a petroleum fuel, indicate the quantity stored on site and how long it will last. Are arrangements and contracts in place for resupply and management of the fuel?	
(d) Backup System	
Is there an independent backup water source to the primary domestic supply system? If yes, specify the type of backup system (such as wells, river, reservoir, tank truck), describe the specific source of the water, indicate the adequacy of the backup supply's capacity, and indicate if it is gravity feed or requires active pumps (generally electric).	
Are the independent backup water source system pumps independent of the asset's/facility's primary electric power supply and distribution system?	
Are there multiple sources of electric supply (such as multiple independent commercial feeds, backup generators, UPSs) explicitly for the backup water source system pumps? If yes, specify them.	
If there is a special UPS, can it support the backup domestic water source pumps at the required levels? Specify for how long it can operate on battery power.	
If there is a special backup generator system, can it support the backup domestic water source system pumps at the required levels? Also indicate the type of fuel or fuels used.	
If the fuel for the dedicated backup generator system for the backup water source system pumps is a petroleum fuel, indicate the quantity stored on site and how long it will last. Are arrangements and contracts in place for resupply and management of the fuel?	

INTERNAL INDUSTRIAL WATER/WASTEWATER SYSTEM

Date: ____________ Facility: ____

This checklist applies to:

Entire Facility

Critical Asset ____

	COMMENTS
(a) Primary Water System	
Does the asset/facility have an industrial water system? If yes, specify the uses of the water (such as wash water, process water, generation of process steam, cooling).	
Does the water supply for the industrial water system come from an external source (such as community, city, or regional water mains) or from an internal system (such as wells, river, or reservoir)? If internal, describe the system.	
(b) External Water Supply System	
What type of external water supply system provides the industrial water? Indicate whether it is public or private and its general size (such as community, city, or regional).	
Are on-site pumps and/or storage tanks used to boost the pressure or provide for periods of peak usage? If yes, briefly describe them and their purpose.	
Are the on-site booster water pumps for the industrial water system independent of the asset's/facility's primary electric power supply and distribution system?	
Are there multiple sources of electric supply (such as multiple independent commercial feeds, backup generators, UPSs) explicitly for the on-site booster water pumps? If yes, specify them.	
If there is a special UPS, can it support the on-site booster pumps at required levels? Specify for how long it can operate on battery power.	
If there is a special backup generator system, can it support the on-site booster pumps at required levels? Also indicate the type of fuel or fuels.	
If the fuel for the dedicated backup generator system for the booster pumps is a petroleum fuel, indicate the quantity stored on site and how long it will last. Are arrangements and contracts in place for resupply and management of the fuel?	
(c) Internal Water Supply System	
Indicate the source of the water (such as wells, river, or reservoir), the adequacy of the supply's capacity, and whether it is gravity feed or requires active pumps (generally electric).	
Are the on-site industrial water system pumps independent of the asset's/facility's primary electric power supply and distribution system?	

Are there multiple sources of electric supply (such as multiple independent commercial feeds, backup generators, UPSs) explicitly for the on-site industrial water system pumps? If yes, specify them.	
If there is a special UPS, can it support the on-site industrial water system pumps at required levels? Specify for how long it can operate on battery power.	
If there is a special backup generator system, can it support the on-site industrial water system pumps at the required levels? Also indicate the type of fuel or fuels.	
If the fuel for the dedicated backup generator system for the on-site industrial water system pumps is a petroleum fuel, indicate the quantity stored on site and how long it will last. Are arrangements and contracts in place for resupply and management of the fuel?	
(d) Backup Water System	
Is there an independent backup water source to the primary industrial water supply system? If yes, specify the type of backup system (such as wells, river, reservoir, tank truck), describe the specific source of the water, indicate the adequacy of the backup supply's capacity, and indicate if it is gravity feed or requires active pumps (generally electric).	
Are the independent backup water source system pumps independent of the asset's/facility's primary electric power supply and distribution system?	
Are there multiple sources of electric supply (such as multiple independent commercial feeds, backup generators, UPSs) explicitly for the backup water source system pumps? If yes, specify them.	
If there is a special UPS, can it support the backup industrial water source pumps at the required levels? Specify for how long it can operate on battery power.	
If there is a special backup generator system, can it support the backup industrial water source system pumps at required levels? Also indicate the type of fuel or fuels.	
If the fuel for the dedicated backup generator system for the backup water source system pumps is a petroleum fuel, indicate the quantity stored on site and how long it will last. Are arrangements and contracts in place for resupply and management of the fuel?	
(e) Primary Industrial Wastewater System	
Does the asset/facility have an on-site industrial wastewater system? If yes, specify the types of wastewater that are processed and the processes used.	
Are the on-site industrial wastewater lift pumps independent of the asset's/facility's primary electric power supply and distribution system?	

Are there multiple sources of electric supply (such as multiple independent commercial feeds, backup generators, UPSs) explicitly for the on-site industrial wastewater lift pumps? If yes, specify them.	
If there is a special UPS, can it support the on-site industrial wastewater lift pumps at required levels? Specify for how long it can operate on battery power.	
If there is a special backup generator system, can it support the on-site industrial wastewater lift pumps at the required levels? Also indicate the type of fuel or fuels.	
If the fuel for the dedicated backup generator system for the on-site industrial wastewater lift pumps is a petroleum fuel, indicate the quantity stored on site and how long it will last. Are arrangements and contracts in place for resupply and management of the fuel?	
f) Backup Wastewater System	
Is there an independent backup system that can be used to handle the industrial wastewater? If yes, specify the type of backup system (such as a redundant system, holding ponds, temporary discharge of unprocessed wastewater), describe the specific process, indicate the adequacy of the backup's capacity and any limitations on how long it can operate, and indicate if it is gravity feed or requires active lift pumps (generally electric).	
Are the independent backup lift pumps independent of the asset's/facility's primary electric power supply and distribution system?	
Are there multiple sources of electric supply (such as multiple independent commercial feeds, backup generators, UPSs) explicitly for the backup wastewater lift pumps? If yes, specify them.	
If there is a special UPS, can it support the backup industrial wastewater system at the required levels? Specify for how long it can operate on battery power.	
If there is a special backup generator system, can it support the backup industrial wastewater lift pumps at required levels? Also indicate the type of fuel or fuels.	
If the fuel for the dedicated backup generator system for the backup wastewater lift pumps is a petroleum fuel, indicate the quantity stored on site and how long it will last. Are arrangements and contracts in place for resupply and management of the fuel?	

INTERNAL PHYSICAL SECURITY SYSTEM

Date: _____________ Facility: _______________________________________

This checklist applies to:

Entire Facility

Critical Asset _______________________________________

	COMMENTS
(a) Electric Power Sources	
Are the asset's/facility's monitoring and alarm systems normally dependent on the asset's/facility's primary electric power supply and distribution system (i.e., is the asset's/facility's primary electric power supply and distribution system the primary electric power source)?	
Are there multiple sources of electric power for the monitoring and alarm systems? This could consist of the asset's/facility's primary electric power supply and distribution system and its backup or redundant systems; or combinations of multiple independent commercial electric feeds, backup generators, UPSs, or batteries dedicated to support the monitoring and alarm systems. Specify what electric power sources are in place.	
If there is a special UPS, can it support all the functions of the monitoring and alarm systems in terms of capacity? Specify for how long it can operate on battery power.	
If there is a special generator system, can it support all the functions of monitoring and alarm systems in terms of capacity? Also indicate the type of fuel or fuels used.	
If the fuel for the special security generator system is a petroleum fuel, indicate the quantity stored on site and how long it will last. Are arrangements and contracts in place for resupply and management of the fuel?	
(b) Communications Pathways	
Are the asset's/facility's monitoring and alarm systems normally dependent upon the asset's/facility's telephone system?	
Are there multiple independent telephone systems or dedicated switches and cables supporting the monitoring and alarm systems? This could consist of the asset's/facility's telephone system and its backup or redundant systems; or combinations of multiple independent telephone systems or dedicated communications lines. Specify the types of systems used and whether they are copper-wire or fiber optic-cable based.	
Are the redundant telephone systems or switches and cables physically separated and isolated to avoid common causes of failure? If not, indicate any potential points of common failure.	

Are the dedicated monitoring and alarm systems telephone switches and data-transfer switches located in a limited access or secured area away from potential damage due to weather or water leaks? If so, specify type of protection.	
Are the asset's/facility's monitoring and alarm systems normally dependent upon the asset's/facility's microwave or radio communications system?	
Are there multiple independent microwave or radio communications systems supporting the monitoring and alarm systems? This could consist of the asset's/facility's primary microwave or radio communications system and its backup or redundant systems; or combinations of multiple independent radios, antennae, and relay towers. Specify the type of radio systems used.	
Are there multiple sources of electric power for the microwave or radio communications systems dedicated to support the monitoring and alarm systems? This could consist of the asset's/facility's electric power supply and distribution system and its backup or redundant systems; or combinations of multiple independent commercial electric feeds, backup generators, UPSs, or batteries dedicated to support the special microwave or radio communications systems. If yes, specify the types and how long they can operate.	
Are the components of the special radio communications system dedicated to the monitoring and alarm systems that are located outside of buildings (such as antennae, on-site towers) protected from vandalism or accidental damage by fences or barriers? If protected, specify the types of protection and level of security they provide.	
Are there remote components to the special radio communications system dedicated to the monitoring and alarm systems (such as relay towers)? If yes, identify the components and their locations.	
Are there backup sources of electric power for the remote components? If used, indicate the type of backup, the fuels used, and the expected length of operations.	
Are there environmental controls required for the remote components of the special monitoring and alarm radio communications system (such as heating, cooling)? If yes, describe them.	
Are there backup environmental controls for the remote components? If yes, indicate the type of backup, the fuel or fuels used, and the expected length of operations.	
(c) Computer Support	
Are the asset's/facility's monitoring and alarm systems normally dependent upon the facility's main computers and servers?	

Are there multiple independent computers supporting the monitoring and alarm systems? This could consist of the asset's/facility's main computers and servers and their backup or redundant systems, or combinations of multiple independent computers. Specify the type of computers used.	
Are there multiple sources of electric power for any computers dedicated to support the monitoring and alarm systems? This could consist of the asset's/facility's primary electric power supply and distribution system and its backup or redundant systems; or combinations of multiple independent commercial electric feeds, backup generators, or UPSs dedicated to support the monitoring and alarm systems. If yes, specify the type and how long they can operate.	
Does the asset's/facility's central HVAC system provide environment control for the separate dedicated computers for the monitoring and alarm systems?	
How long can the separate dedicated computers of the monitoring and alarm systems operate with a loss of all environmental control? Indicate the conditions that could affect the length of time.	
Do the separate dedicated computers for the monitoring and alarm systems have their own independent environmental control system? If yes, specify the type.	
Are there backup environmental controls explicitly for any dedicated computers of the monitoring and alarm systems? If yes, indicate the type of backup and the expected maximum duration of operation.	

INTERNAL HUMAN RESOURCES SUPPORT

Date: ______________ Facility: ______________________________

This checklist applies to:

Entire Facility

Critical Asset ______________________________

	COMMENTS
(a) Electric Power Sources	
Are the asset's/facility's human resources offices and functions normally dependent on the asset's/facility's primary electric power supply and distribution system (i.e., is the asset's/facility's primary electric power supply and distribution system the primary electric power source?)?	
Are there multiple sources of electric supply for the human resources offices and functions? This could consist of the facility's electric power supply and distribution system and its backup or redundant systems; or combinations of multiple independent commercial electric feeds, backup generators, or UPSs, dedicated to support the human resources offices and functions? Specify what electric power sources are in place.	
If there is a special UPS, can it support all the human resources offices and functions? Specify for how long it can operate on battery power.	
If there is a special generator system, can it support all the human resources offices and functions? Also indicate the type of fuel or fuels used.	
If the fuel for the special generator system to support human resources is a petroleum fuel indicate the quantity stored on site and how long it will last. Are arrangements and contracts in place for resupply and management of the fuel?	
(b) Communications Pathways	
Are the asset's/facility's human resources offices and functions normally dependent upon the asset's/facility's telephone system?	
Are there multiple independent telephone systems or dedicated switches and cables supporting the human resources offices and functions? This could consist of the facility's telephone system and its backup or redundant systems; or combinations of multiple independent telephone systems or dedicated communications lines. Specify the types of systems used and whether they are copper-wire or fiber optic-cable based.	
Are the redundant telephone systems or switches and cables physically separated and isolated to avoid common causes of failure? If not, indicate any potential points of common failure.	

Are the dedicated telephone switches and data-transfer switches that support the human resources offices and functions located in a limited access or secured area away from potential damage due to weather or water leaks? If so, specify the type of protection.	
(c) Computer Support	
Are the asset's/facility's human resources offices and functions normally dependent upon the facility's main computers and servers?	
Are there multiple independent computers supporting the human resources offices and functions? This could consist of the asset's/facility's main computers and servers and their backup or redundant systems, or combinations of multiple independent computers. Specify the type of computers used.	
Are there multiple sources of electric power for any computers dedicated to support the human resources offices and functions? This could consist of the asset's/facility's primary electric power supply and distribution system and its backup or redundant systems; or combinations of multiple independent commercial electric feeds, backup generators, or UPSs dedicated to support human resources. If yes, specify the type and how long they can operate.	
Does the asset's/facility's central HVAC system provide environment control for any separate dedicated computers that support the human resources offices and functions?	
How long can the separate dedicated computers that support the human resources offices and functions operate with a loss of any environmental control? Indicate the conditions that could affect the length of time.	
Do the separate dedicated computers that support the human resources offices and functions have their own independent environmental control system? If yes, specify the type.	
Are there backup environmental controls explicitly for any dedicated computers that support the human resources offices and functions? If yes, indicate the type of backup and the expected maximum duration of operation.	

INTERNAL FINANCIAL SYSTEM
(Including Monetary Transactions)

Date: ______________ Facility: ____

This checklist applies to:

Entire Facility

Critical Asset ____

	COMMENTS
(a) Electric Power Sources	
Are the asset's/facility's financial systems and functions normally dependent on the asset's/facility's primary electric power supply and distribution system (i.e., is the facility's electric power supply and distribution system the primary electric power source?)?	
Are there multiple sources of electric power for the financial systems and functions? This could consist of the facility's electric power supply and distribution system and its backup or redundant systems; or combinations of multiple independent commercial electric feeds, backup generators, or UPSs dedicated to support the financial systems and functions? Specify what electric power sources are in place.	
If there is a special UPS, can it support all the financial systems and functions? Specify for how long it can operate on battery power.	
If there is a special generator system, can it support all the financial systems and functions? Also indicate the type of fuel or fuels used.	
Is the fuel for the special security generator system a petroleum fuel? Specify the quantity stored and how long it will last. Are arrangements and contracts in place for resupply and management of the fuel?	
(b) Communications Pathways	
Are the asset's/facility's financial systems and functions normally dependent upon the asset's/facility's telephone system?	
Are there multiple independent telephone systems or dedicated switches and cables supporting the financial systems and functions? This could consist of the facility's telephone system and its backup or redundant systems; or combinations of multiple independent telephone systems or dedicated communications lines. Specify the types of systems used and whether they are copper-wire or fiber-optic cable based.	
Are the redundant telephone systems or switches and cables physically separated and isolated to avoid common causes of failure? If not, indicate any potential points of common failure.	

Are the dedicated telephone switches and data-transfer switches that support the financial systems and functions located in a limited access or secured area away from potential damage due to weather or water leaks? If so, specify the type of protection.	
(c) Computer Support	
Are the asset's/facility's financial systems and functions normally dependent upon the facility's main computers and servers?	
Are there multiple independent computers supporting the financial systems and functions? This could consist of the facility's main computers and servers and their backup or redundant systems, or combinations of multiple independent computers. Specify the type of computers used.	
Are there multiple sources of electric supply for any computers dedicated to support the financial systems and functions? This could consist of the asset's/facility's primary electric power supply and distribution system and its backup or redundant systems; or combinations of multiple independent commercial electric feeds, backup generators, or UPSs dedicated to support the financial systems and functions. If yes, specify the type and how long they can operate.	
Does the asset's/facility's central HVAC system provide environment control for any separate dedicated computers that support the financial systems and functions?	
How long can the separate dedicated computers that support the financial systems and functions operate with a loss of any environmental control? Indicate the conditions that could affect the length of time.	
Do the separate dedicated computers that support the financial systems and functions have their own independent environmental control system? If so, specify the type.	
Are there any backup environmental controls explicitly for the dedicated computers that support the financial systems and functions? If yes, indicate the type of backup and the expected maximum duration of operation.	

EXTERNAL ELECTRIC POWER INFRASTRUCTURE

Date: ______________ Facility: ____

This checklist applies to:

Entire Facility

Critical Asset ____

	COMMENTS
(a) Electric Power Sources	
How many substations feed the area of the asset/ facility and the asset/facility itself? That is, is the area supplied by multiple substations? If more than one, which ones have sufficient individual capac-ities to supply the critical needs of the asset/facility?	
How may distinct independent transmission lines supply the substations? Indicate if an individual substation is supplied by more than one transmission line and which substations are supplied by independent transmission lines.	
(b) Electric Power Pathways	
Are the power lines into the area of the asset/facility and into the asset/facility itself above ground (on utility poles), buried, or a combination of both? If both, indicate locations of portions above ground.	
Do the power lines from these substations follow independent pathways to the area of the asset/facility? If not, specify how often and where they intersect or follow the same corridor.	
Are the paths of the power lines co-located with the rights-of-way of other infrastructures? If yes, indicate how often and where they follow the same rights-of-way and the infrastructures that are co-located.	
Are the paths of the power lines located in areas susceptible to natural or accidental damage (such as overhead lines near highways; power lines across bridges, dams, or landslide areas)? If yes, indicate the locations and types of potential disruptions.	
(c) Electric Power Contracts	
What type of contract does the asset/facility have with the electric power distribution company or transmis sion companies? Specify the companies involved and whether there is a direct physical link (distribution or transmission power line) to each company.	
If there is an interruptible contract (even in part), what are the general conditions placed up interruptions such as the minimum quantity that is not interruptible, the maximum number of disruptions per time period, and the maximum duration of disruptions? Has electric service been interrupted in the past? If yes, describe the circumstances and any effect the outages have had on the critical functions and activities of the asset/facility.	
(d) Historical Reliability	
Historically, how reliable has the commercial electric power been in the area? Quantify in terms of annual number of disruptions and their durations.	
Typically, when power outages occur, are they of significant duration (as opposed to just a few seconds or minutes)? Quantify the duration of the outages.	
Have there ever been electric power outages of sufficient frequency and duration so as to affect the critical functions and activities of the asset/facility?	

EXTERNAL NATURAL GAS INFRASTRUCTURE

Date: ______________ Facility: ____

This checklist applies to:

Entire Facility

Critical Asset ____

	COMMENTS
(a) Sources of Natural Gas	
How many city gate stations supply the natural gas distribution system in the area of the asset/facility and the asset/facility itself? If more than one, which ones are critical to maintaining the distribution system?	
How may distinct independent transmission pipelines supply the city gate stations? Indicate if an individual gate station is supplied by more than one transmission pipeline and which stations are supplied by independent transmission pipelines.	
(b) Pathways of Natural Gas	
Do the distribution pipelines from the individual city gate stations follow independent pathways to the area of the asset/facility? If not, specify how often and where they intersect or follow the same corridor.	
Are the paths of the pipelines co-located with the rights-of-way of other infrastructures? If yes, indicate how often and where they follow the same rights-of-way and the infrastructures that are co-located.	
Are the paths of the pipelines located in areas susceptible to natural or accidental damage (such as across bridges or dams, in earthquake or landslide areas)? If yes, indicate the locations and types of potential disruptions.	
Is the local distribution system well integrated (i.e., can gas readily get from any part of the system to any other part of the system)?	
(c) Natural Gas Contracts	
Does the asset/facility have a firm delivery contract, an interruptible contract, or a mixed contract with the natural gas distribution company or the transmission companies? Specify the companies involved and whether there is a direct physical link (pipeline) to each company.	
If there is an interruptible contract (even in part), what are the general conditions placed up interruptions such as the minimum quantity that is not interruptible, the maximum number of disruptions per time period, and the maximum duration of disruptions? Has natural gas service been interrupted in the past? If yes, describe the circumstances and any effect the outages have had on the critical functions and activities of the asset/facility.	

Does the asset/facility have storage or some other sort of special contracts with natural gas transmission or storage companies? If yes, briefly describe the effect on sustaining a continuous supply of natural gas to the asset/facility.	
In case of a prolonged disruption of natural gas supply, are contingency procedures in place to allow for the use of alternative fuels (such as on-site propane-air, liquefied petroleum gas, or petroleum fuels)? If yes, describe these alternatives and indicate whether they have sufficient capacity to fully support the critical functions and activities of the asset/facility	
(d) Historical Reliability	
Historically, how reliable has the natural gas supply been in the area? Quantify by describing any unscheduled or unexpected disruptions. Were there any effects on the critical functions and activities of the asset/facility?	
If operating under an interruptible service agreement, has natural gas service ever been curtailed? If yes, how often, for how long, and were there any effects on the critical functions and activities of the asset/facility?	

EXTERNAL PETROLEUM FUELS INFRASTRUCTURE

Date: ____________ Facility: ______________________________

This checklist applies to:

Entire Facility

Critical Asset ______________________________

	COMMENTS
(a) Uses of Petroleum Fuels	
Are petroleum fuels used in normal operations at the asset/facility? If yes, specify the types and uses.	
Are petroleum fuels used during contingency or emergency operations such as for backup equipment or repairs? If yes, specify the types of fuels and their uses.	
(b) Reception Facilities	
How are the various petroleum fuels normally delivered to the asset/facility? Indicate the delivery mode and normal frequency of shipments for each fuel type.	
Under maximum use-rate conditions, are there sufficient reception facilities (truck racks, rail sidings, surge tank capacity, barge moorings) to keep up with maximum contingency or emergency demand)? If no, explain where the expected shortfalls would be and their impacts.	
Are the petroleum fuel delivery pathways co-located with the rights-of-way of other infrastructures or located in areas susceptible to natural or accidental damage (across bridges or dams, in earthquake or landslide areas)? If yes, indicate the locations and types of potential disruptions.	
Are contingency procedures in place to allow for alternative mo des or routes of delivery? If yes, describe these alternatives and indicate whether they have sufficient capacity to fully support the critical functions and activities of the asset/facility.	
(c) Supply Contracts	
Are contracts in place for the supply of petroleum fuels? Specify the contractors, the types of contracts, the modes of transport (pipeline, rail car, tank truck), and the frequency of normal shipments.	
Are arrangements for emergency deliveries of petroleum fuels in place? Indicate the basic terms of the contracts in terms of the maximum time to delivery and the minimum and maximum quantity per delivery. Also, indicate if these terms are such that there may be effects on the critical functions and activities of the asset/facility.	

EXTERNAL TELECOMMUNICATIONS INFRASTRUCTURE

Date: ______________ Facility: ________________________________

This checklist applies to:

Entire Facility

Critical Asset ________________________________

	COMMENTS
(a) Telecommunications Carriers	
Are there multiple telecommunications carriers used by the asset/facility (possibly commercial, contracted, or organization-owned)? List them, specify the service they provide or the type of information carried (such as analog telephone voice and FAX, digital telephone voice, internet connections, dedicated data transfer), and the type of media used (copper cable, fiber-optic cable, microwave, satellite).	
(b) Pathways of Telecommunications Cables	
Are the telecommunications cables into the area of the asset/facility and into the asset/facility itself above ground (on utility poles), buried, or a combination of both? If both, indicate locations of portions above ground.	
Do the telecommunications cables follow independent pathways into the area of the asset/facility and into the asset/facility itself? If not, indicate how independent they are (some common corridors, intersect at one or more points).	
Are the paths of the telecommunications cables co-located with the rights-of-way of other infrastructures? If yes, describe the extent of the co-location and indicate the other infrastructures.	
Are the paths of the telecommunications cables located in areas susceptible to natural or accidental damage (such as overhead cables near highways; cables across bridges, dams, or landslide areas)? If yes, indicate the locations and types of potential disruptions.	
Do the various telecommunications carriers and cable pathways use separate independent end offices (EO), access tandems (AT), points of presence (POP), and network access points (NAP) to reach the communications transmission backbones? Briefly describe the extent of this independence.	
(c) Historical Reliability	
Historically, has the public switched network (PSN) telephone system in the area been reliable? Quantify in terms of number of both complete outages and dropped connections.	
Typically, when telephone outages occur, are they of significant duration (as opposed to just a few seconds or minutes)? Quantify in terms of potential effects on the critical functions and activates at the asset/facility.	

Historically, have the internet and dedicated data transfer systems in the area been reliable? Quantify in terms of number of both complete outages and dropped connections.	
Typically, when internet or data transfer connectivity outages or disruptions occur, are they of significant duration (as opposed to just a few seconds or minutes)? Quantify in terms of potential effects on the critical functions and activates at the asset/facility.	
(d) Backup Communications Systems	
Are there redundant or backup telephone systems in place if the primary system is disrupted? Specify the extent to which the secondary systems can support the critical functions and activities at the asset/facility.	
Are there redundant or backup internet and dedicated data transfer systems in place if the primary systems are disrupted? Specify the extent to which the secondary systems can support the critical functions and activities at the asset/facility.	

EXTERNAL WATER AND WASTEWATER INFRASTRUCTURE

Date: ______________ Facility: ____

This checklist applies to:

Entire Facility

Critical Asset ____

	COMMENTS
(a) Water Supply Reliability	
Historically, has the city water supply in the area been reliable and adequate? Quantify the reliability and specify any shortfall in the supply pressure or flow rate.	
Typically, when disruptions in the city water supply occur, are they of significant duration (as opposed to just a few hours)? Quantify in terms of potential effects on the critical functions and activities at the asset/facility.	
(b) Wastewater System Reliability	
Historically, has the public wastewater system in the area been reliable and adequate? Quantify the reliability and specify any shortfall in the capacity of the system.	
Typically, when disruptions in the public wastewater system occur, are they of significant duration (as opposed to just a few hours)? Quantify in terms of potential effects on the critical functions and activities at the asset/facility.	
Are there any contingency plans or procedures in place to handle domestic wastewater from the asset/facility if the public system is temporarily unable to accept the waste? If yes, describe them and mention any limitations on quantity of wastewater and duration of outage that might affect the ability of the asset/facility to carry out critical functions or activities.	

EXTERNAL ROAD TRANSPORTATION INFRASTRUCTURE

Date: ____________ Facility: ____

This checklist applies to:

Entire Facility

Critical Asset ____

	COMMENTS
(a) Road Access	
Are there multiple roadways into the area of the asset/facility from the major highways and interstates? Describe the route or routes and indicate any load or throughput limitations with respect to the needs of the asset/facility.	
Are there any choke points or potential hazard areas along these roadways such as tunnels, bridges, dams, low-lying fog areas, landslide areas, or earthquake faults? Describe the constrictions or hazards and indicate if, historically, closures have occurred somewhat regularly.	
(b) Road Access Control	
Could intruders or others determined to do damage to the asset/facility gain access to the asset/facility or nearby areas by road without being readily identified and controlled? If yes, describe the means of access and indicate any limitations on the number of people, the size and number of vehicles, and the size or quantity of material that could approach the asset/facility by road.	
Are there uncontrolled parking lots or open areas for parking near the facility where vehicles could park without drawing significant attention? If yes, indicate the number of vehicles and the size or types of vehicles that would begin to be noticed.	

EXTERNAL RAIL TRANSPORTATION INFRASTRUCTURE

Date: ______________ Facility: ____

This checklist applies to:

Entire Facility

Critical Asset ____

	COMMENTS
(a) Rail Access	
Are there multiple rail routes into the area of the asset/facility from the nearby rail yards or switchyards? Describe the route or routes and indicate any load or throughput limitations with respect to the needs of the asset/facility.	
Are there any choke points or potential hazard areas along these rail rights-of-way such as tunnels, bridges, dams, landslide areas, or earthquake faults? Describe the constrictions or hazards and indicate if, historically, rail traffic closures have occurred somewhat regularly.	
Is there sufficient rail siding space at or near the asset/facility to accommodate rail cars if the number of incoming cars exceeds normal expectations or if outgoing cars are not picked up as normally scheduled? Indicate the magnitude of this excess capacity in terms of the time period before the critical functions or activities of the asset/facility would be affected.	
(b) Rail Access Control	
Could intruders or others determined to do damage to the asset/facility gain access to the asset/facility or nearby areas by rail without being readily identified and controlled? If yes, describe the means of access and indicate any limitations on the number of people and rail cars that could approach the asset/facility by rail.	
Are there railroad tracks or sidings near the asset/facility where rail cars could be positioned without drawing significant attention? If yes, indicate the number and the types of rail cars that would begin to be noticed.	

EXTERNAL AIR TRANSPORTATION INFRASTRUCTURE

Date: ____________ Facility: ____

This checklist applies to:

Entire Facility

Critical Asset ____

	COMMENTS
(a) Airports and Air Routes	
Are there multiple airports in the area of the site of sufficient size and with sufficient service to support the critical functions and activities at the asset/facility? Enumerate the airports and indicate any limitations.	
Are there any regular air routes that pass over or near the asset/facility that could present a danger to the asset/facility if there were some sort of an air disaster? Record any concerns.	

EXTERNAL WATER TRANSPORTATION INFRASTRUCTURE

Date: ______________ Facility: ____

This checklist applies to:

Entire Facility

Critical Asset ____

	COMMENTS
(a) Waterway Access	
Are there multiple water routes to the ports, harbors, or landings used by the asset/facility from the open ocean or major waterway? Describe the route or routes and indicate any load, draft, beam, or throughput limitations with respect the needs of the organization.	
Are there any choke points or potential hazard areas along these waterways such as bridges, draw or lift bridges, locks and dams, low-lying fog areas, or landslide areas? Describe the constrictions or hazards and indicate if, historically, closures have occurred somewhat regularly.	
Is there sufficient mooring, wharf, or dock space at the ports, harbors, or landings used by the asset/facility to accommodate ships or barges if the number of incoming vessels exceeds normal expectations or if outgoing barges are not picked up as normally scheduled? Indicate the magnitude of this excess capacity in terms of the time period before the critical functions or activities at the asset/facility would be affected.	
(b) Waterway Access Control	
Could intruders or others determined to do damage to the asset/facility gain access to the asset/facility or nearby areas by water without being readily identified and controlled? If yes, describe the means of access and indicate any limitations on the number of people, the size and number of vessels, and the size or quantity of material that could approach the asset/facility by water.	
Are there uncontrolled docks or mooring areas near the asset/facility or the ports, harbors, or landings used by the asset/facility where vessels could moor without drawing significant attention? If yes, indicate the number of vessels and the size or types of vessels that would begin to be noticed.	

EXTERNAL PIPELINE TRANSPORTATION INFRASTRUCTURE

Date: ____________ Facility: ______________________________________

This checklist applies to:

Entire Facility

Critical Asset ______________________________________

	COMMENTS
(a) Pipeline Access	
What materials feedstocks or products (such as crude oil, intermediate petroleum products, refined petroleum products, or liquefied petroleum gas—do not include water, wastewater, or natural gas unless there are special circumstances related to these items) are supplied to or shipped from the asset/facility by way of pipeline transportation?	
Are there multiple pipelines and pipeline routes into the area of the asset/facility from major interstate transportation pipelines? If yes, indicate which pipelines or combinations of pipelines have sufficient capacity to serve the asset/facility.	
List the pipeline owners/operators, indicate the types of service provided (dedicated or scheduled shipments), describe the route or routes, and indicate any capacity limitations with respect the needs of the asset/facility.	
Are there any bottlenecks or potential hazard areas along these pipelines or pipeline routes such as interconnects, terminals, tunnels, bridges, dams, landslide areas, or earthquake faults? Describe the constrictions or hazards and indicate if, historically, outages or delays have occurred somewhat regularly.	
(b) Pipeline Access Control	
Could intruders or others determined to bring down the asset/facility gain access to the pipeline near the asset/facility or elsewhere along the pipeline route? Describe the protective measures that are in place and indicate any pipeline segments or facilities (such as pump stations, surge tanks) of concern.	

OPSEC TABLES

HUMAN RESOURCES SECURITY PROCEDURES

Date: ______________ Facility: ____

	COMMENTS
(a) Responsibilities	
What organization(s) is responsible for dealing with security-related personnel issues?	
(b) Background Checks	
Are background checks done on employees? If yes, for which employees?	
Is the background check done for selected (sensitive) positions? If yes, what are the criteria for identifying sensitive positions?	
How extensive are the background checks? Do they vary with the sensitivity of the position?	
(c) Insider Threats	
Are there current conditions in the company that might create a threat from insiders (e.g., low morale, lay-offs, labor disputes)?	
Are there security procedures for handling disgruntled or at-risk employees? If yes, describe.	
What are the security procedures for handling terminated employees? How many have been terminated in the last year? Have there been any security incidents related to a terminated employee?	
(d) Disciplinary Procedures	
What are the policies and procedures for incidents of security concern?	
What are the policies and procedures for other disciplinary actions?	
(e) Security Training	
Is there a company Security Awareness training program that includes initial and periodic security training? Does it include sections on security contacts, critical assets, threats, sensitive information that needs to be protected, reporting suspicious activities, and employee responsibility?	
(f) Travel	
Are employee travel records (e.g., authorizations, vouchers, trip reports) protected? If yes, describe how.	

FACILITY ENGINEERING

Date: _____________ Facility: ______________________________

This section covers security issues related to the engineering information related to the facility. Included are the facility design, configuration, and layout; utility service systems; building floor plans; etc.

	COMMENTS
(a) Responsibilities	
What organization(s) is responsible for facility engineering?	
(b) Facility Engineering Information	
What facility engineering information (e.g., engineering drawings, site maps, utility service lines, floor plans, entry paths into the facility, etc.) is available?	
What organization(s) has control of this information?	
What other internal organizations are allowed access to this information?	
What external organizations (e.g., fire department, environmental agency) have been given access to this information?	
Is any of the facility engineering information publicly available?	
Can sensitive information be gleaned from commercial overhead imaging (e.g., aerial photography, commercial satellite images)? If yes, describe.	
How is this information protected?	
Is this information on the computer system or network?	
How is the information disposed of when no longer needed?	
(c) Public Access to Facility	
Are tours allowed of any or all of the facility? If yes, describe what portions of the facility are open and who is allowed to tour.	
Is any portion of the facility open to the public or special interest groups? If yes, describe.	
Are periodic meetings held where outsiders are allowed inside the facility? If yes, describe.	
Are there procedures for security escorting of visitors? If yes, describe.	

FACILITY OPERATIONS

Date: ____________ Facility: ____

	COMMENTS
(a) Responsibilities	
What organization(s) is responsible for facility operations?	
(b) Facility Operations Control	
Is the operation of the facility controlled from a central point (or several central points)? Describe.	
Is there an automated process control system, energy management system, SCADA system? If yes, describe. If yes, is it isolated or is there remote access possibility?	
Is facility operations control and information on computer systems? If yes, how is it protected? What other internal organizations have access to operations control capability and information?	
Can sensitive operations information be gathered through the telecommunications system (e.g., microwave, cell phones, RF, pagers, voicemail, teleconferencing)?	
Is access to the control point(s) limited to operations personnel? If no, who else has access (e.g., maintenance, janitors, vendors, etc.) and how is that access controlled?	
(c) Facility Construction, Repair, and Maintenance	
Is construction, repair and maintenance at the facility done by employees, contractors, or both? If contractors are used, describe procedures for screening and monitoring contractor personnel.	
Is cleaning and building maintenance (e.g., janitorial service) at the facility done by employees, contractors, or both? If contractors are used, describe procedures for screening and monitoring contractor personnel.	

ADMINISTRATIVE SUPPORT ORGANIZATIONS

Date: ______________ Facility: ______________________________

	COMMENTS
(a) Procurement	Purchasing and procurement activities including: Generating Need (e.g., requisition or RFP), Selecting Supplier, Documenting the Purchase, Providing Delivery of Item or Service, Payment.
What organization(s) is responsible for reviewing procurement activities from a security perspective?	
What is the process used to review RFPs, contracts, and other procurement documents for security-related information?	
How is the procurement information protected before release? Include documents, files, copiers, facsimiles, computer files?	
Is security-sensitive information uniquely marked, both on paper and electronically? If yes, describe how.	
How is security-sensitive procurement information destroyed?	
How are company credit cards controlled? Who is authorized to have one? How is security-related information from credit card use identified and protected?	
(b) Legal	
What organization(s) is responsible for reviewing legal department activities from a security perspective?	
How are legal documents (e.g., patents, environmental impact statements, safety reports, Securities and Exchange Commission filings, Federal Energy Regulatory Commission filings, etc.) reviewed for security implications?	
How are these documents protected?	
(c) Budget and Finance	
What organization(s) is responsible for reviewing budget and finance activities from a security perspective?	
How are budget and finance documents reviewed for security implications?	
How are these documents protected?	
(d) Marketing	
What organization(s) is responsible for reviewing marketing activities from a security perspective?	
How are marketing materials reviewed for security implications?	
How are these documents protected?	
(e) Internal Information	
Are there policies and procedures for handling "Internal Use Documents" (e.g., memos, notes, newsletters, etc.)? If yes, describe.	
How are these documents protected?	
How are these documents destroyed when no longer needed?	

TELECOMMUNICATIONS AND INFORMATION TECHNOLOGY

Date: ____________ Facility: ______________________________

This checklist covers telecommunications, information technologies, and cyber security. Note that this part of the operations security survey must coordinated with the portions of the interdependencies survey that address the telecommunications and computer equipment.

	COMMENTS
(a) Telecommunications	
What telecommunications equipment is in regular use: __ Telephone __ Cell phones __ Voicemail __ Fax __ Audio/video conferencing __ Radio __ Microwave __ Other__________________	
Does particular equipment carry sensitive traffic? If yes, describe.	
Are particular nodes susceptible to intercept? If yes, describe.	
Is particular equipment restricted to selected users? If yes, describe.	
Are internal telephone lines routed to external switches? If yes, describe.	
Can any telecommunications equipment be operated in reverse as eavesdropping equipment? If yes, describe.	
Are there connections to external radio nets, including paging nets? If yes, describe.	
Is voicemail protected by passwords? Have users changed the vendor-supplied passwords? Is there a master password?	
How are fax machines protected? How is the stored information protected?	
Is encryption used on any telecommunications circuits?	
(b) Information Technology	
Is there a corporate security architecture for the computer network? If yes, describe. Does it include intrusion detection, firewalls, compartmentalization? If yes, describe.	
What computer information is available to outsiders?	
How are computer applications developed (internal, external)? How is software and hardware maintenance performed?	
What are the policies and procedures for passwords?	
Is there dial-up access for operations, maintenance, or other reasons? If yes, describe.	
Are there embedded computer systems in other systems (e.g., HVAC equipment, numerically controlled machines, etc.)? If yes, how are they protected?	
Is there a computer incident response team? If yes, describe.	
Are exercises ("War Dialing") conducted to locate unauthorized modems? If yes, describe.	
Is encryption used for internal files and/or information transmission? If yes, describe.	
Have system administrators been trained to recognize "social engineering attacks" designed to obtain passwords and other security information? If yes, describe.	
Is e-mail monitored? If yes, describe.	

PUBLICLY RELEASED INFORMATION

Date: ______________ Facility: ____

This checklist covers information that is released to the public via corporate communications, press releases, the Internet, and other means.

	COMMENTS
(a) Responsibilities	
What organization(s) is responsible for reviewing information (from a security perspective) that is to be released to the public?	
(b) General Procedures	
What is the process used to review information before release?	
How is the information protected before release? Include documents, files, copiers, facsimiles, computer files.	
(c) Report Release	
Who is responsible for reviewing reports released by the company?	
Who generates the reports?	
What type of information is included?	
What is the distribution and ultimate disposition of company-released reports?	
(d) Press Contacts	
Are specific people designated to interact with the press?	
How are they trained (including training on security issues)? Who trains them?	
(e) Briefing and Presentations	
Are briefings and presentations to be given by company employees reviewed for security issues? If yes, describe how.	
(f) Public Testimony	
Is public testimony that is to be given by company employees reviewed for security issues? If yes, describe how.	
(g) Internet Information	
Is there a policy in place to review information posted on the company Internet site for security issues? If yes, describe.	
Who reviews information before it is posted on the Website?	
Is the Website reviewed and monitored regularly for security-related information? If so, describe how.	

TRASH AND WASTE HANDLING

Date: ______________ Facility: ____

This checklist covers the handling of trash and waste that may have security implications (e.g., documents records, discarded equipment, etc.)

	COMMENTS
(a) Responsibilities	
What organization is responsible for security of trash and waste?	
Are there established policies for trash and waste handling? If yes, describe.	
(b) Trash Handling	
Where is trash accumulated?	
Is the trash accessible to outsiders?	
Who collects the trash?	
Where is the trash taken?	
(c) Paper Waste Handling	
Where is paper waste accumulated?	
Are shredders available and used? If yes, describe.	
Is the paper waste accessible to outsiders?	
Who collects the paper waste?	
Where is the paper waste taken? Is it sent for recycling?	
Is there on-site destruction of paper waste? If yes, describe how it is protected until destroyed.	
(d) Salvage Material Handling	
Does salvage material (e.g., serviceable equipment no longer needed, surplus equipment) potentially contain sensitive information?	
Is salvage material inspected prior to release? If yes, by whom? Describe procedures.	
(e) Dumpster Control	
Are dumpsters (for trash, paper waste, salvage material) that are accessible to the public monitored to prevent "dumpster diving"? If yes, describe how.	
Are publicly accessible dumpsters sampled for sensitive information?	

Appendix C1—Refinery SVA Example

The application of the SVA Methodology to a fictitious refinery is illustrated in the following example. Only the first page of each of the four forms is shown for illustrative purposes. It is assumed that the study is conducted by the refinery company and considers the various interfaces with customers and suppliers. However, the security of the customer and supplier facilities is the responsibility of the owners of those facilities.

The study is conducted in a top-down, systematic manner following the logic flowchart for the SVA as shown in Figure A. The five steps of the process are documented in four forms:

Figure A—SVA Methodology Flow Diagram

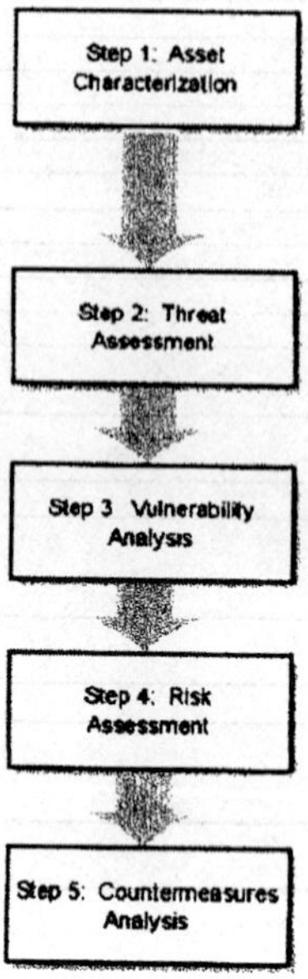

Form 1—Critical Assets/Criticality form

Form 2—Threats Worksheet
Form 3—Attractiveness/Target Ranking Form

Form 4—Scenario Based Vulnerability Worksheet/Risk Ranking/Countermeasures Form

Form 1—Critical Assets/Criticality form

Determine the major assets of the refinery including processes, control rooms, gates and access control points, marine terminals, terminus points for export and import pipelines, utilities, and supporting infrastructure. All entry points should be evaluated as an asset in order to focus the analysis on the need for perimeter security and access control. The team lists all relevant assets on Form 1 in Column 1. Similar facilities with similar geographic locations, common vulnerabilities, and common consequences can be grouped for efficiency and to consider the value of an entire functional set. In Column 2, document the design basis of the asset and the hazards and consequences that would be realized if the asset was damaged, compromised, or stolen. In the Column 3 rank the estimated overall severity of the loss of the asset. Use the five-level Severity Ranking scale for severity or develop an equivalent as required for the particular facility.

Form 2—Threats Worksheet

Document the threats against the facility on Form 2. Include consideration in Column 1 of general types of adversaries that will be considered (usually terrorists, disgruntled employee or contractor, or extreme activist as an example, but more specific or other groups can be considered as required); Column 2 is the source of the attack (EXT—External to the facility, INT—Internal to the facility); Column 3 documents the threat specific to the facility being evaluated; Column 4 documents the specific or general threat of that type of adversary against this or similar assets worldwide; Column 5 documents the potential actions that the adversary could take; Column 6 documents the assumed capabilities, weapons, tactics, and sophistication of the adversary; Column 7 documents their level of motivation; Column 8 provides for an overall ranking assessment per the Threat Ranking scale or equivalent.

Form 3—Attractiveness/Target Ranking Form

Columns 1 – 3 are repeated from Form 1 for reference. Column 4 is a documented rationale for why the particular asset is attractive (or unattractive) and Column 5 is a ranking of that attractiveness on a relative Attractiveness Ranking scale or equivalent. This is repeated for other adversaries. Column 10 is an overall Target Ranking per the same scale, and is normally considered to be the highest attractiveness of any of the individual adversary rankings but also considers that the sum of the different adversary's interests to a common assete may make the asset more attractive. The Target Ranking is used to judge the degree of attractiveness of the target considering all the adversaries.

Form 4—Scenario Based Vulnerability Worksheet/Risk Ranking/Countermeasures Form

Column 1 is the Security Event Type (see Step 3.3—generally one of four security events including loss of containment, degradation of the asset, theft, or contamination); Column 2 is the Threat Category (adversary type such as terrorist, activist, employee); Column 3 is the Type of Adversary Attack (Insider/External); Column 4 is the Undesired Act (the assumed attack scenario, generally taken from the Threats Worksheet Columns 5, 6, 7); Column 5 is the Consequences; Column 6 (S) is the Severity Ranking from the Severity Ranking scale; Column 7 is the Existing Countermeasures, which considers the Deter, Detect, Delay, and Respond philosophy; Column 8 is the Vulnerability, which also considers the weaknesses or missing elements of the security strategy specific to the scenario; Column 9 is the Vulnerability Ranking per the Vulnerability Ranking scale; Column 10 is the Likelihood ranking (L) using the Likelihood scale, which is a judgment of the team considering the factors of Vulnerability, Threat, and Attractiveness; Column 11 is the Risk ranking (R) per the referenced Risk Ranking Matrix values; and Column 12 is the New Countermeasures suggestions (where the risk is considered significant enough to justify the need for change).

Responsibilities

This exa mple includes a sampling of assets that may be owned or operated by various parties. The responsibilities for conducting the SVA and for providing security need to be determined and may not solely be with the refinery owner/operator. It is recommended that the SVA include the appropriate parties to fully analyze the security issues, and that the results are discussed with owner/operators of adjacent facilities and infrastructure providers as required for risk communication and completeness.

Fictitious Refinery Example

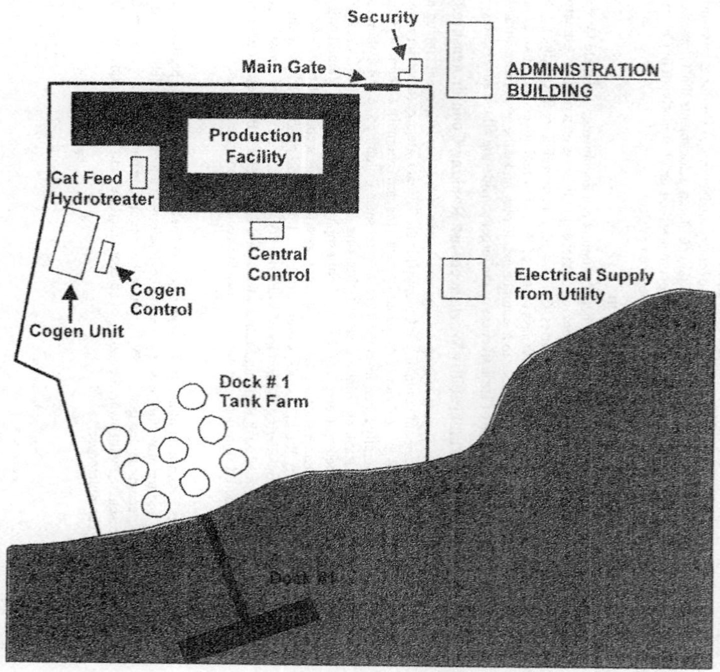

Form 1: Critical Assets/Criticality Form

Facility Name: Fictitious Refinery

Critical Assets Form		
Critical Assets	Criticality/Hazards	Asset Severity Ranking
1. Administration building	Administrative offices including management offices and large number of employees, HR Manager; ordinary office building hazards; personnel exposure to approximately 100 persons; possible loss of personnel and/or critical documents in storage (business sensitive information).	3
2. Central Control Room	Critical security communications and monitoring; Cat, Coker 1, Alkylation, Treating Plant; Crude Units; loss of control function and long time to repair if damaged.	5
3. Cogen Unit and Control Room	Critical steam production and supplemental electrical power generation.	4
4. Dock 1	Loss of logistics for feedstock and products; environmental release; fire and explosion; possible to shutdown channel; coker feed, #2 fuel oil, benzene, toluene, molten sulfur in storage; Coker feed is most critical feedstock.	5
5. Dock 1 Tank Farm—storage in atmospheric tanks north of Dock 1 (crude inT-800; T-802; T-803,T-805; ballast/slop oil tank T-804; lube oils in T-240 to T-244)	Flammable and combustible liquids fire and explosion hazard; possible spill to ship channel; critical to operation of marine terminal.	4
6. Cat Feed Hydrotreater Unit	Significant fire and explosion hazard onsite; possible public impacts from explosion; significant business interruption.	5
7. Electrical supply from Utility to Refinery	Utility supplied; Cat Feed HT, H2 plant, and Units 29 – 35; backup supply from other substations.	3
8. Units 29-35 cooling tower/chlorine containers	Important to operation of units 29 – 35; chlorine toxic hazards may have public impact if damaged.	4

Form 2: Threats Worksheet

Facility Name: Fictitious Refinery

Adversary Types	Source	Site Specific Threat	Threat History	Potential Actions	Assumed Adversary Capability	Adversary Motivation	Threat Ranking
1. Terrorist	EXT	No specific group or threat to the refinery	- General industry terrorist threats only - HSAS Yellow as of the time of the SVA	- Use explosives or small arms to destroy target - May be interested in theft of products of value to terrorist organizations for secondary attack	- Use of improvised explosive device possibly involving a vehicle is most likely scenario - Assume trained, with good information and significant resources to plan and execute attack	Assume highly motivated to cause maximum damage to critical infrastructure and casualties	4
2. Disgruntled employee or contractor	INT	Employees and contractors	- Company facilities have had telephone bomb threats - No actual damage but threats have been made. - Assume general industry experience with insider sabotage is credible	- Might cause intentional overfill of tank or damage to equipment leading to release; might cause product contamination; possible for explosion - Possible for workplace violence - Potential for theft	- Specialized insider knowledge and training - Unrestricted access to entire facility - Not likely to use weapons if sabotage but may use small arms if workplace violence	Potential for disgruntled employee due to disciplinary action; other workplace violence reasons; possibly in collusion with outside terrorist group in extreme case	3
3. Activist	EXT	Citizens for Green Environment has expressed interest	Multiple demonstrations have occurred at the plant	Possibly interested in causing public embarrassment; temporary shutdown of plant; long range goal of elimination of toxic substance in use.	- Highly organized; well funded to cause staged attack of multiple facility operations simultaneously (dock, rail, gate)	Highly politically charged and motivated	4

Form 3: Attractiveness/Target Ranking Form

Facility Name: Fictitious Refinery

Critical Assets	Function/Hazards/ Criticality	Asset Severity Ranking	Asset Attractiveness						
			Foreign/Domestic Attractiveness Rationale	A1	Employee/Contractor Attractiveness Rationale	A2	Activist Attractiveness Rationale	A3	TR
1. Administration building	Administrative offices including management offices and large number of employees, HR Manager; approximately 100 persons; possible loss of personnel and/or critical documents in storage (process and business sensitive information)	3	Management offices and large number of employees and computer systems	2	Management offices and large number of employees; HR Manager	3	Possibly interested in seeking out management for protest but not accessible directly and Business Services building is more accessible	2	TR3
2. Central Control Room	Critical security communications and monitoring; Crude 1, Alkylation, Treating Plant	4	Provides access to control multiple units at the same time	4	Maybe recognizable target; insider information on process control and access; high concentration of processes under single control and large numbers of operators in plant	3	Not easily accessible; does not provide opportunity for media attention and requires trespassing	2	TR4
3. Dock 1	All crude receipts and product transfers occur over Dock 1; hazard of flammable liquids spill and fire and oil spill on water. Possible for disruption to entire refinery and adjacent facilities if waterway is blocked.	5	Immediately accessible; recognizable and importance well understood; critical to refinery operation; long lead time for repair; complicating to adjacent facilities	4	Accessibility; importance well understood; critical to refinery operation; long lead time for repair	4	Could be easily accessible by watercraft; provides opportunity for media attention; activist activity against dock in past.	3	TR4

Form 4—Scenario Based Vulnerability Worksheet/Risk Ranking/Countermeasures

Facility Name: Fictitious Refinery

Critical Assets: 20. Dock 1

Scenario Worksheet Form											
Security Event Type	Threat Category	Type	Undesired Act	Consequences	S	Existing Countermeasures	Vulnerability	Vulnerability Ranking	L	R	New Countermeasures
1.1. Loss of containment	Terrorist	**I/E**	Attack on vessel or dock facility by way of an improvised explosive device	Damage to barge and dock facilities; loss of logistics for feedstock and products; major environmental release; fire and explosion; possible to shutdown channel	S5	1.1. USCG boat patrols of the channel and port 1.2 Roving guardforce	1.1. Lack of access control from water, 1.2 Low lighting 1.3No intrusion detection	5	L4	[redacted]	Consider improving lighting, access control, monitoring by CCTV, and administrative controls per requirements of Enclosure 2 of NVIC 11-02.

Appendix C2—Fictitious Pipeline Example

The application of the SVA Methodology to a fictitious petroleum liquids pipeline system is illustrated in the following example. Only the first page of each of the four forms is shown for illustrative purposes. It is assumed that the study is conducted by the pipeline company and considers the various interfaces with customers and suppliers. However, the security of the customer and supplier facilities is the responsibility of the owners of those facilities.

The general approach is to apply risk assessment resources and, ultimately, special security resources primarily where justified based on the SVA results. The SVA process involves consideration of the pipeline system from both the general viewpoint and specific asset viewpoint. Consideration at the general level is useful for determination of overall impacts of loss, infrastructure and interdependencies at the system level. The benefit of evaluating specific assets is that individual risks can be evaluated and specific countermeasures applied where justified in addition to more general countermeasures.

For example, all facilities will maintain a minimum level of security with general countermeasures such as the pipeline shutdown and control strategies and administrative security controls. Certain assets will justify a more specific level of security based on their value and expected level of interest to adversaries.

The SVA methodology uses this philosophy in several ways. The method is intended to be comprehensive and systematic in order to be thorough. First, it begins with the SVA team gaining an understanding of the entire pipeline system, the assets that comprise the pipeline system, the critical functions of the pipeline, and the hazards and impacts if these assets or critical functions are compromised. This results in an understanding of which assets and functions are "critical" to the business operation. Criticality may be defined both in terms of the potential impact to the workers, community, the environment and the company, as well as to the business importance and continuity of the system. For example, a pumping station or a specific branch along the pipeline system may be a critical part of the operation of the pipeline system due to inability to operate without it or, if attacked, it has the greatest impact. As such it may be given a high priority for further analysis and special security countermeasures.

Based on this first level of screening from all assets to critical assets, a critical asset list is produced. Next, the critical assets are reviewed in light of the threats. Adversaries may have different objectives, so the critical asset list is reviewed from each adversary's perspective and an asset attractiveness ranking is given. This factor is a quick measure of whether the adversary would value damaging, compromising, or stealing the asset, which serves as an indicator of the likelihood that an adversary would want to attack this asset and why.

If an asset is both critical (based on value and consequences) and attractive, then it is considered a "target" for purposes of the SVA. A target may optionally receive further specific analysis, including the development of scenarios to determine and test perceived vulnerabilities. As shown in Figure A, all assets receive at least a general security review. This is accomplished by the basic SVA team's consideration as an asset to begin with, along with a baseline security survey. General security considerations may be found in security references such as the countermeasures checklist provided in Appendix F.

The study is conducted in a top-down, systematic manner following the logic flowchart for the SVA as shown in Figure A. The five steps of the process are documented in four forms:

Figure A— SVA Methodology Flow Diagram

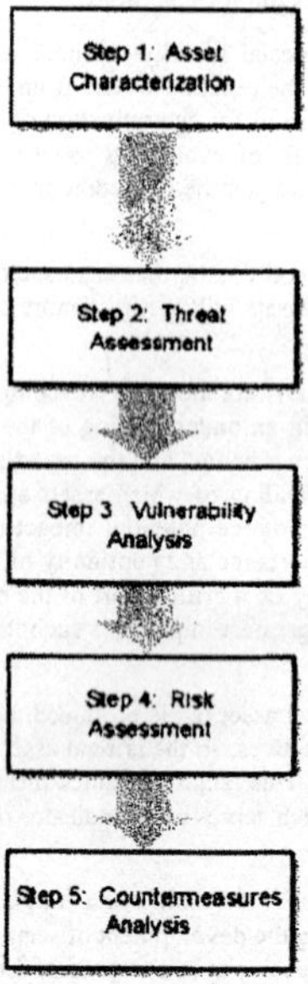

Form 1—Critical Assets/Criticality form

Form 2—Threats Worksheet
Form 3—Attractiveness/Target Ranking Form

Form 4—Scenario Based Vulnerability Worksheet/Risk Ranking/Countermeasures Form

Form 1—Critical Assets/Criticality form

Determine the major assets of the pipeline system including control rooms, gates and access control points, marine terminals, communications networks, terminus points for export and import pipelines, utilities, and supporting infrastructure. All entry points to critical facilities should be evaluated as an asset in order to focus the analysis on the need for perimeter security and access control. The team lists all relevant assets on Form 1 in Column 1. Similar assets within a facility with similar geographic locations on the property, common vulnerabilities, and common consequences can be grouped for efficiency and to consider the value of an entire functional set. In Column 2, document the design basis of the asset and the hazards and consequences that would be realized if the asset was damaged, compromised, or stolen. In the Column 3 rank the estimated overall severity of the loss of the asset. Use the five-level Severity Ranking scale for severity or develop an equivalent as required for the particular facility.

Form 2—Threats Worksheet

Document the threats against the pipeline system or a critical facility on Form 2. Include consideration in Column 1 of general types of adversaries that will be considered (usually terrorists, disgruntled employee or contractor, or extreme activist as an example, but more specific or other groups can be considered as required); Column 2 is the source of the attack (EXT—External to the pipeline/facility, INT—Internal to the pipeline/facility); Column 3 documents the threat specific to the pipeline/facility being evaluated; Column 4 documents the specific or general threat of that type of adversary against this or similar assets worldwide; Column 5 documents the potential actions that the adversary could take; Column 6 documents the assumed capabilities, weapons, tactics, and sophistication of the adversary; Column 7 documents their level of motivation; Column 8 provides for an overall ranking assessment per the Threat Ranking scale or equivalent.

Form 3—Attractiveness/Target Ranking Form

Columns 1 – 3 are repeated from Form 1 for reference. Column 4 is a documented rationale for why the particular asset is attractive (or unattractive) and Column 5 is a ranking of that attractiveness on a relative Attractiveness Ranking scale or equivalent. This is repeated for other adversaries. Column 10 is an overall Target Ranking per the same scale, and is normally considered to be the highest attractiveness of any of the individual adversary rankings but also considers that the sum the different adversary's interests may make the asset more attractive. The Target Ranking is used to judge the degree of attractiveness of the target considering all the adversaries.

Form 4—Scenario Based Vulnerability Worksheet/Risk Ranking/Countermeasures Form

Column 1 is the Security Event Type (generally one of four security events including loss of containment, degradation of the asset, theft, or contamination); Column 2 is the Threat Category (adversary type such as terrorist, activist, employee); Column 3 is the Type of Adversary Attack (Insider/External); Column 4 is the Undesired Act (the assumed attack scenario, generally taken from the Threats Worksheet Columns 5, 6, 7); Column 5 is the Consequences; Column 6 (S) is the Severity Ranking from the Severity Ranking scale; Column 7 is the Existing Countermeasures, which considers the Deter, Detect, Delay, and Respond philosophy; Column 8 is the Vulnerability, which also considers the weaknesses or missing elements of the security strategy specific to the scenario; Column 9 is the Vulnerability Ranking per the Vulnerability Ranking scale; Column 10 is the Likelihood ranking (L) using the Likelihood scale, which is a judgment of the team considering the factors of Vulnerability, Threat, and Attractiveness; Column 11 is the Risk ranking (R) per the referenced Risk Ranking Matrix values; and Column 12 is the New /Countermeasures suggestions (where the risk is considered significant enough to justify the need for change).

Responsibilities

This example includes a sampling of assets that may be owned or operated by various parties. The responsibilities for conducting the SVA and for providing security need to be determined and may not solely be with the pipeline owner/operator. It is recommended that the SVA include the appropriate parties to fully analyze the security issues, and that the results are discussed with owner/operators of adjacent facilities and infrastructure providers as required for risk communication and completeness.

Fictitious Pipeline Example

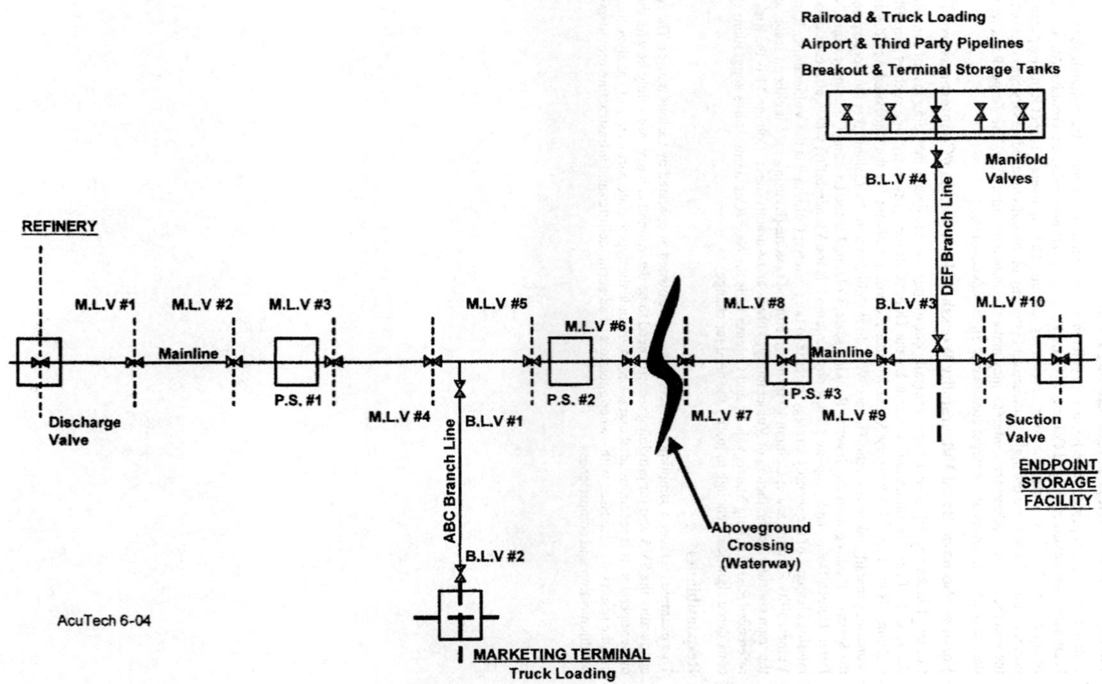

AcuTech 6-04

Form 1: Critical Assets/Criticality Form

Facility Name: 1. Fictitious Pipeline Company

Critical Assets Form		
Critical Assets	Criticality/Hazards	Asset Severity Ranking
1. Main Line, 24-inch Liquids Pipeline System—1000 miles, provides 500,000 b/d. Finished products; Gasoline, Jet Fuel and home heating oil. 35 main-line block valves (approximately every 50 miles), 20 booster (pumping) stations, traverses primarily rural areas.	Main line serves large metropolitan areas. Several million retail customers plus 5 major international airports, and two large military installations. Includes a major above ground river crossing, which provides drinking water to large urban community.	5
2. ABC Branch—10 miles, 8 inch branch line serving mixed products to marketing terminal serving a rural population.	Serves rural customer base. No national defense impact. Remotely located and no major environmental impacts. Alternative delivery sources available.	1
3. DEF Branch and inter-modal terminal—Branch line providing mixed products to multi-modal marketing terminal, breakout facility, interconnection to other pipelines and direct connect to military, commercial airports and power plant.	Possible onsite fatalities. Possible offsite environmental impact. Limited alternative delivery resources to customers.	4
4. Endpoint storage facility—Major tank farm for large metropolitan area, airport and other party pipeline connection.	Serves large metropolitan area. Several million retail customers plus major international airport. Area served by other sources. Located in a sparsely populated industrial area.	2
5. River Span Block Valve	Block valve is upstream from above ground river span (see item 7). Breach could cause release of pipe volume into river and impact public safety and significant contamination to the water supply of a major metropolitan center. Restoration costs significant due to river spill clean up and difficult access to valve. Short timeframe to repair.	5
6. River Span Pipeline (Above Ground)	Above ground river span (see 1 above). Breach could release significant product into river and contaminate public water supply to a major metropolitan center. Block valve used as active mitigation, if not damaged. Significant public safety concern due to frequent recreational and commercial use on river. Long-term repair timeframe and significant repair costs and spill clean up costs. No alternate mode to market. Significant service interruption.	5
7. Inter-modal Terminal	Large inter-modal products terminal with rail, truck and pipeline service. Serves large metropolitan area. Provides gasoline to retail market, jet fuel to 2 major international airports and USAF. Large-scale damage would take months to repair. Repair costs would be significant. Significant disruption to local economy and possible national defense. No significant environmental impact. Limited public safety and employee impact.	4

Form 2: Threats Worksheet

Facility Name: 1. Fictitious Pipeline Company

Adversary Types	Source	Site Specific Threat	Threat History	Potential Actions	Adversary Capability	Adversary Motivation	Threat Ranking
International terrorists	**I/E/C**	No site-specific history of international terrorism.	There have been numerous international terrorist acts against petroleum pipelines in the world to date. Most notably in South America and Middle East. U.S. Homeland Security Advisory System is rated orange presently. According to recent FBI repots, Al Qaeda continues to show interest in the energy sector and large scale operations that have significant impacts to public safety, the national economy, and national symbol of American might and wealth.	Use of stealth or force to cause damage and/or release of hydrocarbons. Possible theft or contamination of product possible but not likely. Degradation of assets and interruption of service biggest concern. Possible environmental release into public water supply and public safety are concerns. Damage to equipment and time to repair are also issues.	High level of organizational support; good resources; good financial backing; network of members; highly developed communication capabilities; weapons including small arms and explosives; possible vehicle bomb based on past events.	Assume adversary is highly motivated, likely extremist, prepared to die for their cause with intent to cause maximum damage to company assets including loss of life and economic disruption.	4
Domestic Terrorist or Activist	**I/E/C**	History at the main-line system of multiple bomb threats over the pas t 2 years. All concluded were fakes.	No confirmed domestic acts of terrorism on the pipeline infrastructure.	Possible for a disruptive event from domestic terrorist such as bombing or disruption of operations.	Low level of organizational support; poor resources and financial backing; small network of members; cell phone/email communication capabilities; weapons including small arms and explosives.	Adversary intent is to cause economic harm through service interruption. If domestic terrorist, intent and motivation could be extreme to cause maximum damage, possibly without personal sacrifice.	3

Form 2: Threats Worksheet

Facility Name: 1. Fictitious Pipeline Company

Adversary Types	Source	Site Specific Threat	Threat History	Potential Actions	Adversary Capability	Adversary Motivation	Threat Ranking
Disgruntled Employee or Contractor	**INT**	No evidence of sabotage has been discovered in the past.	Minimal acts of sabotage or workplace violence.	Sabotage to equipment including SCADA causing possible release of hazardous materials, contamination of products, environmental impact, or major equipment damage and business interruption. Possible for nuisance threats, particularly from contract workers with intent to disrupt operations.	Insider access, knowledge and ability to operate independently with authorization and without question. May have access to keys, computer passwords, gate access codes, communication equipment, records, vehicles, proximity cards for access cards, company process control system.	Nuisance adversary is intent to cause inconvenience and financial impacts to the company or their employer. If very disgruntled or troubled, intent and motivation could be extreme to cause maximum damage, possibly with personal sacrifice as evidenced in various national workplace violence cases.	4

Form 3: Attractiveness/Target Ranking Form

Facility Name: 1. Fictitious Pipeline Company

Critical Assets	Function/Hazards/ Criticality	S	Asset Attractiveness						
			Foreign/Do mestic Attractiveness Rationale	A1	Employee/Contractor Attractiveness Rationale	A2	Activist Attractiveness Rationale	A3	TR
1. Main Line, 24-inch Liquids Pipeline System—1000 miles, provides 500,000 b/d. Finished products; Gasoline, Jet Fuel and home heating oil. 35 main-line block valves (approximately every 50 miles), 20 booster (pumping) stations, traverses primarily rural areas.	Serves rural customer base. No national defense impact. Remotely located and no major environmental impacts. Alternative delivery sources available.	5	Easy access due to length of pipeline and location in a rural area with several above ground - unmanned pumping stations. Minimal disruptions to only a rural customer base no impact to military and minimal potential environmental impact.	1	Some insider insight helpful but not necessary.	2	Limited interest.	2	TR 2
2. ABC Branch—10 miles, 8 inch branch line serving mixed products to marketing terminal serving a rural population.	Main line serves large metropolitan areas. Several million retail customers plus 5 major international airports, and two large military installations. Includes a major above ground river crossing, which provides drinking water to large urban community.	1	Major disruption to residential, air travel and military. Public safety and drinking water contamination. Easy access.	2	Some insider insight helpful but not necessary.	2	Public Image impact due to press/media interest.	3	TR 3

Form 3: Attractiveness/Target Ranking Form

Facility Name: 1. Fictitious Pipeline Company

Critical Assets	Function/Hazards/ Criticality	S	Asset Attractiveness						
			Foreign/Domestic Attractiveness Rationale	A1	Employee/Contractor Attractiveness Rationale	A2	Activist Attractiveness Rationale	A3	TR
3. DEF Branch and inter-modal terminal—Branch line providing mixed products to multi-modal marketing terminal, breakout facility, interconnection to other pipelines and direct connect to military, commercial airports and power plant.	Possible onsite fatalities. Possible offsite environmental impact. Limited alternative delivery resources to customers.	4	Major disruption to air travel, power supply and military. Easy access.	3	Some insider insight helpful but not necessary.	2	Public Image impact due to press/media interest.	3	TR 3
4. Endpoint storage facility—Major tank farm for large metropolitan area, airport and other party pipeline connection.	Serves large metropolitan area. Several million retail customers plus major international airport. Area served by other sources. Located in a sparsely populated industrial area.	2	Hardened facility. Access difficult but impact significant.	3	Insider information very helpful both to gain access and operational.	2	Nuisance issue with trespassing. Public image impact. Operational knowledge needed.	2	TR 3
5. River span block valve	Block valve is upstream from above ground river span (see item 7). Breach could cause release of pipe volume into river and impact public safety and significant contamination to the water supply of a major metropolitan center. Restoration costs significant due to river spill clean up and difficult access to valve. Short timeframe to repair.	5	Public safety and drinking water contamination. Perhaps included with attack on asset--River Span (above ground).	2	Some insider insight helpful but not necessary. Difficult access within minimal success.	1	Limited interest.	2	TR 2

Form 3: Attractiveness/Target Ranking Form

Facility Name: 1. Fictitious Pipeline Company

Critical Assets	Function/Hazards/ Criticality	S	Asset Attractiveness						
			Foreign/Do mestic Attractiveness Rationale	A1	Employee/Contractor Attractiveness Rationale	A2	Activist Attractiveness Rationale	A3	TR
6. River Span Pipeline (Above Ground)	Above ground river span (see 1 above). Breach could release significant product into river and contaminate public water supply to a major metropolitan center. Block valve used as active mitigation, if not damaged. Significant public safety concern due to frequent recreational and commercial use on river. Long-term repair timeframe and significant repair costs and spill clean up costs. No alternate mode to market. Significant service interruption.	5	Public safety and drinking water contamination. Easy access.	3	No insider knowledge needed for breach/access.	1	Public Image impact due to press/media interest.	3	TR 3
7. Inter-modal Terminal	Large inter-modal products terminal with rail, truck and pipeline service. Serves large metropolitan area. Provides gasoline to retail market, jet fuel to 2 major international airports and USAF. Large-scale damage would take months to repair. Repair costs would be significant. Significant disruption to local economy and possible national defense. No significant environmental impact. Limited public safety and employee impact.	4	Hardened facility. Access difficult but impact significant.	3	Insider information very helpful both to gain access and operational.	2	Nuisance issue with trespassing. Public image impact. Operational knowledge needed.	2	TR 3

Form 4—Scenario Based Vulnerability

Facility Name: 1. Fictitious Pipeline Company

Critical Assets: 6. River Span Pipeline (Above Ground)

Scenario Worksheet Form											
Security Event Type	Threat Category	Type	Undesired Act	Consequences	S	Existing Safeguards/ Countermeasures	Vulnerability	V	L	R	Recommendations
1.1. Destruction of span, release of product and loss of containment.	Terrorist	I/E/C	Destruction of river span by bombing.	Damage of river span; release of product into river; contamination of public drinking water supply; loss of service to downstream facilities for an extended period.	S5	1.1. Fencing around cable platform.	1. There are some protective measures; river span remote; easy access - above grade.	4	L3	[redacted]	1. Consider additional hardening to prevent access to river span.
						1.2. Air patrol and ground observation.					2. *Evaluate additional intrusion detectors feasible at this site.
						1.3. Manually operated block valve.					3. *Evaluate if CCTV is feasible.
						1.4. Monitoring pipeline conditions and flow ctrl.					4. Consider additional surveillance of this area.

**Note:* *Additional countermeasures should be based on threat and criticality of the equipment / system under evaluation. Due to remote locations, electric power may not be available or feasible to implement electronic security measures.*

Form 4—Scenario Based Vulnerability

Facility Name: 1. Fictitious Pipeline Company

Critical Assets: 7. Inter-modal Terminal

Scenario Worksheet Form											
Security Event Type	Threat Category	Type	Undesired Act	Consequences	S	Existing Safeguards/ Countermeasures	Vulnerability	V	L	R	Recommendations
1.1. Destruction of inter-modal terminal manifold piping.	Terrorist	I/E/C	Destruction of piping by bombing.	Inability to receive or pump product and possible onsite fatalities.	S4	1.1. Fencing, lighting, access control, CCTV, manned 24/7, security procedures in place.	1. There are multiple protective measures but at least one weakness to gain access.	2	L3	Med	5. Consider improved access control, 24/7 security guards at higher threat levels.
											6. Consider additional countermeasures to achieve "protection in Depth" such as random vehicle inspections, background checks.

Appendix C3—Fictitious Truck Transportation SVA Example

The application of the SVA Methodology to a fictitious products distribution system by truck is illustrated in the following example. Only the first page of each of the four forms is shown for illustrative purposes. It is assumed that the study is conducted by the owner of the trucking company and considers the various interfaces with customers and suppliers. However, the security of the customer and supplier facilities is the responsibility of the owners of those facilities.

The example is of a fictitious hydrocarbon tank truck transportation system, which includes the tank truck, inventory of flammable liquids and the route specific variables such as the type of road, population centers and environmental receptors, and any stops. It is assumed that the shipper and receiver sites will have a separate SVAs. This example is intended to provide some insight on how one might conduct a security vulnerability analysis (SVA) using this methodology on the fictitious truck transportation system. This example is not intended to be all inclusive of every situation or every item that one may consider when conducting an SVA on a tank truck system. It is recognized that not all tank truck systems are the same. Factors such as route length, type of cargo, geographic location and many other factors play a significant role to determine the criticality of the transportation system thereby defining the type and level of analysis that may be appropriate for a particular situation.

The study is conducted in a top-down, systematic manner following the logic flowchart for the SVA as shown in Figure A. The five steps of the process are documented in four forms:

Figure A—SVA Methodology Flow Diagram

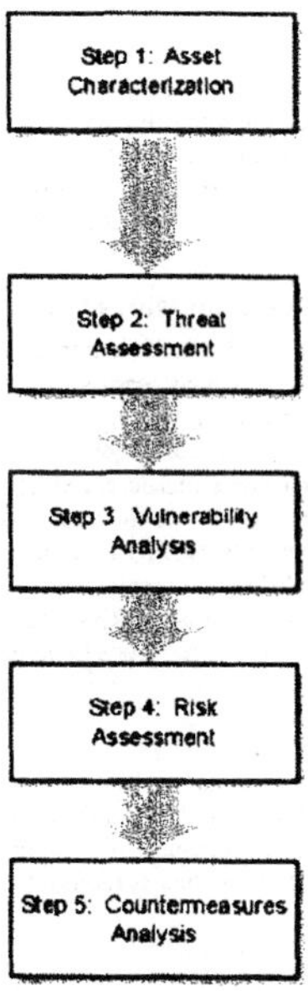

Form 1—Critical Assets/Criticality form

Form 2—Threats Worksheet
Form 3—Attractiveness/Target Ranking Form

Form 4—Scenario Based Vulnerability Worksheet/Risk Ranking/Countermeasures Form

Form 1—Critical Assets/Criticality form

First determine the major assets of the truck transportation system including function, major customers, routes, check points, terminals, utilities, and supporting infrastructure. Next, critical facilities or functions of the transportation system are identified. For all critical facilities, identify critical assets within those functions or within those facilities. All entry points should be evaluated as an asset in order to focus the analysis on the need for perimeter security and access control. The team lists all relevant assets on Form 1 in Column 1. Similar facilities with similar geographic locations, common vulnerabilities, and common consequences can be grouped for efficiency and to consider the value of an entire functional set. In Column 2, document the design basis of the asset and the hazards and consequences that would be realized if the asset was damaged, compromised, or stolen. In the Column 3 rank the estimated overall severity of the loss of the asset. Use the five-level Severity Ranking scale for severity or develop an equivalent as required for the particular facility.

Form 2—Threats Worksheet

Document the threats against the transportation system/facility on Form 2. Include consideration in Column 1 of general types of adversaries that will be considered (usually terrorists, disgruntled employee or contractor, or extreme activist as an example, but more specific or other groups can be considered as required); Column 2 is the source of the attack (EXT—External to the transportation system/facility, INT—Internal to the transportation system/facility); Column 3 documents the threat specific to the transportation system/facility being evaluated; Column 4 documents the specific or general threat of that type of adversary against this or similar assets worldwide; Column 5 documents the potential actions that the adversary could take; Column 6 documents the assumed capabilities, weapons, tactics, and sophistication of the adversary; Column 7 documents their level of motivation; Column 8 provides for an overall ranking assessment per the Threat Ranking scale or equivalent.

Form 3—Attractiveness/Target Ranking Form

Columns 1 – 3 are repeated from Form 1 for reference. Column 4 is a documented rationale for why the particular asset is attractive (or unattractive) and Column 5 is a ranking of that attractiveness on a relative Attractiveness Ranking scale or equivalent. This is repeated for other adversaries. Column 10 is an overall Target Ranking per the same scale, and is normally considered to be the highest attractiveness of any of the individual adversary rankings but also considers that the sum the different adversary's interests may make the asset more attractive. The Target Ranking is used to judge the degree of attractiveness of the target considering all the adversaries.

Form 4—Scenario Based Vulnerability Worksheet/Risk Ranking/Countermeasures Form

Column 1 is the Security Event Type (generally one of four security events including loss of containment, degradation of the asset, theft, or contamination); Column 2 is the Threat Category (adversary type such as terrorist, activist, employee); Column 3 is the Type of Adversary Attack (Insider/External); Column 4 is the Undesired Act (the assumed attack scenario, generally taken from the Threats Worksheet Columns 5, 6, 7); Column 5 is the Consequences; Column 6 (S) is the Severity Ranking from the Severity Ranking scale; Column 7 is the Existing Countermeasures, which considers the Deter, Detect, Delay, and Respond philosophy; Column 8 is the Vulnerability, which also considers the weaknesses or missing elements of the security strategy specific to the scenario; Column 9 is the Vulnerability Ranking per the Vulnerability Ranking scale; Column 10 is the Likelihood ranking (L) using the Likelihood scale, which is a judgment of the team considering the factors of Vulnerability, Threat, and Attractiveness; Column 11 is the Risk ranking (R) per the referenced Risk Ranking Matrix values; and Column 12 is the New /Countermeasures suggestions (where the risk is considered significant enough to justify the need for change).

Responsibilities

This example includes a sampling of assets that may be owned or operated by various parties. The responsibilities for conducting the SVA and for providing security need to be determined and may not solely be with the Shipper. It is recommended that the SVA include the appropriate parties to fully analyze the security issues, and that the results are discussed with owner/operators of adjacent facilities, Receivers, and infrastructure providers as required for risk communication and completeness.

SVA Methodology
Fictitious Truck Transportation Example

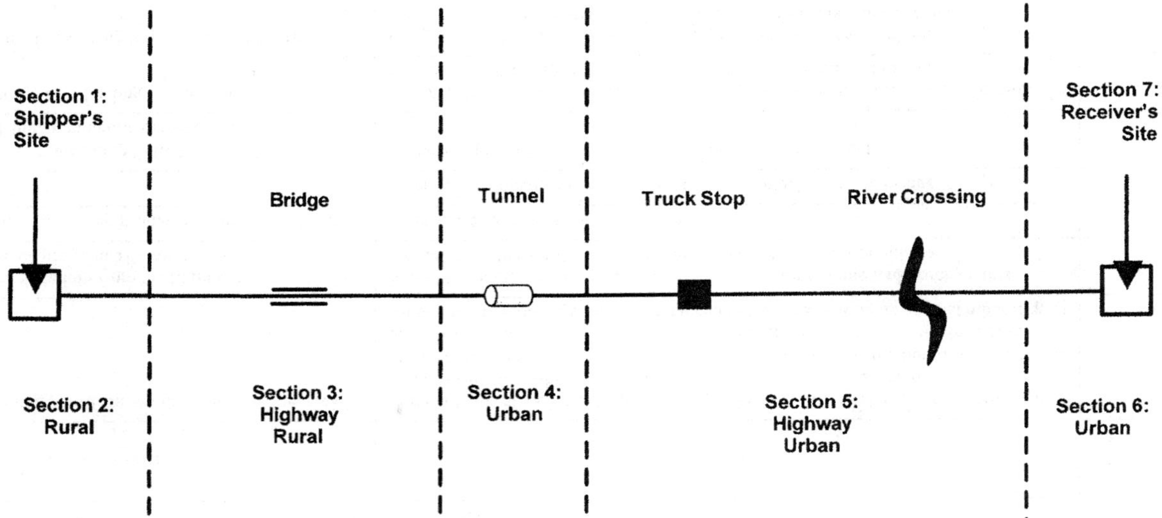

Form 1: Critical Assets/Criticality Form

Facility Name: 1. Fictitious Trucking Company

Critical Assets Form		
Critical Assets	Criticality/Hazards	Asset Severity Ranking
1. Tank Truck containing petroleum products and loading rack operations	Shipper loads 50 tank trucks per day of products and dispatches them to both local receivers and to within nearby neighboring states. Route evaluated is the longest distance transported to a receiver's site. Potential for flammable liquids to be attacked directly to damage the loading rack and operations, to be attacked while en route to cause collateral damage, or to be hijacked and used as a weapon against other targets.	4
2. Rural section of road leading from the Shipper's Site to HWY 100 – 15 miles, traversing primarily rural areas.	Single entrance/exit to supplier's site, but incident involving tank truck on this section of route would result in limited impacts due to low population density.	2
3. HWY 100 (50 miles) traversing primarily through rural areas.	Long stretch across rural section of route.	3
4. Bridge along HWY 100.	Potential to block/damage bridge if tank truck attacked on the bridge.	3
5. Downtown section of route along State Route 5 (15 miles), traversing through high population density area.	Highest population density along route, but shortest segment.	3
6. Tunnel along State Route 5 leading into downtown.	Potential to block/damage tunnel preventing entrance/exit to the city and possible for multiple fatalities/injuries from occupants in other vehicles in tunnel.	4
7. HWY 200 (100 miles) traversing through primarily urban areas.	Longest stretch along the route with a high population density along the segment, potential to not only impact vehicle occupants on road but also surrounding population impacts.	4
8. Truck Stop along HWY 200.	Potential for theft/access to unmanned vehicle.	4
9. River Span along HWY 200.	Potential for environmental impact if product released into river.	3
10. Urban route off HWY 200 to Receiver's site - 10 miles.	Single entrance/exit to receiver's site, with potential for fatalities/injuries due to high population density surrounding the site.	3

Form 2: Threats Worksheet

Facility Name: 1. Fictitious Trucking Company

Adversary Types	Source	Site Specific Threat	Threat History	Potential Actions	Adversary Capability	Adversary Motivation	Overall Assessment	Threat Ranking
International terrorists	I/E/C	1.1. No site-specific history of intentional acts against ACME.	According to information bulletins from DHS there have been suspicious activities involving bulk facilities including surveillance and following trucks. International terrorists have targeted trucks for highjackings and direct attacks.	Use of force to cause damage to vehicles while in transit or at loading/offloading facilities. This could cause a release of hydrocarbons and resulting fire and explosion with possible fatalities and injuries and degradation of transportation assets and environmental release. Terrorists may be interested in 1) weaponization of a tank truck to use fuels as a improvised, field-ready weapon at another location 2) directly damage the truck and cause collateral damage and disruption to the supply chain 3) "Trojan Horse" attack where the truck is used to introduce a weapon into a facility.	Assume a high level of organizational support; good resources; good financial backing; network of members; highly developed communication capabilities; weapons including small arms and explosives; possible vehicle bomb based on past events.	Assume adversary is highly motivated, likely extremist, prepared to die for their cause with intent to cause maximum damage to company assets including loss of life and economic disruption.	Credible threat. Include in analysis. An attempt to cause a violent attack on the truck would be consistent with both the tactics and goals of domestic terrorists.	3

Form 2: Threats Worksheet

Facility Name: 1. Fictitious Trucking Company

Adversary Types	Source	Site Specific Threat	Threat History	Potential Actions	Adversary Capability	Adversary Motivation	Overall Assessment	Threat Ranking
Domestic Terrorist or Activist	I/E/C	2.1. History of bomb threats at ACME Trucking. All concluded were fakes--no bomb or activist found. ACME has had activist protest at the main gate within the past 2 years.	No confirmed domestic acts of terrorism against fuels trucking operations.	Possible for a disruptive event from domestic terrorist such as bombing or disruption of operations, similar to international terrorist objectives but most-likely of a less severe nature. Possible actions would include highjackings, theft, vandalism, and arson.	Assume medium level of organizational support; poor resources and financial backing; small network of members; cell phone/email communication capabilities; weapons including small improvised explosive devices.	Adversary intent is to cause economic harm through service interruption or to emphasize a political cause. If domestic terrorist, intent and motivation could be extreme to cause maximum damage, but more-likely without personal sacrifice.	Credible threat. Included in analysis. An attempt to cause damage or disruption to operation is likely in the future.	3

Form 2: Threats Worksheet

Facility Name: 1. Fictitious Trucking Company

Adversary Types	Source	Site Specific Threat	Threat History	Potential Actions	Adversary Capability	Adversary Motivation	Overall Assessment	Threat Ranking
Disgruntled Employee or Contractor	INT	3.1. No evidence of sabotage has been discovered in the past. Have been several safety systems compromised and incidences of theft.	There have been acts of sabotage, theft and arson to the petroleum trucking operations in the past.	Sabotage to vehicles, including safety systems, arson, and theft of product.	Insider access, knowledge and ability to operate independently with authorization and without question. May have access to vehicles, facilities, gate access codes, communicati on equipment, records, and proximity cards for access cards.	Disgruntled employee is most-likely intent to cause inconvenience and financial impacts to the company or their employer. If very disgruntled or troubled, intent and motivation could be extreme to cause maximum damage, possibly with personal sacrifice as evidenced in various national workplace violence cases.	Credible threat. Include in analysis.	2

Form 3: Attractiveness/Target Ranking Form

Facility Name: 1. Fictitious Trucking Company

Critical Assets	Function/Hazards/ Criticality	Asset Severity Ranking	Asset Attractiveness						
			Foreign/Domestic Attractiveness Rationale	A1	Employee/Contract or Attractiveness Rationale	A2	Activist Attractiveness Rationale	A3	TR
1. Tank truck containing 10,000 gallons of hydrocarbons.	Shipper dispatches 50 trucks per day of gasoline to both local receivers and to within nearby neighboring states. Route evaluated is the longest distance transported to a receiver's site.	3	Potential for release resulting in large fire, potential fatalities and closure/damage to major transportation route.	3	Insider information necessary to gain access to vehicle.	1	Public image impact due to press/media interest.	2	TR3
2. Rural section of road leading from the Shipper's Site to HWY 100 – 15 miles, traversing primarily rural areas.	Single entrance/exit to supplier's site, but incident involving tank truck on this section of route would result in limited impacts due to low population density.	1	Short section of route and limited number of potential impacts.	1	No additional attraction.	1	No additional attraction.	1	TR1
3. HWY 100 (50 miles) traversing primarily through rural areas.	Long stretch across rural section of route.	2	Minimal attraction due to limited impact potential, but length of route provides access to vehicle.	2	No additional attraction.	1	No additional attraction.	1	TR2

Form 3: Attractiveness/Target Ranking Form

Facility Name: 1. Fictitious Trucking Company

Critical Assets	Function/Hazards/ Criticality	Asset Severity Ranking	Asset Attractiveness						
			Foreign/Domestic Attractiveness Rationale	A1	Employee/Contract or Attractiveness Rationale	A2	Activist Attractiveness Rationale	A3	TR
4. Bridge along HWY 100.	Potential to block/damage bridge if tank truck attacked on the bridge.	3	Potential to cause major disruption to US Highway as well as result in potential fatalities and injuries.	3	No additional attraction.	1	Potential to block bridge.	2	TR3
5. Downtown section of route along State Route 5 (15 miles), traversing through high population density area.	Highest population density along route, but shortest segment.	3	High population density and potential to harm a large number of people.	3	No additional attraction.	1	No additional attraction.	1	TR3
6. Tunnel along State Route 5 leading into downtown.	Potential to block/damage tunnel preventing entrance/exit to the city and possible for multiple fatalities/injuries from occupants in other vehicles in tunnel.	3	High population impact potential as well as potential to disrupt local economy by blocking tunnel.	3	No additional attraction.	1	Potential to block tunnel.	2	TR3
7. HWY 200 (100 miles) traversing through primarily urban areas.	Longest stretch along the route with a high population density along the segment, potential to not only impact vehicle occupants on road but also surrounding population.	3	Long section of route provides access to truck highly populated area.	3	No additional attraction.	1	No additional attraction.	1	TR3
8. Truck Stop along HWY 200.	Potential for theft/access to unmanned vehicle.	3	Potential to gain access to truck--theft.	2	No additional attraction.	1	No additional attraction.	1	TR2
9. River Span along HWY 200.	Potential for environmental impact if product released into river.	2	Material not likely to cause sustained environmental impact.	1	No additional attraction.	1	No additional attraction.	1	TR1
10. Urban route off HWY 200 to Receiver's site – 10 miles.	Single entrance/exit to receiver's site, with potential for fatalities/injuries due to high population density s urrounding the site.	2	Limited access due to shortness of route, but high population density makes section attractive.	2	No additional attraction.	1	No additional attraction.	1	TR2

Form 4—Scenario Based Vulnerability

Facility Name: 1. Fictitious Trucking Company

Critical Assets: 1. Tank Truck containing 10,000 gallons of hydrocarbons

Scenario Worksheet Form

Security Event Type	Threat Category	Threat Type	Undesired Act	Consequences	S	Existing Safeguards/ Countermeasures	Vulnerability	V	L	R	Recommendations
1.1. Truck is attacked enroute resulting in a release of hydrocarbons.	Terrorist	I/E/C	Release and ignition of hydrocarbons on a major roadway.	Potential fatalities and injuries from resulting fire. Possible closure of a major transportation route.	S4	1.1. Experienced/ Licensed Drivers--background checks before employment. 1.2. Identification of driver's checked at both the shipper and receiver's sites. 1.3. Drivers trained in HAZMAT. 1.4. Truck is in constant radio contact while enroute.	1. Longest route exposes the truck many hours per shipment; provides the opportunity for surveillance and unexpected attack; route also passes along several areas of high population density, including bridge and tunnel.	4	L3	[redacted]	1. Consider developing company system to alert drivers to DHS/FBI alerts. 2. Consider developing a system to cont act local law enforcement at DHS "red" levels for information prior to traveling. 3. Consider providing security awareness and emergency action training to drivers.
1.2. Truck is highjacked enroute.	Terrorist	I/E/C	Loss of truck and product.	Potential for injury/fatality to driver in an attack by force. Loss of truck and product, but unlikely to be used in subsequent attack.	S4	2.1. Truck is in constant radio contact while enroute. 2.2. Single scheduled truck stop along route. 2.3. Truck is normally locked when driver is at the truck stop. 2.4. Truck has electronic disengagement systems.	1. Long stretches of rural areas along route provide opportunity for surveillance and attack; truck is left unattended while at the truck stop.	3	L2	Med	4. Consider adding GPS tracking system to truck so that they can be tracked/located if stolen. 5. Consider additional radio checks at elevated security levels.

Appendix C4—Fictitious Rail Transportation SVA Example

The application of the SVA Methodology to a fictitious petroleum liquids pipeline system is illustrated in the following example. Only the first page of each of the four forms is shown for illustrative purposes. It is assumed that the study is conducted by the shipper company and considers the various interfaces with customers, suppliers and en-route interfaces. However, the security of the customer and supplier facilities and the en-route interfaces is the responsibility of the owners of those facilities, as well as the general route risk assessment issues. An example may include the switchyard security plan. It is the responsibility of the switchyard operator to ensure the security of the switchyard.

The general approach is to apply risk assessment resources and, ultimately, special security resources primarily where justified based on the SVA results. The SVA process involves consideration of the rail transportation system from both the general viewpoint and specific asset viewpoint. Consideration at the general overall route level is useful for determination of overall impacts of loss, infrastructure and interdependencies at the route level. The benefit of evaluating specific assets is that individual interface risks can be evaluated and specific countermeasures applied where justified in addition to more general countermeasures.

The SVA methodology uses this philosophy in several ways. The method is intended to be comprehensive and systematic in order to be thorough. First, it begins with the SVA team gaining an understanding of the entire rail transportation route that applies to the route that the shipper's products take through the value chain from production facility to various customers and end users. The SVA will analyze the critical assets that comprise the transportation system, the critical functions of the system, and the hazards and impacts if these assets or critical functions are compromised. This results in an understanding of which assets and functions are "critical" to the business operation. Criticality may be defined both in terms of the potential impact to the workers, community, the environment and the company, as well as to the business importance and continuity of the system. For example, a rail loading station or a specific branch along the route may be a critical part of the operation of the system due to inability to operate without it or, if attacked, it has the greatest impact. As such it may be given a high priority for further analysis and special security countermeasures.

Based on this first level of screening from all assets to critical assets, a critical asset list is produced. Next, the critical assets are reviewed in light of the threats. Adversaries may have different objectives, so the critical asset list is reviewed from each adversary's perspective and an asset attractiveness ranking is given. This factor is a quick measure of whether the adversary would value damaging, compromising, or stealing the asset, which serves as an indicator of the likelihood that an adversary would want to attack this asset and why.

If an asset is both critical (based on value and consequences) and attractive, then it is considered a "target" for purposes of the SVA. A target may optionally receive further specific analysis, including the development of scenarios to determine and test perceived vulnerabilities. As shown in Figure A, all assets receive at least a general security review. This is accomplished by the basic SVA team's consideration as an asset to begin with, along with a baseline security survey. General security considerations may be found in security references such as the countermeasures checklist provided in Appendix F.

The study is conducted in a top-down, systematic manner following the logic flowchart for the SVA as shown in Figure A. The five steps of the process are documented in four forms:

Figure A—SVA Methodology Flow Diagram

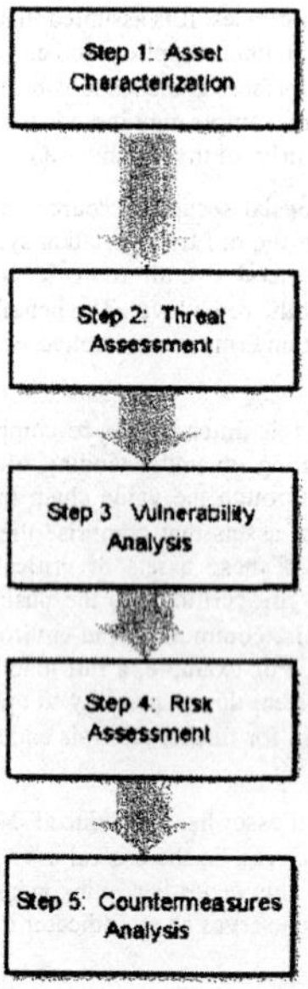

Form 1—Critical Assets/Criticality form

Form 2—Threats Worksheet
Form 3—Attractiveness/Target Ranking Form

Form 4—Scenario Based Vulnerability Worksheet/Risk Ranking/Countermeasures Form

Form 1—Critical Assets/Criticality Form

Determine the major assets of the rail transportation system including loading facilities, switching yards, specific routes, control rooms, gates and access control points, marine terminals, bridges, tunnels, utilities, supporting infrastructure, and other considerations. All entry points to a facility should be evaluated as an asset in order to focus the analysis on the need for perimeter security and access control. The team lists all relevant assets on Form 1 in Column 1. Similar facilities with similar geographic locations, common vulnerabilities, and common consequences can be grouped for efficiency and to consider the value of an entire functional set. In Column 2, document the design basis of the asset and the hazards and consequences that would be realized if the asset was damaged, compromised, or stolen. In the Column 3 rank the estimated overall severity of the loss of the asset. Use the five-level Severity Ranking scale for severity or develop an equivalent as required for the particular facility or transportation system. Conduct the study on the overall general route, followed by more detailed evaluation of critical facilities.

Form 2—Threats Worksheet

Document the threats against the facilities or transportation system on Form 2. Include consideration in Column 1 of general types of adversaries that will be considered (usually terrorists, disgruntled employee or contractor, or extreme activist as an example, but more specific or other groups can be considered as required); Column 2 is the source of the attack (EXT—External to a facility or rail system, INT—Internal to a facility or rail system); Column 3 documents the threat specific to the facility or rail system being evaluated; Column 4 documents the specific or general threat of that type of adversary against this or similar assets and operations worldwide; Column 5 documents the potential actions that the adversary could take; Column 6 documents the assumed capabilities, weapons, tactics, and sophistication of the adversary; Column 7 documents their level of motivation; Column 8 provides for an overall ranking assessment per the Threat Ranking scale or equivalent.

Form 3—Attractiveness/Target Ranking Form

Columns 1 – 3 are repeated from Form 1 for reference. Column 4 is a documented rationale for why the particular asset or operation is attractive (or unattractive) and Column 5 is a ranking of that attractiveness on a relative Attractiveness Ranking scale or equivalent. This is repeated for other adversaries. Column 10 is an overall Target Ranking per the same scale, and is normally considered to be the highest attractiveness of any of the individual adversary rankings but also considers that the sum the different adversary's interests may make the asset more attractive. The Target Ranking is used to judge the degree of attractiveness of the target considering all the adversaries.

Form 4—Scenario Based Vulnerability Worksheet/Risk Ranking/Countermeasures Form

Column 1 is the Security Event Type (generally one of four security events including loss of containment, degradation of the asset, theft, or contamination); Column 2 is the Threat Category (adversary type such as terrorist, activist, employee); Column 3 is the Type of Adversary Attack (Insider/External); Column 4 is the Undesired Act (the assumed attack scenario, generally taken from the Threats Worksheet Columns 5, 6, 7); Column 5 is the Consequences; Column 6 (S) is the Severity Ranking from the Severity Ranking scale; Column 7 is the Existing Countermeasures, which considers the Deter, Detect, Delay, and Respond philosophy; Column 8 is the Vulnerability, which also considers the weaknesses or missing elements of the security strategy specific to the scenario; Column 9 is the Vulnerability Ranking per the Vulnerability Ranking scale; Column 10 is the Likelihood ranking (L) using the Likelihood scale, which is a judgment of the team considering the factors of Vulnerability, Threat, and Attractiveness; Column 11 is the Risk ranking (R) per the referenced Risk Ranking Matrix values; and Column 12 is the New /Countermeasures suggestions (where the risk is considered significant enough to justify the need for change).

Responsibilities

This example includes a sampling of assets that may be owned or operated by various parties. The responsibilities for conducting the SVA and for providing security need to be determined and may not solely be with the Shipper. It is recommended that the SVA include the appropriate parties to fully analyze the security issues, and that the results are discussed with railroad owner/operators, owner/operators of adjacent facilities and infrastructure providers as required for risk communication and completeness.

API/NPRA SVA Methodology
Rail Transportation Example

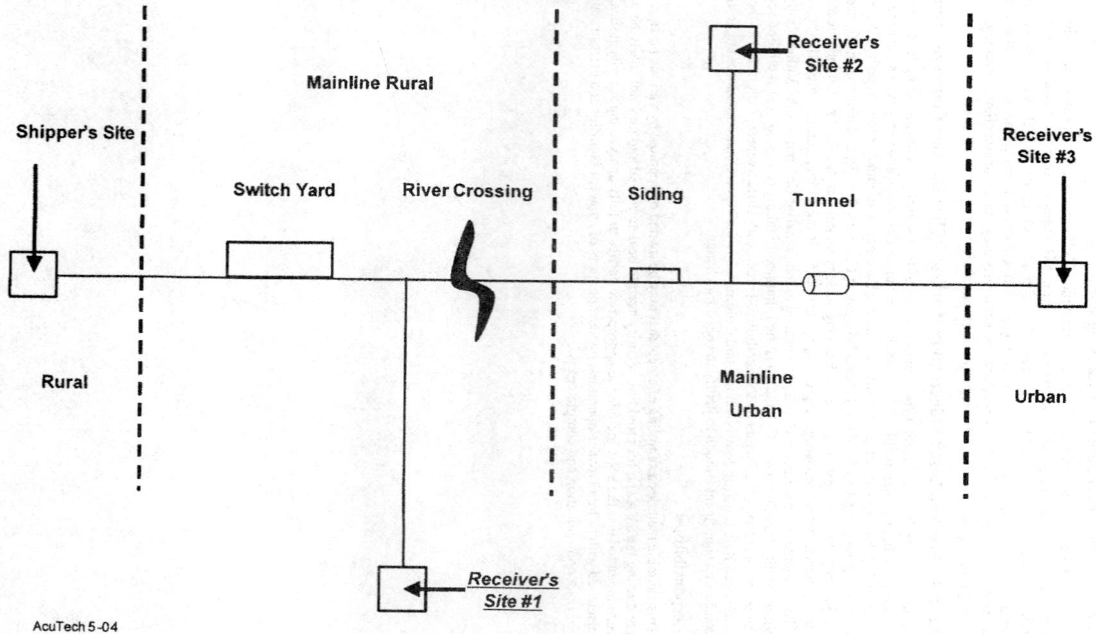

Form 1: Critical Assets/Criticality Form

Facility Name: 1. Fictitious Rail Company

Critical Assets Form		
Critical Assets	Criticality/Hazards	Asset Severity Ranking
1. 25 railcars of petroleum products.	Two trains comprised solely of 25 petroleum products railcars are shipped daily from the shipper's terminal. After leaving the terminal the tankcars are divided into three separate trains at the switchyard and sent to three final receiver's sites. Site #1 - 25 railcars per day. Site #2 - 10 railcars per day. Site #3 - 15 railcars. En route from the switch yard to Site #1 is on a mainline track along a mostly rural area. En route to Site #2 and #3 crosses a river and have access to a siding as needed. The route to Site #2 branches off on an urban mainline, while the route to Site #3 continues through a tunnel before reaching its final destination. Potential hazard for this route is the potential to release one or more railcars resulting in a large environmental impact and or fire and subsequent fatalities and injuries if ignited.	4
2. Rural section of track to switch yard - 25 miles from shipper's site.	Single rail entrance/exit to supplier's site; incident involving railcar on this section of the route would result in limited fatalities/injuries due to low population density, but large fire could damage rail line.	3
3. Mainline section of track in rural area - 200 miles. Including rail spur to Receiver Site #1.	Long stretch across rural section of route.	3
4. Switch Yard	Switch point to individual trains to receiver's sites. Potential to damage site, other railcars and various products if petroleum products released and ignited.	4
5. River crossing	Potential for environmental impact if product released into river.	3
6. Mainline section of track in urban area - 300 miles. Including rail spurs to Site #2 and Site #3.	Long stretch across urban section on route to Site #2 and Site #3.	3
7. Siding in Urban Area (see 6)	Potential for theft/access to unmanned railcars.	4
8. Tunnel in Urban Area (see 6)	Potential to block/damage tunnel.	4

Form 2: Threats Worksheet

Facility Name: 1. ACME Rail Company

Adversary Types	Source	Site Specific Threat	Threat History	Potential Actions	Adversary Capability	Adversary Motivation	Overall Assessment	Threat Ranking
International terrorists	**I/E/C**	1.1. No site-specific history of intentional acts against ACME.	Bombings in Madrid have recently indicated the vulnerability of the rail transportation infrastructure.	Terrorists may be interested in 1) weaponization of a train to use fuels as a improvised, field-ready weapon at another location 2) directly damage the railcar(s) and cause collateral damage and disruption to the supply chain 3) "Trojan Horse" attack where the railcars are used to introduce a weapon into a facility.	Assume a high level of organizational support; good resources; good financial backing; network of members; highly developed communication capabilities; weapons including small arms and explosives; possible vehicle bomb based on past events.	Assume adversary is highly motivated, likely extremist, prepared to die for their cause with intent to cause maximum damage to company assets including loss of life and economic disruption.	Credible threat. Include in analysis. An attempt to cause a violent attack on the railcar/train would be consistent with both the tactics and goals of domestic terrorists.	3
Domestic Terrorist or Activist	**I/E/C**	2.1. History of bomb threats at ACME. No actual bombs found or activist groups claiming responsibility. ACME has had activist protest at the corporate headquarters within the past 5 years.	No confirmed domestic acts of terrorism against fuels rail operations.	Possible for a disruptive event from domestic terrorist such as bombing or disruption of operations, s imilar to international terrorist objectives but most-likely of a less severe nature. Possible actions would include vandalism, blockage of track and arson.	Assume medium level of organizational support; poor resources and financial backing; small network of members; cell phone/email communication capabilities; weapons including small improvised explosive devices.	Adversary intent is to cause economic harm through service interruption or to emphasize a political cause. If domestic terrorist, intent and motivation could be extreme to cause maximum damage, but more-likely without personal sacrifice.	Credible threat. Included in analysis. An attempt to cause damage or disruption to operation is likely in the future.	3

Form 2: Threats Worksheet

Facility Name: 1. ACME Rail Company

Adversary Types	Source	Site Specific Threat	Threat History	Potential Actions	Adversary Capability	Adversary Motivation	Overall Assessment	Threat Ranking
Disgruntled Employee or Contractor	**INT**	3.1. No evidence of sabotage has been discovered in the past.	There have been acts of sabotage, theft and arson to the petroleum railcar operations in the past.	Sabotage to railcars including safety systems, and arson.	Insider access, knowledge and ability to operate independently with authorization and without question. May have access to railcars/train, facilities, gate access codes, communication equipment, records, and proximity cards for access cards.	Disgruntled employee is most-likely intent to cause inconvenience and financial impacts to the company or their employer. If very disgruntled or troubled, intent and motivation could be extreme to cause maximum damage, possibly with personal sacrifice as evidenced in various national workplace violence cases.	Credible threat. Include in analysis.	4

Form 3: Attractiveness/Target Ranking Form

Facility Name: 1. Fictitious Rail Company

Critical Assets	Function/Hazards/ Criticality	S	Asset Attractiveness						
			Foreign/Domestic Attractiveness Rationale	A1	Employee/Contractor Attractiveness Rationale	A2	Activist Attractiveness Rationale	A3	TR
1. 25 railcars of petroleum products.	Two trains comprised of 25 petroleum products railcars are shipped daily from the shipper's terminal. After leaving the terminal the tankcars are divided into three separate trains at the switch yard and sent to three final receiver's sites. Site #1 - 25 railcars per day. Site #2 - 10 railcars per day. Site #3 - 15 railcars. Potential to release one or more railcars resulting in a large environmental impact and or fire and subsequent fatalities and injuries if ignited.	3	Potential for release resulting in large fire, potential fatalities and closure/damage to major transportation route.	3	Insider information necessary to gain access to vehicle.	1	Public image impact due to press/media interest.	2	TR 3
2. Rural section of track to switch yard - 25 miles from shipper's site.	Single rail entrance/exit to supplier's site; incident involving railcar on this section of the route would result in limited fatalities/injuries due to low population density, but large fire could damage rail line.	1	Short section of route and limited number of potential impacts.	1	No additional attraction.	1	No additional attraction.	1	TR 1
3. Mainline section of track in rural area - 200 miles. Including rail spur to Receiver Site #1.	Long stretch across rural section of route.	2	Minimal attraction due to limited impact potential, but length of route provides access to vehicle.	2	No additional attraction.	1	No additional attraction.	1	TR 2

Form 3: Attractiveness/Target Ranking Form

Facility Name: 1. Fictitious Rail Company

Critical Assets	Function/Hazards/ Criticality	S	Asset Attractiveness						
			Foreign/Domestic Attractiveness Rationale	A1	Employee/Contractor Attractiveness Rationale	A2	Activist Attractiveness Rationale	A3	TR
4. Switch Yard	Switch point to individual trains to receiver's sites. Potential to damage site, other railcars and various products if petroleum products released and ignited.	2	Potential to cause major disruption to rail transportation systems.	3	No additional attraction.	1	Potential to block bridge.	2	TR 3
5. River crossing	Potential for environmental impact if product released into river.	2	Potential contamination of drinking water supply and major disruption to rail transportation system.	3	No additional attraction.	1	No additional attraction.	1	TR 3
6. Mainline section of track in urban area – 300 miles. Including rail spurs to Site #2 and Site #3.	Long stretch across urban section on route to Site #2 and Site #3.	3	High population density and potential to harm a large number of people. Ability to disrupt Sites #2/3	4	Ability to disrupt Sites #2/3	4	Ability to disrupt Sites #2/3	2	TR 4
7. Siding in Urban Area (see 6)	Potential for theft/access to unmanned railcars.	3	Siding provides access to unmanned railcars in populated area.	3	No additional attraction.	1	No additional attraction.	1	TR 3
8. Tunnel in Urban Area (see 6)	Potential to block/damage tunnel.	3	Potential to cause major disruption to rail transportation system.	2	No additional attraction.	1	No additional attraction.	1	TR 2

Form 4—Scenario Based Vulnerability

Facility Name: 1. Fictitious Rail Company

Critical Assets: 1. 25 railcars of petroleum products.

Scenario Worksheet Form											
Security Event Type	Threat Category	Type	Undesired Act	Consequences	S	Existing Safeguards/ Countermeasures	Vulnerability	V	L	R	Recommendations
1.1. Train is attacked en route with a bomb resulting in a release of petroleum products.	Terrorist	I/E/C	Release and ignition of petroleum products on a major roadway.	Possible closure/damage to major transport ion rail line and potential fatalities and injuries from resulting fire.	S4	1.1. Major Class I Railroad used to carry materials along the entire route to all receivers' sites.	1. Railcars are exposed many hours per shipment; provides the opportunity for surveillance and unexpected attack; route also passes along several areas of high population density and includes both bridge and tunnel.	4	L3	Med	1. Meet with rail company and develop security plan. Discuss access control and staging of cars at elevated threat levels.
						1.2. Security Plan at both the shipper and receiver's site.					2. Consider providing security awareness and emergency action training to rail personnel.
						1.3. Train is in constant radio contact while en route.					3. Review security procedures/plan at the switch yard; revise plan as necessary to address any security concerns
1.2. Bomb is attached to railcar while in switchyard or while on siding.	Terrorist	I/E/C	Bomb is brought onto receiver's site.	Explosion/fire on the rail spurs of at the receiver's site resulting in fatalities/injuries and potential damage to spur and receivers process equipment.	S4	2.1. Security Plan at both the shipper and receiver's site.	1. Railcars are exposed and vulnerable to placement of hidden bomb on railcar while in yard and while on spur.	5	L5	High	4. Meet with switchyard operator to review security issues.
											5. Review security procedure at receiver's site for accepting and screening railcars for delivery.
											6. Consider adding lighting and CCTV around siding to prevent access to stopped train, while en route.

References

"Chemical Accident Prevention Provisions" (part 68 of Title 40 of the *Code of Federal Regulations (CFR)*).

Chemical Facility Vulnerability Assessment Methodology, NIJ Special Report, U.S. Department of Justice, Office of Justice Programs, National Institute of Justice, July, 2002.

Counterterrorism and Contingency Planning Guide. Special publication from Security Management magazine and American Society for Industrial Security, 2001.

Guidance Document for Implementing 40 *CFR* Part 68, USEPA, 1998.

Guidelines for Chemical Process Quantitative Risk Analysis, Second Ed., Center for Chemical Process Safety, American Institute of Chemical Engineers, 2000.

Guidelines for Consequence Analysis of Chemical Releases, Center for Chemical Process Safety, American Institute of Chemical Engineers, 1999.

Guidelines for Technical Management of Chemical Process Safety, Center for Chemical Process Safety, American Institute of Chemical Engineers, 1998.

Guidelines for Technical Planning for On-Site Emergencies, Center for Chemical Process Safety, American Institute of Chemical Engineers, 1996.

Inherently Safer Chemical Processes – A Life Cycle Approach, Center for Chemical Process Safety, American Institute of Chemical Engineers, 1996.

Layers of Protection Analysis, Center for Chemical Process Safety, American Institute of Chemical Engineers, 2001.

"Site Security Guidelines for the U.S. Chemical Industry", American Chemistry Council, October, 2001.

Bowers, Dan M., "Security Fundamentals for the Safety Engineer", *Professional Safety*, American Society of Safety Engineers, December, 2001, pgs. 31-33.

Dalton, Dennis. *Security Management: Business Strategies for Success*. (Newton, MA: Butterworth-Heinemann Publishing, 1995).

Fischer, Robert J. and Green, Gion. *Introduction to Security*, 6th ed. (Boston: Butterworth-Heinemann, 1998).

Ragan, Patrick T., et al., "Chemical Plant Safety", *Chemical Engineering Progress*, February, 2002 pgs. 62-68.

Roper, C.A. *Physical Security and the Inspection Process* (Boston: Butterworth-Heinemann, 1997).

Roper, C.A. *Risk Management for Security Professionals* (Boston: Butterworth-Heinemann, 1999).

Walsh, Timothy J., and Richard J. Healy, eds. *Protection of Assets Manual* (Santa Monica, CA: Merritt Co.). Four-volume loose-leaf reference manual, updated monthly.

Additional copies are available through Global Engineering Documents at (800) 854-7179 or (303) 397-7956

Information about API Publications, Programs and Services is available on the World Wide Web at http://www.api.org

American Petroleum Institute
1220 L Street, Northwest
Washington, D.C. 20005-4070
202-682-8000

Product No: OSVA02

Third Edition
Security Guidelines
for the Petroleum Industry
Petroleum Refineries
Liquid Petroleum Pipelines
Petroleum Products Distribution and Marketing
Oil and Natural Gas Production Operations
Marine Transportation
Cyber/ Information Technology for the Petroleum Industry
American Petroleum Institute
April 2005

Homeland Security Advisory System

www.dhs.gov

SPECIAL NOTES

API publications necessarily address problems of a general nature. With respect to particular circumstances, local, state, and federal laws and regulations should be reviewed.

API is not undertaking to meet the duties of employers, manufacturers, or suppliers to warn and properly train and equip their employees, and others exposed, concerning health and safety risks and precautions, nor undertaking their obligations under local, state, or federal laws.

Information concerning safety and health risks and proper precautions with respect to particular materials and conditions should be obtained from the employer, the manufacturer or supplier of that material, or the material safety data sheet.

Nothing contained in any API publication is to be construed as granting any right, by implication or otherwise, for the manufacture, sale, or use of any method, apparatus, or product covered by letters patent. Neither should anything contained in the publication be construed as insuring anyone against liability for infringement of letters patent.

Generally, API standards are reviewed and revised, reaffirmed, or withdrawn at least every five years. Sometimes a one-time extension of up to two years will be added to this review cycle. This publication will no longer be in effect five years after its publication date as an operative API standard or, where an extension has been granted, upon republication. Status of the publication can be ascertained from the API Standards department telephone (202) 682-8000. A catalog of API publications, programs and services is published annually and updated biannually by API, and available through Global Engineering Documents, 15 Inverness Way East, M/S C303B, Englewood, CO 80112-5776.

This document was produced under API standardization procedures that ensure appropriate notification and participation in the developmental process and is designated as an API standard. Questions concerning the interpretation of the content of this standard or comments and questions concerning the procedures under which this standard was developed should be directed in writing to the Director of the Standards department, American Petroleum Institute, 1220 L Street, N.W., Washington, D.C. 20005. Requests for permission to reproduce or translate all or any part of the material published herein should be addressed to the Director, Business Services.

API standards are published to facilitate the broad availability of proven, sound engineering and operating practices. These standards are not intended to obviate the need for applying sound engineering judgment regarding when and where these standards should be utilized. The formulation and publication of API standards is not intended in any way to inhibit anyone from using any other practices.

Any manufacturer marking equipment or materials in conformance with the marking requirements of an API standard is solely responsible for complying with all the applicable requirements of that standard. API does not represent, warrant, or guarantee that such products do in fact conform to the applicable API standard.

FOREWORD

This document is intended to offer security guidance to the petroleum industry. Individual companies have assessed their own security needs and have implemented security measures they consider appropriate. This document is not intended to supplant the measures adopted by individual companies or to offer commentary regarding the effectiveness of individual company efforts. With respect to particular circumstances, local, state and federal laws and regulations should be reviewed.

Information concerning security risks and proper precautions with respect to particular materials and conditions should be obtained from individual companies or the manufacturer or supplier of a particular material.

API is not undertaking to meet the duties of employers, manufacturers, or suppliers to warn and properly train and equip their employees, and others exposed, concerning security risks and precautions, nor undertaking their obligation under local, state or federal laws.

To the extent this document contains company specific information such information is to be considered confidential.

API publications may be used by anyone desiring to do so. Every effort has been made by the Institute to assure the accuracy and reliability of the data contained in them; however, the Institute makes no representation, warranty, or guarantee in connection with this publication and hereby expressly disclaims any liability or responsibility for loss or damage resulting from its use or for the violation of any federal, state, or municipal regulation with which this publication may conflict.

Suggested revisions are invited and should be submitted to API, RASA department, 1220 L Street, NW, Washington, DC 20005.

TABLE OF CONTENTS

Page
Executive Summary vii

1.0 Introduction 1
1.1 Scope and Objective 1
1.2 Organization of the Document 1
1.3 Underlying Basis of this Guidance 2
1.4 Other Guidelines and Security References 2

2.0 Overview of Terrorism and the Petroleum Industry 3
2.1 Background on Terrorism and Security 3
2.2 Threat to the Petroleum Industry 3

3.0 Threat Assessment 4
3.1 The Value of Threat Assessment 4
3.2 Threat Assessment Process 4
3.3 Security Alert Level Systems 6
3.3.1 Introduction 6
3.3.2 Department of Homeland Security Alert System (HSAS) 6
3.3.3 U.S. Coast Guard Maritime Security Levels 7
3.3.4 International Ship and Port Facility Security (ISPS) Alert Levels 8

4.0 The Security Management System Process 8
4.1 Initial Screening 9
4.2 Data Gathering 10
4.3 Initial SVA 10
4.4 Example Elements of a Security Plan 12
4.4.1 Security Administration & Organization of the Facility 13
4.4.2 Personnel Training 13
4.4.3 Drills and Exercises 14
4.4.4 Record and Documentation 14
4.4.5 Response to Change in Alert Level 14
4.4.6 Communications 15
4.4.7 Security Systems and Equipment Maintenance 15
4.4.8 Security Measures for Access Control, Including Designated Public Access Areas 15
4.4.9 Protected/Controlled/Restricted Areas 16
4.4.10 Security Measures for Monitoring 16
4.4.11 Security Incident Procedures 16
4.4.12 Audits and Security Plan Amendments 16
4.4.13 Security Vulnerability Analysis (SVA) Report 16

5.0 Security Vulnerability Assessment (SVA) Concepts 17
5.1 Security Vulnerability Assessment Overview 17
5.2 Steps in the SVA Process 18
5.3 Estimating Risk Using SVA Methods 19
5.4 Definition of SVA Terms 19
5.4.1 Risk Definition for SVA 19
5.4.2 Consequences (C) 21
5.4.3 Threat (T) 22
5.4.4 Vulnerability (V) 22
5.4.5 Target Attractiveness (A_T) 22
5.5 Characteristics of a Sound SVA Approach 23
5.6 First Step in the SVA Process 23

5.7 SVA Strength and Limitations 24
5.8 Recommended Times for Conducting and Reviewing the SVA 25
5.9 Risk Control and Mitigation 25
5.10 Risk Screening 26

6.0 Security Conditions and Potential Response Measures 27
6.1 Low Condition—Green 27
6.2 Guarded Condition—Blue 28
6.3 Elevated Condition—Yellow 29
6.4 High Condition—Orange 29
6.5 Severe Condition—Red 30

7.0 Information (Cyber) Security 30
7.1 Introduction 30
7.2 Specific Security Guidelines 31
7.2.1 Security Policies, Standards and Procedures 31
7.2.2 Security Awareness and Education 32
7.2.3 Accountability and Ownership 32
7.2.4 Data/Information Classification 33
7.2.5 Security Vulnerability Assessments 33
7.2.6 Physical and Environmental Security 33
7.2.7 Access Controls and Identity Management 33
7.2.8 Network Security 34
7.2.9 Systems Development 34
7.2.10 Change Control 35
7.2.11 Viruses and other Malicious Code 35
7.2.12 Intrusion Detection and Incident Management 35
7.2.13 Business Continuity, Business Resumption and Disaster Recovery 35
7.2.14 Regulatory Compliance 36
7.2.15 Audit (Compliance and Assurance) 36

Figures

4.1 Security Management System Process 9
4.2 Example Elements of a Security Plan 13
5.1 Security Events Evaluated during the API SVA Process 18
5.2 API/NPRA Security Vulnerability Assessment Methodology 19
5.3 Example Risk Matrix 20
5.4 SVA Risk Definition 20
5.5 SVA Risk Variables 21
5.6 Target Attractiveness Factors 23
5.7 Times for Conducting and Reviewing the SVA 25

Tables

3.1 Homeland Security Alert System 7
4.1 Examples of Petroleum Facility Assets Subject to Potential Security Risk 10
4.2 Examples of Security Risks or Threats in the Petroleum Industry 11
5.1 Questions to Determine SVA Approach Needed 24

Appendix A Security Regulations Affecting the U.S. Petroleum Industry 37
Appendix B Glossary and Terms 41
Appendix C Communication of Security Intelligence 45
Appendix D References 46

EXECUTIVE SUMMARY

Safe and reliable energy is a vital link in the nation's critical infrastructure. Petroleum products play an important role in our national economy, national security and are integral to the American way of life. As such, security has always been and continues to be a priority across the petroleum industry. The American Petroleum Institute is the petroleum industry's primary trade association. API provides a forum for the industry to come together and discuss important issues with Government, develop industry guidelines and share best practices. From developing industry safe operating practices, to assessing vulnerability at facilities, to coordinating emergency response training, API and its members are committed in taking a leadership role to ensure the safety and security of our workers, our surrounding communities and to provide a transparent flow of reliable energy that we have all come to expect in our daily lives.

In order to help petroleum companies evaluate and respond appropriately to their potential and real security threats, the American Petroleum Institute has worked with other industry associations, government and private companies to prepare this security guidance. The risks from terrorist attacks to the U.S. energy supply vary by segment of the petroleum industry, which is broadly defined as petroleum exploration and production, petroleum refining, pipeline transportation (liquids), marine transportation, and petroleum products distribution and marketing. This document provides general security guidance and other reference data on applicable regulatory requirements, which can be tailored to meet the differing security needs of the petroleum industry.

This security guidance is by necessity general in nature. It is intended to provide an overview of security issues in the petroleum industry and provide general guidance on effective policies and practices. Individual companies, working cooperatively with local officials, are best suited for conducting detailed assessments of their own facilities and assets and determining how to protect them. This is because both potential threats and appropriate security measures vary based on size, location, facility type and existing security measures already in place. Due to the sensitive nature of this information, security screenings, site-security plans and vulnerability assessments should be protected under the company's confidentiality program to ensure that detailed information regarding vulnerabilities, threats and countermeasures is available only to those who need such information.

Security Guidelines for the Petroleum Industry

1.0 Introduction

In order to assist petroleum companies evaluate and respond to security threats, the American Petroleum Institute has:

- Assessed the general types of security risks to the public and to petroleum supplies that each sector may face due to terrorism;
- Identified existing standards, recommended practices, guidance and other operational practices, as well as ongoing initiatives that may mitigate these risks;
- Developed guidance on conducting Security Vulnerability Assessments (SVA)[a] in the petroleum and petrochemical industries;
- Developed Recommended Practices for security for offshore oil and gas operations.[b]
- Worked with the Federal Government, other industry associations and petroleum companies to prepare appropriate guidance.

1.1 Scope and Objective

The objective of this document is to provide general guidance to owners and operators of U.S. domestic petroleum assets for effectively managing security risks and provide a reference of certain applicable Federal security laws and regulations that may impact petroleum operations.

Domestic petroleum assets are widely distributed, consisting of over 300,000 producing sites, 4,000 offshore platforms, 600 natural gas processing plants, 160,000 miles of liquid pipelines, numerous crude oil and liquefied natural gas (LNG) offloading ports and terminals, 144 refineries, 1,400 finished product terminals, 7,500 bulk stations and 170,000 gasoline retail stations. The vast majority of these assets are small and geographically remote and do not present a significant security risk to the national economy, national security or public safety. However, the petroleum industry supports taking prudent measures to effectively minimize security risks posed by acts of terrorism where warranted.

Certain petroleum facilities are covered by the Maritime Transportation Security Act of 2002 (MTSA), which was signed into law on November 25, 2002. In compliance with MTSA, the U.S. Coast Guard has promulgated federal rules under 33 *CFR* Subchapter H, Parts 101 – 106 that cover port, OCS and vessel security. These regulations require certain vessels and port facilities that could be involved in a transportation security incident prepare a vessel or facility security plan and submit it to the USCG. See Appendix A for a reference table of Federal security regulations that affect the U.S.

1.2 Organization of the Document

This document is organized into seven chapters plus three Appendix items for reference. Chapter 1.0 describes the objectives, intended audience, and scope of the guidance and the various references for other security regulations. Chapter 2.0 includes an overview of terrorism and the petroleum industry. Chapter 3.0 describes a process for a threat assessment including the use of security intelligence and threat-based countermeasures systems such as the Department of Homeland Security Alert System (HSAS) and the USCG Maritime Security (MARSEC) levels. Chapter 4.0 describes the elements of a

[a] American Petroleum Institute/National Petrochemical and Refiner's Association Guidance "Security Vulnerability Assessment Methodology, October, 2004"

[b] API RP 70 *Security for Offshore Oil and Natural Gas Operations,* First Edition, March 2003 and RP 701 *Security for International Oil and Natural Gas Operations*, First Edition, May 2004.

security plan and provides a plan outline. Chapter 5.0 includes an overview of security vulnerability assessment. Chapter 6.0 includes security conditions and potential response measures. Chapter 7.0 provides an overview of information (cyber) security. The Appendix items provide useful reference information such as a matrix of certain Federal laws and regulations on security and a glossary of terms and references used to develop this document.

1.3 Underlying Basis of this Guidance

Owners and operators in the petroleum industry can enhance the security of their assets and continuity of business operations through the effective management of security risks. By considering site-specific circumstances, security risks can be managed through a risk-based, performance-oriented management systems approach. The foundation of a security management systems approach is to identify and analyze security threats and vulnerabilities, and to evaluate the adequacy of countermeasures provided to mitigate the threats. Security Vulnerability Assessment (SVA) is a management tool that is flexible and adaptable to a wide range of applications and can be used to assist management is identifying and prioritizing security risks and determining the appropriate type and level of protection required at the local asset level.

The need for and type of security enhancements will be determined based on site-specific factors such as the degree of the threat, the degree of vulnerability, the potential consequences of a security event, and the attractiveness of an asset to an adversary. In the case of the terrorist threat, higher-risk sites are those that have critical importance, are attractive targets to the adversary, have a high level of potential consequences, where assets are vulnerable and the threat is great. In these high-risk situations, security enhancements/countermeasures should be considered that reduce one or several of these items to an acceptable level.

Appropriate strategies for managing security risk can vary widely depending on site-specific factors such as the type of facility (fixed or mobile/remote or urban), the operation involved, the type of substances being stored and processed, and the threats facing the facility. As a result, this guidance does not prescribe specific security measures but provides a means of identifying, analyzing, and reducing vulnerabilities based on the unique needs of the location. Each facility should be evaluated individually by management using the best judgment of applicable practices and appropriate security risk management decisions should be made commensurate with the risks. This recognizes that there isn't a uniform approach to security in the petroleum industry, and that resources should be used effectively to reduce high-risk situations on a priority basis. It is recognized that while all security risks cannot be completely eliminated it can be significantly reduced through implementing an effective security risk management program. The security objectives are to employ four basic strategies to manage the risk, including, Deter, Detect, Delay, and Respond.

All owner/operators are encouraged to seek out assistance and coordinate efforts with federal, state, and local law enforcement agencies, and with the local emergency services and Local Emergency Planning Committee as applicable. Owner/Operators can also obtain and share intelligence, coordinate training, and utilize other resources to help deter attacks and to manage emergencies.

1.4 Other Guidelines and Security References

API has developed this guidance for the petroleum industry as a reference to be used with other available sources. This document does not attempt to provide an all-inclusive list of security considerations, but more as a basis for what might be considered when evaluating and implementing security measures. Additionally, it is recognized that certain information included in a security program needs to remain confidential. Petroleum companies should consider a confidentiality program to understand what information can be shared and what should remain confidential. Other available resources on security include:

- American Petroleum Institute RP 70, *Security for Offshore Oil and Natural Gas Operations*, 1st Ed., April 2003.
- American Petroleum Institute RP 70I, *Security for Worldwide Offshore Oil and Natural Gas Operations*, 1st Ed., May 2004.
- American Petroleum Institute Std 1164, *SCADA Security*, 1st Ed., September 2004.
- American Petroleum Institute / National Petrochemical and Refiners Association, "Security Vulnerability Assessment Methodology," October 2004.
- American Chemistry Council, "Site Security Guidelines for the U. S. Chemical Industry," 2001.
- American Chemistry Council, "Implementation Resource Guide for Responsible Care Security Code® of Management Practices: Value Chain Activities," 2003.
- American Chemistry Council, "Transportation Security Guidelines for the U.S. Chemical Industry," 2001.
- American Institute of Chemical Engineers (AIChE) Center for Chemical Process Safety (CCPS®), "Guidelines for Managing and Analyzing the Security Vulnerabilities of Fixed Chemical Sites," August 2002.[1]
- DOT, Office of Pipeline Safety, "Pipeline Security Information Circular, Information of Concern to Pipeline Security Personnel, *Security Guidance for Natural Gas, and Hazardous Liquid Pipelines and Liquefied Natural Gas Facilities*," September 5, 2002.
- Sandia National Laboratories, "Vulnerability Assessment Methodology for Chemical Facilities (VAM-CF)".
- U.S. Coast Guard NVIC 11-02 (and other NVICs).

In addition to these references, owners and operators should be aware of applicable local and national laws and regulations. See the reference table included in Appendix A for a list of final security regulations impacting the petroleum industry that were enacted prior to the release of this document.

2.0 Overview of Terrorism and the Petroleum Industry

2.1 Background on Terrorism and Security

The FBI defines terrorism as, "the unlawful use of force or violence against persons or property to intimidate or coerce a Government, the civilian population, or any segment thereof, in furtherance of political or social objectives." The number of international terrorist incidents has increased in recent years and the potential threat posed by terrorists has increased[2]. All sectors of the U. S. economy are potentially subject to these illicit activities.

2.2 Threat to the Petroleum Industry

Reports from the Department of Homeland Security (DHS), the U. S. Department of State[3], the Federal Bureau of Investigation (FBI), have indicated that the petroleum industry may be a target of terrorism due to the inherent nature of the products used and its importance on the national infrastructure. Specifically, the petroleum industry may be a target for terrorism due to the following characteristics:

- The physical and chemical properties of the products handled at petroleum sites
- The importance of petroleum to the national economy
- The importance of petroleum to national security
- The symbolism of the industry as a cornerstone of capitalism and western culture.

Fortunately there is little experience with actual terrorism in the U.S. However, this fact poses a challenge for domestic petroleum owners/operators. As a result, government and industry are working together to better protect the national infrastructure and our national security. Facility owners and operators should establish a close relationship with various sources of intelligence, both at the local and national levels. Certain key sources of intelligence include: the local law enforcement, regional FBI offices, emergency response organizations, USCG Office of Intelligence and Investigations and the Energy ISAC. By providing certain basic awareness training, employees and members of the pubic can act as the watchful eyes and ears for the company by reporting suspicious activity in and around the facility. Lastly, most domestic petroleum companies operate internationally and in remote regions of the world where security has historically been a significant concern. Domestic firms should where possible, tap that experience to help strengthen its domestic security program.

3.0 Threat Assessment

3.1 The Value of Threat Assessment

Threat assessment is an important part of a security management system. This chapter describes a threat assessment approach as part of a security management system process. In chapter 5.0 the use of threat assessment in the SVA is explained in greater detail.

A threat assessment is used to evaluate the likelihood of an attack against a given asset or group of assets.[4] It is a decision support tool that helps to establish and prioritize security-program requirements, planning and resource allocations. A threat assessment identifies and evaluates each threat on the basis of various factors, including capability, intent, and impact.

Threat assessment is a process that should be systematically performed and kept current to be useful. The determination of these threats posed by different adversaries leads to the recognition of vulnerabilities and to the evaluation of required countermeasures to manage the threats. Without a specific threat in mind, a company cannot effectively develop a cost-effective security management system.

3.2 Threat Assessment Process

In characterizing the threat to a facility or a particular asset for a facility, a company examines the historical record of security events and adversaries and obtains available general and localized threat information from government organizations and other sources. It then evaluates these threats in terms of company assets that represent more likely, higher payout targets to those adversaries.

Certain threats are assumed continuous, whereas others are assumed to be variable. As such, this guidance follows the Department of Homeland Security's Homeland Security Advisory System (HSAS) for management of varying threat levels to the industry, which is further explained in section 3.4. It should be noted that other agencies and groups (e.g., the USCG MARSEC Levels) have established threat levels other than HSAS. While these systems differ in the number and description of the threat levels, they provide essentially the same information and may be correlated. The threat assessment determines the estimated general threat level, which forms a baseline. Then intelligence and threat assessment helps to evaluate situations as they develop. Depending on the increased threat level, different security measures above baseline may be necessary.

While threat assessments are key decision support tools, it should be recognized that, even if updated on a regular basis, threat assessments might not adequately capture emerging threats posed by some terrorist groups. Consequently, a threat assessment must be accompanied by a vulnerability assessment to provide better assurance of preparedness.

Intelligence and law enforcement agencies assess the foreign and domestic threats to the United States. The U.S. intelligence community—which includes the Central Intelligence Agency, the Defense Intelligence Agency, and the State Department's Bureau of Intelligence and Research, among others—monitors the foreign-origin terrorist threat to the United States. The Terrorist Threat Integration Center was established to gather and coordinate information and assess the threat posed by domestic sources of terrorism. [5]

Threat information gathered by both the intelligence and law enforcement communities can be used to develop a company-specific threat assessment. However, it should be understood that much of this information is classified and will not be readily accessible without a security clearance. A company should attempt to identify threats in order to decide how to manage risk in a cost-effective manner. Many companies are exposed to a multitude of threats, including terrorism or other forms of threat. A threat assessment can take different forms, but the key components include:

1. the identification of known and potential adversaries, where such information is available and accurate;
2. the recognition and analysis of their intent, motivation, operating history, methods, weapons, strengths, weaknesses, and intelligence capabilities;
3. the assessment of the threat posed by the adversary factors mentioned above against each asset, and the assignment of an overall criticality ranking for each adversary.

Threats need to be considered from both insiders and outsiders, or a combination of those adversaries working in collusion. An external adversary uses unauthorized access to the facility and systems to destroy or steal a target asset. Insiders are defined as those individuals who normally have authorized access to the asset. Insiders pose a particularly difficult threat, due to the possibility for deceit, deception, training, knowledge of the facilities, and unsupervised access to critical information and assets.

The threat categories that should be considered are those that have the intent and capability of causing major catastrophic harm to the facilities and to the public or environment. Four typical threats that may be included in a SVA are the threat posed by international terrorists, domestic terrorists including disgruntled individuals/'lone wolf' sympathizers, disgruntled employees, and extreme activists. Other adversaries may need to be evaluated as appropriate.

All companies are encouraged to discuss threats with local and Federal law enforcement officials, and to maintain networking with fellow national, regional, and local industrial groups to improve the quality of information relied upon. In particular, owner/operators should coordinate with the Joint Terrorism Task Force offices.

The threat assessment is not necessarily based on precise information. In fact, for most facilities, the best available information is vague or nonspecific to the facility. A particularly challenging part of the analysis can be the absence of site-specific information on threats, particularly the recent concern for international terrorism. A suggested approach is to make a threat assumption that international terrorism is possible at every facility that has adequate attractiveness to that threat.

To be effective, threat assessment must be considered a dynamic process, whereby the threats are continuously evaluated for change. During any given SVA exercise, the threat assessment is referred to for guidance on general or specific threats facing the assets. At that time, the company's threat assessment should be referred to and possibly updated given additional information and assessment of vulnerabilities.

3.3 Security Alert Level Systems

3.3.1 Introduction

Flexibility provides the basis of operational security due to the dynamic threat environment and the need to apply variable security measures are employed accordingly. Alert levels describe a progressive measure of the likelihood of terrorist actions, from normal to imminent risk of attack or action, based on government or company intelligence information. There are three relevant alert level systems that have been developed by the government and international sources to warn of potential acts of terrorism:

1. **Homeland Security Advisory System (HSAS)**—This five-level alert system is based on the National Threat Advisory System developed by the Department of Homeland Security.
2. **Maritime Security Levels (MARSEC)**—This three-level alert system was developed by the U.S. Coast Guard for use by marine vessels, ports and port facilities.
3. **International Ship and Port Facility Security (ISPS) Code**—This three-level alert system is similar to the MARSEC system and applies to foreign flagged vessels and ports.

The purpose of these systems is to provide clear information to both the private and public sectors about the potential for a terrorist action and to help implement appropriate response measures during a threat crisis.

3.3.2 Department of Homeland Security Alert System (HSAS)

The Homeland Security Advisory System (HSAS) was established on July 27, 2002. This five level color-coded threat advisory system was designed to improve coordination and communication at all levels of Government and with the American public in the fight against terrorism. HSAS provides a framework to assign threat conditions, which can apply nationally, regionally, by sector or to a specific target. The following factors that may be used to assess the threat are:

- Is the threat credible?
- Is the threat corroborated?
- Is the threat specific and/or imminent?
- What are the potential consequences of the threat?

Threat conditions characterize the risk of a terrorist attack. Protective measures are the steps to be taken by a potential target to reduce their vulnerabilities. The HSAS establishes five threat conditions with associated general protective measures. It must be emphasized that specific protective measures should be developed by the facility based on the unique characteristics of that particular facility and from the findings from a site-specific SVA. Section 6 of this publication provides an in-depth discussion of specific protective measures that owners/operators of petroleum facilities should consider when the national alert level changes.

Following is the HSAS five level alert system and their general protective measures.

Table 3.1—Homeland Security Alert System
Severe Condition—Red: Severe risk of terrorist attacks. In addition to the previous protective measures, the following may be applied:
o Assign emergency response personnel and pre-position specially trained teams;
o Monitor, redirect or constrain transportation systems;
o Close facilities;
o Increase or redirect personnel to address critical emergency needs.
High Condition—Orange: High risk of terrorist attacks. In addition to the previous protective measures, the following may be applied:
o Coordinate necessary security efforts with armed forces or local law enforcement;
o Take additional precautions at public events;
o Prepare to work at an alternate site or with a dispersed workforce;
o Restrict access to essential personnel only.
Elevated Condition—Yellow: Significant risk of terrorist attacks. In addition to the previous protective measures, the following may be applied:
o Increase surveillance of critical locations;
o Coordinate emergency plans with local jurisdictions;
o Assess further refinement of protective measures within the context of the current threat information;
o Implement, as appropriate, contingency and emergency response plans.
Guarded Condition—Blue: General risk of terrorist attacks. In addition to the previous protective measures, the following may be applied:
o Check communications with designated emergency response locations;
o Review and update emergency response procedures;
o Provide the surrounding community with necessary information.
Low Condition—Green: Low risk of terrorist attacks. The following general protective measures may apply:
o Refine and exercise preplanned protective measures;
o Ensure personnel receive training on HSAS, corporate and facility specific protective measures;
o Regularly assess facility vulnerability and take measures to reduce them.

The National Infrastructure Protection Center, U.S. Coast Guard and other agencies publish guidance on protective measures that are recommended for the different threat levels[6].

3.3.3 U. S. Coast Guard Maritime Security Levels

The U.S. Coast Guard has developed a three-level Maritime Security (MARSEC) alert system for use by marine vessels, certain energy facilities and ports. The MARSEC alert levels are:

- **MARSEC I:** Low or Moderate Threat—this alert is defined as the "new normalcy".
- **MARSEC II:** Heightened Alert—this alert is used when there is credible intelligence suggesting a high threat, but no specific target or delivery method is known.

- **MARSEC III:** Maximum Alert—this alert is issued when there is credible intelligence coupled with a specific threat.

The U.S. Coast Guard will communicate heightened levels of alert using Maritime Security levels (MARSEC) 1, 2, and 3 that essentially align with the graduated color-coded threat condition levels defined by the Homeland Security Advisory System (HSAS). MARSEC is the maritime sector's tool for communicating risk and is linked to the HSAS.

MARSEC Level I generally correspond to the lowest three levels of HSAS: Green (Low), Blue (Guarded), and Yellow (Elevated). MARSEC Level 2 corresponds to HSAS Orange (High), and MARSEC Level 3 corresponds to HSAS Red (Incident Imminent).

Facilities should develop and implement protective measures, to be reflected in their security plans, if necessary, which increase as the MARSEC level increases to reduce the risk of a transportation security incident. MARSEC levels may be assigned for the entire nation, or they may be set for a particular geographic area, industrial sector, or operational activity. It should be noted that it is possible to shift from MARSEC 1 directly to MARSEC 3 without an intermediate shift to MARSEC 2.[7]

Section 6.0 provides in-depth discussion of specific protective measures that owners/operators of petroleum assets may consider when the national alert level changes.

3.3.4 International Ship and Port Facility Security (ISPS) Alert Levels

The ISPS code is a three-level alert system similar to the MARSEC system.

Security level 1: (Normal) The level at which the ship or port facility normally operates. Security level 1 means the level for which minimum appropriate protective security measures shall be maintained at all times.

Security level 2: (Heightened) The level applying for as long as there is a heightened risk of a security incident. Security level 2 means the level where appropriate additional protective security measures shall be maintained for a period of time as a result of heightened risk of a security incident.

Security level 3: (Exceptional) The level applying for the period of time when there is the probable or imminent risk of a security incident. Security level 3 means the level for which further specific protective security measures shall be maintained for a limited period of time when a security incident is probable or imminent, although it may not be possible to identify the specific target.

Setting security level 3 should be an exceptional measure, used only when credible intelligence indicates that a security incident is probable or imminent. Security level 3 should only be set for the duration of the identified security threat or actual security incident. While the security levels may change from level 1, through level 2 to level 3, it is possible that the security levels will change directly from security level 1 to security level 3.

4.0 The Security Management System Process

There is a significant variation in the detail and complexity associated with different SVA methods. Many companies without a formal SVA processes may find that an initial screening level SVA can be beneficial in terms of focusing resources on the most important areas. Companies may find that a screening approach is the most practical means to prioritize facilities for SVA. Depending on the nature of the location and its operations, not all facilities may require a formalized SVA and security plan.

Each owner/office should establish a security management system to effectively manage security risks as appropriate. Since all petroleum operations have unique characteristics, the management system should provide for flexibility and continuous improvement due to changing conditions. However, an effective security management system should have a solid base of several essential elements.

Figure 4.1 illustrates an example of a security management system. The decision flow provides a common process to develop and maintain a site-specific security plan. Owner/operators should consider their unique security risks and then assess those risks to ensure that the plan effectively addresses the highest risks first. There are many different approaches to implementing the elements identified in Figure 4.1, ranging along a continuum from simple to complex. There is no "best" approach that is applicable to all petroleum operations for all situations. This guideline recognizes the importance of flexibility in designing security plans and provides guidance commensurate with this need.

Figure 4.1—Security Management System Process

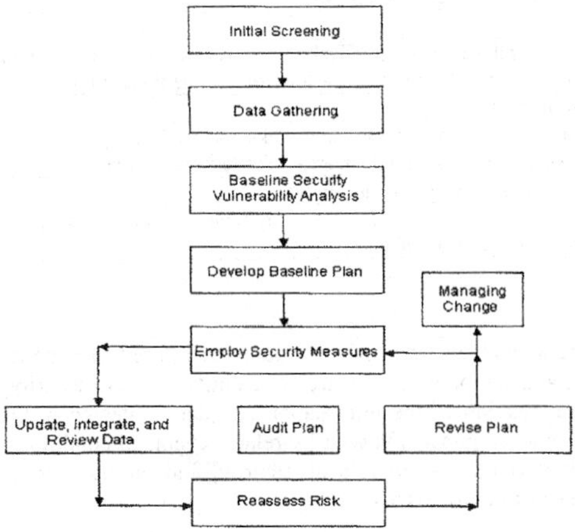

4.1 Initial Screening

An initial evaluation should be conducted prior to launching a formal SVA. The screen should evaluate petroleum facilities at a "systems level" (high level) by considering the potential economic ramifications, public safety and health impacts, national security and the effects on the value chain (interdependencies) as a result of a significant event. If done at a corporate level, screening can be used to help prioritize which facilities would be candidates for further analysis. Screening can also be helpful when evaluating regional impacts. For those facilities that are identified for further evaluation, a formal SVA should be considered that looks at individual assets within the facility and helps to identify and prioritize vulnerabilities that should be addressed.

4.2 Data Gathering

After the initial screening, the first step in an SVA is to assemble information about the location, its assets and any potential threats to those assets. In this element, one performs the initial collection, review, and integration of data that is needed to understand location-specific risks to security. The types of data to support a SVA may include information on the operation, surveillance practices, security measures, and the specific security issues and concerns that are unique. For those that are just formalizing an approach to a security plan, the initial data gathering may be focused on a limited number of assets so that a screening for the most significant security risks can be readily identified.

Table 4.1—Examples of petroleum facility assets subject to potential security risk
Buildings:
Administration offices, corporate offices, control rooms
Equipment:
Process units and associated control systems; product storage tanks; surge vessels, boilers, turbines, process heaters, sewer systems
Support systems:
Utilities such as natural gas lines, electrical power grid and facilities (including back-up power systems), water-supply systems, wastewater treatment facilities
Transportation interface:
Railroad lines and railcars, product loading racks and vehicles, pipelines entering and leaving facility, marine vessels and dock area, off site storage areas
Cyber systems and information technology:
Computer systems, networks, all devices with remote maintenance ports, SCADA systems, laptops, PDAs and cell phones.

4.3 Initial SVA

In this element, the data assembled from the previous step is used to conduct a SVA. The SVA begins with a systematic and comprehensive search to identify possible security risks to the facility. Through the integrated evaluation of the information and data collected in the previous step, the SVA process identifies the location-specific security-related events or conditions, or combinations of events and conditions that could lead to loss of security, and provides an understanding of the likelihood and consequences of these events.

There is a significant variation in the detail and complexity associated with different SVA methods. Some companies without a formal SVA processes may find that an initial screening level SVA can be beneficial in terms of focusing resources on the most important areas. Companies may find a screening approach as the most practical means to prioritize facilities for SVA.

Table 4.2—Examples of security risks or threats in the petroleum industry
• Intentional release (loss of containment) from a process unit or storage tank
• Loss of a critical management team or member
• Destruction or disruption of support systems, such as:
o Electrical power; water supply, sewer systems
o Communications systems, computer systems
o Raw material (crude oil) supply, finished product distribution
• Contamination of raw material or finished product
• Bomb threat or discovery of an Improvised Explosive Device (IEDs) or Vehicle Borne Explosive Devices (VBED)
• Bio-terrorism or eco-terrorism
• Cyber attack
• Vandalism or theft

After identifying the most significant risks next determine what countermeasures should be implemented to reduce or eliminate the risk, and where additional assessment techniques would be of the most value in identifying future risk-threatening issues. The risk control and mitigation process may involve:

- Identification of risk control options that lower the likelihood of an incident, reduce the consequences, or both;
- A systematic evaluation and comparison of those options;
- Selection and implementation of a strategy for risk control.

A SVA may also help to identify and prioritize likely targets and avoid expending resources where the likelihood of attack is remote or where the consequence is less than other targets. A tiered, risk-based approach may be the most effective way to evaluate, identify, and prioritize potential targets. There are, however, a number of methods that can be employed to conduct a SVA and identify risk control activities.

Develop Baseline Security Plan. Using the output of the SVA, a plan is developed to address the most significant risks and assess the security of the facility or asset. This plan should include the mitigation risk control actions, as well as security assessment activities (e.g., inspections and traffic and personnel control).

Employ Security Measures. In this element, the baseline security plan activities are implemented, the results are evaluated, and the necessary changes are made to ensure risks that might lead to system failures are controlled. As noted previously, a SVA may identify other risks that should be addressed.

Examples of physical security elements may include, but are not limited to:

- Controlling access into, within and out of a facility or critical asset areas;
- Perimeter protection including immediately beyond the perimeter;
- Security personnel;
- Redundant systems (electrical, water, computing, communications, sewer, gas);
- Mail and package screening system.

Update, Integrate, and Review Data. After the initial security assessments have been performed, the facility will have improved and updated information about the security of the facility. This information should be retained and added to the database of information used to support future SVAs and security evaluations.

Reassess Risk. SVAs should be performed periodically to factor in recent operating data, consider changes to the facility design, and to analyze the impact of any external changes that may have occurred since the last SVA, e.g., adjacent facilities and changes in traffic flow. The results of security assessments, such as inspections and drills, should also be factored into future SVAs to ensure the process reflects the latest understanding of the security issues.

Revise Plan. The baseline security management plan should be transformed into an on-going security assessment plan that is periodically updated to reflect new information and the current understanding of security risks. As new risks or new manifestations of previously known risks are identified, additional mitigation actions to address these risks should be performed, as appropriate. Furthermore, the updated SVA results should also be used to support scheduling of future security assessments.

Audit Plan. Companies should collect information and periodically evaluate the success of their security assessment techniques and other mitigation risk control activities.

Managing Change. A systematic process should be used to ensure that changes to a facility or its operations are evaluated for their potential risk impacts prior to implementation, and to ensure that changes in the environment in which the facility operates are evaluated. After these changes have been made, they should be incorporated, as appropriate; into future SVAs to be sure the SVA process addresses the facility as it is currently configured. As this final element indicates, managing security is not a one-time process. As implied by the loop in the lower portion of Figure 4.1, a security management system involves a continuous cycle of monitoring conditions, identifying and assessing risks, and taking action to minimize the most significant risks. SVAs should be reviewed and revised to reflect current conditions.

It is important to emphasize that a security plan should be a highly integrated and iterative process. Although the elements depicted in Figure 4.1 are shown sequentially for ease in illustration, there is a significant amount of information flow and interaction between the different steps. For example, the selection of a SVA approach depends in part on what risk related data and information are available. Conversely and while performing a SVA, additional data needs are usually identified to better address potential vulnerability issues.

4.4 Example Elements of a Security Plan

Security plans should address a number of key elements related to an organization's security policies, practices, and procedures as well as describe the physical and cyber security features being employed to protect a particular asset. Figure 4.2 is an example of certain key elements that may be considered as part of a security plan. Figure 4.2 was created to be consistent with the Maritime Transportation Security Act (MTSA) as required under the U.S. Coast Guard regulations, 33 *CFR* 105.405. If you are a MTSA covered facility, your FSP requirements may be significantly more stringent than those outlined in this document in Figure 4.2. You are therefore encouraged to review USCG Regulations 33 *CFR* Parts 101-106 for more detailed information about your obligations. For a more comprehensive reference of federal laws and security regulations, please refer to Appendix A.

Figure 4.2—Example Elements of a Security Plan	
1.	Security Administration & Organization
2.	Personnel Training
3.	Drills and Exercises
4.	Records and Documentation
5.	Response to Change in Alert Level
6.	Communications
7.	Security Systems & Equipment Maintenance
8.	Security Measures for Access Control, Including Designated Public Access Areas
9.	Security Measures for Protected/ Controlled/Restricted Areas
10.	Security Measures for Monitoring
11.	Security Incident Procedures
12.	Audits & Security Plan Amendments
13.	Security Vulnerability Analysis (SVA) Report

In general, the security plan should be customized to support each owner/operator's unique needs therefore, not all of the items listed in Figure 4.2 may be necessary at a particular location. It is up to the company determine its security needs based on a sound risk-based decision making process. For more information about security risk-based decision-making, please refer to section 5.0.

The security plan should be periodically evaluated and updated to account for changes in operation, the environment in which the system operates, new data and other security-related information. Periodic plan review and improvement is helpful to take advantage of new information, improved technology, and changes in the operating plan of a facility. For example, the availability of new threat information may require a change in strategy for access control. An effective security plan should be flexible to account for changes in the operating environment and to meet the goals of an organization's management system.

4.4.1 Security Administration and Organization of the Facility

This section of the security plan should identify the Security Officer and/or the person(s) primarily responsible for administering the security program at the location. Other site/company personnel with security responsibilities should also be identified, along with a description of their duties and responsibilities (e.g., a guard force supervisor, other guards, receptionists that confirm the identification of visitors, etc.).

4.4.2 Personnel Training

This section of the security plan should describe the security-related training provided to the Security Officer(s) and/or the person(s) primarily responsible for administering the security program at the location. Training for other site/company personnel with security responsibilities should also be identified as well as other security awareness training provided to employees at the location.

For efficiency purposes it is noted that many EHS-training topics, have a direct or peripheral relationship to security (e.g., emergency response, particularly in a petroleum handling/processing facility). These topics should also be described as appropriate. For MTSA facilities, the USCG Regulations under 33 *CFR* 105.205 provide a list of qualifications for Facility Security Officers (FSOs), other persons with security duties, and all other employees respectively. Note that these comprehensive lists of skills do not all have to be explicit training topics. They can be obtained through either training and/or experience. The training for all other employees of the site is orientation and security awareness, stressing the notion that all employees need to develop a healthy level of skepticism about what they see and hear on or adjacent to the site while performing their normal duties.

4.4.3 Drills and Exercises

This section of the security plan should describe the planned activities that rehearse aspects of the security plan and any procedures that support the plan. Each location should determine the extent and frequency required to conduct security drills and exercises. Based on a security risk assessment, a specific location may find that no drills or exercises are warranted, others may find that short, focused activities that test one portion of the security program and involve one person or group and their duties (e.g., vehicle searches by main gate guards) will be sufficient, while higher risk sites may require full-scale roll-out or table-top exercises involving multiple groups and offsite responders.

Many of these activities may share the same goals, the same onsite personnel and the same offsite responders as those required for environmental, health, or safety (EHS) related events. Again, efficiency should be considered to minimize any duplication and to leverage existing programs and activities.

For MTSA facilities, the USCG regulations require certain drills and exercises at defined maximum intervals. Many EHS laws and regulations have similar requirements. For example, a petroleum processing facility may be covered by the Oil Pollution Act, SARA Title III regulations, and possibly OSHA and EPA requirements. It is suggested that the EHS and security staffs at the site and corporate levels reconcile these requirements and devise a drill and exercise plan that meets all regulatory requirements simultaneously, including documentation. This plan should then be incorporated into or referenced by the security plan. The security plan should describe, in general terms, the follow-up process for drill and exercise critique action items. If this is the same process that used to resolve EHS-related recommendations and action items, this information can be referenced to the appropriate procedures, databases, or other documents.

In addition to facility drills and exercises, the company's crisis management plan (CMP), if applicable, should also be described in this section of the security plan, to the extent that the security program of the site will rely on the CMP as part of its security program, and what information and support the CMP describes will be provided by the individual site(s). The site emergency response plan(s) and the company CMP are also described and referenced in the security incident procedures section of the security plan.

4.4.4 Records and Documentation

This section of the security plan should describe what security-related records will be kept and how they will be protected from unauthorized disclosure. To the extent possible, existing EHS, quality, and other recordkeeping systems should be utilized to avoid duplication and overlap. Many petroleum facilities have thorough recordkeeping systems already in place for EHS and/or ISO purposes. Therefore, this section of the security plan should describe how the existing documentation systems will be modified to include security-related matters, and who has the responsibility for maintaining the security records, as well as record retention policies for security-related records. MTSA facilities have eight (8) specific types of records that must be kept.

4.4.5 Response to Change in Alert Level

This section of the security plan should describe the security alert system in use at the site or company, whether it is the Department of Homeland Security (DHS) Homeland Security Advisory

System color-coded system, U.S. Coast Guard Maritime Security (MARSEC) levels, International Ship & Port Security (ISPS) Code Security Levels, or a company-specific system. Specifically, the security plan should describe what the site would do at each level in the alert system. For example, if the site uses the DHS HSAS alerts, the plan should describe what additional security measures will be employed if the alert level is elevated from Yellow to Orange. Since most of the alert systems are maintained by external government organizations, the security plan should also describe how changes in alert levels are recorded and the time taken to achieve the declared level. Even in the absence of direct regulatory requirements (e.g., the MTSA 12 hour limit to achieve declared level), the site or company might be asked to report this time interval to external organizations. Refer to section 3.4 of this guidance for a more thorough discussion of alert levels. Refer to section 6.0 for certain example response measure related to changes in the alert level.

4.4.6 Communications

This section of the security plan should describe the necessary communications capabilities of the facility with respect to implementing the security plan. Certain elements to consider are:

- Communications capabilities between employees (e.g., radio, telephone, etc.).
- Communications between the facility and offsite responders or support (e.g., 911).
- Communications between vessels and the facility, if applicable.
- Communication of data, including which computer systems and networks are critical to security (e.g., process control systems; electronic access control systems, etc.), including a general description of the cyber security provisions for these systems.

It should be noted that not all of these elements might be appropriate for a specific location. For example, a small low-risk, unmanned, remote facility may require periodic checks on a weekly or monthly basis.

4.4.7 Security Systems and Equipment Maintenance

This section of the security plan should describe the inspection, test, and preventive maintenance program for security equipment (e.g., camera systems, lighting fencing, etc.).

4.4.8 Security Measures for Access Control, Including Designated Public Access Areas

This section of the security plan should include the policies, practices, and procedures that are important to effectively implement the security plan. The following is a list of items to consider. It should be cautioned that not all of these elements may be appropriate for a specific location.

- Identification requirements for employees, visitors, contractors, truck drivers, railroad crews, government employees/law enforcement and other who may seek access.
- Sign-in or documentation of access procedures.
- Escorting policies for visitors, contractors, and government employees. (Circumstances when escorts are required and the procedures to be followed under each situation.)
- Screening and searching procedures for vehicles, baggage (accompanied and unaccompanied), hard-carried articles.
- Physical security measures applicable to access control (Fencing/barriers, locks, lighting, intrusion detection, etc.).
- Physical barriers that prevent vehicles from being used as weapons.
- The escalation in the implementation of access control procedures as alert levels escalate (How vehicle search procedures change as alert levels rise).

4.4.9 Protected/Controlled/Restricted Areas

If the location designates certain areas as protected, controlled or restricted, then the physical security measures pertinent to those areas should be described this section of the plan.

4.4.10 Security Measures for Monitoring

This section of the plan should describe how the facility is monitored for unauthorized access. Monitoring can be done through a variety of methods to meet the needs of a particular location. For remote facilities that are considered less attractive, frequency of operational checks may be sufficient. For more sophisticated facilities, a combination of personnel monitoring (gaurds and dogs) and technology (intrusion detection) may be more appropriate. As with access control measures, the security plan should describe how the monitoring equipment, personnel, and procedures change as alert levels escalate. For example, if the facility employs off-duty law enforcement officers at "Orange" alert, then this arrangement should be described in the security plan.

4.4.11 Security Incident Procedures

This section of the plan should define what events constitute a breach of security, who is to be notified and the order of such notification. Additionally, the plan should describe the procedure to conduct an investigation of security breaches and incidents (note that this procedure may require some modification to include security related incidents within its scope and to define unique requirements for such investigations). This section should also generally describe or reference the site emergency response plan and the company crisis management plan, if applicable.

4.4.12 Audits and Security Plan Amendments

This section of the security plan should describe how the plan should be audited, including periodicity, audit team leadership/membership, documentation, and follow-up of findings. For MTSA facilities, the USCG regulations contain specific provisions for security plan audits. Non-MTSA facilities may wish to develop their own or use existing HES auditing.

Following an audit, or for other reasons, the security plan may require amending. The process for generating security plan amendments, how they are approved (both internally, and possibly by external organizations) should be described. The USCG regulations contain a defined interface process between the Coast Guard and the facility to amend a security plan. If the facility is not USCG regulated and is ISO-9000 certified, the ISO process for maintaining controlled documents, or an equivalent may be used.

4.4.13 Security Vulnerability Analysis (SVA) Report

This section of the plan may include the SVA report as an attachment, a summary of the SVA, or reference the SVA report. The SVA contains the basis for many of the other items described in the security plan and hence becomes a part of the plan. This includes the need to keep the SVA current, as well as the security plan itself. If the facility is Coast Guard regulated, the SVA is referred to as a Facility Security Assessment (FSA) and accomplishes the same purpose as a SVA. Additionally, if the facility is Coast Guard-regulated, the completed Facility Vulnerability and Security Measures Summary (Form CG-6025) must also be included in the security plan. (Refer to Chapter 5.0 for more information on security vulnerability assessment.)

5.0 Security Vulnerability Assessment (SVA) Concepts

5.1 Security Vulnerability Assessment Overview

Security Vulnerability Assessment (SVA) is a systematic process that evaluates the likelihood that a threat against a facility or asset will be successful and considers the potential severity of consequences to the facility itself, to the surrounding community and on the energy supply chain. One purpose of an SVA is to identify countermeasures that may reduce the risk of an attack and its potential consequences.

There are several SVA techniques and methods available, all of which share common elements. Ultimately, it is the responsibility of the owner/operator to choose the SVA method and depth of analysis that best meets the facility's needs. Differences in geographic location, type of operations, and on-site quantities of hazardous substances, if any, all play a role in determining the level of SVA and the approach taken. Examples include:

1. ***Characterize the facility*** to understand what critical assets need to be secured, their importance and their interdependencies and supporting infrastructure, and the consequences if they are damaged or stolen.
2. ***Identify and characterize threats*** against those assets and evaluate the assets in terms of attractiveness of the targets.
3. ***Identify potential security vulnerabilities*** that threaten the system's service or integrity.
4. ***Determine the risk*** represented by these events or conditions by determining the likelihood of a successful event and the consequences of an event if it were to occur.
5. ***Rank the risk*** of the event occurring and, if high risk, make recommendations for lowering the risk.
6. ***Identify and evaluate risk mitigation options*** and re-assess risk.

The objective of conducting an SVA is to identify security hazards, threats, vulnerabilities and countermeasures that will provide for the protection of the public, workers, national interests, the environment, and the company.

Owner/operators may use any appropriate security vulnerability assessment methodology that effectively achieves this objective. Following are a few published methodologies that are currently available for this use:

- API RP 70 *Security for Offshore Oil & Natural Gas Operations*, 1st Ed., March, 2003
- API RP 70I *Security for International Oil and Natural Gas Operations*, 1st Ed., April 2004
- API/NPRA *Security Vulnerability Assessment Methodology*, September 2004
- American Institute of Chemical Engineers (AIChE) Center for Chemical Process Safety (CCPS®) "Guidelines for Managing and Analyzing the Security Vulnerabilities of Fixed Chemical Sites, August 2002"[8]
- Sandia National Laboratories Vulnerability Assessment Methodology for Chemical Facilities (VAM-CF)
- USCG NVIC 11-02

This guidance should also be considered in light of any applicable governmental security regulations and other guidance as outlined in Appendix A, Regulatory Matrix.

The SVA process may be used to assess a wide range of security issues such as those listed in Figure 5.1.

Figure 5.1—Security Events Evaluated During the API SVA Process
1. *Loss of containment* of toxic substances or flammable hydrocarbons at the facility from intentional damage of equipment or the malicious release of these materials, which may cause multiple casualties, severe damage, and public or environmental impact. 2. *Theft* of toxic substance or flammable hydrocarbons with the intent to cause severe harm at the facility or offsite. 3. *Contamination* or spoilage of products to cause workers or public harm on or offsite. 4. *Degradation* of assets or infrastructure or the business function or value of the facility or the entire company through destructive malevolent acts.

If a facility is covered under USCG regulations 33 *CFR* 101 through 106, there are specific security events that need to be evaluated as part of the SVA. Please refer to the applicable parts of the regulation and U.S. Coast Guard NVIC 11-02 for details on these events, as they are specific to the type of vessel/facility/operation.

5.2 Steps In the SVA Process

Figure 5.2 presents the SVA process flow diagram from the API/NPRA Security Vulnerability Assessment Methodology. It should be noted that this approach to conducting security vulnerability assessments has been developed specifically for the petroleum and petrochemical industries. Other valid approaches, such as outlined in API RP 70 and RP 70I, have been developed and are being used successfully within the petroleum industry as mentioned in Section 5.1 above. To obtain a copy of the "API/NPRA SVA Methodology" contact:

American Petroleum Institute 1220 L. Street, N.W. Washington, DC 20005 (202) 682-8000 www.api.org	National Petrochemical and Refiners Association 1899 L. Street, N.W. Washington, D.C. 20036 (202) 457-0480 Attn: Maurice McBride

Figure 5.2—API/NPRA Security Vulnerability Assessment Methodology

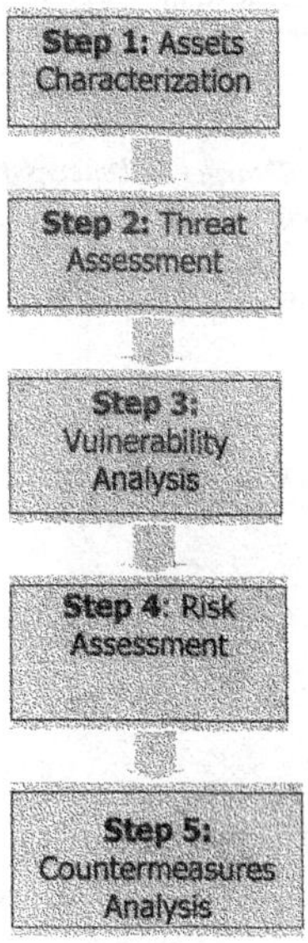

1.1 Identify critical assets and infrastructure.
1.2 Evaluate existing countermeasures.
1.3 Evaluate severity of impacts.

2.1 Adversary identification.
2.2 Adversary characterization.
2.3 Target attractiveness.
2.4 Select targets for further analysis.

3.1 Define scenarios and evaluate specific consequences.
3.2 Evaluate effectiveness of existing security measures.
3.3 Identify vulnerabilities and estimate degree of vulnerability.

4.1 Estimate likelihood of attack by vulnerability, threat, and attractiveness.
4.2 Evaluate risk and need for additional countermeasures.

5.1 Identify and evaluate countermeasures options.
5.2 Prioritize potential enhancements by cost, effectiveness, and other factors.

5.3 Estimating Risk Using SVA Methods

Risk management principles recognize that risk generally cannot be eliminated, however by enhancing protection from known or potential threats it can be reduced. It is important to make risk decisions about these threats using a systematic method. SVA methods are tools that provide management with risk information based on a thorough, defensible process. However, the quality of the study is dependent on the quality of the inputs and the soundness of the logical relationships inherent in the SVA method used to evaluate the input and output conditions. Much of the threat information that the Government possesses is classified and is not generally available to the public.

5.4 Definition of SVA Terms

5.4.1 Risk Definition for SVA

Security risks are different from safety risks. The concept of threat needs to be understood as a combination of an adversary's capability plus their intent. One without the other, and there is no threat.

The petroleum industry has a great deal of experience in managing risks in the safety arena. In that context, risk is usually expressed as a product of probability and consequences. Traditional risk management has focused on the likelihood of an accidental event. In the security realm, this traditional model begins to break down. In the absence of specific intelligence, it is impossible to be specific about the likelihood of an attack. One conclusion of this reasoning is that there is no risk – a potentially misleading and incorrect conclusion.

For this reason, surrogates to likelihood of attack are necessary. Due to the uncertainty of estimating the likelihood of an attack on any particular location, it is recommended to use several variables to compose an estimate. These are a function of an assumed threat, for example, a terrorist. For the purposes of a SVA, the definition of risk is:

"Risk is an expression of the likelihood that a defined threat will target and successfully exploit a specific vulnerability of an asset and cause a given set of consequences."[9]

Figure 5.3 provides a simple depiction of risk, and Figure 5.4 defines risk for the SVA process.

Figure 5.3—Example Risk Matrix

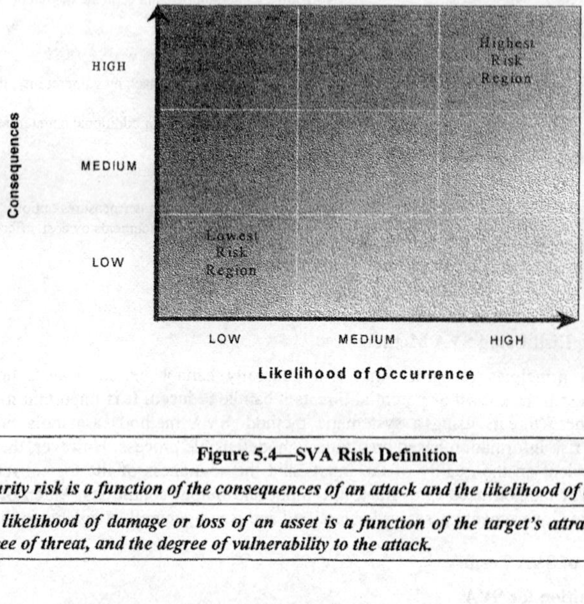

Figure 5.4—SVA Risk Definition
Security risk is a function of the consequences of an attack and the likelihood of the attack.
The likelihood of damage or loss of an asset is a function of the target's attractiveness, the degree of threat, and the degree of vulnerability to the attack.

The risk variables are defined as shown in Figure 5.5.

Figure 5.5—SVA Risk Variables[10]	
Consequences	Consequences are the potential impacts of the event.
Likelihood	The chance of being targeted for attack, and the conditional chance of mounting a successful attack (both planning and executing) given the threat and existing security measures. This is a function of the three variables below.
Threat	Threat is a function of the adversary intent, motivation, capabilities, and known patterns of potential adversaries. Different adversaries may pose different threats to various assets within a given facility.
Vulnerability	Vulnerability is a weakness that can be exploited by an adversary to gain access and damage or steal an asset or disrupt a critical function. This is a variable that indicates the likelihood of a successful attack given the intent to attack an asset.
Target Attractiveness	Target Attractiveness is a surrogate measure for likelihood of attack. This factor is a composite estimate of the perceived value of a target to the adversary and their degree of interest in attacking the target.

A high-risk event is represented by a high likelihood of a successful attack against a given critical target asset. Likelihood is determined by its attractiveness to the adversary, the degree of threat, and the degree of vulnerability. Criticality is determined by the asset's importance or value, and the potential consequences if attacked. If the likelihood of a successful attack is high, then the risk is considered high and appropriate countermeasures would be required for a high-risk asset.

For the SVA, the risk of the security event is estimated qualitatively. It is based on the consensus judgment of knowledgeable people as to how the likelihood and consequences of an undesired event scenario compares to other scenarios. The assessment is based on best available information, using experience and expertise to make sound risk management decisions. The company may use a risk matrix, which is a graphical representation of the risk factors, as a tool for risk assessment decisions.

5.4.2 Consequences (C)

The severity of the consequences of a security event at a facility is generally expressed in terms of the degree of injury or damage that would result if there was a successful attack. They may involve effects that are more severe than expected with accidental risk. Several examples of relevant consequences in a SVA include:

- Injuries to the public or to workers.
- Severe environmental damage (such as contamination of drinking water).
- Direct and indirect significant financial losses to the company.
- Disruption to the national, regional, or local operations and economy.
- Loss of business viability.

The estimate of consequences may be different in magnitude or scope than is normally anticipated for accidental releases. In the case of security events, adversaries are determined to maximize damage, so a worst case credible security event should be defined. Critical infrastructure may have dependencies and interdependencies that need careful consideration.

In addition, theft of hazardous materials should be included in SVAs as applicable. Terrorists may be interested in theft of hazardous materials to either cause direct harm at a later date or possibly to make chemical weapons using the stolen materials as constituents.

Consequences are used as one of the key factors in determining the criticality of the asset and the degree of security countermeasures required. During the initial screening, consequences and attractiveness are used to screen low value assets from further consideration.

5.4.3 Threat (T)

Threat can be defined as any indication, circumstance, or event with the potential to cause loss of, or damage, to an asset.[11] It can also be defined as the intention and capability of an adversary to undertake actions that would be detrimental to valued assets. Sources of threats may be categorized as:

- Terrorists (international or domestic),
- Activists, pressure groups, single-issue zealots,
- Disgruntled employees,
- Criminals (e.g., white collar, cyber hacker, organized, opportunists).

Adversaries may be categorized as occurring from three general groups:

- Insider threats,
- External threats,
- Insiders working as colluders with external threats.

Threat information is gathered and used during the SVA process as an important reference point. To assess an adversary's capability and intent, one must understand what may motivate them. A company should consider a range of threats and then look at their system's vulnerabilities to each type of threat. That assessment will determine the areas where an company will need additional help and information from federal, state, and local governments.

5.4.4 Vulnerability (V)

Vulnerability is any weakness that can be exploited by an adversary to gain unauthorized access and subsequent destruction or theft of an asset.[12] Vulnerabilities can result from, but are not limited to, weaknesses in current management practices, physical security, or operational security practices. In a SVA, vulnerabilities are evaluated either by broadly considering the threat and hazards of the assets they could attack or affect, or analyzed by considering multiple potential specific sequences of events (a scenario-based approach).

5.4.5 Target Attractiveness (A_T)

Not all targets are of equal value to adversaries. A basic assumption of the SVA process is that target attractiveness is one factor that influences the likelihood of a security event. Target attractiveness is an estimate of the real or perceived value of a target to an adversary based on such factors as shown in Figure 5.6.

During the SVA, the attractiveness of each asset should be evaluated based on the adversary's intents or anticipated level of interest in the target if known. Security strategies can be developed around the estimated targets and potential threats.

Figure 5.6—Target Attractiveness Factors
Type of effect:
• Potential for causing maximum casualties
• Potential for causing maximum damage and economic loss to the facility and company
• Potential for causing maximum damage and economic loss to the geographic region
• Potential for causing maximum damage and economic loss to the national infrastructure
Type of target:
• Usefulness of the process material as a weapon to cause collateral damage
• Proximity to a national asset or landmark
• Difficulty of attack including ease of access and degree of existing security measures
• High company reputation and brand exposure
• Iconic or symbolic target
• Chemical or biological weapons precursor chemical
• Target recognition

5.5 Characteristics of a Sound SVA Approach

It is important to distinguish between a security risk management process and a SVA method. Security risk management is the overall process that includes the SVA, development and implementation of a security plan, and reintegration of data into subsequent SVAs. SVA is the estimation of risk for the purposes of decision-making. SVA methods may be very powerful analytical tools to integrate data and information, and help understand the nature and locations of risks of a system. However, SVA methods alone should not be relied upon to establish risk, nor solely determine decisions about how risks should be addressed. SVA methods should be used as part of a process that involves knowledgeable and experienced personnel that review the input, assumptions, and results. This review should integrate the SVA output with other factors, the impact of key assumptions, and the impact of uncertainties created by the absence of data or the variability in assessment inputs before arriving at decisions about risk and actions to reduce risk.

5.6 First Step in the SVA Process

After obtaining management approval and authorization to proceed, a typical first step in all SVA approaches is to collect a representative group of company experts, and outside experts if needed, to identify potential security related events or conditions, the consequences of these events, and the risk reduction activities for the company's system. These experts draw on the years of experience, practical knowledge, and observations from experienced field operations and maintenance personnel in understanding where the security risks may reside and what can be done about them. Such a company group typically consists of representation from: company security, risk management, operations, engineering, safety, environmental, regulatory compliance, logistics/distribution, IT and other team members as required. This group of experts will focus on the potential problems and risk control activities that would be effective in a facility security plan. The primary goal of this group is to capture and build into the SVA method the experience of this diverse group of individual experts so that the SVA process will capture and incorporate information that may not be available in typical operator databases.

There are a number of techniques employed by these expert teams that have proven useful in assuring a systematic and thorough review. These include:

- Free-form brainstorming of issues and potential risks.
- Conducting an asset-by-asset review.

- Using checklists or structured question sets designed to solicit information on a comprehensive list of potential risks and integrity issues, and
- Using simple risk matrices to qualitatively portray and communicate the likelihood and consequences of different security related events.

For each potential security threat or risk factor, the characteristics or variables that potentially could impact risk (both beneficially and adversely) are identified. During the SVA process, specific risk increasing characteristics of the system are either external variables (e.g., outside influences acting on the system), or operation variables (e.g., characteristics associated with the physical properties). In either case, these variables are features of the in-service system and are not easily altered. Variables should be considered individually based on how they impact a specific risk factor. This means that variables could be used in different ways and with potentially contradictory influences within the SVA.

5.7 SVA Strengths and Limitations

Each of the SVA methods commonly used has its strengths and limitations. Qualitative methods are well suited for making good sound security management decisions at the local asset level. In selecting an appropriate SVA method, there are a number of questions that should be considered. Some of the more significant ones are summarized below.

Table 5.1—Questions to determine SVA Approach Needed
• Does the scope of the SVA method identify significant security related events and risks of the facility or along the system? If not, how can the risks that are not included in the SVA method be assessed and integrated in the future?
• Will all data be assessed, as it really exists along the system? Data should be location specific so that additive effects of the various risk variables can be determined. Can the assessment resolution be altered, e.g. station-by-station or mile-by-mile, dependent on the evaluation needs?
• Does the SVA method use numerical weights and other empirical factors to derive the risk measures and priorities? Are these weights based on the experience of the system, operator, industry, or external sources?
• Do the basic input variables of the SVA method require data that is available to the company? Do data systems and industry data updating procedures provide sufficient support to apply the SVA method effectively? What is the process for updating the SVA data to reflect changes in the system, the infrastructure, and new security related data? How is the input data validated to ensure that the most accurate, up-to-date depiction of the system is reflected in the SVA?
• Does the SVA output provide adequate support for the justification of risk-based decisions? Are the SVA results and output documented adequately to support justification of the decisions made using this output?

5.8 Recommended Times for Conducting and Reviewing the SVA

Figure 5.7—Times for Conducting and Reviewing the SVA	
1	An initial review of all relevant facilities and assets per a schedule set by the an initial planning process
2	When an existing process or operation is proposed to be substantially changed and prior to implementation (revision or rework)
3	When a significant new process or operation is proposed and prior to implementation (revision or rework)
4	When the threat substantially changes, at the discretion of the owner/operator of the facility (revision or rework)
5	After a significant security incident, at the discretion of the owner/operator of the facility (revision or rework)
6	Periodically to revalidate the SVA (revision or rework)

5.9 Risk Control and Mitigation

SVA methods are also important tools to help owner/operators make cost effective and sound decisions to control security risks on their systems. Once a potential risk has been identified, SVA methods can be used to estimate the expected risk reduction or benefits that will be achieved. Potential capital and maintenance improvement activities may be prioritized to support management decision-making. This section provides an overview of this process.

After the results of the SVA are available, the next step is to examine the most significant risks on the system, as well as other opportunities to more efficiently control risks and determine what mitigation actions might be desirable. The risk control and mitigation process involves:

- Identification of risk control options that lower the likelihood of a security related event, reduce the consequences, or both, i.e., mitigation activities.
- A systematic evaluation and comparison of those options to quantify the risk reduction impact of the proposed project, and
- Selection and implementation of the optimum strategy for risk control.

Typically there are many ways to address a particular risk. For example, improvements or modifications can be made to the system hardware or equipment configuration, operation and maintenance practices, assessment practices, personnel training, control and monitoring methods, emergency response, and interface with the public and other external organizations. This guideline provides a discussion of risk control options that are frequently used to reduce different petroleum sector security risks. In order to find the optimum approach to risk control, it is important that a variety of options, and perhaps a combination of activities be considered rather than just taking the first idea that is proposed or doing what has always been standard practice. This allows management to consider innovative solutions and perhaps new technologies that may be more effective in addressing risk.

After identifying the risk control options available, the next step is to evaluate and compare the effectiveness of the different alternatives. This evaluation and comparison is often performed at more than one level. For example, a company may desire to select the best approach among several options to address a specific risk. In each case, the basis for comparison and ranking should consider both the magnitude of risk reduction benefits expected as well as the resources expended. Many owner/operators use a benefit-to-cost ratio where the benefit is the expected risk reduction to evaluate and rank potential risk control projects. This can provide a simple, easy-to-understand metric that allows projects with diverse benefits to be compared.

When conducting a ranking of projects based on a benefit-to-cost approach, a comprehensive evaluation and comparison process should also include a review of the system risks to be sure that relatively high risks are not overlooked simply because the risk control projects proposed don't have a high benefit-to-cost ratio. This may signal the need to consider other risk control options.[c] The process should also consider the amount of risk reduction being achieved to be sure the most effective projects are being proposed. There are many other practical factors that are typically considered when evaluating and prioritizing activities. These can include:

- Uncertainties in both the risk reduction and cost estimates.
- Technological value of a particular option, e.g., employing a new security camera.
- Human resource and equipment constraints.
- Logistical and implementation issues, e.g., delay in ability of vendor to supply necessary equipment.
- Concerns of government organizations and other external constituencies.

When establishing a SVA program, an operator should consider the many features that are unique to its systems and operations to determine which approach is most appropriate. SVA is a "fact finding", not a "fault finding" system analysis. The ultimate goal of SVA is to identify and prioritize significant security risks in the system so the operator can determine how, where, and when to allocate risk mitigation resources to improve system security. The operator must decide what information could be useful in performing the assessment and how that information can be used to maximize the accuracy and effectiveness of the SVA.

5.10 Risk Screening

Security issues potentially exist at every facility managed by the petroleum industry, but the threat of malevolent acts is likely to be differentiated across the industry. This is captured by the factor known as 'target attractiveness', whereby certain assets are considered to be more likely to be of interest to terrorists than others. Based on many reported threat assessments, intelligence reports, and actual events around the world, these factors can be used to evaluate target attractiveness.[13]

It is likely that most facilities have no specific threat history. A screening process may contain the following factors:

1. Target attractiveness or target value,
2. Degree of threat,
3. Difficulty of attack (function of adversary, current security and vulnerabilities),
4. Potential consequences (casualties, environmental, infrastructure and economic).

These are the same factors as are used for evaluating an individual asset risk, but the difference is that this is done at a generalized facility level for the risk screening.

Note that target attractiveness itself includes the other factors of consequences and difficulty of attack/vulnerability.

Arguably target attractiveness is the dominant factor in determining terrorist risk. Priority should be given to the Attractiveness Ranking when making assessments. In this way resources can be appropriately applied to assets where they are most likely to be important.

---

[c] Although summarized in a linear fashion for this guideline, the risk control and mitigation process, like the risk assessment process, can be highly iterative in nature.

6.0 Security Conditions and Potential Response Measures

This section describes a progressive level of protective measures that may be implemented in response to the possibility of a terrorist threat directed at a petroleum facility, facility assets, and personnel (including contractors) consistent with the Homeland Security Advisory System (HSAS) developed by the Department of Homeland Security. The purpose of the HSAS is to establish standardized alert and response measures for a broad range of threats and to help disseminate appropriate and timely information for the coordination and implementation of the response measures by management and operator personnel prior to and during a threat crisis. The associated response measures may be implemented for each security alert level at a facility.

In addition to HSAS, there are several other threat level systems used by both industry and other agencies. While the MARSEC levels utilize only a 3 Tier system, it may essentially be compared to HSAS with:

- MARSEC 1 equivalent to HSAS Green, Blue and Yellow.
- MARSEC 2 equivalent to HSAS Orange.
- MARSEC 3 equivalent to HSAS Red.

If a system other than HSAS or MARSEC has been implemented by an individual company it most likely has been developed based on HSAS, MARSEC or both and specific guidance contained below should be considered where appropriate.

Each company should be able to advise and communicate to company personnel and others as warranted the security condition at the facility. The potential measures associated with each alert level are not always prioritized but those implemented should be initiated concurrently where practical and as applicable. Facility management should maintain a record of specific actions taken for each alert level. Less attractive facilities, remote facilities, unmanned facilities may employ less stringent measures. Following is a detailed explanation for each alert level and the potential response measures associated with each level:

6.1 Low Condition—Green

This condition exists when there is a low risk of possible terrorist activity or civil unrest. **Green** condition is for normal operating conditions. All measures under **Green** should be maintained indefinitely. Potential measures to consider implementing include:

<u>Access Control/Perimeter Protection</u>

- Have all contractors and visitors check or sign in and out of the facility at designated location(s).
- Ensure existing security measures are in place and functioning such as fencing, locks, camera surveillance, intruder alarms, and lighting as appropriate.

<u>Communications</u>

- Establish emergency communications and contact information with appropriate agencies. Consider redundant emergency communications in both the hardware and the means for contacting agencies.

<u>Training/Policies/Procedures/Plans</u>

- Develop terrorist and security awareness information and provide relevant education to employees on security standards and procedures. Caution employees not to talk with outsiders concerning their facility or related issues.

- Advise all facility personnel to report the presence of unknown personnel, unidentified vehicles, aircraft or watercraft, vehicles, watercraft or aircraft operated out of the ordinary, abandoned packages, and other suspicious activities.
- Incorporate security awareness and information into public education programs and notifications to emergency response organizations as appropriate.
- Survey surrounding areas to determine those activities that might increase the security risks that could affect the facility (e.g., airports, government buildings, other industrial facilities).
- Ensure contingency and business continuity plans are current and include a response to terrorist threats.
- Review existing emergency response plans and modifying them, if required, in light of potential threats.

IT Security

- Develop and implement hardware, software, and communications security for computer-based operating systems.

6.2 Guarded Condition—Blue

This condition exists when there is an increased general threat of possible terrorist activity against the facility or facility personnel, the nature and extent of which are unpredictable, and circumstances do not justify full implementation of higher alert measures. It may be necessary to implement certain selected measures from higher alert levels to address information received or to act as a deterrent. All measures under **Blue** should be maintained as long as the **Blue** threat exists. In addition to the measures suggested by **Green**, the following measures could be considered:

Perimeter Protection/Access Control

- Secure all facilities, buildings and storage areas not in regular use, if possible. Increasing frequency of inspections and patrols within the facility, including the interior of buildings and along the facility perimeter.
- Inspect perimeter fencing and repairing all fence breakdowns. Review all outstanding maintenance and capital projects that could affect the security.
- Reduce the number of access points for and spot-check the contents of vehicles, aircraft, watercraft and personnel. Be alert to vehicles or watercraft parked or moored for an unusual length of time in or near a facility.
- Check designated unmanned sites at more frequent intervals for signs of unauthorized entry, suspicious packages, or unusual activities. Increase surveillance in designated areas.
- Require visitors to check in at a facility office and verifying their identification. Be especially alert to repeat visitors or outsiders who have no apparent business at the facility and are asking questions about the facility or the facility's personnel. Familiarizing facility personnel with vendors who service the facility and investigate unusual changes in vendor personnel.
- Inspect all packages/equipment coming into the facility. Do Not open suspicious packages. Consider reviewing the USPS "Suspicious Mail Alert" and the "Bombs by Mail" publications with all personnel involved in receiving packages.

Communications

- Inform personnel of the change in alert status. Review with employees the operations plans, personnel safety, and security details and logistic requirements that pertain to the increased security level. Implement procedures to provide periodic updates to employees on security measures being implemented that are considered confidential.
- Test security and emergency communications procedures and protocols as appropriate.

Training/Policies/Procedures/Plans

- Review all operations plans, personnel details, and logistics requirements that pertain to implementing higher alert levels.
- Review communications procedures and back-up plans with all concerned.

6.3 Elevated Condition—Yellow

This condition exists when there is an elevated risk of terrorist activity against the facility or facility personnel. All measures under **Yellow** should be maintained as long as the **Yellow** threat exists. In addition to the measures suggested by **Blue**, the following measures could be considered:

Perimeter Protection/Access Control

- Close and lock gates and barriers except those needed for immediate entry and egress. Inspect perimeter and perimeter fences on a regular basis. Ensure that other security systems are functioning and are available.
- Inspect on a more frequent basis the interior and exterior of all critical buildings and around all storage tanks and other designated critical areas.
- Dedicate personnel to assist with security duties to monitor personnel entering the facility and to inspecting the area on a regular basis, reporting to facility management as issues surface.
- Limit visitors and confirm that the visitor has a need to be and is expected at the facility. Escort visitors while at the facility pursuant to the specifics outlined in the security plan.

Communications

- Inform personnel of the change in alert status. Review with employees the operations plans, personnel safety, and security details and logistic requirements that pertain to the increased security level as appropriate. Implement procedures to provide periodic updates to employees on security measures being implemented.
- Check to ensure all emergency telephone, radio, and satellite communication devices are in place and they are operational.

Training/Policies/Procedures/Plans

- Confirm availability of security resources that assist with extended coverage.
- Identify areas where explosive devices could be potentially hidden.
- Instruct employees working alone to check-in on a periodic basis.

6.4 High Condition—Orange

This condition applies when there is a high risk of terrorist attacks or an incident occurs or information is received indicating that some form of terrorist action against the facility or facility personnel is imminent. Implementation of measures in this alert for more than a short period will probably create hardship and affect the routine activities of the facility and its personnel. In addition to the measures suggested for **Yellow**, the following measures could be considered:

Perimeter Protection/Access Control

- Reduce facility access points to the absolute minimum necessary for continued operation.
- Increase security patrol activity such as perimeter patrols and inspections.
- Check security systems such as lighting and intruder alarms to ensure they are functioning. Install additional, temporary lighting if necessary to adequately light all suspect areas or decreasing lighting to detract from the area.
- Prohibit unauthorized or unidentified vehicles/personnel entrance to the facility.

- Inspect vehicles entering the facility, including the cargo areas, undercarriage, glove boxes, and other areas where dangerous items could be concealed pursuant to the specifics outlined in the security plan. Inspect all packages and cargo being delivered by aircraft or watercraft in the same manner.
- Limit access to the facility to those personnel who have a legitimate and verifiable need to enter. Implementing positive identification of all personnel.

Communications

- Advise appropriate agencies that the facility is at an **Orange** alert level and advise of the measures being employed—requesting an increase in the frequency of their patrol of the facility.
- Consider consultation with local authorities about control of public roads and accesses by waterway that might make the facility more vulnerable to terrorist attack if they were to remain open.

Training/Policies/Procedures/Plans

- Continue **Green, Blue** and **Yellow** measures or introduce those that have not already been implemented.
- Develop procedures for shutting down and evacuation of the facility, if considered necessary, in case of imminent attack.
- Ensure that employees not work alone in remote areas or increasing the frequency of call-ins from remote locations.

6.5 Severe Condition—Red

This condition applies when there is a severe risk of terrorist attacks, an attack has occurred in the immediate area which may affect the facility, or when an attack is initiated on the facility and its personnel. Normally, this alert is declared as a localized condition at the facility. In addition to the measures suggested for **Orange**, the following measures could be considered:

Perimeter Protection/Access Control

- Augment security forces. Establish surveillance points and reporting criteria and procedures. Solicit assistance from appropriate agencies in securing the facility and access, if possible. Cooperate with authorities if they take control of security measures.

Training/Policies/Procedures/Plans

- Continue **Orange** and **Yellow** measures or introduce those that have not already been implemented.
- Consider shutting down the facility and operations in accordance with security contingency plans and evaluating security prior to resuming operations if they are temporarily shut down.
- Implement business contingency and continuity plans as appropriate.

7.0 Information (Cyber) Security

7.1 Introduction

The petroleum industry is a worldwide industry that is highly dependent on technology for its communications and operations. Technological advances that promote better efficiency and more automation within the petroleum industry also make information security an increasingly important issue. Technology is an important component of information security but without the integration of policies, procedures, processes and people, technology alone can not provide adequate information security.

It is widely understood that information security is important for office computing systems such as desktop PCs, laptops, servers, software programs, etc. What is less recognized is that computer technology has become pervasive throughout the entire organization, including network access to plant equipment to allow vendors to maintain systems remotely, and remote access connections to process control systems (SCADA) to allow engineers to trouble-shoot problems. In all of these environments, improper controls could allow unauthorized individuals to accidentally or intentionally harm the information assets of the petroleum industry.

To ensure that adequate and appropriate resources are allocated within the information security program, information security activities should be based on a thorough analysis of risks to the confidentiality, integrity and availability of the information assets. A comprehensive information technology security program implemented by member companies improves the security of the petroleum industry as a whole by effectively:

- Identifying and analyzing actual and potential precursor events that could result in cyber security-related incidents;
- Identifying the likelihood and consequence of potential cyber security-related events;
- Providing a comprehensive and integrated means for examining and comparing the spectrum of risks and risk reduction activities;
- Providing a structured, easily communicated means for selecting and implementing risk reduction activities;
- Monitoring program performance with the goal of improving that performance;
- Establishing alert and response measures for a broad range of security threats.

Additionally, the establishment of a communication program between federal agencies and the industry to share threat information also improves the security of the industry by providing an early warning mechanism so appropriate action can be taken in a timely manner.

ISO/IEC International Standard 17799, ***Information technology—Code of practice for information security management***, describes a framework for creating an information security program and forms the basis of this guideline. ISO/IEC 17799 attempts to ensure preservation of confidentiality, integrity and availability of user access, hardware, software and data. The standard describes eight steps of an information security process: create an information security policy; select and implement appropriate controls; obtain upper management support; perform security vulnerability assessments (SVAs), create statements of applicability for all employees; create an information security management system; educate and train staff; and perform regular audits.

This framework has been endorsed by API's Information Technology Security Forum (ITSF) as voluntary guidance to protect the petroleum industry's information assets. The guidance contained herein and in ISO/IEC International Standard 17799 does not attempt to provide an all-inclusive list of information security considerations, but rather a framework for the evaluation and implementation of information security measures. The concepts mentioned in this Introduction are expounded upon in the following section.

7.2 Specific Security Guidelines

7.2.1 Security Policies, Standards and Procedures

Information Security policies, standards and procedures that focus on protecting a company's information technology assets are the foundation of a Security Management process. Policies are a prerequisite for defining the acceptable behaviors that a company desires to promote in protecting its critical information technology assets. Since policies set the tone for the company's culture relative to protecting information and information technology, a policy must have executive management sponsorship, clearly articulate accountabilities and responsibilities, and be communicated to every employee and system user in the company. Company policies should address topics such as:

- Assignment of management responsibilities
- Business conduct and appropriate system use
- E-mail and internet use
- Remote access & third party connectivity
- System monitoring and compliance (audit)
- Physical security (laptops, computer rooms, etc)
- Incident reporting and response
- Data retention
- Business continuity and disaster recovery

The company Information Security Officer or Manager is generally accountable for the development, implementation and maintenance of a company's information security policies. However, it is recommended that this be accomplished by working in "partnership" with representatives from the functional areas of IT Audit, Human Resources, Legal, Corporate Security and Information Technology.

Each policy should be accompanied by a set of standards and procedures that provide guidance for the operational implementation and compliance assessment of the policies. The standards and procedures should be derived from industry technology standards and/or "best practices" and where appropriate, clearly define "mandatory" requirements to which adherence is not an option. Security policies should be tested from time to time to ensure adequate protections are in place. When new information assets are introduced, policies should be updated to reflect any changes that may be necessary.

7.2.2 Security Awareness and Education

Companies should invest time and resources on an Information Security Awareness Program. To help safeguard company assets, employees must have the knowledge to understand the significance of their actions. A Security Awareness Program should designate responsibility for security training, clarify why security is important, identify who should attend Security Awareness Training, explain employee responsibilities, discuss existing security controls being taken to protect personnel and assets, and serve as a forum to discuss security questions.

Security awareness education should include "new hire" orientations, multi-media campaigns, and ongoing refresher activities. Incentive programs may also be utilized to bolster awareness and training efforts. Comprehensive security awareness programs will include both physical and cyber security initiatives.

7.2.3 Accountability and Ownership

It is important to establish an owner for all policies, procedures, hardware, software and information assets. Having identifiable responsibility for these assets within a company is fundamental to the control process. The responsibility for many owner tasks can be delegated to custodians, but the owner remains accountable for the asset. Some of the key responsibilities of an owner include:

- Defining the business requirements for which the asset is needed,
- Establishing the value, criticality and sensitivity of the asset,
- Establishing, maintaining, documenting and verifying cost effective controls commensurate with the risk,

- Establishing policies and procedures to deal with issues related to the asset.

Since the business unit is typically in a better position to effectively assess business requirements, value, and sensitivity of an asset, it is recommended that ownership be placed within the business unit under most circumstances, not in the IT function. However, it would be appropriate for the IT function to own computing infrastructure and services that support the entire company, such as the company's network, etc.

7.2.4 Data/Information Classification

Information classification is the process of assigning protection categories or labels to information materials such as hardcopy documents and computer files. Classification of assets is generally based on the impact to the business if the information is lost, disclosed, corrupted or made unavailable. It is important to identify an organization's most critical information assets so that protection efforts and budget can be focused on those resources.

Typical components of a classification program include a policy that defines the classification program, identification of asset owners, definitions for various classifications, guidelines for handling, storing, transmitting and accessing information with various classifications, and an education program for employees. An information classification framework was developed by the API IT Security Forum. For more information call 202-682-8590.

7.2.5 Security Vulnerability Assessments

Security Vulnerability Assessments (SVA) are a cost-effective method to identify risks and reduce them to acceptable levels. SVAs should be performed on information technology assets on a routine basis to identify significant exposures that could lead to negative consequences. SVAs should evaluate the potential business and financial impacts of loss of information integrity, disclosure of sensitive information, loss of processing capability, violation of regulations, and the impact on health, safety or the environment. Key outcomes of an SVA are the documentation of the owner's judgment of exposures and risks in the absence of controls, and the documentation of follow-up action plans or the justification for accepting residual risks.

7.2.6 Physical and Environmental Security

It is important to prevent unauthorized access, theft or damage to computing systems and information assets. Critical or sensitive information processing equipment should be housed in secure facilities, protected by a defined security perimeter. The nature of this perimeter should be commensurate with the identified risks and value of the business assets. Protection should be extended to supporting facilities such as electrical supply and cabling infrastructure. Placement of systems should take into account environmental risks and should provide protection and detection from hazards such as fire. Policies should be implemented when feasible that require desks to be left clear of sensitive documents and media, and computer screens to be locked when unattended.

7.2.7 Access Controls and Identity Management

The implementation of appropriate access controls and the management of user identities are essential for the preservation of confidentiality, integrity and availability. These processes are typically applied to network, host, application and physical assets. The resulting audit trails should be monitored to detect anomalies.

Access control systems must allow authorized use of systems and resources, while preventing direct access by unauthorized users. Authorized users may be employees, contractors, third parties, or the general public, but should be defined. Access controls include administrative controls such as policies, procedures, training, background checks and supervision; logical or technical controls such as passwords, two-factor authentication mechanisms, encryption, system hardening and protected protocols; and physical controls such as locks, cables, security cameras, guards and fences.

Identity Management or User Management systems maintain system user identities for the purpose of authenticating individuals to multiple systems. Identity management processes create, remove or modify an individual's access to systems in compliance with company policy. When an Identity Management system is functioning properly, a change to an individual's status will automatically and appropriately modify the access permitted to that individual throughout the environment.

7.2.8 Network Security

Many controls are required to achieve and maintain the security of computer networks. Network controls should be implemented based on a clear policy that defines:

- The networks and network services which are allowed to be accessed.
- Authorization procedures for determining who is allowed to access which networks and networked services.
- Management controls and procedures to protect the access to network connections and network services.
- The degree of testing, monitoring and intrusion detection that is required to ensure required security levels are maintained.

Access to networks by remote users, access to network management facilities, and access to remote diagnostic ports on network equipment should require an appropriate level of authentication, such as two-factor authentication. Additional controls within the network to segregate information systems or groups of users should be considered when different levels of trust or security requirements exist. Shared networks and those linked to third parties require particular access control policies, traffic filtering, and routing controls to ensure that computer connections and information flows do not breach the access control policy of business applications. Security patches should be maintained on all network devices.

7.2.9 Systems Development

Information security controls should be integrated into the initial phases of any application, data or system development process because it is much more effective to design information security requirements early in a development process rather than attempting to retrofit them after the system is operational. Security controls should be designed according to a risk mitigation strategy that attempts to reduce risk to levels acceptable to the business unit, based on the value of the asset and the likelihood of threats against it.

Periodic design reviews should be conducted during development and modification processes to assure that the design satisfies the specified security requirements. Production data should not be used to test application software until software integrity is assured. Application software should not be placed into production until the system tests have been successfully completed and the application has been properly certified and accredited. (See Change Control)

Infrastructure that supports applications that process or maintain sensitive data must be protected as well. Specific security controls such as intrusion detection/prevention and anti-virus should be implemented on hardware platforms and operating systems utilized during application development phases. Vulnerability assessment and patch management processes should be implemented to reduce or eliminate known or recently released vulnerabilities. Development and production environments should be continuously monitored to verify controls such as identity management and access control are functioning as intended.

7.2.10 Change Control

It is important to establish a methodology to evaluate system changes and configuration controls to ensure the secure operation of the networking infrastructure and the continued confidentiality, integrity and availability of information systems. A change control process should be chartered and empowered to manage change within the information technology environment. This change control process should include features such as submission and evaluation of change requests, recovery and back-out procedures, and a mechanism to monitor and protect the organization's capacity to ensure uninterrupted availability.

7.2.11 Viruses and other Malicious Code

Increasingly complex and sophisticated malicious code continues to be prevalent, making it essential to implement effective controls to mitigate this risk. Recent versions of malicious code combine different infection techniques, carry new payloads, and steal or expose information rather than just destroying it. To reasonably mitigate this risk, multiple solutions should be deployed. Standard anti-virus software should be installed throughout the enterprise, on personal computers, data file servers, centralized application servers such as e-mail and web servers, and in the firewall complex. Anti-virus solutions should scan all protocols that could contain malicious code. To the extent possible, anti-virus software should be centrally administered to ensure desktops are updated quickly and uniformly.

Consideration should be given to the deployment of desktop (personal) firewalls and anti-spyware systems. Operating system and application security patches should be evaluated based on the risk they mitigate and installed as appropriate to reduce the effectiveness of malicious code. Finally, it is important to maintain employee awareness efforts since users are typically the first to receive malicious code and most often the cause of its distribution.

7.2.12 Intrusion Detection and Incident Management

Systems should be implemented and qualified personnel should be assigned to log and monitor inappropriate or unauthorized network activities. Electronic firewalls and other systems should be installed and configured to detect and prevent hostile activity at all external network access points, and between certain internal networks as appropriate. An incident response plan should be developed to ensure the timely and effective response to relevant exploits and report information of concern to appropriate Information Technology and business contacts, including internal public relations staff and government or law enforcement agencies. An incident response team should be assigned to respond to security events such as virus outbreaks, network penetration attempts, denial of service, intrusions and data theft or compromise. A computer security incident response plan was developed by the API IT Security Forum. For more information call 202-682-8590.

7.2.13 Business Continuity, Business Resumption and Disaster Recovery

Business Continuity, Business Resumption and Disaster Recovery are somewhat interchangeable terms. The intent of these plans is to enhance an organization's ability to counteract interruptions to normal operations. Business Impact Assessments should be performed by each department or function to determine the length of time they can operate without critical systems or processes before the business unit would incur a material loss. Appropriate business resumption plans, including well defined and tested data backup processes, should then be developed and implemented that would have a reasonable probability of preventing such a material loss. These plans should be documented to form the Business Resumption Plan for the entire business unit. It is critical that Companies regularly test their Business Continuity Plans and revise the documentation as necessary to ensure the long-term effectiveness of their overall business continuity strategy.

7.2.14 Regulatory Compliance

Companies should establish a regulatory baseline to measure and provide corporate wide visibility to legal compliance requirements. To establish this baseline, all applications, systems and infrastructures should be identified and documented. Communication between corporate information security planners and other corporate functional sponsors or business owners should be established to ensure proper attention, visibility and guidance is obtained.

All relevant statutory, regulatory and contractual requirements should be identified, defined and documented for each information system. Major legislation has been passed in the following areas and should be addressed:

- Intellectual property (business information and copyrighted materials)
- Records retention (safeguard organizational records)
- Data protection and privacy of personal information
- Import/Export regulation (such as laws related to the use of encryption)
- Law enforcement (Rules of evidence)
- HIPPA, Sarbanes-Oxley, Graham-Leach-Bliley and others

7.2.15 Audit (Compliance and Assurance)

Security standards and policies can be very effective at safeguarding information assets and employees. However, in order to be effective, the standards and policies must be enforced. One way to ensure adequate protections are in place is by means of a standards compliance and assurance audit.

A company's executive management and Audit Committee have become increasingly interested in how well the company is protecting its critical information technology assets from unauthorized access and inappropriate use. One of the key assurance methods used by management is audit. Unsatisfactory audit reviews are discussed with management and/or the Audit Committee. These reviews typically require a clear definition of actions to be taken to prevent reoccurrence and a clear accountability for ensuring the actions are executed in a timely manner.

Other metrics that can be routinely evaluated and reported as indicators of the quality of health of the Information Security Management process and the associated policies, standards and procedures are the following:

- Appropriate use of Internet and e-mail systems
- Intrusion Detection reporting
- Password strength
- User account administration (modifications, additions, deletions)
- Change Management compliance

Appendix A—Security Regulations Affecting the U.S. Petroleum Industry

Security Regulations Affecting the U. S. Petroleum Industry					
Operating Sector	**Federal Agency**	**Issue**	**Requirement**	**Deadline**	**Authority References**
Marine, Upstream, Downstream	USCG, DHS	Area Maritime Security Improvements – General Provision	Establishes framework for vessels and facilities located under, in, on or adjacent to U.S. waters to implement security plans developed under Parts 104, 105 and 106, to deter transportation security incidents; provides for civil and criminal penalties for noncompliance; provides for Coast Guard approval of Alternative Security Programs.		Final rule – 10/22/03 [68 FedReg 60448] 33 *CFR* Subchapter H, Part 101. See also Interim Final Rule 7/1/03 [68 FedReg 39240]
	USCG, DHS	Area Maritime Security	Integrates port security-related requirements in the Maritime Transportation Security Act of 2002 with International Ship and Port Security Code (ISPS) and amendments to International Convention for Safety of Life at Sea (SOLAS). Establishes Area Maritime Security (AMS) Committee, directs the Committee to develop a risk-based AMS Assessment and an AMS Plan to respond to maritime security threats. (See J and K.)		Final rule – 10/22/03 [68 FedReg 60448] 33 *CFR* Subchapter H ,Part 103. See also Interim Final Rule 7/1/03 [68 FedReg 39240]
	USCG, DHS	Vessel Security	Requires owners or operators of vessels calling on U.S. ports to designate security officers for vessels, develop a Vessel Security Assessment, develop and submit to the USGS for approval a Vessel Security Plan that addresses components outlined in the rule, implement security measures specific to the vessel's operation, and comply with Maritime Security Levels. (See G.)	Plans to be submitted on or before 12/29/03. Compliance required on or before 6/30/04. Foreign vessels must have certificate of compliance with SOLAS and ISPS on or before 7/1/04.	Final rule – 10/22/03 [68 FedReg 60448] 33 *CFR* Subchapter H, Part 104. See also Interim Final Rule 7/1/03 [68 FedReg 39240]
	USCG, DHS	Port Facility Security	Requires owners or operators of certain facilities at U.S. ports to designate security officers for facilities, develop a Facility Security Assessment, develop and submit to the USCG for approval a Facility Security plan that addresses components outlined in the rule, implement security measures specific to the facilities' operations, and comply with Maritime Security Levels. (See H.) See also updated regulations for handling of Class I (explosives) or other dangerous cargoes within or contiguous to waterfront facilities.	Plans to be submitted on or before 12/29/03. Compliance required on or before 6/30/04.	Final rule – 10/22/03 [68 FedReg 60448] 33 *CFR* Subchapter H, Part 105. See also Interim Final Rule 7/1/03 [68 FedReg 39240] Final Rule – 9/26/03 [68 FedReg 55436]
	USCG, DOT	Port/Facility Access: Identification Credentials	Clarifies the identification credentials that are acceptable to allow access to waterfront facilities and to port and harbor areas, including the vessels in them.	Clarification effective 9/6/02.	Clarification of Regulation – 8/7/02 [67 FedReg 51082] See also 33 *CFR* 6.10-5, 125.09(f), 125.15 and 125.53

Security Regulations Affecting the U. S. Petroleum Industry

Operating Sector	Federal Agency	Issue	Requirement	Deadline	Authority References
	USCG, DHS	Vessel Communication	Establishes technical and performance standards for an Automatic Identification System (AIS) and implements the AIS carriage requirements of the Maritime Transportation Security Act (MTSA) and the International Maritime Organization requirements adopted under International Convention for Safety of Life at Sea (SOLAS), 1974, as amended. Requires AIS on all vessels subject to SOLAS, Vessel Traffic Service Users and certain other commercial vessels. (See I and J.)	Varies by type of ship.	Final rule – 10/22/03 [68 FedReg 60448] 33 *CFR* Parts 26, 161, 164, 165. See also Interim Final Rule 7/1/03 [68 FedReg 39240]
	USCG, DOT	Vessels: Notification of Arrival (NOA) in US Ports	For vessels bound for or departing US ports: Specifies information required in a NOA including additional crew and passenger information, consolidates and centralizes NOA submissions, requires earlier NOA submission times, provides exemptions for certain vessels, and creates exceptions to submission times for cargo declaration.	Requirements effective 4/1/03.	Final Rule – 2/28/03 [68 FedReg 9537]
Upstream	USCG, DHS	Outer Continental Shelf Facility Security	Requires certain offshore mobile drilling units and fixed oil and gas platforms to develop Facility Security Plans and Facility Security Assessment reports (See A,B, and E), designate security officers for OCS facilities, implement security measures specific to the facility's operation, and comply with Maritime Security Levels. Criteria based on production or number of personnel. Smaller facilities are not required to have assessments and plans but are encouraged to use industry standards such as API RP 70 (See F.) Coast Guard will review need for further security requirements and then consider separate rule making that would require compliance with industry standards.	Plans to be submitted on or before 12/29/03. Compliance required on or before 6/25/04. Facilities built after 7/1/04 must file for approval 60 days prior to beginning operations.	Final Rule – 10/22/03 [68 FedReg 60448] 33 *CFR* Subchapter H, Part 106. See also Interim Final Rule 7/1/03 [68 FedReg 39240]
Transportation	RSPS, DOT	Hazmat transportation: Generally	Shippers and carriers of certain hazardous materials must develop and adhere to security plans. (See I.) Includes personnel security, unauthorized access information and en route security. Shippers and transporters of certain hazardous materials are required to comply with Federal security regulations that apply to motor carrier and vessel transportation.	Plans must be developed by 9/25/03. Compliance by 10/27/03.	Final rule – 3/25/03 [68 FedReg 14509] 49 *CFR* Part 172 Final rule – 9/26/03 [68 FedReg 55436] 33 *CFR* Part 126

Security Regulations Affecting the U. S. Petroleum Industry

Operating Sector	Federal Agency	Issue	Requirement	Deadline	Authority References
	RSPS, DOT	Hazmat transportation: Employee Training	Shippers and carriers of certain hazardous materials must ensure that employee training includes a security awareness component. In-depth training required for shippers which have security plans. See 3.1	Compliance required no later than the date of the first scheduled recurrent training after March 25, 2003, and in no case later than March 24, 2006. New employees must receive training within 90 days of hire. Compliance by 12/22/03.	Final rule – 3/25/03 [68 FedReg 14509] Final rule – 3/25/03 [68 FedReg 14509]
	FMCSA, DOT	Hazmat transportation: Employee security	Applicants for a commercial driver's license (CDL) to transport hazardous materials must pass a security screening/background check by the Transportation Security Administration. States required to change procedures for issuing licenses, including collecting fingerprints and biographical and criminal history information of applicants for a hazmat endorsement for a CDL. Security threat assessment standards established to review applicants for hazmat endorsement to commercial driver licenses (CDL). Appeal and waiver procedures established. Certain individuals barred from shipping explosives. Exemption process provided.	State compliance on 4/1/04 (extended from 11/03/04). Limitations imposed beginning 9/2/03. After 4/01/04 (extended from 11/3/03), no renewals or issuances without TSA review Extension of licenses until 4/29/04 while TSA conducts reviews. Effective 3/11/04.	Delay of compliance date – 11/7/03 [68 FedReg 63030] Interim final rule – 5/5/03 [68 FedReg 23844] 49 *CFR* Parts 383, 384 Delay of compliance date – 11/7/03 [68 FedReg 63030] Interim final rule -- 5/5/03 [68 FedReg 23852] 49 *CFR* Parts 1570,1572 Final rule – 2/10/04 [69 FedReg 6195] Interim final rule -- 5/5/03 [68 FedReg. 23832] 49 *CFR* 107.105(c) 18 USC 842, 845
	USCG, DHS	Hazmat transportation: Facility security	Requires improved security and procedures related to the handling of dangerous cargoes and to and from vessels at such facilities, including fire extinguishing equipment, fire appliances, warning signs, outdoor lighting, international shore connection meeting for facilities involved with foreign-flag vessels, limited personnel access, certified material handling and other vehicles, and adequate equipment, materials and standards. Applicable also to waterfront facilities.	Compliance by 10/27/03.	Final rule – 9/26/03 [68 FedReg 55436] 33 *CFR* 126

Security Regulations Affecting the U. S. Petroleum Industry

Operating Sector	Federal Agency	Issue	Requirement	Deadline	Authority References
	RSPS, FMCSA, DOT	Hazmat transportation: Security measures for motor carriers	Imposes specific security measures, e.g., escorts, vehicle tracing and monitoring systems, remote shutoffs, anti theft devices. Research and Special Programs Administration assumed the lead role from the Federal Motor Carrier Safety Administration for rulemaking addressing security of motor carrier shipments of hazardous materials.		ANPRM 7/16/02 [67 FedReg 46622] Notice – 3/19/03 [698 FedReg 13250]
Terminals	FERC, USCG, OPS, RSPA, DOT	LNG Terminal Siting	Applications for authorization to build LNG terminals to FERC (land based) or Coast Guard (offshore) must include security assessment and security plan. (See O.)	With application.	Title 49 *CFR* Part 193, Subpart J – Security 33 *CFR* Part 127
Pipelines	TSA, DHS	Security Assessment and Plan	OPS Pipeline Security Information Circular (non-public distribution) directs pipelines to identify critical facilities and develop, implement and annually review a security plan, utilizing industry association guidelines. OPS will audit to verify company response to circular. (See A, B, C, D and E.)	Written confirmation of compliance with the PSIC due 3/5/03.	Guidance with expectations and recommendations but not statutorily mandated. Pipeline Security Information Circular 9/5/02.
All Sectors	DHS	Procedures for handling Critical Infrastructure Information	Establishes procedures by which DHS will manage confidential data voluntarily submitted by companies. Implements Homeland Security Act of 2002 Sec. 214, also known as the Critical Infrastructure Act of 2002. Addresses how FOIA requests for physical and cyber vulnerability information will be handled.	Interim rule effective 2/20/04. Comments are due on 5/20/04	Interim rule – 2/20/04 [69 FedReg 8074] 6 *CFR* 29.1 et seq. Proposed rule -- 4/15/03 [68 FedReg 18523]

<u>Statutory Authority:</u>

- Homeland Security Act of 2002—Signed into law 11/25/02. Public Law 107-296.
- Pipeline Safety Improvement Act of 2003—Signed into law 12/17/02. Public Law 107-355
- Maritime Transportation Security Act of 2002—Signed into law 11/25/02. Public Law 107-295
- USA PATRIOT Act—Signed into law 10/26/01. Public Law 107-56
- Safe Explosives Act—Signed into law 11/25/02. Public Law 107-296

Appendix B—Glossary and Terms

Adversary: Any individual, group, organization, or government that conducts activities, or has the intention and capability to conduct activities detrimental to critical assets. An adversary could include intelligence services of host nations, or third party nations, political and terrorist groups, criminals, rogue employees, and private interests. Adversaries can include site insiders, site outsiders, or the two acting in collusion.

Alert Levels: Describe a progressive, qualitative measure of the likelihood of terrorist actions, from negligible to imminent, based on government or company intelligence information. Different fixed or variable security measures may be implemented based on the level of threat to the facility.

Asset: An asset is any person, environment, facility, material, information, business reputation, or activity that has a positive value to an owner. The asset may have value to an adversary, as well as an owner, although the nature and magnitude of those values may differ. Assets in the SVA include the community and the environment surrounding the site.

Asset category: Assets may be categorized in many ways. Among these are:

- Activities/Operations
- Environment
- Equipment
- Facilities
- Hazardous materials (used or produced)
- Information
- People

Computer incident: refers to an adverse event in an information system and/or network, or the threat of such an occurrence, which could cause loss of data confidentiality, disruption of data or system integrity, or disruption or denial of availability. Examples include: unauthorized use of another user's account, unauthorized use of system privileges, or execution of malicious code that destroys data. Adverse events such as natural disasters and power-related disruptions, though certainly undesirable incidents, are not generally within the scope of incident response teams and should be addressed in the business continuity (contingency) and Disaster Recovery plans. For the purpose of *Incident Response*, therefore, the term "computer incident" refers to an adverse event that is related to Information Security.

Consequences: The amount of loss or damage estimated to result from a successful attack against an asset. This should include consideration of casualties, facility damage, environmental impacts, and business interruption as appropriate.

Control center: A location from where a pipeline system is remotely monitored and operated. A control center is typically staffed on a 24/7 basis and is the location for continuous and centralized control of a pipeline system.

Countermeasures: An action taken or a physical capability provided whose principal purpose is to reduce or eliminate one or more vulnerabilities. The countermeasure may also affect the threat(s) (intent and/or capability) as well as the asset's value. The cost of a countermeasure may be monetary, but may also include non-monetary costs such as reduced operational effectiveness, adverse publicity, unfavorable working conditions, and political consequences.

Damage: Impairment of the usefulness or value of information or computer resources (e.g., when a virus scrambles a file or makes a hard disk inoperable).

Delay: A countermeasures strategy that is intended to provide various barriers to slow the progress of an adversary in penetrating a site to prevent an attack or theft, or in leaving a restricted area to assist in apprehension and prevention of theft.

Detection: A countermeasures strategy to that is intended to identify an adversary attempting to commit a security event or other criminal activity in order to provide real-time observation as well as post-incident analysis of the activities and identity of the adversary.

Deterrence: A countermeasures strategy that is intended to prevent or discourage the occurrence of a breach of security by means of fear or doubt. Physical security systems such as warning signs, lights, uniformed guards, cameras, bars are examples of countermeasures that provide deterrence.

Energy ISAC: The Energy Information Sharing and Analysis Center is an industry organization that provides a secure database, analytic tools, and information gathering and distribution facilities designed to allow authorized individuals to submit either anonymous or attributed reports about information security threats, vulnerabilities, incidents and solutions.

Event: any observable occurrence in a system and/or network. Examples of events include the system boot sequence, a system crash and packet flooding within a network. Events sometimes provide indication that an incident is occurring. In reality, events caused by human error (e.g., unintentionally deleting a critical directory and all files contained therein) are the most costly and disruptive. Computer security-related events are attracting an increasing amount of attention among Information Security Professionals and within the general computing community.

Hazard: A situation with the potential for harm.

Intelligence: Information to characterize specific or general threats including the motivation, capabilities, and activities of adversaries.

Intent: A course of action that an adversary intends to follow.

Likelihood of adversary success: The potential for causing a catastrophic event by defeating the countermeasures. Likelihood of adversary success is an estimate that the security countermeasures will thwart or withstand the attempted attack, or if the attack will circumvent or exceed the existing security measures. This measure represents a surrogate for the conditional probability of success of the event.

MOC (Management of Change): An internal company management system to define, document, and communicate changes to a process as applicable.

Operator: A person or company who owns and/or operates petroleum facilities. For a person or company who owns or operates pipeline segments and/or facilities, the definition of operator is based on Title 49 *CFR* Part 195.

Pipeline security plan: Documentation that describes an operator's plan to address security issues and related events including security assessment and mitigation options and includes security condition levels and protective measures to security threats.

Pipeline system: Pipeline or pipeline segment and pipeline facilities such as a terminal, pump station, or other remote site plus the control center.

Response: The act of reacting to detected criminal activity either immediately following detection or post-incident via surveillance tapes or logs.

Risk: A measure of loss in terms of both the incident likelihood of occurrence and the magnitude of the consequences.

Risk management: An overall program consisting of: identifying potential threats to an area or equipment; assessing the risk associated with those threats in terms of incident likelihood and consequences; mitigating risk by reducing the likelihood, the consequences, or both; and measuring the risk reduction results achieved.

Risk mitigation: Those security measures employed at a facility to reduce the security risk to that facility.

Safeguard: Any device, system or action that either would likely interrupt the chain of events following an initiating event or that would mitigate the consequences.[1]

SCADA: Supervisory Control and Data Acquisition used for the remote control and monitoring of a pipeline system

Security plan: A document that describes an operator's plan to address security issues and related events including security assessment and mitigation options and includes security alert levels and response measures to security threats.

Security risk management: An overall plan consisting of: identifying potential security threats to pipeline segments and facilities; assessing the risks associated with those threats in terms of incident likelihood and consequences; mitigating the risk by reducing the likelihood, the consequences, or both; and evaluating the risk reduction results achieved.

Security risk mitigation: Those security measures employed on a pipeline system to reduce the security risk to the pipeline system.

Security Vulnerability Assessment (SVA): A systematic, analytical process in which potential security threats and vulnerabilities to facility or system operations are identified and the likelihood and consequences of potential adverse events are determined. SVAs can have varying scopes and can be performed at varying levels of detail depending on the operator's objectives - see Section 5.

Segment: an aspect of the petroleum industry that represent one of the steps needed to find, produce, process and transport petroleum from where they are found deep below the earth's surface to where they will be consumed. For purposes of this guidance document, the petroleum segments are defined as petroleum exploration and production (Upstream), petroleum refining, pipeline transportation (liquids), marine transportation, and petroleum products distribution and marketing.

Should: The term "should" is used in this document to indicate those practices which are preferred, but for which Owner/Operators may determine that alternative practices are equally or more effective or those practices for which engineering judgment is required.

Terrorism: "The unlawful use of force or violence against persons or property to intimidate or coerce a Government, the civilian population, or any segment thereof, in furtherance of political or social objectives" - (FBI).

Threat: Any indication, circumstance, or event with the potential to cause the loss of, or damage to an asset. Threat can also be defined as the intention and capability of an adversary to undertake actions that would be detrimental to critical assets.

Threat categories: Adversaries may be categorized as occurring from three general areas:

- Insiders
- Outsiders
- Insiders working in collusion with outsiders

Vulnerability: Any weakness that can be exploited by an adversary to gain access to and damage or steal an asset. Vulnerabilities can include but are not limited to building characteristics, equipment properties, personnel behavior, locations of people, equipment and buildings, or operational and personnel practices.

Appendix C—Communication of Security Intelligence

One important key to mitigate acts of terror and to protect facilities is good intelligence, and the quick dissemination of information to the large number of Owner/Operators that may need the information.

Information Sharing and Analysis Centers (ISACs) were created to serve as information dissemination organizations to provide government intelligence to industry concerning potential acts of terrorism. An ISAC consists of a secure database, analytic tools, and information gathering and distribution facilities that allow authorized individuals to submit either anonymous or attributed reports about information and physical security threats, vulnerabilities, incidents, and solutions. ISAC members also have access to information and analysis related to information provided by other members and obtained from other sources, such as the US government and law enforcement agencies, technology providers, and security associations such as CERT. The ENERGY-ISAC is exclusively for, and designed by, professionals in the energy industries. No U.S. government agency, regulator, or law enforcement agency can access the ENERGY-ISAC. Other critical industries, such as finance and telecommunications, also have ISACs in place.

Organizations wishing to apply for membership in the ISAC may obtain membership information at (http://www.energyisac.com/) or by calling 202-682-8286. Membership requests should be mailed to the ISAC administrator at:

ENERGY-ISAC
1220 L. Street N.W., Suite 900
Washington, D.C. 20005
USA

Appendix D—References

[1] American Institute of Chemical Engineers (AIChE) Center for Chemical Process Safety (CCPS) "Guidelines for Managing and Analyzing the Security Vulnerabilities of Fixed Chemical Sites", August, 2002.
[2] "The Sociology And Psychology Of Terrorism: Who Becomes A Terrorist And Why?," A Report Prepared under an Interagency Agreement by the Federal Research Division, ,Rex A. Hudson, et. al. Library of Congress, September,1999.
[3] "Patterns of Global Terrorism" 2001, May, 2002, U. S. State Department.
[4] Testimony Before the Senate Committee on Governmental Affairs, United States General Accounting Office, October 31, 2001, "A Risk Management Approach Can Guide Preparedness Efforts", Statement of Raymond J. Decker, Director, Defense Capabilities and Management.
[5] CCPS, 2002.
[6] The National Infrastructure Protection Center ,"Suggested Guidance on Protective Measures," Information Bulletin 03-002, February 7, 2003.
[7] COMDTPUB P 16700.4, U.S. DOT, USCG, NVIC 11-02, 13 January 2003.
[8] American Institute of Chemical Engineers (AIChE) Center for Chemical Process Safety (CCPS) "Guidelines for Managing and Analyzing the Security Vulnerabilities of Fixed Chemical Sites", August, 2002.
[9] Ibid, AIChE.
[10] Ibid, AIChE.
[11] Ibid, AIChE.
[12] Ibid, AIChE.
[13] "National Infrastructure Protection Center, Homeland Security Information Update, Potential Al-Qa'ida Operational Planning," Information Bulletin 03-001, February 7, 2003.

API Creative Services: 25045 | 04.05

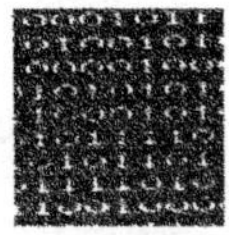

Petroleum Refineries

Liquid Petroleum Pipelines

Petroleum Products Distribution and Marketing

Oil and Natural Gas Production Operations

Marine Transportation

Cyber/Information Technology for the Petroleum Industry

Additional copies are available through Global Engineering Documents at 1-800-854-7179 or 303-397-7956.

Information about API Publications, Programs and Services is available on the web at www.api.org.

API®
American Petroleum Institute

1220 L Street, NW
Washington, DC 20005-4070
USA
202-682-8000

Product No. OS0002

Unnecessary Dangers: Emergency Chemical Release Hazards at Power Plants

Paul Orum

Working Group on Community Right-to-Know

July 2004

www.crtk.org

Acknowledgements:

The author acknowledges helpful comments from Carol Andress, Environmental Defense; Tom Natan, National Environmental Trust; Janelle Cunningham, Waterkeeper Alliance; Fred Millar, chemical safety consultant; Jim Young, Work Environment Council; George Sorvalis, Working Group on Community Right-to-Know.

The Working Group on Community Right-to-Know receives support from the Bauman Foundation and the Beldon Fund. The opinions expressed in this report do not necessarily reflect the views of our reviewers or supporters.

For copies of this report, visit www.crtk.org or send $15 (including postage) to:

Working Group on Community Right-to-Know
218 D Street, SE
Washington, DC 20003
(202) 544-9586
www.crtk.org

Table of Contents **Page**

I. Executive Summary.. 3
Scope and Limitations... 4

II. Background.. 4
The Clean Air Act Cleans Up Dirty Power Plants........................... 4
Pollution Control Chemicals Pose Unnecessary Dangers................. 4
Ammonia is Hazardous to Health... 5
Chlorine Gas is Hazardous to Health.. 6
Emergency Releases Occur Repeatedly... 6

III. Findings... 7
The Technology Determines the Hazard... 7
Risk Management Program Tracks Hazard Reduction..................... 8
Public Pressure Prompts Safer Practices.. 9

IV. Safer Alternatives Are Available.. 10
Safer Technologies Reduce Hazards – Anhydrous Ammonia............. 10
Safer Technologies Reduce Hazards – Chlorine Gas........................ 11
Strict Pollution Requirements Improve Technologies...................... 12

V. Despite Warnings, Prevention is Not Required.............................. 12
Terrorism Warnings Abound.. 12
Site Security Alone is Insufficient.. 13
No Federal Rules Protect Power Plants.. 14
The White House Blocks EPA Action.. 14
Congress Fails to Act... 14

VI. Conclusion and Recommendations.. 15

Appendix A: Power Plants That Submit Risk Management Plans, by State.... 16

Endnotes.. 23

Tables and Figures

Table 1: Average Residents in Danger Near Power Plants, by Technology in Use...... 8
Table 2: The Ten States With Over 100,000 Residents in Danger Near Power Plants. 8
Figure 1: Enhancing Security: Barriers to Access vs. Safer Chemicals................. 13

Executive Summary

Across the country, some 3.5 million people live near any of 225 power plants (other than nuclear) that could suddenly release extremely hazardous chemicals and cause serious injury or death. Power plants could all but eliminate these hazards by using safer chemicals.

Ammonia poses the principal emergency chemical release danger to workers and communities at these power plants. Power plants use ammonia in air pollution control equipment. Chlorine gas, used in cooling water systems at power plants, can also endanger communities in the event of a sudden release.

Under certain conditions, both ammonia and chlorine gas can form toxic and lethal clouds if released in large quantities. Frequent industrial accidents involving ammonia or chlorine and the threat that terrorists could use stored chemicals as weapons demonstrate the need to use safer alternatives at power plants and other industrial facilities. Gaseous ammonia and gaseous chlorine can both burn the eyes and lungs; high levels can cause fluid in the lungs, leading to death.

Power companies' choice of technology determines the danger to communities:

- Some 166 power plants report using anhydrous ammonia, endangering an average of 21,506 people around each facility.
- An additional 69 power plants report using aqueous ammonia, endangering an average of 205 people off-site.
- Forty power plants report chlorine gas as their greatest emergency release hazard, endangering an average of 4,618 nearby residents.
- Power plants can readily substitute safer chemicals that work just as well to control pollution. The simplest changes are to substitute aqueous ammonia or urea for anhydrous ammonia, and chlorine bleach or bromine for chlorine gas. Under development are other alternatives that do not rely on ammonia or chlorine gas.

The data in this report also show that:

- Just two-dozen power plants account for two-thirds of the people in danger. By using readily available safer chemicals these two-dozen plants could all but eliminate the danger to 2.4 million people.
- In ten states more than 100,000 people live in danger of emergency chemical releases from power plants. These states are California, Texas, Florida, Illinois, Minnesota, Pennsylvania, Missouri, Rhode Island, Virginia, and New Jersey.

Community members, regulators, news reporters, power plant managers, and employees all have a role in investigating or addressing chemical hazards in our communities. We urge readers to use this report and its recommendations to seek changes at high hazard power plants that eliminate the possibility of a catastrophic chemical release.

Scope and Limitations

This report examines 275 current and 36 deregistered power plants that reported using extremely hazardous substances under the Environmental Protection Agency's Risk Management Planning program as of September 2003. Most of these power plants burn coal or other fossil fuels, and a few burn solid waste or produce electricity from cogeneration or other sources. Not included are nuclear power plants or power distribution and transmission facilities. This report addresses the potential for sudden chemical releases rather than routine pollution and stack emissions that include mercury (a potent nerve toxin), cancer-causing substances, fine particles, smog-forming nitrogen oxides, and acid-rain from sulfur dioxide. Power plants are just one industry with vulnerabilities to chemical release hazards and terrorism.[1]

II. Background

The Clean Air Act Cleans Up Dirty Power Plants

Since its inception, the Clean Air Act has required states to reduce emissions that form smog – the brownish haze that hangs over cities and whole regions, especially in the summertime. The primary component of smog is ground-level ozone, a gas that forms when oxides of nitrogen (NOx) react with other air pollutants. Ozone is a potent respiratory irritant that can burn the lungs and cause breathing problems that range from chest pain and coughing to asthma attacks and pulmonary inflammation.[2] Nationwide, power plants are the second-largest source of ozone-creating NOx emissions (after automobiles).[3, 4]

The Clean Air Act requires all new power plants and some older plants to control NOx emissions. These pollution controls are important for healthy air. However, power plants often use ammonia in pollution control equipment to react with and control NOx emissions. The most hazardous forms of ammonia can pose unnecessary dangers to workers and communities.

Power plants can prevent NOx formation during combustion or can control NOx emissions with add-on equipment.[5] Control technologies most commonly include selective catalytic or non-catalytic reduction systems that rely on ammonia.[6] These control technologies are used at hundreds of power plants and are generally the most effective commercially widespread method of controlling NOx emissions. Pollution controls that rely on alternatives to anhydrous ammonia are generally just as effective at cleaning up NOx pollution, and some developing technologies may prove more effective.

Pollution Control Chemicals Pose Unnecessary Dangers

Hundreds of fossil fuel power plants in the U.S. use chemicals that are dangerous to transport and store. In particular, anhydrous ammonia and chlorine gas are extremely hazardous chemicals that can form lethal toxic clouds in an emergency release. For this reason, many power plants already use readily available safer alternatives that effectively control pollution without posing the same dangers to workers and nearby communities.

Anhydrous Ammonia – Many power plants use hazardous anhydrous ammonia to remove NOx in smokestack pollution control systems. Available alternatives to anhydrous ammonia do not pose the same dangers. Dilute aqueous ammonia presents lower health and safety hazards, but does not eliminate the danger entirely, because aqueous ammonia retains limited ability to form a toxic cloud. Solid urea poses less danger because it allows utilities to generate ammonia for pollution control systems on-demand. Power plants can easily convert from anhydrous to aqueous ammonia or urea. Under development are pollution prevention and control technologies that do not rely on ammonia at all.

Chlorine Gas – Some power plants use chlorine gas as a biocide to prevent fouling of water used in cooling or to generate steam. Fouling can be in the form of algae and slime in cooling towers or mussels and clams in intake water pipes. Such fouling can aggravate corrosion, increase mineral deposition, and reduce heat transfer. Common alternatives to chlorine gas at power plants include chlorine bleach or bromine. Additional approaches include ultraviolet light, pulsed electric power, filtration, and anti-fouling surface coatings. Power plants frequently use more than one method to filter or treat water.

Ammonia is Hazardous to Health

Ammonia is a corrosive colorless gas with a strong odor. While used safely most of the time, ammonia can endanger workers and surrounding communities by forming a lethal toxic cloud in the event of an emergency release. Acute ammonia exposure can irritate the skin, burn the eyes, cause temporary or permanent blindness, and cause headaches, nausea, and vomiting. High levels can cause fluid in the respiratory system (pulmonary or laryngeal edema), which may lead to death. Chronic exposure damages the lungs; repeated exposure can lead to bronchitis with coughing or shortness of breath.

Aqueous ammonia solution (ammonia in water) is inherently safer than anhydrous ammonia (without water) because it is less capable of forming a dense ground-hugging plume that can drift downwind. Urea is a solid form of ammonia that releases ammonia slowly; it is inherently safer than either anhydrous or aqueous ammonia because it does not form toxic plumes under normal conditions of use.[7]

Ammonia's danger depends on its form:

- Anhydrous ammonia is a *gas* that poses significant hazards to workers and communities in the event of an emergency release.
- Aqueous ammonia is a *solution* of ammonia mixed with water at concentrations of typically 19 percent or 29 percent ammonia. The lower concentration of ammonia poses lesser hazards but requires more volume and shipments when compared to anhydrous ammonia.

- Urea is a *solid* that can be used to form ammonia as needed on-site. Producing ammonia on-site from urea reduces transportation and storage dangers associated with anhydrous or aqueous ammonia.

Chlorine Gas is Hazardous to Health

Chlorine gas is greenish-yellow with a strong, irritating odor. Chlorine gas can form a dangerous toxic plume in an emergency release. Chlorine in water harms fish and other marine life. Acute human exposure can cause permanent damage by severely burning the eyes and skin, and can cause throat irritation, tearing, coughing, nose bleeds, chest pain, fluid build-up in the lungs (pulmonary edema), and even death. Chronic exposure can damage the teeth and irritate the lungs, causing bronchitis, coughing, and shortness of breath. A single high exposure can permanently damage the lungs. Chlorine's danger depends on its form; chlorine bleach is a solution that does not have the potential of chlorine gas to form a dangerous toxic cloud.[8]

Emergency Releases Occur Repeatedly

National data show frequent ammonia and chlorine spills at industrial facilities. The National Response Center received reports of 6,400 ammonia spills over the 10 years ending March 1, 2004, and 2,200 releases involving chlorine gas.[9] Spills reported to the National Response Center range from minor to very large.[10]

Ammonia and chlorine were the first and second most commonly reported extremely hazardous substances involved in serious non-transportation industrial accidents – those involving death, injury, or evacuation – from 1994 to 1999. Some 656 of these serious incidents involved ammonia and 518 involved chlorine gas.[11]

A sampling of recent serious accidents illustrates ammonia dangers and underscores the desirability of substituting safer alternatives where feasible:

- On January 18, 2002, a train derailment leaked a large anhydrous ammonia plume over portions of Minot, North Dakota, killing one person, hospitalizing 15, and sending over 1,600 to hospitals and emergency medical centers.[12]

- In July 2002, an ammonia spill killed at least 13 people when a pipe burst at a fertilizer factory in Shandong province, China.[13]

- In July 2002, a planned release of ammonia to water from a power plant inadvertently killed 100,000 fish along ten miles of the Vermillion and Salt Fork Rivers near Urbana, Illinois.[14]

III. Findings

The Technology Determines the Hazard

Across the country, some 275 power plants report that they use large amounts of ammonia or chlorine gas. These power plants report the information under the Environmental Protection Agency's Risk Management Planning (RMP) program. Each facility's Risk Management Plan indicates among other things how many nearby residents live in danger of exposure to a worst-case chemical release. **Power plant RMP reports indicate that 225 of these 275 plants could harm people off-site in an emergency chemical release. These 225 power plants use enough ammonia or chlorine gas to collectively endanger any of 3,568,658 people who live in nearby communities.[15] Just two-dozen power plants are responsible for two-thirds of the population in danger.**

By switching to readily available and inherently safer pollution control options these power plants could eliminate or significantly reduce dangers that accidents or acts of terrorism pose to surrounding communities. Safer alternatives to anhydrous ammonia and chlorine gas are commercially available for use at power plants. Aqueous ammonia can replace anhydrous ammonia, and chlorine bleach or bromine can replace chlorine gas. Diverse technologies are available or under development that do not rely on ammonia or chlorine. The population figures do not include potential accidents during transport of the hazardous materials to the power plants. Safer technologies can reduce transportation hazards as well.

Out of the 275 power plants that submit RMPs:

- Some 166 power plants report *anhydrous ammonia* as their most dangerous chemical in the event of an emergency release. These 166 power plants endanger an average of 21,506 people per facility. All but one of these facilities endanger people off-site.

- An additional 69 power plants report *aqueous ammonia* (at concentrations of 20 percent or more) as the most serious emergency release hazard. These 69 plants endanger an average of 205 people per facility. Only one-third of these facilities, or 23 plants, endanger people off-site.

- Forty power plants report *chlorine gas* as their greatest emergency release hazard. These 40 plants endanger an average of 4,618 people per facility. All but three of these facilities endanger the surrounding community.

- Just two-dozen power plants in 11 states account for two-thirds of the people in danger. These two-dozen power plants are in California, Texas, Florida, Illinois, Minnesota, Pennsylvania, Missouri, Rhode Island, Virginia, New Jersey, and Delaware.

Table 1. Average Residents in Danger Near Power Plants, by Technology in Use

Technology in use	RMP power plants	Avg. residents in danger[16]	Avg. vulnerability in miles[17]
Anhydrous ammonia	166	21,506	3.56
Aqueous ammonia	69	205	0.35
Chlorine gas	40	4,618	2.22

In a worst-case release, ammonia or chlorine gas would drift downwind in a ground-level toxic plume. Such a release would generally endanger only a portion of the people who live in the downwind "vulnerability zone" because the wind generally only blows in one direction. The figures are not forecasts of potential casualties.

In ten states more than 100,000 people live in danger of emergency releases from power plants (Table 2).[18]

Table 2. The Ten States With Over 100,000 Residents in Danger Near Power Plants

State	Number of RMP power plants	Residential population in danger
California	75	803,311
Texas	23	607,067
Florida	17	438,655
Illinois	10	220,728
Minnesota	3	215,075
Pennsylvania	17	206,833
Missouri	4	173,000
Rhode Island	3	152,500
Virginia	8	126,258
New Jersey	5	107,646

Risk Management Program Tracks Hazard Reduction

The EPA's RMP program enables people to identify industrial facilities that switch to safer chemicals and processes, in particular when the facilities "deregister" from the program. At least three-dozen power plants have deregistered from the RMP program since 1999. These 36 power plants in some cases no longer use extremely hazardous substances and in other cases no longer use the chemicals above RMP threshold amounts. Before they deregistered, these 36 facilities collectively endangered any of 1,337,144 people from potential emergency chemical releases.

The RMP program requires industrial facilities that use extremely hazardous substances to carefully examine and describe the hazards they pose. The process can prompt facilities to select safer technologies, but nothing in the program *requires* facilities to use or even review available safer alternatives. Communities can track facilities' progress in reducing chemical hazards by reviewing RMP information through designated federal "reading rooms" (see endnote).[19]

The RMP planning process and other safety concerns have helped prompt a number of power plants to switch to safer chemicals, including:

- GWF Power Systems switched from anhydrous ammonia to aqueous ammonia at a half-dozen California power plants, eliminating potential dangers to thousands of nearby residents.

- Central and South West Corporation switched several power plants in Texas and Oklahoma from chlorine gas to safer chlorine bleach for cooling water treatment. The change reduces hazards to workers as well as people off-site.

- Wisconsin Power's Pulliam Plant (Green Bay, Wis.) switched from liquid anhydrous sulfur dioxide, used to capture particulates in pollution control equipment, to a safer solid form of the chemical. The change eliminated potential off-site injury to 180,000 people.

Public Pressure Prompts Safer Practices

Where neighbors have insisted on safer alternatives, some power plants have responded by substituting safer technologies that reduce the possibility of a lethal toxic cloud. The utility industry generally only spends money for safer alternatives, even if barely more costly, when there is a pressing need – such as public pressure or regulation.

- Constellation Power Source's Brandon Shores Power Plant near Baltimore agreed to use aqueous ammonia after four months of protests by local residents over the possible use of anhydrous ammonia.[20] Aqueous ammonia poses lesser dangers to the community from both storage and truck transportation.

- American Electric Power's Gavin power plant in Cheshire, Ohio selected a urea-based pollution control technology after residents raised concerns about the dangers of an emergency ammonia release. "Our neighbors in and around Cheshire told us they were very concerned about the impact of a serious accident involving a major release of anhydrous ammonia," said John Norris, senior vice president of operations and technical services, in a press release. "We took those concerns to heart."[21] Subsequently, however, larger pollution conflicts lead the company to buy out and relocate the entire immediate Cheshire community.

IV. Safer Alternatives Are Available

"If we make fewer toxic products, use milder manufacturing conditions, and produce less toxic waste, we reduce the opportunities for terrorists." – *National Research Council, Making the Nation Safer: The Role of Science and Technology in Countering Terrorism, 2002.*

Safer Technologies Reduce Hazards – Anhydrous Ammonia

Power plants have readily available alternatives to significantly reduce or eliminate the danger that ammonia poses to workers and nearby communities.

- At least 69 U.S. power plants use aqueous ammonia (at concentrations of 20 percent or more) rather than anhydrous ammonia, which significantly reduces but does not eliminate ammonia dangers (see Table 1, page 8). Switching from anhydrous to aqueous ammonia is a straightforward conversion that does not require significant new equipment.

- Ammonia-on-demand technology provides ammonia from urea as needed for pollution control systems, eliminating off-site emergency release hazards. Urea based pollution control systems have recently proven effective at large power plants. As of 2003, more than a dozen units had been installed at five commercial power stations.[22]

Power plants address NOx emissions with a combination of approaches.[23] The effectiveness of each technology depends on the combustion unit, the fuel, and operating requirements. Performance and cost-effectiveness are rapidly improving and changing as power plants gain experience with new technologies.[24]

Newer technologies are in various stages of commercial application. A sampling is listed below. Each has advantages and disadvantages in terms of pollution and cost, but all avoid the dangerous use of anhydrous ammonia. These technologies include:

- Goal Line Environmental Technology's SCONOx™ (also EmeraChem EMx™) system uses a catalyst to remove NOx and other pollutants without the use of ammonia as an additional reagent. This technology is used commercially at natural gas-fired power plants in California and Massachusetts.[25] It is more expensive than selective catalytic reduction systems.

- Catalytica Combustion Systems' XONON™ technology uses a catalyst in natural gas combustion at temperatures below which most NOx emissions form. Since the lower temperatures do not generate NOx, this system does not require add-on, ammonia-based pollution control equipment; it prevents rather than controls pollution.[26] XONON™ is used at one plant in California. Catalytica is adapting the technology to more types of natural gas turbines.

- EnviroScrub Corporation's Pahlman™ Process is an emerging add-on technology that captures almost all NOx, sulfur, and mercury from coal-fired plants. This technology uses an oxide of manganese sorbent to capture these pollutants reducing air emissions to very low levels. EnviroScrub is currently bidding on large-scale commercial power projects.[27]

- PowerSpan Corporation's Electro-Catalytic Oxidation™ system uses ultra-violet light and conventional pollution equipment to control multiple pollutants, including NOx emissions. The system works with aqueous ammonia. PowerSpan is testing the system for commercial application.[28]

Safer Technologies Reduce Hazards – Chlorine Gas

Chlorine gas is only one commercially available treatment to prevent biological fouling of cooling water. In fact, while still widely used, gaseous chlorine is less used than previously due to its safety hazards.[29] There are safer substitutes for chlorine gas that do not form dangerous chemical plumes if inadvertently or deliberately released. These options include:

- Switching from chlorine gas to chlorine bleach at power plants would all but eliminate catastrophic release dangers. However, this would not resolve certain workplace hazards and environmental problems associated with routine discharge of chlorine.

- Bromine is another anti-fouling biocide that is widely used in power plants, most commonly in the form of dry bromine tablets. Dry bromine nonetheless poses hazards to workers, although it is thought to be somewhat less residually toxic than chlorine.[30]

Additional technologies can be used at power plants, or are in limited use, whether alone or in combination:

- Clearwater Systems' Dolphin™ pulsed power device has successfully limited scaling, biological activity, and corrosion in some smaller gas turbine plants that recirculate water, eliminating the need for chemicals on-site.[31]

- Trojan Technologies™ markets ultraviolet light systems that limit fouling of cooling water, as does LightStream Technologies™, which uses mercury-free pulsed ultraviolet light. Ultraviolet light is an emerging application for power plant turbines and cooling water, and may be more suitable for smaller plants that recirculate water (rather than once through water use).

- Potassium permanganate controls zebra mussels and algae in water intake pipes. It is widely used to pre-treat water at drinking water treatment plants. It does pose some hazards to workers, but not catastrophic emergency release hazards.[32]

Another strategy to address mussels is "non-stick" fouling release coatings. However, some antifouling coatings harm the environment. For example tributylin is a persistent toxic chemical that bioaccumulates in the food chain; it has been banned from use as hull paint on many ships, but remains registered for use in cooling towers.[33] Also toxic are products that contain other biocides such as atrazine. Copper-based anti-fouling paints also harm the marine environment. Some environmentally safer fouling release products include:

- SealCoat™ protects against corrosion and barnacles without poisons by providing a slippery micro fiber surface that imitates how marine mammals such as seals naturally protect themselves from barnacles.[34]

- The Department of Defense has validated environmentally safer silicon-based anti-fouling coatings through testing on boats and utility water intakes.[35]

More restrictive chlorine discharge standards will increase the use of alternate anti-fouling methods.

Strict Pollution Requirements Improve Technologies

In addition to emergency release hazards, some ammonia from pollution control escapes via the smokestack, contributing to routine air pollution. This smokestack ammonia that does not combine with nitrogen oxides can contribute to the release of fine particulate matter. Communities close to power plants have complained about sore throats, breathing problems, and other local health issues.[36]

Selective catalytic reduction (SCR) systems may not meet future restrictions on stack emissions. For example, regulators in Massachusetts and California are proposing stricter pollution requirements that may in effect limit the use of SCR and thereby force the development of non-ammonia pollution control or prevention alternatives (sometimes called Zero Ammonia Technology). Several non-ammonia pollution technologies listed above are under development or in early commercialization at power plants that burn natural gas. These technologies are designed not only to control or prevent NOx emissions, but also avoid release of added ammonia as a routine pollutant in stack emissions.

V. Despite Warnings, Prevention is Not Required

Terrorism Warnings Abound

Government agencies and other experts have warned that terrorists can target facilities that use extremely hazardous substances. Agencies that have issued such warnings include the Department of Homeland Security,[37] Department of Justice,[38] Environmental Protection Agency,[39] General Accounting Office,[40] Congressional Research Service,[41] Agency for Toxic Substances and Disease Registry,[42] Naval Research Laboratory,[43] and

Army Surgeon General.[44] The chemical industry has also warned about these dangers in its report, "The Terrorist Threat in America."[45] The non-governmental Brookings Institute,[46] Rand Corporation,[47] and Center for Strategic and International Studies[48] have described the threat. Public interest groups including Environmental Defense, The Safe Hometowns Initiative, and U.S. Public Interest Research Group have documented industrial chemical hazards and safer alternatives to those unnecessary dangers.[49]

Site Security Alone is Insufficient

Add-on physical site security measures such as barriers, lights, and guards do not help utilities produce power, and because they leave communities vulnerable may ultimately fail to address terrorism (figure 1, left). Indeed, investigative reporters have walked into dozens of chemical facilities, finding open gates, holes in fences, and other lax security.[50] To more fully address the threat of terrorism, safer technologies reduce or altogether remove chemical dangers to workers and communities (figure 1, right).

Figure 1. Enhancing Security: Barriers to Access vs. Safer Chemicals

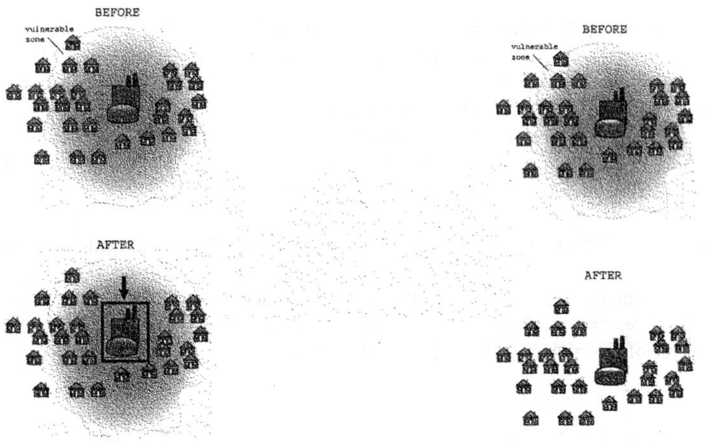

Before and after improving barriers to access – the community is still vulnerable

Before and after using safer chemicals – the community is no longer in danger.

No Federal Rules Protect Power Plants

More than two and a half years after September 11, 2001, there are still no substantial federal security standards for industries that use extremely hazardous substances. No federal standards require power plants and other industries to protect dangerous chemicals from theft or release by potential intruders. Nor does federal law require such companies to consider safer chemicals and processes that can reduce or eliminate chemical dangers.

The Clean Air Act, section 112(r) requires industrial facilities that use large amounts of certain extremely hazardous substances to disclose the dangers to workers and communities and to document measures that prevent a catastrophic chemical release. This law prevents pollution, saves lives, and protects property by stimulating safety steps *before* there is ever a major chemical release. As noted, under this law some 275 power plants report Risk Management Plans. These plans tell how the power plants will manage the risks of using anhydrous ammonia, chlorine gas, or aqueous ammonia – but the plans do not directly address chemical security.

The White House Blocks EPA Action

The Environmental Protection Agency (EPA) arguably has authority under the Clean Air Act section 112(r) general duty clause to compel companies to reduce chemical hazards in order to prevent accidents and improve security. Companies that reduce chemical dangers have less need for add-on safety controls (monitors, alarms, containment) and add-on physical security (guards, fences, lights). However, EPA has not used this authority and has not established any other systematic program to make communities safer by reducing chemical hazards at industrial plants.

In June 2002 EPA prepared options for an active chemical security program, including a draft press release and talking points, but the White House blocked these steps and transferred emphasis to the Department of Homeland Security, which is not a regulatory agency.[51] Nonetheless, lack of clear federal authority prompted both EPA and the Department of Homeland Security to call for legislation addressing chemical site security.[52]

Congress Fails to Act

To remedy these deficiencies, Senator Corzine (D-N.J.) and Congressman Pallone (D-N.J.) introduced the Chemical Security Act (S.157, H.R.1861). These bills require high priority chemical facilities to consider safer technologies and use them where practicable. Where safer technologies are not available, these bills set appropriate federal security standards.

Senator Corzine's bill unanimously passed the Senate Environment and Public Works Committee in July 2002. However, chemical manufacturers quickly organized opposition. As a result, seven Senators who had voted for the bill in committee raised objections that effectively scuttled chemical security legislation for the remainder of 2002.[53]

In 2003, Senator Inhofe (R-Okla.) introduced a very limited bill, the Chemical Facilities Security Act (S.994). This bill directs the Department of Homeland Security to endorse voluntary industry security initiatives rather than requiring industries to meet strict federal standards. This bill passed the Senate Environment and Public Works Committee on October 23, 2003 with a few strengthening amendments. These amendments require covered facilities to review safer chemicals and processes, and to report vulnerabilities to the federal government.

VI. Conclusion and Recommendations

Some power plants that use anhydrous ammonia or chlorine gas pose significant dangers to workers and communities. Just two-dozen power plants that use anhydrous ammonia impose two-thirds of the danger in terms of residential population at risk. Safer alternatives to anhydrous ammonia and chlorine gas are readily available for use at power plants, and new technologies are emerging. Power plants that use anhydrous ammonia or chlorine gas in populated areas pose unnecessary dangers.

Recommendations

The **power industry** should curtail unnecessary dangers by: converting high hazard power plants in populated areas to readily available safer alternatives to anhydrous ammonia and chlorine gas.

Community members should investigate solutions to local hazards by: contacting high hazard power plants with basic questions, including 1) has the facility explored alternatives to dangerous chemicals, and 2) when will the facility implement changes that eliminate catastrophic release dangers?

The **U.S. Environmental Protection Agency** should address power plant pollution and chemical hazards by: 1) using Clean Air Act general duty authority to reduce chemical dangers to communities around high hazard power plants, and 2) requiring new power plants and those that significantly upgrade to use safer control technologies.

Supporting Recommendations

Lawmakers should improve homeland security by: enacting an aggressive program of hazard reduction at high priority power plants and other facilities that endanger communities with extremely hazardous substances.

The **U.S. Department of Homeland Security** should reduce terrorism opportunities by: 1) developing a standard methodology for high hazard facilities to identify and switch to safer technologies that do not endanger workers and surrounding communities in the event of a terrorist-caused chemical release, and 2) allocating homeland security funding to convert high hazard facilities to safer technologies.

The **U.S. Chemical Safety Board** should reduce chemical hazards by: 1) working with federal agencies for comprehensive national public reporting and verification of chemical spills and emergencies, and 2) including safer technology conversion opportunities as a standard element in Board recommendations to industry and government.

The **U.S. Department of Labor** should improve worker safety by: developing training grants to non-profit organizations, through the National Institute for Environmental Health Sciences, to train and educate employees and first responders on preventing chemical vulnerabilities by means of inherently safer technologies.

State and local pollution prevention programs should leverage existing resources by: incorporating technology options analyses for inherent safety into pollution prevention technical assistance and site visits, and involving workers in site inspections.

Appendix A: Power Plants That Submit Risk Management Plans, by State

Power Plant Name	City	State	County	Toxic Chemicals*
Tenaska Central Alabama Generating Station	Billingsley	AL	Autauga	Ammonia (anhydrous)
Tenaska Lindsay Hill Generation Station	Billingsley	AL	Autauga	Ammonia (anhydrous)
E. B. Harris Electric Generating Plant	Prattville	AL	Autauga	Ammonia (anhydrous)
James A. Vann, Jr. Power Plant	Gantt	AL	Covington	Ammonia (aqueous)
TVA - Widows Creek Fossil Plant	Stevenson	AL	Jackson	Ammonia (anhydrous)
J. H. Miller Electric Generating Plant	Quinton	AL	Jefferson	Chlorine
Plant Franklin Combined Cycle Units	Smiths	AL	Lee	Ammonia (anhydrous)
Barry Combined Cycle Electric Generating Facility	Bucks	AL	Mobile	Ammonia (anhydrous)
Theodore Cogeneration Plant	Theodore	AL	Mobile	Ammonia (anhydrous)
Hog Bayou Energy Center	Mobile	AL	Mobile	Ammonia (aqueous)
Decatur Energy Center	Decatur	AL	Morgan	Ammonia (anhydrous)
Calpine Morgan Energy LLC	Decatur	AL	Morgan	Ammonia (anhydrous)
E. C. Gaston Electric Generating Plant	Wilsonville	AL	Shelby	Chlorine
Gorgas Electric Generating Plant	Parrish	AL	Walker	Ammonia (anhydrous)
Hot Spring Power	Malverne	AR	Hot Spring	Ammonia (aqueous)
Coronado Generating Station	St. Johns	AZ	Apache	Chlorine
APS West Phoenix Power Plant	Phoenix	AZ	Maricopa	Ammonia (aqueous)
Mesquite Generating Station	Arlington	AZ	Maricopa	Ammonia (anhydrous)
Pinnacle West Energy Redhawk Power Plant	Arlington	AZ	Maricopa	Ammonia (aqueous)
Griffith Energy	Golden Valley	AZ	Mohave	Ammonia (anhydrous)
South Point Energy Center	Mohave Valley	AZ	Mohave	Ammonia (anhydrous)

Pacific Oroville Power, Inc.	Oroville	CA	Butte	Chlorine
Contra Costa Power Plant	Antioch	CA	Contra Costa	Ammonia (aqueous)
Calpine Pittsburg	Pittsburg	CA	Contra Costa	Ammonia (anhydrous)
Delta Energy Center	Pittsburg	CA	Contra Costa	Ammonia (anhydrous)
Los Medanos Energy Center	Pittsburg	CA	Contra Costa	Ammonia (aqueous)
Pittsburg Power Plant	Pittsburg	CA	Contra Costa	Ammonia (aqueous)
Coalinga Cogeneration Company	Coalinga	CA	Fresno	Ammonia (anhydrous)
Kingsburg Cogeneration Facility	Kingsburg	CA	Fresno	Ammonia (anhydrous)
AES Mendota, L.P.	Mendota	CA	Fresno	Ammonia (anhydrous)
Rio Bravo Fresno	Fresno	CA	Fresno	Ammonia (anhydrous)
Heber Geothermal Company	Heber	CA	Imperial	Chlorine
Second Imperial Geothermal Company	Heber	CA	Imperial	Chlorine
AES Delano, Inc.	Delano	CA	Kern	Ammonia (anhydrous)
Mid-Set Cogeneration Company	Fellows	CA	Kern	Ammonia (anhydrous)
Sunrise Power Company, LLC	Fellows	CA	Kern	Ammonia (anhydrous)
Berry Cogen-18 Facility	Maricopa	CA	Kern	Ammonia (anhydrous)
Berry Cogen-38 Facility	Taft	CA	Kern	Ammonia (anhydrous)
DAI Oildale, Inc.	Bakersfield	CA	Kern	Ammonia (anhydrous)
Rio Bravo Poso	Bakersfield	CA	Kern	Ammonia (anhydrous)
Rio Bravo Jasmin	Bakersfield	CA	Kern	Ammonia (anhydrous)
Mt. Poso Cogeneration Company	Bakersfield	CA	Kern	Ammonia (anhydrous)
Texaco South East Kern River Cogeneration Facility	Bakersfield	CA	Kern	Ammonia (anhydrous)
Mojave Cogeneration Company	Boron	CA	Kern	Ammonia (anhydrous)
HL POWER COMPANY	Wendel	CA	Lassen	Ammonia (anhydrous), Chlorine
El Segundo Generating Station	El Segundo	CA	Los Angeles	Ammonia (aqueous)
AES Redondo Beach, L.L.C.	Redondo Beach	CA	Los Angeles	Ammonia (aqueous)
Scattergood Generating Station	Playa Del Rey	CA	Los Angeles	Ammonia (aqueous)
Praxair - Wilmington, CA	Wilmington	CA	Los Angeles	Ammonia (anhydrous)
Harbor Cogeneration Company Wilmington Plant	Wilmington	CA	Los Angeles	Ammonia (anhydrous)
Harbor Generating Station	Wilmington	CA	Los Angeles	Ammonia (aqueous)
AES Alamitos, L.L.C.	Long Beach	CA	Los Angeles	Ammonia (aqueous)
Haynes Generating Station	Long Beach	CA	Los Angeles	Ammonia (aqueous)
Pasadena Water & Power Broadway Power Plant	Pasadena	CA	Los Angeles	Ammonia (aqueous)
Grayson Power Plant	Glendale	CA	Los Angeles	Chlorine
Berry Cogen-42 Facility	Sant Clarita	CA	Los Angeles	Ammonia (anhydrous)
Valley Generating Station	Sun Valley	CA	Los Angeles	Ammonia (aqueous)
City of Burbank Public Service Department	Burbank	CA	Los Angeles	Chlorine
Sargent Canyon Cogeneration Company	San Ardo	CA	Monterey	Ammonia (anhydrous)
Salinas River Cogeneration Company	San Ardo	CA	Monterey	Ammonia (anhydrous)
King City Power Plant	King City	CA	Monterey	Ammonia (anhydrous)
SOLEDAD ENERGY LLC	Soledad	CA	Monterey	Ammonia (anhydrous)
Moss Landing Power Plant	Moss Landing	CA	Monterey	Ammonia (aqueous)
Rio Bravo Rocklin	Lincoln	CA	Placer	Ammonia (anhydrous)
Blythe Energy Project	Blythe	CA	Riverside	Ammonia (anhydrous), Ammonia (aqueous)
Colmac Energy, Inc.	Mecca	CA	Riverside	Ammonia (anhydrous)

Carson Energy Cogeneration Plant	Sacramento	CA	Sacramento	Ammonia (anhydrous)
SCA Congeneration Plant II	Sacramento	CA	Sacramento	Ammonia (aqueous)
OLS Energy-Chino Cogeneration Facility	Chino	CA	San Bernardino	Ammonia (anhydrous)
Colton Plant	Colton	CA	San Bernardino	Ammonia (aqueous)
High Desert Power Plant	Victorville	CA	San Bernardino	Ammonia (aqueous)
ACE Cogeneration Facility	Trona	CA	San Bernardino	Ammonia (anhydrous)
South Bay Power Plant	Chula Vista	CA	San Diego	Ammonia (aqueous)
Goal Line, LP	Escondido	CA	San Diego	Ammonia (aqueous)
University of California, San Francisco	San Francisco	CA	San Francisco	Ammonia (aqueous)
POSDEF Power Company L.P.	Stockton	CA	San Joaquin	Ammonia (anhydrous)
Stockton Cogen Company, Inc.	Stockton	CA	San Joaquin	Ammonia (anhydrous), Chlorine
Northern California Power Agency	Lodi	CA	San Joaquin	Ammonia (anhydrous)
Ripon Cogeneration, Inc.	Ripon	CA	San Joaquin	Ammonia (anhydrous)
OLS Energy Agnews	San Jose	CA	Santa Clara	Ammonia (anhydrous)
Wheelabrator Shasta Inc	Anderson	CA	Shasta	Ammonia (anhydrous)
Burney Forest Power Cogeneration Plant	Burney	CA	Shasta	Ammonia (anhydrous)
Stanislaus County Resource Recovery Facilty	Crows Landing	CA	Stanislaus	Ammonia (anhydrous)
Turlock Irrigation District	Ceres	CA	Stanislaus	Ammonia (anhydrous), Chlorine
Modesto Energy Limited Partnership	Westley	CA	Stanislaus	Ammonia (anhydrous)
Greenleaf 2 Power Plant	Yuba City	CA	Sutter	Ammonia (anhydrous)
Sutter Energy Center	Yuba City	CA	Sutter	Ammonia (anhydrous)
Pacific-Ultrapower Chinese Station	Jamestown	CA	Tuolumne	Ammonia (anhydrous)
OLS Energy-Camarillo Cogeneration Facility	Camarillo	CA	Ventura	Ammonia (anhydrous)
E. F. Oxnard, Inc.	Oxnard	CA	Ventura	Ammonia (anhydrous)
Ormond Beach Generating Station	Oxnard	CA	Ventura	Ammonia (aqueous)
Mandalay Generating Station	Oxnard	CA	Ventura	Ammonia (aqueous)
Woodland Biomass Power Ltd.	Woodland	CA	Yolo	Ammonia (anhydrous)
Craig Station	Craig	CO	Moffat	Chlorine
Nucla Station	Montrose County	CO	Montrose	Chlorine
Fort St. Vrain Station	Platteville	CO	Weld	Chlorine, Ammonia (aqueous)
Bridgeport Energy LLC	Bridgeport	CT	Fairfield	Ammonia (aqueous)
Conectiv - Hay Road Power Complex	Wilmington	DE	New Castle	Ammonia (anhydrous)
Gulf Power Co. Lansing Smith Elec Generating Plant	Southport	FL	Bay	Chlorine
FPL-Cape Cavaneral	Cocoa	FL	Brevard	Chlorine
Cedar Bay Generating Facility	Jacksonville	FL	Duval	Ammonia (aqueous)
St. Johns River Power Park	Jacksonville	FL	Duval	Chlorine
Payne Creek Generating Station	Bowling Green	FL	Hardee	Ammonia (aqueous)
Bayside Power Station	Tampa	FL	Hillsborough	Ammonia (anhydrous)
Lee County Solid Waste Resource Recovery Facility	Fort Myers	FL	Lee	Ammonia (anhydrous)
Indiantown Cogeneration Company, L.P.	Indiantown	FL	Martin	Ammonia (aqueous)
Stanton Energy Center	Orlando	FL	Orange	Ammonia (anhydrous), Chlorine
Curtis H. Stanton Energy Center Unit A	Orlando	FL	Orange	Chlorine, Ammonia (anhydrous)

Riviera Power Plant	Riviera Beach	FL	Palm Beach	Chlorine
Pinellas County Waste to Energy Facility Facility	St. Petersburg	FL	Pinellas	Chlorine
McIntosh Power Plant/Northside WWTP	Lakeland	FL	Polk	Ammonia (anhydrous)
Mulberry Cogeneration Facility	Bartow	FL	Polk	Chlorine
Orange Cogeneration Facility	Bartow	FL	Polk	Chlorine
Hines Energy Complex	Bartow	FL	Polk	Ammonia (aqueous)
Plant Bowen	Cartersville	GA	Bartow	Ammonia (anhydrous)
Plant Hammond	Rome	GA	Floyd	Ammonia (anhydrous)
Plant Wansley Combined Cycle Units	Franklin	GA	Heard	Ammonia (anhydrous)
Chattahoochee Energy Facility	Franklin	GA	Heard	Ammonia (aqueous)
Mid-Georgia Cogeneration Facility	Kathleen	GA	Houston	Ammonia (aqueous)
Plant Scherer	Juliette	GA	Monroe	Chlorine
Hamakua Energy Partners	Honokaa	HI	Hawaii	Ammonia (anhydrous)
AES Hawaii Inc.	Kapolei	HI	Honolulu	Ammonia (anhydrous)
Power Generation Station	Muscatine	IA	Muscatine	Chlorine, Sulfur dioxide (anhydrous)
Kincaid Generation, L.L.C.	Kincaid	IL	Christian	Ammonia (anhydrous)
Ameren Energy Generating Newton Plant	Newton	IL	Jasper	Chlorine
Kendall County Generating Facility	Minooka	IL	Kendall	Ammonia (aqueous)
AEG Coffeen Power Station	Coffeen	IL	Montgomery	Ammonia (anhydrous), Chlorine
Ameren CILCo Edwards Power Plant	Bartonville	IL	Peoria	Ammonia (anhydrous)
Dynegy Midwest Generation, Baldwin Complex	Baldwin	IL	Randolph	Ammonia (anhydrous), Chlorine
Cordova Energy Company, LLC	Cordova	IL	Rock Island	Ammonia (aqueous)
CWLP's Dallman Power Station	Springfield	IL	Sangamon	Ammonia (anhydrous)
Holland Energy LLC	Beecher City	IL	Shelby	Ammonia (anhydrous)
Marion Generating Station	Marion	IL	Williamson	Ammonia (anhydrous), Chlorine
Tanners Creek Plant	Lawrenceburg	IN	Dearborn	Ammonia (anhydrous)
PSI Energy Gibson Generating Station	Owensville	IN	Gibson	Ammonia (anhydrous)
Whiting Clean Energy Cogeneration Facility	Whiting	IN	Lake	Ammonia (anhydrous)
Merom Generating Station	Sullivan	IN	Sullivan	Ammonia (anhydrous)
East Bend Generating Station	Rabbit Hash	KY	Boone	Ammonia (anhydrous)
Owensboro Municipal Utilities	Owensboro	KY	Daviess	Chlorine
Mill Creek Station	Louisville	KY	Jefferson	Ammonia (anhydrous)
Spurlock Power Station	Maysville	KY	Mason	Ammonia (anhydrous)
Kentucky Utilities-E.W. Brown Station	Burgin	KY	Mercer	Ammonia (anhydrous)
TVA - Paradise Fossil Plant	Drakesboro	KY	Muhlenberg	Ammonia (anhydrous)
Western Kentucky Energy - D. B. Wilson Station	Centertown	KY	Ohio	Ammonia (anhydrous), Chlorine
Trimble County Station	Bedford	KY	Trimble	Ammonia (anhydrous)
Western Kentucky Energy-Reid/Henderson/Green	Sebree	KY	Webster	Chlorine
Acadia Power Station	Eunice	LA	Acadia	Ammonia (anhydrous)

AEP-Plaquemine Cogeneration Facility	Plaquemine	LA	Iberville	Ammonia (anhydrous), Chlorine
Perryville Power Station	Sterlington	LA	Ouachita	Ammonia (anhydrous)
Big Cajun 2	New Roads	LA	Pointe Coupee	Chlorine
AES Warrior Run	Cumberland	MD	Allegany	Ammonia (anhydrous)
Brandon Shores Power Plant	Baltimore	MD	Anne Arundel	Ammonia (aqueous)
Montgomery County Resource Recovery Facility	Dickerson	MD	Montgomery	Ammonia (anhydrous)
Westbrook Energy Center	Westbrook	ME	Cumberland	Ammonia (anhydrous), Ammonia (aqueous)
Androscoggin Energy Center	Jay	ME	Franklin	Ammonia (aqueous)
Rumford Power Associates	Rumford	ME	Oxford	Ammonia (anhydrous), Ammonia (aqueous)
Maine Independence Station	Veazie	ME	Penobscot	Ammonia (aqueous)
Erickson Station	Lansing	MI	Eaton	Chlorine
J. B. Sims Generating Station	Grand Haven	MI	Ottawa	Chlorine
Zeeland Generating Plant	Zeeland	MI	Ottawa	Ammonia (aqueous)
Black Dog Generating Plant	Burnsville	MN	Dakota	Ammonia (aqueous)
Covanta Hennepin Energy Resource Company	Minneapolis	MN	Hennepin	Ammonia (anhydrous)
Rochester Meat Co.	Rochester	MN	Olmsted	Ammonia (anhydrous)
Aries Power Plant	Pleasant Hill	MO	Cass	Ammonia (anhydrous)
Sibley Generating Station	Sibley	MO	Jackson	Ammonia (aqueous)
KCPL - Hawthorn Generating Facility	Kansas City	MO	Jackson	Ammonia (anhydrous)
New Madrid Power Plant	Marston	MO	New Madrid	Ammonia (anhydrous)
Choctaw County Generation Station	French Camp	MS	Choctaw	Ammonia (aqueous)
Daniel Electric Generating Plant	Escatawpa	MS	Jackson	Ammonia (anhydrous)
PPL Montana	Colstrip	MT	Rosebud	Chlorine
Butler Warner GPlant	Fayetteville	NC	Cumberland	Ammonia (anhydrous)
Roxboro Steam Electric Plant	Semora	NC	Person	Ammonia (anhydrous)
Cliffside Steam Station	Cliffside	NC	Rutherford	Ammonia (anhydrous)
Belews Creek Steam Station	Belews Creek	NC	Stokes	Ammonia (anhydrous)
Coal Creek Station	Underwood	ND	Mclean	Chlorine
Salt Valley Generating Station	Lincoln	NE	Lancaster	Ammonia (anhydrous)
Lincoln Electric System Rokeby Station	Lincoln	NE	Lancaster	Ammonia (anhydrous)
PSNH Merrimack Generating Station	Bow	NH	Merrimack	Ammonia (anhydrous)
Newington Energy	Newington	NH	Rockingham	Ammonia (anhydrous)
American Reffuel of Essex County	Newark	NJ	Essex	Ammonia (aqueous)
Logan Generating Co., L.P.	Swedesboro	NJ	Gloucester	Ammonia (aqueous)

Bayonne Plant Holding, L.L.C.	Bayonne	NJ	Hudson	Ammonia (anhydrous)
Carneys Point Generating Co., L.P.	Carneys Point	NJ	Salem	Ammonia (aqueous)
Cogen Technologies Linden Venture, LP	Linden	NJ	Union	Ammonia (aqueous)
Plains Escalante Generating Station	Prewitt	NM	Mckinley	Chlorine
El Dorado Energy, L.L.C.	Boulder City	NV	Clark	Ammonia (aqueous)
Saguaro Power Company	Henderson	NV	Clark	Ammonia (anhydrous)
Bighorn Electric Generating Station	Primm	NV	Clark	Ammonia (aqueous)
Apex Generating Station	North Las Vegas	NV	Clark	Ammonia (anhydrous)
TRI-Center Power Plant	Mccarran	NV	Storey	Ammonia (anhydrous)
Saranac Power Partners, L.P.	Plattsburgh	NY	Clinton	Ammonia (aqueous)
CH Resources, Beaver Falls	Beaver Falls	NY	Lewis	Ammonia (aqueous)
AES Somerset L.L.C.	Barker	NY	Niagara	Ammonia (anhydrous)
Onondaga County Resource Recovery Facility	Jamesville	NY	Onondaga	Ammonia (anhydrous)
Onondaga Cogeneration Facility	Syracuse	NY	Onondaga	Ammonia (aqueous)
CH Resources, Syracuse	Solvay	NY	Onondaga	Ammonia (aqueous)
Massena Energy Facility	Massena	NY	St. Lawrence	Ammonia (anhydrous)
Ogdensburg Energy Facility	Ogdensburg	NY	St. Lawrence	Ammonia (aqueous)
AES Cayuga LLC	Lansing	NY	Tompkins	Ammonia (anhydrous)
DP&L - J.M. Stuart Generating Station	Manchester	OH	Adams	Ammonia (anhydrous)
Conesville Power Plant	Conesville	OH	Coshocton	Ammonia (anhydrous)
Grand River Dam Authority Coal Fired Complex	Chouteau	OK	Mayes	Chlorine
AECI CC Power Plant -- Chouteau Power Plant	Pryor	OK	Mayes	Ammonia (anhydrous)
PSO Northeastern Station	Oologah	OK	Rogers	Chlorine
PSO Riverside Power Station	Jenks	OK	Tulsa	Chlorine
Klamath Cogeneration Project	Klamath Falls	OR	Klamath	Ammonia (anhydrous)
Coyote Springs Plant	Boardman	OR	Morrow	Ammonia (anhydrous)
Hermiston Power Project	Hermiston	OR	Umatilla	Ammonia (anhydrous)
Hermiston Generating Plant	Hermiston	OR	Umatilla	Ammonia (aqueous)
Hunterstown Combined-Cycle Power Plant	Gettysburg	PA	Adams	Ammonia (aqueous)
Keystone Station	Shelocta	PA	Armstrong	Ammonia (anhydrous), Ammonia (aqueous)
FirstEnergy Bruce Mansfield Plant	Shippingport	PA	Beaver	Ammonia (aqueous)
Ontelaunee Energy Center	Reading	PA	Berks	Ammonia (anhydrous)
Colver Power Project	Colver	PA	Cambria	Ammonia (anhydrous)
Air Products, Cambria Cogen Company	Ebensburg	PA	Cambria	Ammonia (aqueous)
Panther Creek Energy Facility	Nesquehoning	PA	Carbon	Ammonia (anhydrous)
Shawville Station	Shawville	PA	Clearfield	Ammonia (anhydrous)
NorCon Power Partners, L.P.	North East	PA	Erie	Ammonia (anhydrous)
EME Homer City Generating, L.P.	Homer City	PA	Indiana	Ammonia (anhydrous)
Conemaugh Station	New Florence	PA	Indiana	Ammonia (anhydrous)

Seward Station	New Florence	PA	Indiana	Ammonia (aqueous)
Montour Steam Electric Station	Washingtonville	PA	Montour	Ammonia (anhydrous)
Conectiv Bethlehem Plant	Bethlehem	PA	Northampton	Ammonia (anhydrous)
Northampton Generating Company, LP	Northampton	PA	Northampton	Ammonia (aqueous)
Shamokin-Coal Township Joint Sewer Authority	Shamokin	PA	Northumberland	Chlorine
Ecoelectrica, L.P.	Penuelas	PR	Penuelas	Ammonia (anhydrous)
Tiverton Power Associates	Tiverton	RI	Newport	Ammonia (anhydrous)
Ocean State Power	Harrisville	RI	Providence	Ammonia (aqueous)
Pawtucket Power	Pawtucket	RI	Providence	Ammonia (anhydrous), Ammonia (aqueous)
Santee Cooper Cross Generating Station	Pineville	SC	Berkeley	Ammonia (anhydrous)
South Carolina Electric and Gas, Wateree Station	Eastover	SC	Richland	Ammonia (anhydrous)
TVA - Allen Fossil Plant	Memphis	TN	Shelby	Ammonia (anhydrous)
TVA - Cumberland Fossil Plant	Cumberland City	TN	Stewart	Ammonia (anhydrous)
Lost Pines 1 Power Plant	Bastrop	TX	Bastrop	Ammonia (anhydrous)
CPL La Palma Power Station	San Benito	TX	Cameron	Chlorine
Baytown Energy Center	Baytown	TX	Chambers	Ammonia (anhydrous)
Cedar Bayou Electric Generating Station	Eldon	TX	Chambers	Ammonia (aqueous)
W. A. Parish Electric Generating Station	Thompsons	TX	Fort Bend	Ammonia (aqueous)
P. H. Robinson Electric Generating Station	Bacliff	TX	Galveston	Ammonia (aqueous)
CPL Coleto Creek Power Plant	Fannin	TX	Goliad	Chlorine
Channel Energy Center	Houston	TX	Harris	Ammonia (anhydrous)
AES Deepwater Cogeneration Plant	Pasadena	TX	Harris	Ammonia (anhydrous)
Pasadena Cogeneration, L.P.	Pasadena	TX	Harris	Ammonia (anhydrous)
Pasadena P2 Power Plant	Pasadena	TX	Harris	Ammonia (anhydrous)
Reliant Energy Channelview, L.P.	Channelview	TX	Harris	Ammonia (aqueous)
Magic Valley Generation	Edinburg	TX	Hidalgo	Ammonia (anhydrous)
AES Wolf Hollow, L.P.	Granbury	TX	Hood	Ammonia (aqueous)
Blackhawk Station	Borger	TX	Hutchinson	Ammonia (anhydrous)
Tenaska IV Texas Partners, LTD.	Cleburne	TX	Johnson	Ammonia (anhydrous)
Lewis Creek Plant	Willis	TX	Montgomery	Ammonia (anhydrous)
CPL Lon C. Hill Power Station	Corpus Christi	TX	Nueces	Chlorine
SRW Cogeneration Limited Partnership - SRWCLP	Orange	TX	Orange	Ammonia (aqueous)
Monticello Steam Electric Station	Mt. Pleasant	TX	Titus	Ammonia (anhydrous)
WTU San Angelo Power Station	San Angelo	TX	Tom Green	Chlorine
CPL Victoria Power Station	Victoria	TX	Victoria	Chlorine
Intermountain Generating Station	Delta	UT	Millard	Chlorine
Altavista Power Station	Altavista	VA	Campbell	Ammonia (anhydrous)
Chesapeake Energy Center	Chesapeake	VA	Chesapeake (City)	Ammonia (anhydrous), Chlorine
Chesterfield Power Station	Chester	VA	Chesterfield	Ammonia (anhydrous)

Doswell Combined Cycle Facility	Ashland	VA	Hanover	Ammonia (anhydrous)
Birchwood Power Facility	King George	VA	King George	Ammonia (anhydrous)
Gordonsville Energy L.P.	Gordonsville	VA	Louisa	Ammonia (aqueous)
Hopewell Power Station	Hopewell	VA	Prince George	Ammonia (anhydrous)
Bellemeade Power Station	Richmond	VA	Richmond (City)	Ammonia (anhydrous)
Moses Lake Generating	Moses Lake	WA	Grant	Ammonia (aqueous)
Frederickson Power LP	Tacoma	WA	Pierce	Ammonia (aqueous)
March Point Cogeneration Company	Anacortes	WA	Skagit	Ammonia (anhydrous)
Spokane Regional Waste-to-Energy Facility	Spokane	WA	Spokane	Ammonia (anhydrous)
Encogen Northwest Cogeneration Plant	Bellingham	WA	Whatcom	Ammonia (anhydrous)
Tenaska Washington Partners, LP	Ferndale	WA	Whatcom	Ammonia (anhydrous)
Sumas Cogeneration L.P.	Sumas	WA	Whatcom	Ammonia (anhydrous)
Pleasant Prairie Power Plant	Pleasant Prairie	WI	Kenosha	Ammonia (aqueous)
WEPCO Germantown Turbine Inlet Cooling System	Germantown	WI	Washington	Ammonia (anhydrous)
Mt. Storm Power Station	Mt. Storm	WV	Grant	Ammonia (anhydrous)
Wygen 1	Gillette	WY	Campbell	Ammonia (anhydrous)

Endnotes

[1] Industries other than power plants with significant chemical release dangers include some chemical manufacturers, oil refineries, and drinking water and wastewater treatment plants, among others.
[2] American Lung Association, State of the Air: 2004, April 2004.
[3] U.S. Environmental Protection Agency 1986 National Air Pollution Emission Estimates, 1940-1984, EPA-450/4-85-014 (January), Office of Air Quality Planning and Standards, U.S. Environmental Protection Agency (Research Triangle Park, North Carolina).
[4] Fossil fuel-fired power plants produce NOx emissions, regardless of whether coal, oil, or natural gas is burned to generate power.
[5] Prevention technologies include low NOx burners, flue gas recirculation, fuel reburning, and steam or water injection. These technologies may, however, decrease combustion efficiency or increase particulate pollution.
[6] NOx Control White Paper, Plant Automation Services, 2001.
[7] Information sources include: New Jersey Hazardous Substance Fact Sheets (www.state.nj.us/health/eoh/rtkweb/rtkhsfs.htm) and National Library of Medicine Hazardous Substance Data Bank (www.toxnet.nlm.nih.gov).
[8] Ibid.
[9] These figures exclude transportation. National Response Center, www.nrc.uscg.mil/foia.html.
[10] A cursory review of chemical spills reported to the National Response Center found 32 incidents involving chlorine gas at power plants.
[11] Belke, James C., Chemical accident risks in U.S. industry – A preliminary analysis of accident risk data from U.S. hazardous chemical facilities, U.S. Environmental Protection Agency, September 25, 2000.
[12] Radig, Scott, Catastrophic Anhydrous Ammonia Release, Minot, North Dakota, North Dakota Department of Health, Division of Water Quality, 2002.
[13] World Watch: Deadly Chemical Spill, St. Petersburg Times, July 9, 2002.
[14] River back to normal after summer ammonia spill, The Daily Illini, August 29, 2002.
[15] Where two or more power plant vulnerability zones overlapped we eliminated the lesser of the vulnerability zones when calculating residential population in danger.

[16] "Residents in danger" indicates the residential population within the vulnerability distance. Others may work, travel, or go to school within the vulnerability distance.

[17] "Vulnerability in miles" indicates the distance within which exposure to the high chemical concentration of a toxic cloud may lead to severe health effects or death for people who are unable to readily escape.

[18] Where two or more power plant vulnerability zones overlapped we eliminated the lesser of the vulnerability zones when calculating residential population in danger.

[19] For a list of federal reading rooms that hold Risk Management Plan information see http://yosemite.epa.gov/oswer/ceppoweb.nsf/content/readingroom.htm or call the Environmental Protection Agency at 800-424-9346 or the Department of Justice at 888-442-9267.

[20] "Power company changes plan after local outcry," *Baltimore Sun*, October 7, 2000.

[21] Public statement, American Electric Power, December 19, 2000.

[22] Walker, Hamilton G., James J. Ferrigan, Ammonia-on-Demand™ Installations at American Electric Power's Clifty Creek and Kyger Creek Plants Provide 100% Availability and Safe Operation, Hera LLC, undated.

[23] EPA cites six general methods of reducing NOx: reducing combustion temperatures, reducing time at peak temperatures, chemical reduction, oxidation, removing nitrogen from combustion, and sorption. Technical Bulletin: Nitrogen Oxides (NOx), Why and How They Are Controlled, Clean Air Technology Center, U.S. Environmental Protection Agency, EPA-456/F-99-006R, November 1999.

[24] Bitler, John R., Alternatives to Ammonia Dependent NOx Control Technologies for Cogeneration and DHC Applications, Levitan & Associates, undated.

[25] Slocomb, Stephen H., and Dale T. Raczynski, Comparison of the Cost Effectiveness of New NOx Control Technologies and Conventional Selective Catalytic Reduction for Combined Cycle Combustion Turbine Power Plants, Epsilon Associates, Inc., 2001.

[26] Slocomb and Raczynski, op. cit.

[27] EnviroScrub Homepage, www.enviroscrub.com, accessed April 2004, and communication with EnviroScrub's Curt Steinbergs, March 15, 2004.

[28] PowerSpan website (www.powerspancorp.com) and communication with PowerSpan's Chris McLarnon, April 16, 2004.

[29] Technifax: Cooling Water Chlorination, Nalco Chemical Company, 1998.

[30] Technical Topic 030-0898: Testing and Control of Bromine-Based Biocides, and Technical Topic 029-0798: Bromine Safety, International Chemtex Corporation, 1998.

[31] Lane, John and Gerald Kutner, A Non-Chemical Water Treatment for Cooling Towers, February 2000, and personal communication with John Lane, April 12, 2004.

[32] Potassium Permanganate; EPA Guidance Manual, Alternative Disinfectants and Oxidants, U.S. Environmental Protection Agency, April 1999.

[33] Wounded Waters: The Hidden Side of Power Plant Pollution, Clean Air Task Force, February 2004.

[34] SealCoat website, www.sealcoats.com/fr_cpro.htm, accessed April 2004.

[35] Advanced Nontoxic Anti-Fouling Coatings, Environmental Security Technology Certification Program, Department of Defense, www.estcp.org/projects/pollution/199502v.cfm.

[36] National Response Center, Incident Report 575686, report taken August 7, 2001.

[37] Press Release: Statement by the Department of Homeland Security on Continued Al-Quada Threats, Department of Homeland Security, November 21, 2003.

[38] Assessment of the Increased Risk of Terrorist or Other Criminal Activity Associated With Posting Off-site Consequence Analysis Information on the Internet, U.S. Department of Justice, April 18, 2000; and, A Method to Assess the Vulnerability of U.S. Chemical Facilities, National Institute of Justice, U.S. Department of Justice, November 2002.

[39] Strategic Plan for Homeland Security, U.S. Environmental Protection Agency, September 2002.

[40] Homeland Security: Voluntary Initiatives Are Under Way at Chemical Facilities, but the Extent of Security Preparedness is Unknown, U.S. General Accounting Office, GAO-03-439, March 14, 2003.

[41] CRS Report to Congress: Chemical Plant Security, Congressional Research Service, January 2003.

[42] Industrial Chemicals and Terrorism: Human Health Threat Analysis, Mitigation and Prevention, Agency for Toxic Substances and Disease Registry, 1999; and, Terrorist Use of Expedient Chemical Agents: Health Risk Assessment and Las Vegas Case Study, Agency for Toxic Substances and Disease Registry, undated.

[43] Testimony of Dr. Jay Boris of the Naval Research Laboratory before the Committee on Public Works and the Environment of the Council of the District of Columbia, January 23, 2004.

[44] Study Assesses Risk of Attack on Chemical Plant, *Washington Post*, March 12,2002.

[45] The Terrorist Threat in America, Chemical Manufacturers Association (American Chemistry Council), April 1998.
[46] Protecting the American Homeland, Brookings Institution, March 2002.
[47] Toxic Warfare, RAND Corporation, 2002.
[48] News Release: Chemical Facilities Vulnerable, Center for Strategic and International Studies, December 23, 2003.
[49] See Eliminating Hometown Hazards: Cutting Chemical Risks at Wastewater Treatment Facilities, Environmental Defense, December 2003; and, Needless Risk: Oil Refineries and Hazard Reduction, U.S. Public Interest Research Group, October 2003; and, The Safe Hometowns Guide, The Safe Hometowns Initiative, 2002.
[50] Chemical Plant Security Breaches in the News, Working Group on Community Right-to-Know, February 18, 2004.
[51] Chemical Site Security: EPA's Strategy for Improving Site Security and Preventing Releases Caused by Criminal Attacks at Hazardous Chemical Facilities (Pre-Decisional Draft), U.S. Environmental Protection Agency, June 2002; and, "EPA Drops Chemical Security Effort," *Washington Post*, October 3, 2002.
[52] Letter to Editor from Tom Ridge, Director, Office of Homeland Security and Christine Whitman, Administrator, Environmental Protection Agency, *Washington Post*, October 6, 2002.
[53] Senate Dear Colleague letter signed by Senators Inhofe (R-Okla.), Smith (R-N.H.), Specter (R-Pa.), Voinivich (R-Ohio), Domenici (R-N.M.), Crapo (R-Idaho), Bond (R-Mo.), Chemical Security Bill Misses the Mark, September 10, 2002.

NRDC
THE EARTH'S BEST DEFENSE

NATURAL RESOURCES DEFENSE COUNCIL

STATEMENT OF
JON P. DEVINE, JR.
SENIOR ATTORNEY
NATURAL RESOURCES DEFENSE COUNCIL

BEFORE THE
U.S. SENATE COMMITTEE ON HOMELAND SECURITY
AND GOVERNMENTAL AFFAIRS

FOR THE RECORD OF THE HEARING ENTITLED:

"CHEMICAL FACILITY SECURITY:
WHAT IS THE APPROPRIATE FEDERAL ROLE?"

HEARING DATE: JULY 13, 2005
SUBMITTED: JULY 27, 2005

www.nrdc.org
1200 New York Avenue, NW, Suite 400
Washington, DC 20005
TEL 202 289-6868 FAX 202 289-1060

NEW YORK · LOS ANGELES · SAN FRANCISCO

My name is Jon Devine, and I am a Senior Attorney in the Health and Environment Program of the Natural Resources Defense Council. NRDC is a national, non-profit organization of scientists, lawyers and environmental specialists dedicated to protecting public health and the environment. Founded in 1970, NRDC has more than 1.2 million members and e-activists nationwide. My work at NRDC focuses on reducing the public's exposure to toxic chemicals. At NRDC and in my prior position as an attorney in the Environmental Protection Agency's Office of General Counsel, I have worked on implementing a number of provisions of the Clean Air Act, including the risk management planning requirements applicable to chemical facilities.

Thank you for inviting me to testify about the importance of securing a vital part of our nation's critical infrastructure – facilities that produce, store, or use large amounts of highly toxic chemicals. I commend you for your serious consideration of this issue and for your commitment to developing legislation to ensure that there is a fair and consistent set of requirements to help these plants guard against acts of terrorism. I also want to thank your staff for their dedication to this crucial issue and for their willingness to meet with me and my colleagues from the public health community to explore the details of a potential chemical security bill.

The dangers of chemical plants are by now well known. It has often been repeated, for instance, that over 100 facilities could endanger 1 million or more people if attacked, but focusing upon that statistic masks the scope of the problem. Thousands of plants each put thousands of people at risk; according to the Congressional Research Service, an analysis by an EPA specialist "found that . . . at least 3,000 facilities could threaten 10,000 people in the vicinity" and the Department of Justice "analyzed EPA data and concluded that among facilities submitting risk management plans to EPA, more than 7,000 facilities projected worst case scenarios for toxic substances that could potentially affect more than 1,000 people."[1]

It is reasonable to think that these plants will be targeted. The tragic events in London are a striking reminder that terrorists are continuing to seek opportunities to cause widespread harm and mayhem. Moreover, we must heed the lessons of September 11 – the terrorists will not hesitate to turn our own infrastructure against us. As Stephen Flynn of the Council on Foreign Relations told this Committee:

> There are hundreds of chemical facilities within the United States that represent the military equivalent of a poorly guarded arsenal of weapons of mass destruction. Terrorists do not need to produce or procure chemical weapons and smuggle them into the United States. Just as on 9/11 they converted domestic airliners into missiles that destroyed the twin towers, they can target facilities that manufacture or conveyances that transport such lethal chemicals as chlorine, anhydrous ammonia, boron triflouride, cyanide, and nitrates. These facilities are found around the country in industrial parks, in seaports, and near the major population centers.[2]

In other words, in light of the terrorists' cunning and deadly purpose, chemical facilities must be considered a prime target.

Because chemical plants are attractive to terrorists, and because an attack could be catastrophic, there is not a serious debate today that these facilities need to be secured. Indeed, a Sense of the Senate resolution on chemical plant security passed unanimously this month, concluding that "Congress should pass legislation establishing enforceable Federal standards to protect against a terrorist attack on chemical facilities within the United States."[3] Moreover, this committee has heard expert testimony that leaving plant security to the voluntary efforts of the industry will not be sufficient to ensure that facilities are adequately protected, and that the

[1] Linda-Jo Schierow, Congressional Research Service, *Chemical Plant Security*, at 9 (updated Jan. 20, 2004) (citations omitted).
[2] Stephen E. Flynn, Ph.D., "Ending the Post 9/11 Security Neglect of America's Chemical Facilities," Testimony before the Senate Homeland Security & Government Affairs Committee, at 2 (Apr. 27, 2005).
[3] H.R. 2360, 109th Cong., 1st Sess., § 533 (July 14, 2005).

nation needs national standards. From the Government Accountability Office to the Department of Homeland Security (DHS) and even to the American Chemistry Council (ACC) itself, there is widespread agreement that there should be industry-wide standards.

Accordingly, the real debate is about how secure these facilities must be made:

- Will legislation deprive DHS of the authority to enforce one particular kind of security measure – using alternative chemicals, processes, storage volumes, and operating conditions so as to make the facility a less attractive target?
- Will legislation allow some companies to do less than others simply because they belong to an industry association or because they are subject to other security requirements?
- Will legislation simply assume that companies will do what the rules specify, or will it require DHS to receive and review facilities' vulnerability assessments and security plans to ensure they are sufficient?
- Will legislation prohibit experts at EPA from assisting DHS in its security mission?

These issues have been debated in the development of prior legislative proposals, and my testimony focuses on how best to answer these crucial questions.

Inherently Safer Design

Perhaps the most contentious issue that has arisen during the debate about chemical plant security legislation has been the role of "inherently safer technology," in the words of Senator Corzine's bill (S. 157, 108th Congress), or "alternative approaches," in the words of the bill that was reported out of the Environment and Public Works Committee last Congress (S. 994). What is an "inherently safer" approach? Simply put, it is a way that a company can do the same thing it has been doing, but in a way that makes it less likely or impossible to experience a catastrophic chemical release. For instance, the attached report, *New Strategies to Protect America: Securing our Nation's Chemical Facilities*, was authored by my NRDC colleague, Dr. Linda Greer, and it discusses four kinds of inherently safer approaches: "materials substitution, just-in-time manufacturing, inventory reduction, and hardened storage."[4]

Because it can eliminate the threat to the public posed by terrorism at chemical facilities, materials substitution is the most effective alternative approach. Materials can be substituted in a number of common industrial processes so as to avoid the use of an acutely toxic chemical altogether or in amounts that may be attractive to terrorists:

- Wastewater treatment plants do not need to use chlorine gas to disinfect water, but can use sodium hypochlorite or ultraviolet light.[5]

- Petroleum refineries can, with some equipment changes, substitute less hazardous sulfuric acid for hydrofluoric acid as a process catalyst, and an emerging technology, the solid acid catalyst, can eliminate the need for either hydrofluoric acid or sulfuric acid.[6]

- Power plants that use anhydrous ammonia in their pollution control systems can use safer substitutes, including aqueous ammonia or solid urea.[7]

- Though more difficult, polycarbonate plastics can be manufactured using a transesterification process instead of relying on phosgene gas. Similarly, polyurethane foams can be produced with alternatives to toluene isocyanate.[8]

Replacing hazardous chemicals with less dangerous alternatives with a lower potential to cause harm to workers or nearby residents will make a facility a less attractive target for terrorists, and will minimize the harm that a successful terrorist attack could cause.

But replacing chemicals is only a part of inherent safety; numerous other measures make a plant less of a hazard to its neighbors if attacked. Companies that lack substitutes for a dangerous chemical can synthesize the chemical immediately before it is needed in a process, rather than storing a large amount on site. This approach is known as just-in-time

[4] Dr. Linda Greer, *New Strategies to Protect America: Securing our Nation's Chemical Facilities*, at 5.
[5] Carol Andress, *Eliminating Hometown Hazards: Cutting Chemical Risks at Wastewater Treatment Facilities*, at 7 (2003).
[6] Meghan Purvis & Warren Claflin, *Needless Risk: Oil Refineries and Hazard Reduction*, at 16-17 (Oct. 2003).
[7] Paul Orum, *Unnecessary Dangers: Emergency Chemical Release Hazards at Power Plants*, at 5 (July 2004).
[8] Greer, *New Strategies to Protect America* at 6.

manufacturing. Likewise, plants that need to have large quantities of a hazardous chemical in storage can reduce the size of the storage vessels they use, which would make it more difficult to cause a catastrophic release. Reducing hazardous temperatures or pressures in a company's chemical process also can make a release less likely or less deadly should there be an act of terrorism. Beyond these options, Senator Corzine's bill would define inherent safety to include "reduc[ing] the possibility and potential consequences of equipment failure and human error" and "improv[ing] inventory control and chemical use efficiency".[9]

There has been significant agreement that these strategies are – like guards, fences, and security cameras – effective at reducing the potential hazard posed by terrorism at chemical plants. For instance, the industry security code developed by ACC provides that member companies should, as part of implementing security measures, "tak[e] into account inherently safer approaches to process design. . . ."[10] Similarly, a February 2000 EPA alert about site security notes that "[c]onsidering inherent safety in the design and operation of any facility will have the benefit of helping to prevent and/or minimize the consequences of any release."[11] And both bills that have passed the Environment and Public Works Committee directed companies to evaluate inherently safer approaches as part of designing their site security plans.[12]

Notwithstanding the widespread acceptance of safer technologies' role in facility security, the notion that chemical security legislation might give DHS the authority to require companies to implement such strategies has been demonized. For instance, after Senator Corzine's bill passed the Senate Environment and Public Works Committee unanimously, numerous industry groups told the Senate that an inherent safety requirement would "grant[] new

[9] S. 157, 108th Cong., 1st Sess. §§ 3(11)(B)(iv) &(v) (Jan. 14, 2003).
[10] American Chemistry Council, Responsible Care® Security Code of Management Practices, at 2-3.
[11] U.S. EPA, Chemical Accident Prevention: Site Security, at 5 (Feb. 2000).
[12] S. 994, 108th Cong. 1st Sess. § 3(a)(2)(C); S. 1602, 107th Cong., 2d Sess. § 4(a)(3)(B).

authority to allow government micromanagement in mandating substitutions of all processes and substances."[13] Others have suggested that inherent safety is an idea promoted by environmentalists trying to sneak their agenda into homeland security legislation.[14] Although it is undeniably true that facilities making changes to reduce their potential hazards will make communities safer from *both* terrorism and accidents, it is odd to consider this fact a strike against the requirement. As advocates of including inherent safety in security legislation, we are concerned about our members, our neighbors, and our families, and we understand that purely environmental goals are not paramount in the present debate. For that reason, we hail a wastewater treatment plant that converts from chlorine gas to sodium hypochlorite, even though the replacement is still a hazardous chemical and even though we might prefer that it use an environmentally benign technology like ultraviolet light, because making the change eliminates the danger that the plant could release a deadly gas if attacked.

In addition, we do not oppose reasonable limits on the application of safer technologies. For instance, Senator Corzine's bill would have required facilities to implement safer design elements "to the extent practicable,"[15] and the Committee report explained that "[t]he term 'practicable,' as used in the definition of 'safer design and maintenance,' is intended to incorporate consideration of both technical feasibility and cost."[16] This approach is appropriate; facilities should do what is cost-effective, and DHS should retain the authority to insist on more if there is an emergency or if the Department has specific intelligence indicating that a given facility may be targeted. Similarly, we believe that an alternative approach is not truly "safer" if

[13] Letter from Agricultural Retailers Assn. et al., to Senators (Aug. 29, 2002) (on file with NRDC).
[14] *See, e.g.,* Paul Rosenzweig, Heritage Foundation, *The Chemical Security Act: Using Terrorism as an Excuse to Criminalize Productive Economic Activity* (Sept. 12, 2002) ("some backers of the bill are attempting to marry the long-held environmentalist agenda of chemical-use reduction to the war on terrorism"), available online at http://www.heritage.org/Research/HomelandDefense/em833.cfm (visited July 21, 2005)
[15] S. 1602, 107th Cong., 2d Sess. § 3(6).
[16] S. Rep. 107-342, 107th Cong., 2d Sess., at 5 (Nov. 15, 2002).

it creates a collateral risk that outweighs the anti-terrorism benefit (for instance, if reducing on-site storage requires increased transportation, and if that transportation is more dangerous than having the storage volumes at the plant), so companies should not be forced to alter their processes if doing so would increase risk. Finally, we believe that facilities' security requirements should reflect the danger they pose, so those facilities that endanger a small number of people should face less onerous obligations – and this also applies to the adoption of safer technologies.

Some have suggested that companies should have the final word on whether safer approaches will be used. This has taken at least two forms – the bill that passed the Environment and Public Works Committee last Congress would have required a company to implement a safer option if the approach is "practicable *in the judgment of the owner or operator* of the chemical source,"[17] and Dr. Richard Falkenrath has argued that a tiered conventional-security-only regime would lead plants to make changes to reduce their security obligations and thereby "create market-based incentives for the chemical industry to reduce the inherent danger of their facilities and practices."[18] However, denying DHS enforcement authority for alternative approaches will not ensure the security improvements that our nation needs chemical plants to make. As we have seen in the years between September 11th and now, companies have been reluctant – even when there are recognized and cost-effective safer solutions – to make investments to reduce the dangers of their operations,[19] and thus turning over complete authority to decide what is practicable to company managers will not guarantee a full assessment of available options and the adoption of those approaches that are available and affordable. With regard to Dr.

[17] S. 994, 108th Cong. 1st Sess. § 3(a)(2)(C) (emphasis added).
[18] Testimony of Richard A. Falkenrath, Visiting Fellow, Brookings Institution, Before the Senate Committee on Homeland Security & Governmental Affairs, at 17 (Apr. 27, 2005).
[19] *See, e.g.*, Andress, *Eliminating Hometown Hazards*, at 7 (in 2003, "an estimated 19 million people remain at risk from the 45 wastewater facilities that use toxic chlorine gas in populated areas").

Falkenrath's suggestion, we agree that tiered standards are appropriate and will encourage the adoption of safer technologies, but we believe that relying entirely on this incentive is insufficient to secure some plants. We have seen too many examples of lax physical security at chemical plants to rely only on "guards, gates, and guns" to protect against the possibility of an attack – we need to also take steps to make sure that plants are less dangerous if an attack is successful. According to a recent account, "Georgia State University professor Sal DePasquale, a former security director for the U.S. Department of Energy who helped draft the American Chemistry Council's voluntary standards, told House leaders that 'night security is better at most liquor stores in our cities than what you'll find at our chemical plants.'"[20] Moreover, a number of facilities are likely to be extraordinarily hard to guard, because they are located along public roadways or rivers or are sprawling, with lengthy perimeters.

Consequently, DHS should be given the authority to enforce a requirement to use cost-effective and safer approaches, just as legislation will surely give the Department the power to force companies to make physical security upgrades. Because DHS would only need to step in when a company refused to adopt an approach that was clearly practicable (for instance, if it had been employed elsewhere in the industry and there were no site-specific barriers to its use), we do not expect this authority to be needed frequently. After all, in enforcing the Marine Transportation Security Act (MTSA) of 2002, the Coast Guard took 312 enforcement actions against facility owners for noncompliance, out of over 2,900 plants subject to the Act.[21] More to the point, the government only "imposed operational controls on 29 MTSA facilities, such as

[20] Carl Prine, *Radical changes in chemical plant security urged*, Pittsburgh Tribune-Review (June 16, 2005).
[21] U.S. Government Accountability Office, *Protection of Chemical and Water Infrastructure: Federal Requirements, Actions of Selected Facilities, and Remaining Challenges*, at 8 (Mar. 2005).

suspending certain facility operations,"[22] which indicates to us that DHS is hardly eager to meddle with ongoing plant processes.

Exemptions for Other Security Programs

A credible security bill should not allow companies to opt out of the federal requirements simply because they belong to an industry with a self-designed security scheme or because they are subject to separate federal, state, or local security requirements. For example, one Senate bill would have allowed a company or industry trade association to petition to have the Secretary of Homeland Security "endorse" its security program if the Secretary found it to be "substantially equivalent" to the federal requirements, after which point companies could comply with the substitute program instead of the Department's rules.[23] This idea is flawed in several ways.

First, this idea is an overreaction to a problem the industry identified when the notion of chemical security legislation was initially being debated in the Environment and Public Works Committee. The industry argued that new legislation might delay work that facilities had begun doing in the wake of September 11th – the idea was that companies would wait to see what the legislation would require rather than make improvements immediately.[24] Now, however, several years have elapsed and the industry's security code has been implemented by many, if not all, ACC member companies,[25] so the concern that the legislation will slow down voluntary efforts should have significantly abated.

---

[22] *Id.*
[23] S. 994, 108th Cong. 1st Sess. §§ 3(c) & (d).
[24] *See* Eric Pianin, *U.S. Faulted on Chemical Plants' Security: Government Inaction Leaves Industry Vulnerable Target to Terrorists, Critics Say*, Washington Post, at A10 (June 13, 2002) ("Chris VandenHeuvel, a spokesman for the American Chemistry Council, an Arlington-based trade group that represents firms such as Dow Chemical Co. and ExxonMobil Corp., said new legislation or government mandates would merely 'slow down our efforts.'").
[25] Testimony of Martin J. Durbin, Managing Director, Security & Operations, American Chemistry Council, Before the Senate Committee on Homeland Security & Governmental Affairs, at 4 (July 13, 2005) ("All 2,040 ACC member company facilities have completed their vulnerability assessments, and virtually all have completed their enhancement verifications.").

Second, the "substantially equivalent" standard is vague and could permit DHS to approve a program that is weaker than the federal requirements. For instance, is an industry program that requires guards to be on duty from 6 A.M until midnight "substantially equivalent" to a federal 24-hour guard requirement? Is it "substantially equivalent" to have companies evaluate safer technologies, as ACC urges its members to do, if the federal regulations require companies to implement those alternative approaches that are practicable? Or is it "substantially equivalent" to have a program where third parties, rather than DHS, review facility vulnerability assessments and security plans, as they do under ACC's Responsible Care® Security Code, even if those third parties are not required "to verify that a vulnerability assessment was conducted appropriately or that actions taken by a facility adequately address security risks"?[26] We fail to see how it is a good security policy to allow some facilities – including ones that could endanger huge populations — to do less than others for no other reason than their participation in an industry program.

Third, it would add unnecessary complication to have different facilities subject to different standards, and to have DHS enforcing a variety of industry, state, or local security requirements. The Department has concluded that the current state of affairs is unacceptable; as Robert Stephan, Acting Undersecretary for Information Analysis and Infrastructure Protection, reported to this Committee, "the existing patchwork of authorities does not permit us to regulate the industry effectively."[27] In the same way, the Committee should not replace the current approach with legislation that could lead to a similarly balkanized set of requirements for the industry.

---

[26] *See* GAO, ***Protection of Chemical and Water Infrastructure*** at 18.
[27] Testimony of Robert Stephan, Acting Undersecretary for Information Analysis and Infrastructure Protection, Department of Homeland Security, Before the Senate Homeland Security & Governmental Affairs Committee, at 2 (June 15, 2002)

Finally, providing this new loophole is particularly unnecessary, as there is an obvious solution to the one legitimate complaint that industry has raised on this point – that it would be inappropriate to force companies to re-do work they have already done to comply with an industry security program or some other requirement. A bill could allow individual companies to submit pieces of work performed to comply with other laws or industry programs (for instance, a vulnerability assessment done under the Bioterrorism Act or MTSA), as part of meeting their obligations under chemical security legislation, and simply require them to supplement those submissions with any additional information required by the chemical security bill.

Ensuring DHS Review of Company Assessments and Plans

Responsible oversight of chemical plants is an essential element of credible chemical security legislation. Although the real work of assessing plant vulnerabilities, developing security plans, and making necessary improvements will be done by the plant managers and other company experts who know the most about their unique processes, there must be federal supervision to guard against noncompliance, to ensure consistency across the industry, and to monitor trends at chemical facilities. Accordingly, the legislation needs to require facilities to develop vulnerability assessments and security plans, but also submit them to DHS for review and approval.

In this respect, the bill introduced last Congress by Senator Inhofe, S. 994, was a complete failure. First, the bill only required companies to submit vulnerability assessments and security plans upon request by DHS.[28] Such an approach creates an obvious "chicken-and-egg" problem; the government would not know if a facility is vulnerable to terrorism without seeing its vulnerability assessment (assuming, of course, that DHS will not assess the vulnerabilities of

[28] S. 994, 108th Cong., 1st Sess., § 4(a)(1)(C) (May 5, 2003) (version as introduced).

each of the approximately 15,000 chemical plants), and it will not know which plants' vulnerability assessments and security plans it needs to review without knowing which facilities were most dangerous. Second, the bill would direct DHS to establish "deadlines" for submitting facility assessments and plans, but did not specify when those deadlines would need to fall, thereby allowing indefinite delay in companies' compliance.[29] Third, the bill would instruct DHS to "ensure and evaluate" sources' compliance with the act, but only "at such times and places as the Secretary determines to be appropriate,"[30] which would allow facility assessments and plans to go completely unreviewed unless DHS took affirmative steps to look at each of them. Last Congress, the Environment and Public Works Committee corrected the first two of these flaws, and we implore this Committee to make sure that all plans are submitted, reviewed, and assessed for adequacy against each of the bill's standards.

Using Other Agencies' Expertise

Finally, we want to stress that, even though DHS may have the formal lead within the government on issues of chemical plant security, EPA retains significant expertise that should be brought to bear in designing and implementing chemical security regulations. Indeed, in light of EPA's experience with chemical facilities and processes and its authority in implementing the requirements of the accidental release program of the Clean Air Act, Senator Corzine's bill would have had EPA administer the program, and EPA itself had developed a plan to require plants to upgrade security several years ago.[31]

---

[29] *Id.* § 4(a)(2).
[30] *Id.* § 4(b)(3).
[31] Alexander Lane, *Whitman: GOP foiled security efforts: Book says legislators helped lobbyists defeat rules for chemical plants*, Star-Ledger (Jan. 28, 2005) ("Whitman wrote that she and Homeland Security Secretary Tom Ridge crafted rules requiring the 15,000 most high-risk plants to 'take reasonable steps to address those vulnerabilities, and report to the EPA that they had complied.' * * * She said she grew so frustrated she formally asked the White House to 'relieve EPA of its lead responsibility for reducing the vulnerability of the chemical sector to attack.'").

Unfortunately, as the debate on chemical security legislation has gone on, EPA's experience and expertise has been undercut. For instance, the committee-passed bill last Congress would bar other federal agencies from performing "field work" to support DHS in implementing chemical security legislation.[32] It is unclear what exactly "field work" entails, but it would seem that EPA facility inspectors would not be authorized to report to DHS on obvious security lapses or safer techniques that could be employed at the plant.

We believe that EPA should be directly involved in the administration of chemical plant security legislation. At the same time, we recognize that DHS will have expertise on matters of physical security and intelligence about likely terrorist targets, so it is reasonable to have the Department take the lead. DHS should consult with EPA on matters on which the agency has greater expertise, such as the use of safer technologies, and the particular hazards posed by different kinds of processes. This should, we believe, include giving EPA a formal role in reviewing the relevant portions of company assessments and plans. To do otherwise would force DHS to develop expertise similar to EPA's in chemical toxicity and safer technologies, which could be a waste of government anti-terror resources. At a minimum, the nonsensical prohibition on "field work" (the need for which has never been explained, as far as we know) should not be included in this Committee's proposed legislation.

Conclusion

Thank you for this opportunity to present our views on this crucial issue. I applaud your serious and thorough attention to chemical security, and I hope that these thoughts, along with our experience in working on the issue in previous Congresses, will be of assistance to you as you go forward. We are eager to continue to work with you to make our nation more secure and to combat the possibility of terrorism at chemical plants.

---

[32] S. 994, 108th Cong. 1st Sess. § 5(1).

Center for American Progress

Critical Infrastructure Security Series

New Strategies to Protect America: Securing our Nation's Chemical Facilities

by Dr. Linda Greer

CRITICAL INFRASTRUCTURE SECURITY SERIES

New Strategies to Protect America: Securing our Nation's Chemical Facilities

EXECUTIVE SUMMARY

More than three years after the attacks of 9/11, our nation's chemical manufacturing and transport facilities remain extremely vulnerable to terrorism. The Bush administration's reliance on voluntary actions by the chemical industry has failed to produce sufficient change at the nation's 15,000 facilities that use or produce deadly chemicals. Yet there are readily available hazard-reduction techniques including replacing the most dangerous chemicals with less toxic substances, reducing the amount of chemicals we store, and hardening facilities to both deter and protect against potential attacks.

Voluntary approaches have failed to accomplish what our national security requires, leaving us with no choice: the administration must put aside its ideological resistance to federal action, overcome private sector inertia, force a change in the status quo, and put into place new safety measures. Tax incentives, up-front low interest loans or homeland security grants can be used to speed the process and relieve some financial burden on the industry. Action cannot wait.

In this, the first in a series of papers on protecting our critical infrastructure, the Center for American Progress sets out a 12-month action plan to reduce the risks posed by the nation's chemical facilities. We recommend using existing government and industry data to create a priority list of the most vulnerable facilities that produce or use the most dangerous toxic chemicals, making them prospective terrorist targets. We then call for immediate leadership by the Environmental Protection Agency to write new, effective guidelines on reducing hazards. This would be followed by audits of these priority facilities and the creation of plans to use state-of-the-art techniques to increase safety. Facility operators who fail to aggressively implement these plans would be subject to strong enforcement action and significant penalties. Finally, we recommend that the government fund a new program devoted to the longer term task of developing safer alternatives to today's deadly chemicals. Taken together, these steps form a specific, concrete and actionable plan to protect our communities and make our country more secure.

BACKGROUND

The attacks of 9/11 tragically revealed our nation's vulnerability to terrorism. In response, Congress and the Bush administration committed to secure "critical infrastructure and key assets" under the Patriot Act.[1] The 2002 National Strategy for Homeland Security established critical infrastructure protection as a "critical mission area."[2] But today, despite this obligation, there has been insufficient progress in protecting the public from potential attacks on the nation's most dangerous chemical facilities, which remain vulnerable high-impact targets for terrorists intent on damaging our nation.[3]

The U.S. Environmental Protection Agency (EPA) estimates that more than 100 chemical facilities in 24 states threaten *more than a million people each*,[4] and the industry's own assessments do not disagree.[5] These are only the biggest targets among the nation's more than 15,000 chemical plants. "No one needed to convince us that we could be – and indeed would be – a target at some future date," the president of the American Chemistry Council (a Washington-based industry association) said a few weeks after 9/11. "If you are looking for a big bang, obviously you don't have to go far in your imagination to think about what the possibilities are."[6] Recent accidents underscore the severity of the risk. For example, in January, a train carrying chlorine gas derailed near Graniteville, South Carolina. The resulting release of the gas killed nine, but might have caused more than 100,000 deaths in a major urban area.

Several dozen acutely dangerous chemicals are used in significant quantities in and around large population centers. Chlorine gas is commonly used at wastewater treatment facilities, putting 19 million Americans at risk.[7] The nerve gas phosgene is a key ingredient for manufacturing plastics. Concentrated ammonia – the major ingredient used in the Oklahoma City bombing – is used by scores of manufacturers to make fertilizer. And cyanide compounds are used to manufacture nylon.

Despite this danger, security at the nation's chemical plants has historically been treated as a secondary concern, the province of a small cadre of environmental and public safety professionals. As part of the 1990 amendments to the Clean Air Act, Congress directed each chemical facility to develop a "Risk Management Plan" (RMP) (see box). These plans include, among other things, five-year accident histories, measures to prevent an accidental release, response plans to mitigate damage should one occur, and assessments of potential dangers to surrounding communities, including worst-case scenarios. Yet up until now, companies have not been required to assess and consider inherently safer methods of operation. The private sector has yet to understand or embrace the increased security requirements that now exist beyond traditional safety and environmental concerns. We have lost valuable time.

Immediately after 9/11, there was an attempt to jump-start an aggressive program to reduce the attractiveness of chemical plants to terrorists. Congress held hearings and, just six weeks after the attacks, Sen. Jon Corzine (D-NJ) introduced the Chemical Security Act of 2001. The bill, which quickly received substantial bipartisan support, focused on sites across the country where hazardous chemicals were produced or stored and called for the chemical industry to switch to less dangerous processes "to the extent practicable."

However, industry strongly resisted government efforts to decrease the nation's vulnerability to terrorist attacks at its facilities. Thirty trade associations, including the American Chemistry Council, American Petroleum Institute, American Farm Bureau, Edison Electric Institute, National Association of Manufacturers, and U.S. Chamber of Commerce, opposed the Corzine bill. The bill died in Congress without even a vote.

For appearances sake, the administration worked with Sen. James Inhofe (R-OK) to develop a far weaker bill. It provided no government authority to enforce safety requirements or require emergency action by companies. Incredibly, companies were not even obligated to submit self-assessment plans for government review and approval. As Rena Steinzor of the Center for Progressive Regulation put it, the Inhofe bill "was like giving your class an open-book take home exam – and telling them you're not going to collect it."

Excessive Secrecy Threatens Accountability

More than two years before the 9/11 terrorist attacks, Congress decided to restrict public access to worst-case scenario assessments contained in the Risk Management Plans (RMP) of chemical companies. These assessments estimate the number of people in the surrounding area who would be killed or injured from a catastrophic chemical release.

As a result of Congress's action, the public can only obtain this information in 50 "reading rooms" around the country. EPA and other federal agencies are prohibited from disseminating it through the Internet (though parties who obtain the information from the reading rooms can disseminate it as they see fit).

Congress took this action after the chemical industry – a longtime opponent of such disclosure – convinced the FBI that worst-case scenario data created an increased risk of terrorism. At the time, the FBI determined there was no increased risk associated with the rest of the information contained in RMPs, including accident histories, prevention measures, and disaster plans. Nonetheless, despite the FBI assessment, EPA immediately yanked this information from its web site following 9/11. To date, all RMP information remains off-line, without any detailed explanation.

Security is important, and goes hand-in-hand with accountability. In the past, community groups and environmental organizations, as well as the media and everyday citizens, have used such information to hold corporations and government accountable. This public pressure has achieved significant safety improvements. For example, since facilities began publicly reporting toxic releases in 1988, releases have declined by nearly 50 percent.[21] This is no less true when it comes to security. Sufficient information should be available for the public to judge that clear security standards are being established and action taken. Risk Management Plans continue to have an important role to play and need to more accessible to the public.

Reece Rushing

Chemical Transport Part of the Equation

When industrial plants switch to non-explosive and non-lethal chemical alternatives, they will no longer need to transport these hazardous materials. Thus, improving chemical plant security can also improve transportation security.

As it stands, railroad tank cars carrying deadly chemicals make for inviting targets. They are not guarded like chemical plants; indeed, many tank cars are covered by graffiti, testifying to their vulnerability. Making matters worse, more than half of the nation's 60,000 tank cars that carry hazardous materials were built before 1989 and are not up to current industry standards, making them less resistant to rupture, according to the National Transportation Safety Board.[22]

Unfortunately, the Bush Administration has failed to show leadership in the transportation area, just as it has with chemical plants. "The federal government has the authority to regulate the security of chemicals as they are being transported on roads, railways and waterways," Richard A. Falkenrath, President Bush's former deputy homeland security adviser, pointed out in a recent *Washington Post* op-ed. "With only one minor exception, the administration has not exercised this authority in any substantial way since Sept. 11. There has been no meaningful improvement in the security of these chemicals moving through our population centers."[23]

A number of measures are needed to address the transport of hazardous chemicals. This includes improved physical security and surveillance, real-time tracking of trucks and rail cars hauling dangerous cargoes, routing hazardous material away from target cities, and halting storage of hazardous chemicals in rail cars outside a plant's perimeter. The Center for American Progress intends to offer more detailed recommendations on chemical transport in a forthcoming report.

Reece Rushing

Even the Inhofe bill never went to the Senate floor for debate and a vote.

While this was dying in Congress, EPA initially stepped up to the plate, asserting that it could address the problem using its existing authority under the Clean Air Act. The Clean Air Act authorizes EPA to issue regulations to prevent any "unanticipated emission of a regulated substance or other extremely hazardous substance into the air" and imposes a "general duty" of precaution on sources, directing them to "design and maintain a safe facility" in order to prevent dangerous releases.[8] Because few actions are as unanticipated as terrorism and because operating a safe facility includes reducing its vulnerability to attack, this language gives EPA ample authority to create strong, mandatory security standards.

Unfortunately, EPA had little political support within the administration to move forward, and the Bush White House ultimately transferred lead responsibility for chemical plant safety to the Department of Homeland Security (DHS), which lacks clear statutory authority to require industry action. In her new book, then-EPA Administrator Christine Todd Whitman explains that she was frustrated by "the lack of support we were receiving in meeting our responsibility."[9]

At the time of the transfer, Whitman and then-DHS Secretary Tom Ridge issued a joint statement explaining that voluntary measures by the chemical industry were not enough. Ridge also testified before Congress that post-9/11 security deficiencies had been validated at dozens of chemical facilities across the country.[10] Nonetheless, the Bush administration failed to show the leadership necessary to overcome Congressional opposition to passing robust and mandatory protections.

On the industry side, in June 2002, the American Chemistry Council (ACC) issued new standards that called on its members to add security assessments, timelines and independent validation of security improvements. However, this third-party certification is limited to goals set by the companies themselves.[11] Indeed, external reviewers, who are selected by the company being evaluated, do not consider safer chemicals and process changes that could eliminate the need for add-on physical security (such as fences, alarms, and lights). Nor do they need to have expertise in design engineering for reducing hazardous chemicals. This approach, as documented by the General Accounting Office, does not go far enough to meet today's security requirements.[12]

Thus far, industry steps have focused on enhancing physical security around their plants. This has meant higher fences and increased surveillance, and in some cases, updated warning and evacuation plans. These techniques are not focused on reducing the chances of a deadly attack but rather on detecting and frustrating attacks about to occur and protecting residents after an attack.

The lack of risk reduction has left the public at substantial risk. Journalist Carl Prine of the *Pittsburgh Post Gazette* attempted to enter 30 facilities and found "almost non-existent security in a lot of places," including one Chicago facility where he sat on top of a chemical tank and waved at security personnel. "I began to wonder," Prine told *60 Minutes,* "what would it take for me to get arrested at one of these plants? Would I have to come in carrying an AK-47? What would it take for someone to say, 'Why is this guy walking around taking pictures of our tanks?'"

It is not hard to identify the primary reason behind the anemic effort to reduce the hazards posed by deadly chemicals: the Bush administration's decision to rely only on voluntary industry efforts. As in other major economic sectors, the White House has essentially allowed the chemical industry to police itself – even in the face of widespread consensus by its own experts that these plants are easy, deadly targets. Security has been trumped by a reflexive ideology that rejects government regulation out of hand. Recently, the Bush administration listed 15 potential disaster scenarios to focus homeland security preparedness. One of these scenarios involved the deliberate explosion of a chlorine tank at an industrial facility. Yet the administration has never required steps that would make such a catastrophe less likely to occur. It also joined with the railroad and chemical industries to contest a new ordinance passed by the City of Washington, D.C. that restricts the routing of rail cars carrying hazardous material through the city to other destinations (see box page 3). Warren Rudman, co-chair of the U.S. Commission on National Security, pointed out the problem with following industry's line: "With all due respect, and I'm a great admirer of private business, private business does not necessarily always have the public interest uppermost in their minds."[13]

Existing Solutions: Inherent Hazard Reduction

Ironically, chemical manufacturers and users in the United States already have the techniques and tools they need to better protect public safety and reduce the terrorism risk to their facilities. Physical security upgrades should be encouraged and supported, but alone are inadequate. The chemical industry must rapidly reduce reliance on potentially lethal chemicals that are most likely to attract terrorist interest in the first place.

The best answer lies in a process known as "inherent hazard reduction," which sets forth a hierarchy of effective, readily available techniques. By eliminating or greatly minimizing the quantities of acutely hazardous chemicals stored in any one plant or site, this approach has the potential to significantly reduce the number of vulnerable targets and the risks associated with those targets. Up to this point, however, the chemical industry has overlooked or ignored these techniques. Inherent hazard reduction includes four primary elements: materials substitution, just-in-time manufacturing, inventory reduction, and hardened storage.

Materials Substitution

The goal of materials substitution is clear: replace acutely toxic substances with less dangerous alternatives. This technique sits at the top of the hazard reduction hierarchy and should be the option of choice wherever possible.

Take the example of water treatment and disinfection, where there are already non-toxic, non-explosive alternatives for widely used and deadly toxic compounds. The current system combines chlorine gas and sulfur dioxide gas. Each of these chemicals is highly toxic; chlorine gas poses an acute and deadly risk to populations wherever it is in heavy use. Using the substitution approach, chlorine is replaced with sodium hypochlorite (industrial bleach), which won't explode, or ultraviolet light, which avoids chemicals altogether. Sulfur dioxide is readily replaced with a variety of alternative chemical-reducing agents, including thiosulfate.

Elimination of these two highly toxic chemicals is an extremely practical option that is already gaining acceptance. Many water/wastewater facilities have already undertaken this option. In a December 2003 report, Environmental Defense examined each wastewater facility that in 1999 reported that a chemical accident at its plant would endanger 100,000 or more people.[14] Of the 62 such plants, a dozen had since switched from chlorine to sodium hypochlorite or ultraviolet light. This includes plants in California, Florida, Georgia, Louisiana, Michigan, Ohio, Pennsylvania, Utah, Washington, and Washington, D.C. (see box).

Substitution is more difficult, but not impossible, where a chemical is used as a critical feedstock for industrial manufacturing. For example, polycarbonate plastics, which are currently manufactured by some companies using the dangerous nerve gas phosgene, can be made using a much more benign transesterification process. Polyurethane foams, which are currently manufactured by some companies using toluene isocyanate (a cyanide derivative) can also be manufactured with alternative chemical compounds.

Material substitution also has a clear side benefit for transportation, particularly rail and truck security. Because of material substitution, fewer rail cars carrying hazardous materials are today moving through on lines that pass literally within yards of critical areas, including government offices like the U.S. Capitol.

Of course, companies that use dangerous chemicals in a manufacturing or business process are most likely to adapt less toxic alternatives if they are relatively cost effective. Substitution benefits safety and security, which in turn reduces long-term risk. Less dangerous substances can be

What's Good for Washington, D.C...[24]

For years, the Blue Plains Wastewater Treatment Plant in Washington, D.C. stored hazardous chlorine gas in 90-ton rail cars. A rupture of just one of these rail cars would have put 1.7 million people at risk, and would cover the White House, Congress, and Bolling Air Force Base.

These risks had been known for almost two decades, prompting repeated complaints from the Department of Defense and the City of Washington – which commissioned a study in 1991 that recommended industrial bleach as a safer substitute for the more dangerous chlorine. Yet the Blue Plains facility refused to change, no government action was taken, and the danger persisted.

Then came 9/11. Suddenly, the threat of a terrorist attack on the plant, setting off a deadly release of chlorine, became very real. Indeed, the *Washington Post* reported that trade publications from the U.S. chemical industry were found in a hideout of Osama bin Laden.[25] In short order, the Blue Plains facility removed its 90-ton rail cars, and began to use sodium hypochlorite bleach, which does not have the potential to drift off-site, as a substitute for chlorine.

Initial construction costs associated with the switch were about $500,000; subsequent capital improvements were completed in 2003 at a cost of $15 million, adding about 25 cents to the average customer's monthly bill.[26] These costs have been substantially offset by a reduction in costs for security, maintenance (which have declined $300,000 annually), and hazardous substance rule compliance.

Reece Rushing

cheaper to transport. Less risk reduces liability, which may also translate into lower commercial insurance premiums.

Substitution may not be immediately economically feasible for companies that manufacture acutely toxic chemicals as products. However, the government should develop financial incentives – including tax breaks, upfront low-interest loans or homeland security grants – for research and development of safer alternatives.

Just-in-Time Manufacturing

Just-in-time manufacturing aims to limit storage of acutely toxic substances by adopting manufacturing processes that reduce the need to store deadly chemicals. Where industry simply cannot substitute fewer acutely toxic substances, this process calls for synthesizing immediately before using the molecules of a chemical needed for a reaction, rather than synthesizing them earlier and storing them in reserve.

Just-in-time manufacturing is an eminently practical option in many situations and has been undertaken by many companies. For example, at many of its plants today, Dow Chemical produces phosgene using this process. DuPont adopted this technique for producing methyl isocyanate in the immediate wake of the Bhopal disaster in 1984.[15] With the advent of nanotechnology that enables more efficient use of chemicals, the potential for just-in-time manufacturing could increase exponentially,[16] adding further impetus for the chemical industry to abandon old school rules and adopt inherently safer and more secure practices.

Inventory Reduction and Separation

Where substitution and just-in-time manufacturing are not feasible, chemical plants should aim to decrease total storage inventories of acutely toxic substances, or should that prove impossible, take steps to separate inventory into smaller tanks and containment vessels. Both of these techniques will decrease the likely impact of a terrorist attack, while also offering terrorists less accessible targets.

Reduction in on-site storage and so-called "fractionation" involve marginal up-front costs, but risks can be substantially diminished. For example, in the late 1990s, Bayer redesigned a cooling system at one of its plants to eliminate about 30 percent of the ammonia inventory;[17] around the same time, Kodak stopped using large one-ton containers to store chlorine gas and began using 150-pound cylinders instead, eliminating the potential off-site impact from a worst-case release.[18] Industry needs to adapt its facilities and operations based on the real possibility that intruders will deliberately try to kill or injure as many people as possible, destroy as much property as possible, and instill panic and economic disruption within major communities.

Hardened Storage

In cases where all of the first three techniques are determined to be infeasible, the best short-term solution is to harden or hide storage vessels of acutely toxic chemicals to decrease their vulnerability as targets. One way to do this is to store chemicals underground. A number of large flammable gas storage facilities are underground caverns, including a Marathon facility in Woodhaven, Michigan, and an AmeriGas facility in Waddell, Arizona. Tanks and other storage facilities need to be less visible, less accessible, and less vulnerable to deliberate attacks or sabotage. The federal government has the power to raise minimum facility standards.

A Strategic Action Plan

New Secretary of Homeland Security Michael Chertoff spoke recently about a "hierarchy of risks" and the need to "put our resources to work in a way that most closely approximates the most serious risks with the worst consequences and the greatest vulnerabilities."[19] The administration's hands-off approach to chemical facilities is at odds with this stated objective.

The issue is not whether voluntary approaches are inherently good and government regulation is inherently bad. The only question that matters is whether our critical infrastructure is adequately protected. When it comes to chemical facilities, the answer is no. Voluntary approaches have not worked. It's time to change our strategy.

We strongly urge the administration to set aside its general hostility toward regulation of industry and properly address the catastrophic potential that an attack on a chemical plant poses to public safety. The required federal authority already exists and none of the recommendations that follow requires the enactment of new legislation. Genuine progress in assessing and addressing the risks posed by chemical plants to the American public is achievable within 12 months. A fast-track action plan should include the following steps.

Set Priorities

Neither the private sector nor the federal government can solve overnight the problem posed by the more than 15,000 chemical facilities around the country. Our goal should be to stay focused on the facilities that pose the greatest threat, and to set priorities based on which ones use the deadliest chemicals in significant quantities—those that cause immediate, catastrophic effects such as creation of windborne toxic plumes that can move without warning into nearby communities. Many, if not most, of the chemicals used in daily commerce do *not* meet these criteria.

Fortunately, information is readily available to identify and prioritize the most dangerous chemicals that make priority facilities vulnerable to terrorism. EPA has been in the business of collecting information about chemical plants for more than 30 years; various technical guidance documents and computer programs are already available to interpret this information and set priorities. Furthermore, for many years, industry has been required to provide the government with chemical storage data for every facility.[20] There are no legal or regulatory obstacles to quickly identifying the highest priority industrial facilities. We can develop a priority list through four simple steps:

1. *Identify acutely toxic chemicals.* Acute toxicity values for a chemical indicate the concentration that could be fatal. Several government agencies, including EPA, the Occupational Safety and Health Administration and the Department of Transportation, provide toxicity values of a long list of chemicals, often tiered by the severity of their toxic effect. Toxicity values from the National Advisory Committee for Acute Exposure Guideline Levels for Hazardous Substances and the American Industrial Hygiene Institute's Emergency Response Planning Guidelines are particularly comprehensive and well-suited for the task of assessing the level of toxicity of facilities, though several other compilations could also be used.

2. *Focus on chemicals that will create a toxic plume in the event of an attack.* The volatility of the selected acutely toxic chemicals must be evaluated in order to properly identify those that would disperse off site if a storage tank were hit in close proximity to a major population center. For this analysis, we recommend EPA's Computer Aided Management of Emergency Operations (CAMEO) and Risk Management Planning Program (RMPComp), two computer programs which have been used for many years by local emergency planning agencies and industry for assessing potential off-site consequences of chemical accidents. The volatility values assigned in these programs take into account the vapor pressure of a chemical and its molecular weight to predict the extent to which a chemical will create a vapor plume.

Table 1 provides a list of 20 chemicals in commerce that rise to the top of the list using an assessment methodology that combines toxicity and volatility. Table 2 summarizes the adverse health effects caused by exposure to these chemicals and their predominant use in industry.

3. *Research quantities of toxic material stored at nation's plants.* We should identify the facilities that store significant quantities of the top priority acutely toxic chemicals. The chemical storage data that industry is required to routinely report to the government provides the basic information necessary for this task. Because peculiarities of the current reporting requirements can cause "double-counting" errors – making plants appear more dangerous than they actually are – we recommend that the government check against these errors by contacting key facilities that rise to the top of the priority list. Longer term, the existing reporting requirements should be modified to eliminate the possibilities of double-counting.

4. *Rank danger of plants by creating inherent hazard score.* The next step is to create a simple quantitative equation to analyze these factors and develop an inherent hazard score of the chemical facilities that pose the greatest risks to the nation. We recommend the equation presented below, which gives a straightforward quantification based on the key characteristics of concerns discussed above, although other similar quantifications could also be developed.

(Toxic Plume Factor/Toxicity) X Storage Quantity = Facility Score

The government could also include additional risk factors, including population density, into its ranking of priority sites. Because necessary information is readily accessible and the technical issues key to assessment are trivial, we conservatively estimate that the government should be able to create its list of target chemicals of concern and the most important facilities that store them in less than 30 days.

Strengthen Existing Government Authorities

On a parallel track, the government should set up an expert task force, through the National Academy of Sciences or a similar organization, to write guidelines for aggressive and effective hazard-reduction measures. EPA should adopt these recommendations as an update to its guidance under section 112(r) of the Clean Air Act.

EPA should issue emergency regulations under section 112(r) of the Clean Air Act and section 553(b) of the Administrative Procedure Act requiring priority facilities to arrange for immediate and rigorous independent audits of their management of priority chemicals. These audits would help determine every possible option for adopting inherent hazard reduction techniques. We recommend that these audits culminate in the filing of a sworn statement that every feasible option for inherent hazard reduction has been identified, and that the recommended hazard reduction program be implemented as quickly as possible, with a schedule and milestones agreed to by the audit team. In every instance where inherently safer technologies cannot be adopted, facilities should implement site security that includes 24/7 monitoring of all facility perimeters to prevent potential attacks, criminal screening of employees, and use of the best available technologies to otherwise secure the facility.

False statements, or failure to submit a sworn statement, should be enforceable by both civil and criminal penalties as violations of the general duty clause in section 112(r) of the Clean Air Act and/or 18 U.S.C. 1001.

A 12-Month Plan of Action

Within 30 days, the EPA should develop a list of plants that pose the greatest risk to citizens based on the criteria identified above. This plan is fully consistent with both the nature of the threat from global terrorist networks, such as al Qaeda, and the Department of Homeland Security's repeated pledge to devote its energies and resources where the risk is most significant.

Within 60 days, joint DHS/EPA teams trained in both *inherent hazard reduction* and site security should inspect the 25 highest priority facilities in the country, moving on to the next 25 within the following 60 days. Inspections should specify areas that make priority facilities vulnerable to terrorist attack and recommend steps that could and should be taken to employ inherent hazard reduction techniques and improve site security.

These findings and recommendations should be transmitted to each facility's owner and operator in the context of a letter defining their "general duty" under section 112(r) of the Clean Air Act. Industry should be required to respond within 90 days. The government should take appropriate enforcement action against those who do not agree to implement the recommendations.

Assuming that the rate of inspections will accelerate as more experience is gained, joint EPA/DHS teams should complete evaluations of the top 500 priority facilities within a year. During that same timeframe, the government should establish a program at the National Institute for Standards and Technology to conduct research on substitutes for certain industrial applications of acutely toxic chemicals. In conjunction with this effort, EPA should conduct an inventory of hazard-reduction efforts – both in the United States and in other countries – and identify best practices. This information should be used to educate facilities and promote quick adoption of safer manufacturing processes.

Conclusion

Homeland security is an integral dimension of national security. If we are to make our citizens, communities, and economy less vulnerable to terrorist attacks, the status quo is unacceptable. In the eyes of global terrorists, chemical facilities are "potential weapons of mass destruction." We have to be just as serious and determined to make them more secure as we are confronting the very real dangers posed by weapons of mass destruction abroad. Since 9/11, the Bush administration has engaged in preventive war abroad, but has not yet taken preemptive steps to address imminent threats in our midst.

Where the threat is real and the risk significant, the security of the United States – its people, its communities and its economy – should trump special interests. Voluntary approaches have been tried by the Bush administration. The industry's response has been insufficient. The government has the authority and power to identify the most vulnerable sites and insist on action. The government must lead. Fortunately, there is a strategy that can be immediately implemented to make our country and our communities more secure.

ABOUT THE AUTHOR

Dr. Linda Greer is a Senior Scientist and Director of NRDC's public health program. She specializes in issues related to toxic chemicals and hazardous waste, currently focusing on mercury pollution and the risk assessment methodologies for toxic chemicals. Linda is also the author of numerous technical and policy articles on environmental matters, and she has frequently testified before Congress. She received her doctorate in Environmental Toxicology from the University of Maryland and her master's degree in environmental sciences and engineering from the University of North Carolina's School of Public Health. Mr. Steve Anderson, an independent consultant now working for the New Jersey Department of Environmental Protection, contributed substantially to the original underlying analysis that led to this report.

ACKNOWLEDGEMENTS

This paper was commissioned by the Center for American Progress as part of its Critical Infrastructure Security Series. Editorial direction and production assistance were provided by Robert O. Boorstin, senior vice president for national security; Peter Rundlet, vice president for national security and international affairs; P.J. Crowley, senior fellow and director of national defense and homeland security; Reece Rushing, associate director for regulatory policy; Adam Hinds, research associate; and Matt Brown, production manager. The Center also wishes to thank Paul Orum, Sean Moulton of OMB Watch, and George Sorvalis of the Working Group on Community Right-to-Know, a project of OMB Watch, for their assistance with this project.

ENDNOTES

[1] USA PATRIOT ACT, Pub. L. No. 107-56 (2001), § 1016(e).
[2] Office of Homeland Security, National Strategy for Homeland Security, at 29 (July 2002), *available at* http://www.whitehouse.gov/homeland/book/nat_strat_hls.pdf.
[3] This paper focuses on chemical manufacturing facilities and the substances stored on their premises. The second in the Center's series on chemical infrastructure will explore in-depth the issue of transporting chemicals and other toxic substances across the country.
[4] Joby Warrick, *An Easier, but Less Deadly, Recipe for Terror*, Washington Post, December 31, 2004, at A1.
[5] EPA's assessments are based on industry-generated Risk Management Plans.
[6] Eric Pianin, *Toxic Chemicals' Security Worries Officials*, Washington Post, November 12, 2001, at A14.
[7] Environmental Defense, Eliminating Hometown Hazards: Cutting Chemical Risks at Wastewater Treatment Facilities, at 7 (December 2003), *available at* http://www.environmentaldefense.org/documents/3357_EliminatingHometownHazards.pdf.
[8] 42 U.S.C. 7412(r)
[9] Christine Todd Whitman, It's My Party Too, 163-165 (Penguin, 2005).
[10] Testimony before the Senate Committee on Environment and Public Works, July 10, 1002.
[11] Paul Orum, *Responsible Care Still Lacks Teeth*, Careline, Spring 2003.
[12] GAO, Voluntary Initiatives are Underway at Chemical facilities, but the Extent of Security Preparedness is Unknown, GAO 03-493, March 2003, *available at* http://www.gao.gov/new.items/d03439.pdf.
[13] Speaking on "Now with Bill Moyers" about Bush administration policies on possible terrorist attacks at chemical plants, March 21, 2003.
[14] Environmental Defense, Eliminating Hometown Hazards: Cutting Chemical Risks at Wastewater Treatment Facilities, at 6 (December 2003), *available at* http://www.environmentaldefense.org/documents/3357_EliminatingHometownHazards.pdf.
[15] On Dec. 3, 1984, a Union Carbide plant in Bhopal, India, released 40 tons of the toxic chemical methyl isocyanate into the surrounding community, killing 3,800 and sickened or injured more than 170,000, tens of thousands of which still suffer long-term effects.
[16] For a discussion of nanomanufacturing, *see* The Royal Society, Nanoscience and Nanotechnologies: Opportunities and Uncertainties, Chapter 4, July 29, 2004, *available at* http://www.nanotec.org.uk/finalReport.htm. As this report notes, some applications of nanotechnology may actually pose new health and environmental risks; such potential risks should be considered before nanotechnology is used in manufacturing.
[17] This example is drawn from a letter from Helge H. Wehmeier, president and chief executive officer of Bayer, to a coalition of environmental and public interest organizations (August 27, 1999).
[18] This example is drawn from a letter from David M. Kiser, director of health, safety and environment at the Kodak Park Site, to a representative at the National Environmental Trust (August 9, 1999).
[19] Testimony before the Senate Homeland Security and Governmental Affairs Committee, March 9, 2005.
[20] 42 U.S.C. 7412(r)
[21] *See* EPA, *The Toxics Release Inventory(TRI) and Factors to Consider When Using TRI Data*, at 2 (2002), *available at* http://www.epa.gov/tri/2002_tri_brochure.pdf.
[22] National Transportation Safety Board, Derailment of Canadian Pacific Railway Freight Train 292-16 and Subsequent Release of Anhydrous Ammonia Near Minot, North Dakota, at 50 (March 9, 2004), *available at* http://www.ntsb.gov/publictn/2004/rar0401.pdf.

[23]Richard A. Falkenrath, *We Could Breathe Easier*, Washington Post, March 19, 2005, at A15.
[24]This sidebar is adapted from a previous report by the Center for American Progress and OMB Watch, *Special Interest Takeover*, May 2004, available at http://www.americanprogress.org/site/pp.asp?c=biJRJ8OVF&b=81986.
[25] James V. Grimaldi and Guy Gugliotta, *Chemical Plants Are Feared as Targets*, Washington Post, December 16, 2001, at A1.
[26]Environmental Defense, Eliminating Hometown Hazards: Cutting Chemical Risks at Wastewater Treatment Facilities, at 4 (December 2003), *available at* http://www.environmentaldefense.org/documents/3357_EliminatingHometownHazards.pdf.

Table 1
Twenty Acutely Toxic Chemicals and their Quantities Used in Commerce

Chemical Name	Production (million lbs. per year)
Ammonia (anhydrous)	31,697
Chlorine	30,836
Nitric acid (concentration of 80% or greater)	22,469
Ethylene oxide	9,190
Propylene oxide	4,980
Hydrochloric acid (anhydrous)	4,942
Phosgene	4,294
Vinyl acetate	3,730
Acrylonitrile	3,650
Hydrocyanic acid	1,937
Sulfur dioxide (anhydrous)	964
Hydrofluoric acid (concentration of 50% or greater)	835
Phosphorus trichloride	718
Bromine	715
Phosphorus oxychloride	88
Dimethyldichlorosilane	N/A
Hydrogen sulfide	N/A
Methyl mercaptan	N/A
Methyltrichlorosilane	N/A
Trichlorosilane	N/A

N/A = Not available. The databases we consulted did not include this information.

Table 2
Chemical Summaries

The following chart summarizes the adverse health effects caused by exposure to the 20 worst chemicals, along with the predominant uses of these chemicals in commerce.

Information for these chemical summaries was obtained from several public sources including: (1) the chemical profiles published by Schnell Publishing available at www.chemexpo.com, (2) information developed by Cambridgesoft.com available at www.chemfinder/cambridgesoft.com (3) product focus report published by Chemical Week available at http://www.chemweek.com/marketplace/prod_focus.html, (4) ATSDR fact sheets, (4) New Jersey Department of Health (NJDOH) chemical fact sheets http://www.state.nj.us/health/eoh/rtkweb/rtkhsfs.htm#C, (5) EPA profiles at http://www.epa.gov/ceppo/ehs/ehsalpha.html.

Chemical Description and Acute Health Effects	Uses
Ammonia: Colorless gas with a penetrating, pungent suffocating odor detectable at 17 ppm; can be a liquid when under pressure, also as an aqueous solution. It dissolves easily in water and evaporates quickly. Exposure to high concentrations of ammonia in the air may cause severe burns on skin, eyes, throat, and lungs. In extreme cases, blindness, lung damage, heart attack, or death occur. Higher concentrations can cause a build up of fluids in the lungs (pulmonary edema), a medical emergency with severe shortness of breath.	Fertilizer 72%; Ammonium nitrate explosives 4%; Fiber and plastic intermediates 7%; Miscellaneous, 14%.

Chemical Summaries (continued)

Chemical Description and Acute Health Effects	Uses
Chlorine: Amber liquid or greenish-yellow gas with characteristic irritating, bleach-like odor detectable at 0.02 to 3.4 ppm. Chlorine is corrosive and may be converted to hydrochloric acid in the lungs. Signs and symptoms of acute exposure to chlorine may include tachycardia (rapid heart rate), hypertension (high blood pressure) followed by hypotension (low blood pressure), and cardiovascular collapse. High exposures can cause a build up of fluids in the lungs (pulmonary edema), a medical emergency with severe shortness of breath.	Polyvinyl chloride, 37 %; Other organic chemicals, 25 %; Inorganic chemicals, 17 %; Water treatment, 6 %; Pulp and paper, 5 %; Miscellaneous, 10 %.
Nitric Acid: Colorless, yellow, or red fuming liquid with an acrid, suffocating odor detectable at <5.0 ppm. High exposures can cause a build up of fluids in the lungs (pulmonary edema), a medical emergency with severe shortness of breath.	Ammonium nitrate, 75 %; Adipic acid, 9 %; Nitrobenzene, 4 %; Toluene diisocyanate (TDI), 4 %; Miscellaneous, 8 %.

Chemical Summaries (continued)

Chemical Description and Acute Health Effects	Uses
Ethylene Oxide: Colorless liquid or gas with an ether-like odor detectable at 257 to 690 ppm; irritating at high concentrations. Higher levels of exposure cause vomiting, memory loss and numbness, as well as a build up of fluids in the lungs (pulmonary edema), a medical emergency with severe shortness of breath.	Ethylene glycol, 57 %; Ethoxylates (surfactants), 11 %; Ethanolamines, 11 %; Diethylene, triethylene and polyethylene glycols, 9 %; Glycol ethers, 7 %; Miscellaneous, 5 %.
Propylene oxide: Colorless liquid with an ether-like odor. Repeated exposure can damage lungs and/or lead to pneumonia.	Flexible and rigid urethane foams, 58 %; Propylene glycols, 22 %; P-series glycol ethers, 5.5 %; Di- and tripropylene glycols, 3.5 %; Miscellaneous, 11%.
Hydrochloric acid [Hydrogen chloride]: Colorless gas with an irritating, pungent odor. Breathing can severely irritate the lungs. Higher exposures can cause a build up of fluids in the lungs (pulmonary edema), a medical emergency with severe shortness of breath.	Chemicals production, 30 %; Steel pickling, 20 %; Oil well acidizing, 19 %; Food processing, 17 %; Miscellaneous, including water treatment, 14%.

Chemical Summaries (continued)

Chemical Description and Acute Health Effects	Uses
Phosgene: Colorless liquid or gas with a sweet odor like hay at low concentrations, sharp pungent odor at high concentrations; detectable at 0.1 to 5.7 ppm. A chemical warfare agent, causing severe pulmonary edema but not always immediately irritating.	Toluene diisocyanate, 40 %; MDI/polymeric isocyanates, 40 %; Polycarbonate resins, 13 %; Miscellaneous, 7 %.
Vinyl acetate: Clear, colorless liquid with pleasant, sweet to sharp irritating odor. Irritates nose, throat and lungs causing coughing and shortness of breath. High levels can cause fatigue, irritability, disturbed sleep, dizziness and lightheadedness. May affect heart, nervous system and liver.	Polyvinyl acetate emulsions and resins, 56 %; Polyvinyl alcohol (PVOH), 18 %; Polyvinyl butyral (PVB), 11 % Ethylene-vinyl acetate (EVAc) resins, 8 %; Miscellaneous, 7 %.
Acrylonitrile: Colorless to pale yellow liquid with a mild pyridine- onion- or garlic-like odor at 2 to 22 ppm. It can be dissolved in water and evaporates quickly. Exposure to high concentrations in the air will cause nose and throat irritation, tightness in the chest, difficulty breathing, nausea, dizziness, weakness, headache, impaired judgment, and convulsions. Higher concentrations can cause a build up of fluids in the lungs (pulmonary edema), a medical emergency with severe shortness of breath.	Adiponitrile (nylon production), 33 %; Acrylic fibers, 25 %; ABS/SAN resins, 23 %; Acrylamide, 9 %; Nitrile elastomers, 3 %; Miscellaneous, 7 %.

Chemical Summaries (continued)

Chemical Description and Acute Health Effects	Uses
Hydrocyanic acid [Hydrogen cyanide]: Colorless or pale blue liquid or gas with a bitter almond odor detectable at 1 to 5 ppm. Dizziness, headache, pounding of the heart, trouble breathing and nausea. These can rapidly lead to convulsions and death.	Adiponitrile (nylon 6/6), 47 %; Acetone cyanohydrin, 27 %; Sodium cyanide, 8 %; Methionine, 6 %; Chelating agents, 2 %; Cyanuric chloride, 2 %; Miscellaneous, 8%.
Sulfur Dioxide: Colorless liquid or gas with a characteristic, pungent odor detectable at 0.3 to 5 ppm, can be a liquid at < 14 degrees F. Preservative, disinfectant, bleaching agent and antioxidant. Irritates nose, throat and lungs causing coughing and shortness of breath. Higher exposures can cause a build up of fluids in the lungs (pulmonary edema), a medical emergency with severe shortness of breath.	Chemicals, 40 %; Pulp and paper, 23 %; Food and agriculture (mainly corn processing), 14 %; Water and waste treatment, 9 %; Metal and ore refining, 6 %; Miscellaneous, 8 %.
Hydrofluoric acid [Hydrogen fluoride]: Colorless, fuming liquid or gas with a strong, irritating odor. Breathing can irritate the lungs and cause shortness of breath. Higher exposures can cause a build up of fluids in the lungs (pulmonary edema), a medical emergency with severe shortness of breath.	Fluorocarbons, 60 %; Chemical derivatives, 18 %; Aluminum manufacturing, 6 %; Stainless steel pickling, 5%; Petroleum alkylation catalysts, 4 %; Miscellaneous, 7 %.

Chemical Summaries (continued)

Chemical Description and Acute Health Effects	Uses
Phosphorus trichloride: Colorless to yellow, fuming liquid with an odor like hydrochloric acid. High exposures can cause a build up of fluids in the lungs (pulmonary edema), a medical emergency with severe shortness of breath.	Pesticide intermediates, (particularly glyphosate) 65 %; Phosphorus oxychloride, 15%; Phosphorous acid (primarily for water treatment chemicals), 10 %; Miscellaneous, 10%.
Bromine: Heavy, red-brown, fuming liquid with a choking, irritating odor, causes tears. Odor can be detected at concentrations as low as 0.05 ppm. High exposures can cause a build up of fluids in the lungs (pulmonary edema), a medical emergency with severe shortness of breath.	Flame retardants, 40 %; Drilling fluids, 24 %; Pesticides (mostly methyl bromide), 12%; Water treatment chemicals, 7 %; Miscellaneous, 17 %.
Phosphorus oxychloride: Volatile, colorless to pale yellow, strongly fuming liquid.	Phosphate esters, 85 % (flame retardants and plasticizers for plastics and urethanes); Miscellaneous, including pesticides and lube oil additives, 15 %.
Dimethyldichlorosilane: Violent reaction with water and alcohols. Breathing can irritate the lungs and cause shortness of breath. Higher exposures can cause a build up of fluids in the lungs (pulmonary edema), a medical emergency with severe shortness of breath.	Chemical intermediate for the manufacture of silicone.

Chemical Summaries (continued)

Chemical Description **and Acute Health Effects**	**Uses**
<u>Hydrogen sulfide</u>: Colorless gas with a strong odor of rotten eggs detectable at 0.001 to 0.1 ppm; liquid at high pressure, low temperature. Nausea, dizziness, confusion headache. High concentrations cause immediate death, through the nervous system resulting in paralysis of respiratory centers.	Used in the manufacturing of chemicals; in metallurgy; analytical reagent; agricultural disinfectant; intermediate for sulfuric acid, elemental sulfur, sodium sulfide, and other inorganic sulfides; additives in extreme pressure lubricants and cutting oils; and as an intermediate for organic sulfur compounds.
<u>Methyl mercaptan:</u> Colorless gas with a disagreeable odor like garlic: can be liquid at <43 ° F. Breathing high concentrations causes headache, nausea, vomiting, dizziness, muscle weakness, and loss of coordination. Higher concentrations can cause loss of consciousness and death by respiratory paralysis and pulmonary edema.	Used in manufacturing of pesticides, plastics and pharmaceuticals, and as a gas odorant to serve as a warning property for odorless but hazardous gases.
<u>Methyltrichlorosilane:</u> Colorless liquid with acrid odor. High exposures can cause a build up of fluids in the lungs (pulmonary edema), a medical emergency with severe shortness of breath.	Chemical intermediate for the manufacture of silicone.

Chemical Summaries (continued)

Chemical Description **and Acute Health Effects**	**Uses**
Trichlorosilane: Colorless liquid with acrid odor fumes in air, supports combustion. Irritates nose, throat and lungs causing coughing and shortness of breath. Higher exposures can cause a build up of fluids in the lungs (pulmonary edema), a medical emergency with severe shortness of breath.	N/A

N/A = Not available. The databases we consulted did not include this information.

Center for American Progress

ABOUT THE CENTER FOR AMERICAN PROGRESS

The Center for American Progress is a nonpartisan research and educational institute dedicated to promoting a strong, just and free America that ensures opportunity for all. We believe that Americans are bound together by a common commitment to these values and we aspire to ensure that our national policies reflect these values. We work to find progressive and pragmatic solutions to significant domestic and international problems and develop policy proposals that foster a government that is "of the people, by the people, and for the people."

Center for American Progress
1333 H Street, N.W., 10th Floor
Washington, D.C. 20005
(202) 682-1611
www.americanprogress.org

Written Testimony for Homeland Security and Governmental Affairs Committee

"Chemical Facility Security: What is the Appropriate Federal Role?"
July 27, 2005

Meghan Purvis
Environmental Health Advocate
U.S. Public Interest Research Group

Thank you for inviting me to submit testimony to the Senate Homeland Security and Governmental Affairs Committee hearing "Chemical Facility Security: What is the Appropriate Federal Role?" on July 13, 2005.

I am the Environmental Health Advocate for the U.S. Public Interest Research Group, where I focus my work on protecting the public from exposure to toxic chemicals. U.S. PIRG is a nation-wide network of state-based public interest advocacy organizations, with a thirty-year history of working on behalf of the public to protect the environment, protect consumers, and defend our democracy. The PIRGs have worked for decades to reduce hazardous chemical accidents by advocating for the use of safer alternatives.

I would like to focus my testimony on the appropriate federal role and address issues that arose at the hearing, including the use of safer chemicals to prevent terrorist attacks, the Maritime Transportation Security Act, voluntary industry programs, and state preemption and the role of federal legislation and state legislation.

Chemical Security Overview

The events of September 11, 2001 reminded us that our infrastructure makes appealing terrorist targets. Chemical plants, refineries, and other facilities using and storing large amounts of chemicals are not only sitting ducks, they make ideal targets because of the potential for large scale injuries.

The best way to reduce the likelihood of a terrorist attack at a chemical facility is to make it a less appealing target—by requiring chemical facilities to use safer chemicals and processes wherever possible.

First in 2003, and again this summer U.S. PIRG examined the petroleum refining industry. Not only are oil refineries critical pieces of infrastructure that the U.S. relies on to meet its energy needs, an attack on a refinery that uses hydrofluoric acid could have devastating consequences for neighboring communities. Petroleum refineries stand as a stark example of the safety hazards posed by using toxic chemicals in the manufacturing process and the opportunities to switch to safer alternatives.

I will review our research, and have included our report on petroleum refineries, *Needless Risk: Oil Refineries and Hazard Reduction,* for the record.

Hydrofluoric Acid and the Petroleum Industry

Petroleum refineries are responsible for nearly 11% of all of the high-risk processes in EPA's Risk Management Program.[i] Most notably, many of these refineries use hydrofluoric acid, also known as hydrogen fluoride, as a catalyst to produce an additive to gasoline. Currently there are 148 petroleum refineries in the United States, 50 of which use hydrofluoric acid as a catalyst to produce alkylate.[ii]

Hydrofluoric Acid: A Threat to Health and Safety

Hydrofluoric acid is highly toxic, and has many acute consequences for human health and safety. Even slight contact with hydrofluoric acid may cause a variety of acute symptoms, including skin and deep tissue burns, which may not be felt for up to 24 hours after exposure.[iii]

Hydrofluoric Acid: A Terrorist Target

According to Neil Livingstone, board chairman of Global Options, a security firm in Washington, DC, hydrofluoric acid is a "known quantity to some terrorists," particularly those from oil-producing countries where hydrofluoric acid is commonly used.[iv] Terrorist attacks over the past ten years have consistently targeted petroleum facilities throughout the world because of their vulnerability, value to economies, and high volumes of toxic chemicals stored onsite.

Furthermore, it is relatively simple for individuals to gain access to chemical plants and refineries. In January 2002, a robber carrying a shotgun made his way into a Citgo Petroleum Corporation facility. Citgo was one of the companies that claimed to have dramatically increased security measures after September 11, 2001.[v]

Report Findings: Communities at Risk

In our report, U.S. PIRG analyzed the Risk Management Plans submitted by industry to determine a measure of community risk due to petroleum refineries.

We found that more than 17 million Americans (17,040,905) live inside a vulnerability zone for an oil refinery using hydrofluoric acid.[vi] In some cases, the vulnerability zones overlap, posing an even greater danger to people who live and work within the overlapping areas. In Philadelphia, for example, the vulnerability zones of two refineries overlap across the Delaware River, encompassing Philadelphia International Airport, the sports stadiums, and many city neighborhoods.[vii]

Seven petroleum refineries with hydrofluoric acid alkylation facilities have accidental toxic release "worst-case" scenarios where more than one million people are endangered by potential exposure to a cloud of toxic hydrofluoric acid gas.

Based on conservative estimates, refineries in Pennsylvania using hydrofluoric acid endanger the most people, at 4.4 million. A single New Jersey refinery endangers more than 3.1 million people living within the vulnerability zone. Three Illinois-based refineries also endanger 3.1 million people living within the vulnerability zone.

How Policymakers Should Protect Communities: A Preventive Approach

Reducing or eliminating chemical hazards offers the best strategy to fully protect American communities from both accidents and terrorist attacks involving industrial chemicals. Hazard reduction means making a chemical process *inherently* safer by eliminating the use of highly toxic, volatile, or flammable chemicals or using chemicals in safer quantities or conditions where feasible. The concept of inherent safety leads to a hierarchy to guide decisions on the use and management of chemicals:

The first option in this hierarchy is to reduce or eliminate the *possibility* of a chemical release by choosing inherently safer materials and technologies.

Prevent the Possibility: Solid Acid Catalyst

In the case of oil refineries, preventing the possibility of an attack means switching from the use of hydrofluoric acid to a solid acid catalyst, completely eliminating the need to use either hydrofluoric acid or sulfuric acid to produce alkylate. Industry experts report that a variety of solid acid catalysts will be available for use in alkylation facilities within the next three years. Critics, however, insist this technology has been available since the late 1990s, and simple industry inertia has kept solid acid catalysts from becoming the popular choice for refinery alkylation processes. [viii]

Solid acid catalysts have tremendous safety advantages because they are neither corrosive nor particularly hazardous to people or the environment. Furthermore, in the event that the container housing the catalyst is breached, no off-site damage would result.

Reduce the Probability: Hydrofluoric Acid Modifiers

The second option in the hierarchy is to reduce the *probability* of a chemical release through secondary prevention measures such as safety valves and double-walled vessels.

In the example of the petroleum industry, this option calls for an investment in alkylation modifiers and install active mitigation units, such as water spray systems. This option, however, does not remove the possibility of a terrorist threat, and an adversary could thwart mitigation systems. For example, a terrorist could simply disable the power supply to the water spray system.

Modified hydrofluoric acid reduces the ability of the acid to form an aerosol cloud by a certain percentage, thereby reducing the impact the toxic cloud will have on the surrounding community.

In 2004, public pressure succeeded in persuading the Valero Energy Corporation to switch to modified hydrofluoric acid at its Wilmington, California refinery, near Los Angeles. Since an explosion in 1987, the local community and government have pushed to shut down two refineries that used hydrofluoric acid and required a third facility to change to modified hydrofluoric acid. The community was able to negotiate an agreement with the South Coast Air Quality Management District with regards to the Valero facility; Valero will pay a fine up to $1 million if the renovation is not complete by the end of 2005. The change is expected to cost Valero about $30 million.[ix],[x]

Prevent the Severity: Sulfuric Acid As an Option

The third option in the hierarchy is to reduce the *potential severity* of the impacts of a chemical release through mitigation measures or emergency response plans.[xi]

In oil refineries, preventing the severity of a release calls for the use of sulfuric acid in the alkylation process. When released, sulfuric acid will not readily form an aerosol cloud, but instead is released as a liquid form, making it much easier to contain and prevent exposure to those offsite. As a result, sulfuric acid does not pose as much of a threat to life outside of the facility.[xii]

However direct exposure to sulfuric acid can cause many detrimental health effects at concentrated levels, such as burns or severe irritation to the eyes, skin, and respiratory tract if concentrated fumes are inhaled.

In addition, because sulfuric acid is often regenerated offsite, there is an increased risk of an accident involving sulfuric acid due to increased transportation time. With the current risk identified today in the transportation sector, it is clear this is not the best option. Once onsite, however, sulfuric acid can be regenerated indefinitely if the refinery builds a regeneration facility.

Again, the first option—inherent safety, or the use of solid acid catalysts at oil refineries —provides the best prevention of the threat of chemical releases caused by acts of terrorism because it eliminates the potential hazard.

Voluntary Industry Programs

In 1999, and again in 2002, PIRG and other public interest organizations surveyed the chemical industry to determine the industry's progress and effort to implement safer chemicals and processes to protect communities. These surveys show that whatever the intention of voluntary industry programs, such as the American Chemistry Council's "Responsible Care," the industry does not, with few exceptions, have measurable public goals and timelines to reduce the number of people who live within the vulnerability zones of ACC member companies. Frankly, this is appalling. If voluntary programs have not caused incremental progress to use safer chemicals and processes in order to protect chemical plant neighbors then Congress must step in. The 2002 survey has been attached for the record.

Maritime Transportation Security Act of 2002 and Refineries

In 2002 Congress passed the Maritime Transportation Security Act which requires 3,150 port facilities and more than 9,000 vessels to develop and implement security plans. The MTSA gives the Coast Guard the authority to implement the Act.

We have many concerns with MTSA, and the legislation as a whole should not serve as verbatim model policy for this committee.

First, although many of the facilities covered by MTSA use and store large quantities of hazardous chemicals, the security plans do not require facilities that do use or store extremely hazardous substances to review safer alternatives.

In the section below, on how policymakers should protect communities, I outline why this would be a critical mistake to overlook this preventive approach. The fact that the MTSA does not require even basic planning for safer alternatives and processes means that these vulnerabilities will remain a threat to surrounding communities well after MTSA requirements are met.

A second problem with the legislation was outlined by the GAO Report from June 2004, "Maritime Security: Substantial Work Remains to Translate New Planning Requirements into Effective Port Security." Under MTSA, there are two options for submitting security plans for review to the Coast Guard. The first option is for each facility to submit individual plans. The second option is called Option B, or "alternative security program." Under this option, the facilities sent a letter to the Coast Guard telling them which industry organization's alternative security program they were going to subscribe to. The Coast Guard simply checked to make sure that the owners or operators of the facility were members of the industry group. The Coast Guard did not review any of these facilities plans.

Nearly half of the individual plans were submitted under option B. Therefore, the Coast Guard did not review half of the plans developed to improve security at maritime facilities, although the Coast Guard does plan to review how these plans are implemented at on-site facility inspections in the next phase of MTSA.

The GAO report found that according to the Coast Guard's MTSA security plan program manager, some of the users of option B thought that if they belonged to an industry trade organization with developed voluntary standards they were then in compliance with the law. In one case, a facility reportedly joined an industry organization solely to avoid submitting an individual security plan after the Coast Guard issued a notice that they were late in submitting a security plan.

State Preemption

During the July 13th hearing, at least one industry stakeholder testified they would encourage the Committee to develop legislation that preempts the rights of states to set stricter security standards. U.S. PIRG strongly opposes efforts to preempt the ability of state and local legislatures to set innovative and even stronger chemical security standards.

States have long been the laboratories for innovative public policy. State and local legislatures, being smaller and often more nimble than the federal government, can develop and test innovative policies to address problems identified by local constituents. If a certain policy works, other states can try it. If the policy fails, the state or local government can quickly modify the policy without having affected residents in all 50 states.

Federal preemption of state laws, and particularly chemical security laws, presents a serious challenge in many ways. Federal preemption of state law threatens to limit the states' traditional role as the laboratories of democracy; suppressing innovation and limiting the policy jurisdiction of state and local governments.

Federal preemption of state law also threatens to compromise consumer and environmental protection overall. Congress often only acts following a crisis, such September 11th or the Bhopal chemical accident, or after several states experiment with a policy and Congress follows suit with a federal policy. By preempting state law and effectively removing states from the equation, we are then left to rely on Congress to act without impetus from the states – or to wait until systemic abuses or a major attack at a chemical facility cause widespread harm and rise to the level of scandal or crisis.

In order to preserve states as the laboratory for innovative policy to protect the public from chemical terrorism, federal law should be a floor, not a ceiling. Federal chemical security legislation should be the floor to improve security at chemical plants across the country, but it should not limit states that have innovative ideas or an added incentive to increase security for their residents.

Conclusion

In conclusion, U.S. PIRG urges this committee to prevent terrorist attacks by limiting the number of attractive targets by requiring the use of safer technologies at high priority facilities when it is both technologically and economically feasible, and when it will not create a greater risk to public health and the environment. In addition, the Maritime Transportation Security Act should not serve as a model for chemical security legislation. Finally, we urge the committee to ensure that states will have the continued right to set the level of safety adequate for their citizens and not preempt state laws.

---

[i] James Belke, U.S. Environmental Protection Agency, "Chemical accident risks in U.S. industry – A preliminary analysis of accident risk data from U.S. hazard facilities," September 25, 2000.
[ii] Energy Information Association, Annual Refinery Report, "Petroleum Supply Annual 2004, Volume 1, Table 36. Number and Capacity of Operable Petroleum Refineries by PAD District and State as of January 1, 2005." released February 2003.
[iii] "Hydrofluoric Acid: Chemical Safety Information," Environment Health and Safety at University of North Carolina. Available at http://ehs.unc.edu/pdf/HydrofluoricAcid.pdf.
[iv] Adam Fifield, "In the Shadow of Danger: The Chemical Plant Peril How Safe, How Secure?" *Philadelphia Inquirer*. April 20, 2003.
[v] John Tedesco, "Thief breaks refinery security: Two workers lose wallets," *San Antonio Express-News*, February 16, 2002.
[vi] Risk Management Plan, EPA Database, December 2004.
[vii] Adam Fifield, "In the Shadow of Danger: The Chemical Plant Peril How Safe, How Secure?" *Philadelphia Inquirer*. April 20, 2003.
[viii] Milton Lapkin and Sanford Lewis. "Boosting the First Line of Defense: Report of the Good Neighbor Project for Sustainable Industries." Available at http://gnp.enviroweb.org /hfexec.htm.
[ix] Bill Walsh, "Toxins Make Local Plants Possible Target for Terrorists," *Times-Picayune*, July 6, 2003
[x] South Coast Air Quality Management District, "Highly Toxic Chemical to be Phased Out at Valero Refinery," February 7, 2003, available at http://www.aqmd.gov /news1/hfvalero.htm.
[xi] Adapted from N.A. Ashford, Gobbel, J.V., Lachman, J., Matthiesen, M., Minzner, A., and R.F. Stone, *The Encouragement of Technological Change for Prevention Chemical Accidents: Moving Firms from Secondary Prevention and Mitigation to Primary Prevention.* Cambridge, Massachusetts: Center for Technology, Policy, and Industrial Development, Massachusetts Institute of Technology, Boston, 1993
[xii] Randy Petrson, Alkylation Division at STRATCO, telephone interview on July 21, 2003.

August 2005

Needless Risk

Oil Refineries and Hazard Reduction

U.S. PIRG
Education Fund

NEEDLESS RISK:

OIL REFINERIES AND HAZARD REDUCTION

MEGHAN PURVIS
MARGARET HERMAN

U.S. PIRG EDUCATION FUND

AUGUST 2005

ACKNOWLEDGMENTS

Cover photo of oil refinery obtained from the Hawaii State Department of Business, Economic Development & Tourism, at http://www.hawaii.gov/dbedt/ert/refinery.html. Note that this photo is for illustrative use only; this refinery does not use hydrofluoric acid. Photo of children playing courtesy of Denny Larson of the National Refinery Reform Campaign, www.refineryreform.org. This photo shows children from the Carver Terrace federal housing project playing in the shadow of the Premcor refinery in West Port Arthur, Texas. Photo of homes near an oil refinery obtained from EcoIQ.com Online Images. Photo of emergency medical technicians obtained from the City of Norfolk at www.norfolk.gov/NFR/images/emspic.JPG.

The authors would like to acknowledge Alison Cassady of U.S. PIRG Education Fund for coordinating the research and distribution of this report, as well as her endless editorial support and cover design.

In addition, we would like to thank Paul Orum, senior advisor to the Working Group on Community Right-to-Know, for his research leadership and editorial review. We would also like to thank the U.S. PIRG Education Fund interns for their help in collecting the data.

The U.S. PIRG Education Fund's Toxics and Environmental Health Program is grateful to the Bauman Foundation, the Beldon Fund, and individual contributors for their support.

The authors alone are responsible for any factual errors. The recommendations are those of the U.S. PIRG Education Fund. The views expressed in this report are those of the authors and do not necessarily reflect the views of our funders or those who provided editorial review.

The U.S. PIRG Education Fund is the research and public education center for the U.S. Public Interest Research Group (PIRG), the national advocacy office of the state PIRGs. The state PIRGs are a nationwide network of nonprofit, nonpartisan, state-based public interest advocacy organizations. The state PIRGs' mission is to deliver persistent, result-oriented activism that protects the environment, encourages a fair marketplace for consumers and fosters responsive, democratic government.

For additional copies of this report, send $20 (including shipping) to:

U.S. PIRG Education Fund
218 D Street, SE
Washington, DC 20003
202-546-9707
uspirg@pirg.org
www.uspirg.org

For more information about the state PIRGs, visit our Web site at http://www.pirg.org.

TABLE OF CONTENTS

EXECUTIVE SUMMARY 4
CHEMICAL INSECURITY: HAZARDS LEAVE COMMUNITIES EXPOSED 6
HYDROFLUORIC ACID AND THE PETROLEUM INDUSTRY 8
HYDROFLUORIC ACID: A THREAT TO HEALTH AND SAFETY 8
USE OF HYDROFLUORIC ACID IN THE PETROLEUM INDUSTRY 8
CHEMICAL ACCIDENTS INVOLVING HYDROFLUORIC ACID 9
HYDROFLUORIC ACID: A TERRORIST TARGET 10
REPORT FINDINGS: COMMUNITIES AT RISK 12
HOW POLICYMAKERS AND INDUSTRY SHOULD PROTECT COMMUNITIES 15
A PREVENTIVE APPROACH 15
INHERENT SAFETY AT REFINERIES: ALTERNATIVES TO HYDROFLUORIC ACID 16
PREVENT THE POSSIBILITY: SOLID ACID CATALYST 16
REDUCE THE SEVERITY: SULFURIC ACID AS AN OPTION 17
REDUCE THE PROBABILITY: HYDROFLUORIC ACID MODIFIERS 18
REDUCING CHEMICAL HAZARDS THROUGH POLICY MEASURES 21
INADEQUACIES OF EXISTING POLICIES 21
THE RIGHT-TO-KNOW AS A SAFETY TOOL 22
PROTECTING COMMUNITIES THROUGH INHERENTLY SAFER TECHNOLOGY 23
METHODOLOGY 24
APPENDIX: OIL REFINERIES USING OR STORING HYDROFLUORIC ACID ON-SITE 25
END NOTES 27

EXECUTIVE SUMMARY

Across the country, petroleum refineries, chemical plants and other industrial facilities use and store large amounts of hazardous chemicals that could be released in the event of an accident or terrorist attack. Such releases could endanger thousands or even millions of people who live in communities in close proximity to these facilities. According to the Environmental Protection Agency (EPA), 106 facilities would each endanger at least one million people in the event of a worst-case chemical release. Another 3,000 facilities would endanger at least 10,000 people. Nearly 5,000 facilities store more than 100,000 pounds of at least one EPA-classified "extremely hazardous substance."

Many of these facilities, however, present an unnecessary risk to their surrounding communities. Industries often have multiple options for carrying out similar processes, and some of these options are inherently safer than others. Facilities that use fewer or smaller quantities of hazardous chemicals, or even make changes to storage pressure or other processes, can eliminate the possibility of on-site chemical accidents and make themselves less appealing terrorist targets.

Petroleum refineries stand as a stark example of the needless risk posed by such facilities in the event of an attack or accident as well as the opportunity to mitigate this risk by using safer alternatives to toxic chemicals.

Many petroleum refineries use hydrofluoric acid in their processing, which poses a great public safety risk both because of its extreme toxicity to humans as well as its propensity to form a toxic aerosol cloud when released. A catastrophic event at one of these facilities could cause a potentially lethal release of hydrofluoric acid, forming a stable aerosol cloud above the facility and surrounding neighborhoods. Exposure to hydrofluoric acid results in devastating burns, and pain associated with the exposure may be delayed for up to 24 hours. If the burn is not addressed, tissue destruction may continue for days. Inhalation of fumes can cause symptoms ranging from severe throat irritation to pulmonary edema.

To estimate the number of Americans at needless risk of exposure to hydrofluoric acid in the event of a catastrophic accident or terrorist attack at a petroleum refinery, we examined the Risk Management Plans submitted by oil refineries to EPA. These plans estimate how far a chemical could travel off-site in the event of a release and report the number of people living within the "vulnerability zone," the area potentially affected by the release. Based on this data, we found that petroleum refineries using hydrofluoric acid endanger millions of people.

Specifically:

- Of the 148 petroleum refineries in the United States, 50 use hydrofluoric acid in their processing or store it on-site. The remainder use safer alternatives, such as sulfuric acid.

- These 50 refineries, using and storing 10.6 million pounds of hydrofluoric acid, endanger more than 17 million people living in surrounding communities in 20 different states.

- With 12 refineries using hydrofluoric acid, Texas has more than any state. Louisiana has five oil refineries that currently use hydrofluoric acid, and Montana has four.

- Refineries using hydrofluoric acid in Pennsylvania endanger more than 4.4 million people residing in their vulnerability zones, according to conservative estimates. Refineries using hydrofluoric acid in both Illinois and New Jersey endanger more than 3.1 million people living within the vulnerability zones.

- Seven petroleum refineries using hydrofluoric acid reported toxic release "worst-case" scenarios in which more than one million people could be affected. Furthermore, 15 refineries could place more than 500,000 people in harm's way, and 28 refineries could endanger more than 100,000 people in the event of a worst-case hydrofluoric acid release.

- The companies operating refineries using hydrofluoric acid with the most people residing in their vulnerability zones include Sunoco, Valero Energy Corporation, Marathon Ashland Petroleum, ConocoPhillips, CITGO, and ExxonMobil, each endangering at least two million people in total.

- Many companies owning refineries using hydrofluoric acid also operate refineries without that technology. ConocoPhillips, ExxonMobil, Valero Energy Corporation, and Marathon Ashland, for example, own refineries using hydrofluoric acid as well as refineries that use other technologies.

Fortunately, hydrofluoric acid is not the only material oil refineries can use in their refining processes. Many other refineries already use sulfuric acid, a safer alternative, in the alkylation process. This cost-effective and widely-used alternative diminishes the appeal of refineries as a terrorist target and mitigates the public health and safety consequences of an accident. In addition, a new technology, solid acid catalysts, will soon be available for widespread commercial use, offering an even safer option than the use of sulfuric acid.

Petroleum refineries are but one example of the facilities that pose an immediate risk to public health in the event of a terrorist attack or chemical accident. Refineries also are not the only example of facilities that could make cost-effective changes to manufacturing processes to reduce or eliminate the use of hazardous chemicals.

Unfortunately, most industrial facilities have not responded to the increased awareness of terrorism by switching to inherently safer technologies. Instead, industry organizations such as the American Chemistry Council, the American Petroleum Institute, and the National Petrochemical and Refiners Association have emphasized increasing physical security at facilities. Hiring more guards, building higher fences, and placing more lights may all be part of a strong security plan, but this does not actually reduce the threat to the community. Switching chemicals and processes to something less volatile not only reduces the chemical hazard to the community, but also reduces the need for costly add-on security measures and the attractiveness of the facility as a target for attack.

CHEMICAL INSECURITY: HAZARDS LEAVE COMMUNITIES EXPOSED

Across the United States, thousands of industrial facilities use and store hazardous chemicals in large quantities that pose major risks to their neighbors. According to the Environmental Protection Agency (EPA), 106 facilities would each endanger at least one million people in the event of a worst-case chemical release. Another 3,000 facilities would endanger at least 10,000 people. Nearly 5,000 facilities store more than 100,000 pounds of at least one EPA-classified "extremely hazardous substance."[1]

Accidents at chemical and industrial facilities are common. Each year tens of thousands of spills and gas leaks occur as a result of the ongoing use of hazardous chemicals.[2] As recently as March 2005, an explosion at a BP oil refinery in Texas City, Texas killed 15 employees. In Gulfport, Mississippi in February 2003, a cloud of toxic ammonia leaked from a chemical plant after an intruder broke into a fenced compound, attempting to steal the chemical to make illegal drugs. The toxic cloud led to the evacuation of seven hotels and the closure of Gulfport-Biloxi International Airport and a 10-mile stretch of Interstate 10 for seven hours.[3]

Since September 11, 2001, it has become increasingly apparent that these facilities pose a more sinister threat, as they may become the target of a terrorist attack. A report by the Army Surgeon General ranked an attack on a chemical plant second only to a widespread biological attack in magnitude of the hazard to the public.[4] On February 12, 2003, the National Infrastructure Protection Center warned, "Al Qa'ida operatives…may attempt to launch conventional attacks against the U.S. nuclear/chemical-industrial infrastructure to cause contamination, disruption, and terror."[5]

Even before September 11, 2001, the Agency for Toxic Substances and Disease Registry (ATSDR) addressed the weak security at chemical facilities. In 1999, ATSDR published a study of chemical site security at two key chemical communities – the Kanawha Valley in West Virginia and Las Vegas, Nevada. The study found the industry poorly prepared for terrorist attacks, noting that industrial chemicals provide terrorists with "effective and readily accessible materials to develop improvised explosives, incendiaries and poisons."[6]

Across the country, select facilities have made major progress by switching to safer chemicals and processes that pose less of a threat to surrounding communities in the event of a major chemical release. Soon after September 11th, for example, the Blue Plains Sewage Treatment Plant in Washington, DC switched from using and storing chlorine and sulfur dioxide on-site to using sodium hypochlorite bleach in its processes. For weeks after September 11th, workers at Blue Plains removed up to 900 tons of liquid chlorine and sulfur dioxide. Whereas chlorine gas from the Blue Plains facility could have enveloped downtown Washington, DC, Anacostia, Reagan National Airport, and Alexandria in a toxic cloud, sodium hypochlorite bleach is far more benign if accidentally released.[7]

Unfortunately, too few other chemical facilities have followed the lead of the Blue Plains facility, changing the processes and chemicals they use to make their facilities inherently safer. Instead, industry groups have emphasized increasing physical security at facilities, such as guards and fences. The American Chemistry Council, the chemical industry's lobbying organization, has issued site security guidelines for its member companies and requires these companies take part in its Responsible Care program in order to continue membership in the organization. This set of guidelines, however, focuses mainly on site security and does not include minimum standards that facilities must follow. For example, it does not require that facilities provide for protection against an armed intruder. In addition, Responsible Care does not require that facilities use inherently safer technology to reduce the threat these facilities pose to surrounding communities. The industry does not even require members to set public goals and timelines to reduce chemical dangers.

Hiring more guards, building higher fences, and placing more lights may all be part of a good security plan, but this does not actually reduce the threat to the community. Switching chemicals and processes to something less volatile not only reduces the chemical hazard to the community, but also reduces the cost of physical security and the attractiveness of the facility as a target for attack.

Furthermore, some in industry and the government have proposed limiting the public's access to information on the potential impact of a major chemical release on public health and safety. Limitations have been placed on the information any individual can obtain to protect themselves and their families from chemical releases at neighboring facilities. Instead of safeguarding these facilities from terrorists, these efforts merely deny public accountability measures that encourage industry reform.

Hydrofluoric Acid and the Petroleum Industry

Petroleum refineries stand as a stark example of the safety hazards posed by using toxic chemicals in the manufacturing process and the opportunities to switch to safer alternatives.

Petroleum refineries are responsible for nearly 11% of all of the high-risk processes in EPA's Risk Management Program, the agency's primary chemical accident prevention program established under the 1990 Clean Air Act Amendments.[8] Most notably, many of these refineries use hydrofluoric acid, also known as hydrogen fluoride, as a catalyst to produce an additive to gasoline. This additive is then used to increase the octane levels in gasoline, which addresses the problem of engine "knocking." Currently, 148 petroleum refineries operate in the United States, 50 of which use hydrofluoric acid as a catalyst to produce alkylate.[9]

Hydrofluoric Acid: A Threat to Health and Safety

Highly toxic, hydrofluoric acid has many acute consequences for human health, as well as the ability "to kill people on the spot," according to Jonathan Ward, director of toxicology at the University of Texas.[10] Even slight contact with hydrofluoric acid may cause a variety of acute symptoms, including skin burns and deep tissue burns, which may not be felt for up to 24 hours after exposure. In addition, hydrofluoric acid exposure commonly causes eye irritation and can lead to permanent damage. If inhaled, the acid can cause irritation of the nose, throat, and lungs, causing coughing and dyspnea, or shortness of breath. Severe exposure can cause cyanosis, an indicator of hypoxemia, lung injury and a build up of fluid in the lungs, known as pulmonary edema.[11]

Once hydrofluoric acid penetrates body tissue, it can react with the calcium and magnesium in the blood stream, causing abnormally low calcium concentrations and a condition known as hypocalcaemia.[12]

Case studies have shown hydrofluoric acid to be fast acting, reporting instances of respiratory irritation in less than one minute after acute inhalation of 122 parts per million (ppm).[13] The Occupational Safety and Health Administration estimates that the lowest lethal concentrations range from 50-250 ppm for a five-minute exposure.[14]

Use of Hydrofluoric Acid in the Petroleum Industry

Petroleum refiners use hydrofluoric acid as a catalyst in the alkylation process. With tighter standards on fuel emissions, alkylate has become a key additive in gasoline because it offers a high octane with very low sulfur and nitrogen content.

Alkylation is the chemical process by which high-octane gasoline is made. Light, gaseous hydrocarbons, made up of compounds like propylene and butylenes, feed into a reactor, where contact with the acid catalyst results in a reaction that produces a heavy mixture of hydrocarbons, along with excess heat, which must be removed. It is the liquid portion of this mixture that is known as alkylate, and it is responsible for the anti-knocking property of unleaded gasoline.[15] Lastly, because some of the acid catalyst leaves the reaction chamber along with the alkylate, further treatment to neutralize the acid is required as well in order to prevent corrosion further down the production line.

CHEMICAL ACCIDENTS INVOLVING HYDROFLUORIC ACID

Hydrofluoric acid has a boiling point of 67° F (19.4° C) at atmospheric pressure, and most reactions take place at about 100° F (37.8° C) in petroleum refineries. Therefore, the acid will vaporize whenever the container is penetrated.[16] Furthermore, when the outside temperature is 67° F or greater, the containment unit housing hydrofluoric acid becomes pressurized because of the acid's high volatility. Therefore, if hydrofluoric acid is released from containment, under certain atmospheric conditions it will form a stable aerosol cloud.

Amoco, Mobil, Allied Chemical and DuPont tested the possibility of a release of hydrofluoric acid from one of their refineries then under construction in a Nevada desert in 1986. Under conditions similar to those in an alkylation unit, lethal concentrations of hydrofluoric acid aerosol were present up to five miles (8 km) from the release points, at levels much higher than anticipated.[17] The amount of hydrofluoric acid released in the test was relatively small—1,000 gallons in two minutes.[18]

Hydrofluoric acid refineries have long had a history of accidental releases that prove the potentially devastating effects of a release caused by simple human error. The National Response Center recorded more than 400 incidents at refineries and other facilities involving hydrofluoric acid or hydrogen fluoride from 1990 to 2005.[19] Simple malfunctions, ruptured valves or pipes, or valves accidentally opened often cause significant leaks that can harm both the people and the environment surrounding the refinery.

- On October 30, 1987, a crane at Marathon Oil's Texas City refinery dropped its load on a storage tank, rupturing a pipe and releasing 30,000 pounds of hydrofluoric acid, the largest known release. The resulting vapor cloud sent 1,037 people to the hospital suffering from respiratory problems and skin rashes and forced 3,000 residents out of their homes for three days. "There were houses right up against the fence," said Ronald Koopman of Lawrence Livermore. "The only thing that saved people was that the [hydrofluoric acid] plume shot 200 feet up in the air, and it went about 900 meters downwind before it actually came down into the neighborhood. If it had squirted out sideways, it would have killed hundreds, if not thousands."[20]

- On March 2, 2003, 13 electricians working at the Marathon Ashland oil refinery in St. Paul Park, Minnesota were hospitalized after being exposed to hydrofluoric acid. They had been hired to repair damage to the facility caused by a fire a few days earlier. While working, a pump leaked one cup of hydrocarbons with trace amounts of hydrofluoric acid, which immediately vaporized and entered their lungs.[21]

- An accidental hydrofluoric acid release in 1991 killed two workers and injured five others at Southwestern Refining Co. in Texas.[22]

- Between 1995 and 1997, four separate one-pound releases of hydrofluoric acid at the Sunoco refinery in Philadelphia injured one worker in each accident.[23]

- On October 2, 2001, 150 pounds of hydrofluoric acid in low concentration leaked within the Valero refinery in Paulsboro, New Jersey. Because the wind was blowing toward neighboring Greenwich Township, a nearby elementary school was forced to secure all the children and staff in the gym by sealing the doors and windows with duct tape and plastic.[24]

- On March 19, 1988, at the Sun Co. refinery in Tulsa, Oklahoma, an accidental release of 210 pounds of hydrofluoric acid sent a cloud of hydrofluoric acid drifting five miles through downtown. Had the accident occurred on a weekday, more people would have been injured. One resident, living in the downtown area, said, "I didn't realize it was hazardous until I could see it leave orange particles everywhere…And then my eyes burned, my throat burned and my head ached."[25]

- On March 23, 2005, an explosion at a BP oil refinery in Texas killed 15 employees and injured more than 100. This was the third major accident in a year at this facility. Although it did not involve a release of hydrofluoric acid, the presence of this chemical on-site was a major cause for concern.[26]

HYDROFLUORIC ACID: A TERRORIST TARGET

According to Neil Livingstone, board chairman of Global Options, a security firm in Washington, DC, hydrofluoric acid is a "known quantity to some terrorists," particularly those from oil-producing countries where hydrofluoric acid is commonly used.[27] Terrorist attacks over the past 10 years have consistently targeted petroleum facilities throughout the world because of their vulnerability, value to economies, and high volumes of toxic chemicals stored onsite. A few examples of coordinated attacks on petroleum facilities over the past 10 years include:

- During the Croatian war, Serbian armies attacked a natural gas refinery in eastern Slovenia that stored ethane, propane, and butane with rockets and cluster bombs.[28]

- Serbian forces attacked large fuel storage tanks along the highway from Belgrade to the outskirts of Zagreb and started large fires at Osijek, Sisak, and Karlovak.[29]

- A refinery in Sisak, which produced liquefied petroleum gas, fuel, petroleum coke, and solvents, was attacked with thousands of Serbian artillery rounds, which hit 38 petroleum storage tanks. If these attacks had destroyed existing stored chemical containers, lethal concentrations of chemicals would have covered a wide area.[30]

- In October 2001, a group of Tamil Sea Tigers attacked and set ablaze the oil tanker MV Silk Pride. It was carrying 225 tons of low-sulfur diesel, 160 tons of kerosene oil, and 275 tons of auto diesel.[31]

An organized attack on a U.S. oil refinery, similar to the attack on the Sisak refinery by Serbian forces, would be nearly impossible for security guards to prevent.

Furthermore, it is relatively simple for individuals to gain access to plants with the level of security common at refineries. As recently as January 2002, a robber carrying a shotgun made his way into a CITGO facility in Texas. CITGO was one of the companies that claimed to have dramatically increased security measures after September 11, 2001.[32] In addition, activists and reporters have breached security at refineries and chemical facilities across the country. Greenpeace activists entered Dow Chemical's Plaquemine, Louisiana facility in February 2001 through an unlocked gate and gained access to the control panel that regulates wastewater discharges into the Mississippi River.[33] A reporter in Pennsylvania skirted inadequate security at more than 30 chemical facilities and found that he "could walk or drive right up to tanks, pipes and control rooms considered key targets for terrorists."[34]

Assuming that on-site security was able to prevent all unauthorized access to a refinery and water mitigation systems[a] were installed throughout the refinery, it would still not be difficult to release hydrofluoric acid into surrounding communities. According to Carol Coy, a California regulator whose agency pushed for an end to hydrofluoric acid use in southern California, saboteurs could deactivate the mitigation systems simply by shutting off the electricity.[35] Furthermore, munitions fired from an offsite location could puncture the storage tanks holding hydrofluoric acid.

National security experts recognize that a terrorist attack on chemical facilities and refineries is more likely than an attack with a conventional chemical weapon. Creating chemical weapons is a complex and expensive process, whereas industrial facilities provide relatively easy access to large amounts of chemicals from which a significant chemical release could harm considerable numbers of people. In a Senate committee hearing on June 15, 2005, even the Department of Homeland Security (DHS) expressed concern about the consequences of a terrorist attack on a chemical facility.[36] Robert Stephan, an undersecretary in the DHS, noted that, as an aspect of America's "critical infrastructure," chemical facilities indeed represent a dangerous terrorist target. In addition, chemical facilities have many easy access points, making facility security more difficult. According to Stephan's testimony:

> "DHS has identified five areas as the focus of our primary preparedness work with the industry: access and access control, operational security, process control, facility systems operations, and local first responder and external response and recovery coordination. These preparedness planning variables must be refined with reference to potential methods of attack. These include perhaps most importantly: insider threats or sabotage; cyber attack; and attacks using explosives or other weaponry."[37]

Amy Smithson, director of the Chemical and Biological Weapons Non-Proliferation Project at the Henry L. Stimson Center, testified to this in a House of Representatives committee hearing:

> "Although assembling from scratch an unconventional weapons capability that could cause mass casualties is not that elementary, there are tangible routes whereby terrorists could inflict considerable harm with chemical and biological substances. One shortcut involves foul play with industrial chemicals...Logic dictates that if the same result [mass casualties from a chemical release] can be achieved through a less arduous route, terrorists intent on causing mass casualties with chemicals would probably engineer the intentional release of industrial chemicals rather than wrestle with the complexities of making large quantities of the classic chemical warfare agents."[38]

[a] Water mitigation systems are designed to cool nearby liquid petroleum gas containing vessels and columns to insure their structural integrity and to reduce hydrogen fluoride vapors if a leak were to occur. For water mitigation systems to effectively reduce hydrogen fluoride vapors in the event of a leak, sensors to detect the leak must be installed.

REPORT FINDINGS: COMMUNITIES AT RISK

Under the Clean Air Act's chemical accident prevention requirements, industrial plants that use large volumes of certain toxic materials must file accident prevention plans (Risk Management Plans) with EPA that include worst-case accident scenarios. These estimate how far a chemical could travel off-site and still maintain toxic concentrations in certain weather conditions and report the number of people living within that distance, named the "vulnerability zone."[b] We examined the most recent Risk Management Plans for the 50 petroleum refineries in the United States using hydrofluoric acid in their processing or storing it on-site. These 50 refineries comprise one-third of the 148 refineries operating in the U.S.; meaning, two-thirds of all refineries use technology other than hydrofluoric acid.

These 50 refineries, using and storing 10.6 million pounds of hydrofluoric acid, endanger more than 17 million people living in surrounding communities in 20 states (Table 1). See the Appendix for a list of all 50 refineries using hydrofluoric acid in the U.S.

Texas is home to the most hydrofluoric acid refineries, with 12; Louisiana is second with five facilities; and Montana is third with four facilities. These states are not home to the refineries that collectively endanger the most people, however; often a single refinery can endanger millions. Just two refineries using hydrofluoric acid in Pennsylvania endanger more than 4.4 million people residing in their vulnerability zones, according to conservative estimates. In New Jersey, a single refinery using hydrofluoric acid endangers more than 3.1 million people living within the vulnerability zone. Three Illinois-based refineries also endanger 3.1 million people living within the vulnerability zones.

Table 1. Oil Refineries by State and Population at Risk in Each State in Event of Worst-Case Release of Hydrofluoric (HF) Acid

State	# of Oil Refineries	# of Refineries Using HF Acid	Population Endangered By HF Release
AL	3	0	-
AK	6	0	-
AR	2	0	-
CA	21	2	360,000
CO	2	0	-
DE	1	0	-
GA	1	0	-
HI	2	0	-
IL	4	3	3,183,000
IN	2	1	8,000
KS	3	3	63,800
KY	2	1	300,000
LA	17	5	1,436,411
MI	1	0	-
MN	2	1	2,200,000
MS	4	0	-
MT	4	4	183,704
NV	1	0	-
NJ	6	1	3,170,000
NM	3	2	18,829
ND	1	1	68,013
OH	4	1	940,000
OK	5	3	113,388
OR	1	0	-
PA	5	2	4,400,000
TN	1	1	791,888
TX	26	12	1,930,805
UT	5	3	680,000
VA	1	0	-
WA	5	1	120,000
WV	1	0	-
WI	1	1	180,000
WY	5	2	63,067
Total	**148**	**50**	**17,040,905**[c]

Data on number of oil refineries by state obtained from Energy Information Administration.[19] *Data on oil refineries using hydrofluoric acid and the population at risk obtained from company Risk Management Plans (see methodology for more details).*

[b] It is important to note that not *all* people living within a vulnerability zone would be affected by a single chemical release; those living downwind during a chemical release are most likely to be affected.

[c] This is not the aggregate total of the state totals, as it takes into account overlapping vulnerability zones between states. As such, this is a conservative estimate.

Table 2. Ten Refineries Endangering the Most People in Event of Worst-Case Release of Hydrofluoric Acid	
Facility Name	**Location**
Sunoco Philadelphia Refinery	Philadelphia, PA
Valero Refining Co.	Paulsboro, NJ
PDV Midwest Refining (CITGO)	Lemont, IL
ConocoPhillips Trainer Refinery	Trainer, PA
Marathon Ashland Petroleum, MN Refining Div.	St. Paul Park, MN
Chalmette Refining	Chalmette, LA
Murphy Oil USA, Inc. Meraux Refinery	Meraux, LA
ExxonMobil Oil Corporation Joliet Refinery	Channahon, IL
Marathon Ashland Petroleum, Ohio Refining Div.	Canton, OH
ConocoPhillips Alliance Refinery	Belle Chasse, LA

Data on oil refineries using hydrofluoric acid obtained from company Risk Management Plans (see methodology for more details).

In some cases, however, the vulnerability zones of two or more refineries overlap, posing an even greater danger to people who live and work within the overlapping areas. In Philadelphia, for example, the vulnerability zones of two refineries overlap across the Delaware River, encompassing the airport, sports stadiums, and many neighborhoods.[40] In Corpus Christi, Texas, four refineries using hydrofluoric acid are located near each other; similarly, three refineries using hydrofluoric acid are situated close together near New Orleans, Louisiana.

Single refineries can endanger thousands or millions of people. Seven petroleum refineries with hydrofluoric acid alkylation facilities reported toxic release "worst-case" scenarios in which more than one million people could be affected. Furthermore, 15 refineries could place more than 500,000 people in harm's way, and 28 refineries could endanger more than 100,000 people in the event of a worst-case hydrofluoric acid release.

The 10 facilities putting the most people at risk in the event of a worst-case release of hydrofluoric acid are found in Illinois, Louisiana, Minnesota, New Jersey, Ohio, and Pennsylvania (Table 2).

Many parent companies put millions of people at risk from the hydrofluoric acid they use and store at their refineries. The companies with the most people residing in the vulnerability zones of refineries using hydrofluoric acid include Sunoco, Valero Energy Corporation, Marathon Ashland Petroleum, ConocoPhillips, CITGO, and ExxonMobil (Table 3). These companies each put at least two million people at risk.

Table 3. Parent Companies With at Least One Million People Living in Vulnerability Zones of Their Refineries Using Hydrofluoric Acid	
Parent Company	**Population at Risk**
Sunoco, Inc.	4,400,000
Valero Energy Corporation	4,366,193
Marathon Ashland Petroleum	4,132,993
ConocoPhillips	3,532,763
CITGO	3,415,420
ExxonMobil	2,393,847
Murphy Oil Corporation	1,236,000
Premcor Inc.	1,121,888

Data on oil refineries using hydrofluoric acid and the population at risk obtained from company Risk Management Plans (see methodology for more details).

Many companies that utilize the dangerous hydrofluoric acid technology also use refineries without that technology.

ConocoPhillips, ExxonMobil, Valero Energy Corporaton, and Marathon Ashland, for example, own refineries that use hydrofluoric acid as well as refineries that use other technologies.

- ConocoPhillips owns six refineries that report to EPA's Risk Management Program for their use of hydrofluoric acid, but runs 12 refineries in the United States. The other six facilities presumably use a safer technology than hydrofluoric acid.

- Valero Energy Corporation owns seven refineries in the United States that use hydrofluoric acid and five others that do not use that technology. One Valero refinery in Wilmington, California has responded to concerned residents and is switching from hydrofluoric acid to a safer technology.[41]

- ExxonMobil owns or co-owns four refineries that use hydrofluoric acid and three other refineries in the United States that do not.

- Marathon Ashland owns six hydrofluoric acid refineries in the U.S. and one refinery that does not use that technology.

Companies owning facilities that use different types of chemicals in their alkylation processes illustrate not only that alternative processes are possible to implement, but that some companies have knowledge of safer alternatives and choose not to implement them.

How Policymakers and Industry Should Protect Communities

A Preventive Approach

The actions of American industry—and American regulatory policy—have historically focused on preparing for or managing chemical accidents and releases rather than preventing them. The continuing legacy of chemical accidents and un-addressed vulnerabilities in the United States is evidence that this strategy has failed to protect public safety. Furthermore, the events of September 11th make plain the need for preventive action. Safety valves may mitigate the effects of an accidental release, and employee training may reduce the chances of an accident, but neither can protect public safety if a terrorist parks a truck bomb at a chemical plant or refinery. Designed to protect only against accidental releases, many accident mitigation technologies could be foiled by a deliberate saboteur.

Reducing or eliminating chemical hazards offers the best strategy to fully protect American communities from both accidents and terrorist attacks involving industrial chemicals. Hazard reduction means making a chemical process *inherently* safer by eliminating the use of highly toxic, volatile, or flammable chemicals or using chemicals in safer quantities or conditions. The concept of inherent safety leads to a hierarchy to guide decisions on the use and management of chemicals:

First, reduce or eliminate the *possibility* of a chemical release by choosing inherently safer materials and technologies.

Second, reduce the *probability* of a chemical release through secondary prevention measures such as safety valves and double-walled vessels. In preventing terrorism, increasing site security is an additional secondary prevention measure (although inadequate in the context of modern terrorists' tactics).

Third, reduce the *potential severity* of the impacts of a chemical release through mitigation measures (containment dikes, sprinkler systems) or emergency response plans.[42]

Again, the first option—inherent safety—provides the best response to the threat of chemical releases caused by acts of terrorism because it eliminates the potential hazard. Add-on security and mitigation measures could make minor contributions toward preventing an act of terrorism, but traditional tools of terrorists—truck bombs, suicide bombers, and now airplanes—would likely render such measures useless. Site security measures could prevent a terrorist from entering the grounds of a facility, but in the embassy bombings in Africa, the trucks containing bombs were parked near, not inside, facility grounds. Increasing physical security would have been of little help.

Inherent Safety at Refineries: Alternatives to Hydrofluoric Acid

Three options are available to petroleum refineries using hydrofluoric acid as a catalyst for alkylates in order to make this process inherently safer.

- Change the alkylation process to use a solid acid catalyst;
- Convert the hydrofluoric acid alkylation unit into a sulfuric acid unit; or
- Add modifiers to the hydrofluoric acid that decrease the gaseous nature of hydrofluoric acid and install mitigation systems.

Prevent the Possibility: Solid Acid Catalyst

The best option available to oil refineries is switching from using hydrofluoric acid to a solid acid catalyst, which completely eliminates the need to use either hydrofluoric acid or sulfuric acid to produce alkylate. Industry experts report that a variety of solid acid catalysts will be available for use in alkylation facilities within the next four years. Critics, however, insist this technology has been available since the late 1990s, and simple industry inertia has kept solid acid catalysts from becoming the popular choice for refinery alkylation processes.[43]

Solid acid catalysts have tremendous environmental and safety advantages over both hydrofluoric acid and sulfuric acid because they are neither corrosive nor particularly hazardous to people or the environment. Furthermore, in the event that the container housing the catalyst is breached, no further damage would result. Using a solid acid catalyst provides many benefits, including reduced waste disposal costs and hazards, possibility to expand capacity at a lowered capital cost, and less corrosion of the equipment, resulting in lower maintenance costs.[44]

Considerable research is being conducted on solid acid catalysts, and three main options currently exist: Albemarle Corp./ABB Lummus Global/Fortum Oil's AlkyClean™, UOP's Alkylene™, and Exelus Inc.'s ExSact™.

AlkyClean™'s demonstration phase at Fortum's facilities in Porvoo, Finland was completed in January 2005 and is currently commercially available. According to the manufacturers, the operating and capital costs of the AlkyClean™ process rival those of the liquid acid catalyst processes and produce an alkylate of similar high quality.[45]

This solid acid catalyst also can replace existing hydrofluoric acid or sulfuric acid alkylation units, reusing the existing feedstock pretreatment and product distillation/recycle facilities. Converting to a solid acid catalyst would save the facility money over a sulfuric acid plant because this technology does not require refrigeration for the reaction to take place. Furthermore, there is no need to remove acid from the finished product, thereby eliminating yet another step in the process.[46]

UOP's Alkylene™ operates similarly, with the primary difference between the two available processes consisting of how the solid acid is regenerated. The biggest difference is that Alkylene™ does not have the energy savings associated with AlkyClean™. Furthermore, the feedstock does have to be treated to remove impurities. Yet, even with these differences, Alkylene™ is still very cost competitive with hydrofluoric and sulfuric acid facilities.[47] Recent industry publications report this technology has "overall economics superior to sulfuric acid technology."[48] Alkylene™ also has recently become commercially available, though there is limited interest so far in the United States and Europe given the need to construct new solid acid units.[49] In February 2005, however, the Baku Heydar Aliyev Refinery of Azerbaijan awarded UOP a contract for the design of a new high-octane gas facility, which included, among other technologies, the world's first Alkylene™ unit. Production is estimated to start by 2008.[50]

Exelus Inc. also has a new solid acid catalyst technology (ExSact™) that is currently in its six to nine month demonstration phase that precedes commercialization. This technology not only includes a stable and active catalyst, but also a fixed-bed reactor that enhances the efficiency of the reaction and reduces the cost.[51] According to a representative of Exelus Inc, U.S. petrochemical companies, not refineries, have shown interest in this technology; oil refineries in Norway and Eastern Europe have demonstrated interest.[52]

All three of these alkylation processes cost significantly less than sulfuric acid alkylation units and virtually the same as hydrofluoric acid units. Manufacturers expect the market for solid acid catalyst technology to increase significantly as oil refinery alkylation units age and have to be replaced. The mounting costs of physical site security also add an incentive for refineries to select alternatives to dangerous hydrofluoric acid.

REDUCE THE SEVERITY: SULFURIC ACID AS AN OPTION

Sulfuric acid is often used in the alkylation process instead of hydrofluoric acid because when released, it will not readily form a toxic aerosol cloud. Instead, sulfuric acid is released as a liquid form, making it much easier to contain and prevent exposure to those offsite. As a result, sulfuric acid does not pose as much as a threat to life outside of the facility.

Converting a hydrofluoric acid alkylation unit into a sulfuric acid alkylation unit requires several equipment changes. First, a sulfuric acid alkylation unit requires a refrigeration process.[d] Second, if the hydrofluoric acid facility contains the metal monel, it must be replaced because it reacts with sulfuric acid. Finally, an acid regeneration facility also should be constructed onsite so as to decrease the need to frequently transport sulfuric acid, which poses the risk of offsite consequences.

The fundamental difference between using hydrofluoric acid and sulfuric acid in the production of alkylate is in how the acid is regenerated, once the acid becomes too contaminated with impurities. Hydrofluoric acid can be distilled to remove impurities. However, sulfuric acid must go through a series of steps that first break down the acid into sulfur dioxide and then mix the sulfur dioxide with water to create regenerated sulfuric acid.[53] Sulfuric acid catalysts inevitably cost about fifty cents per barrel of alkylate more than hydrofluoric acid catalysts because of the elaborate regeneration process that sulfuric acid requires.[54]

Although sulfuric acid is less hazardous than hydrofluoric acid, direct exposure to sulfuric acid can cause many detrimental health effects at concentrated levels, such as burns or severe irritation to the eyes, skin, and respiratory

[d] The catalytic reaction using hydrofluoric acid takes place around 100° F, which only necessitates the use of water cooling towers to maintain an optimum reaction temperature. The reaction to produce alkylate using sulfuric acid is optimized at 45°-50° F (7°-10° C), requiring refrigeration to maintain the optimum reaction temperature.

tract if concentrated fumes are inhaled. Because sulfuric acid is often regenerated offsite, there is a risk of an accident involving sulfuric acid during transportation. Once onsite, however, it can be regenerated indefinitely if the refinery builds a regeneration facility.

Two options exist to switch the alkylation process from hydrofluoric acid to sulfuric acid: using a conversion system or building a new sulfuric acid alkylation unit. One system for switching from hydrofluoric acid to sulfuric acid is the Alkysafe™ conversion/expansion process, offered by STRATCO®. Alkysafe™ reuses both the reaction and distillation sections of the alkylation facility.[e] Due to the short downtime and the amount of equipment that is reused, STRATCO claims that Alkysafe™ is cost-competitive with mitigation systems being installed on hydrofluoric acid alkylation units. These mitigation systems, on average, will cost a refinery between $20 million and $30 million, with costs reaching at most $50 million.[55]

The cost of building a sulfuric acid alkylation unit varies according to the amount of alkylate the refinery produces each day. As a general rule, a new alkylation unit will cost about $5,000 per barrel of alkylate produced, per day. Therefore, a new alkylation unit capable of producing 10,000 barrels of alkylate per day would cost about $50 million. STRATCO's Alkysafe™ process is estimated to cost one-half to two-thirds the cost of installing a new sulfuric acid alkylation unit.[56]

ExxonMobil also offers a sulfuric acid alkylation process that can replace hydrofluoric acid alkylation systems, although the cost of the process is higher than using the STRATCO designs. However, this process also would be competitive with building a new hydrofluoric acid alkylation unit.[57] It is currently in use in approximately 12 refineries.[58]

REDUCE THE PROBABILITY: HYDROFLUORIC ACID MODIFIERS

The final option available to decrease the threat of a hydrofluoric acid release is to invest in alkylation modifiers and install active mitigation units, such as water spray systems. This option, however, does not remove the possibility of a terrorist threat, and an adversary could thwart mitigation systems.

Modified hydrofluoric acid reduces the ability of the acid to form an aerosol cloud by a certain percentage, thereby mitigating the impact the toxic cloud will have on the surrounding community. UOP and Texaco estimate the cost of modifying a hydrofluoric acid alkylation refinery using Alkad, a passive mitigation system, at $7 million. This estimate, however, does not include the cost of active mitigation systems, which would be necessary in order to truly reduce the potential severity of a release.[59] Mitigation systems cost, on average, between $20 million and $30 million to install.[60]

ConocoPhillips and ExxonMobil have devised their own alkylation modifier under the name ReVAP, which has the ability to reduce hydrofluoric acid aerosol formation when leaked by 60% to 90%. ReVAP is currently licensed at five refineries, including the Woods Cross Refinery in Utah and the Torrance Refinery in California.[61]

In 2003, public pressure succeeded in persuading the Valero Energy Corporation to switch to modified hydrofluoric acid at its Wilmington, California refinery, near Los Angeles. Since an explosion that caused an accidental release of hydrofluoric acid at a neighboring Torrance refinery in 1987, the local community and government have

[e] The distillation section of the alkylation unit neutralizes any acid leaving the reaction chamber with the alkylate.

pushed to shut down two refineries that used hydrofluoric acid and required a third facility to change to modified hydrofluoric acid. The community was able to negotiate an agreement with the South Coast Air Quality Management District with regards to the Valero facility; Valero decided on ConocoPhillips' ReVap technology and will pay a fine up to $1 million if the renovation is not complete by the end of 2005. The change is expected to cost Valero about $30 million.[62,63]

Since both modification systems cannot completely reduce the threat of a hydrofluoric acid release, active mitigation units also must be installed. Active mitigation for hydrofluoric acid requires sensors to detect a hydrofluoric acid release; acid pumps to quickly move the acid to remote locations and decrease the amount of acid that is released; and water spray systems that knock the acid to the ground and prevent it from affecting surrounding communities. If the sensors are working properly and detect a leak immediately, water spray systems can knock up to 90% of the hydrofluoric acid to the ground; however, tests have shown that it requires at least 40 volumes of water for each volume of hydrofluoric acid released. Effective water spray systems often include water spray curtains and remotely operated water cannons.[64]

TABLE 4. COMPARISON OF INHERENTLY SAFER TECHNOLOGIES FOR PETROLEUM REFINERIES USING HYDROFLUORIC ACID

	Sulfuric Acid	Solid Acid Catalyst	HF modifier
How Catalyst Regenerated	Decomposed into sulfur dioxide, then regenerated by mixing with water vapor	Frequent regeneration with dissolved hydrogen every two to four weeks.	Distilled
Required Facility Modification	Some versions require a new refrigeration facility and design change and modifications of current equipment; other versions cannot be fitted to existing alkylation processes.	Alkylene™ and ExSact™ require entirely new alkylation facilities, while AlkyClean™ only requires a new regeneration system.	Additional separation equipment needed to remove leaching acid from alkylate.
Advantage over HF Alkylation Unit	Will not form an aerosol cloud and will not pose a threat to life outside the facility.	No corrosive acid can leave the reaction chamber. *No threat of a chemical spill to anyone.*	Decrease HF aerosol formation from 60% to 90% in the event of a leak, depending on version used.
Additional Mitigation Systems Needed	None	None	Active mitigation systems needed.
Safety Concerns	Acid spill during transport to the refinery and to the regeneration facility if the regeneration facility is not part of the refinery.	None	Active mitigation systems can fail.
Cost of Conversion	Varies depending on the daily output required. Estimated to cost between $2,500 to $3,333 per barrel of alkylate produced per day for one version, $5,000 per barrel for another.	Cost competitive with installing a new hydrofluoric acid unit, less than installing a sulfuric acid alkylation unit.	Valero is paying an estimated $30 million to convert its Wilmington, CA refinery; most modifiers cost $7 million plus the cost of active mitigation systems.

Reducing Chemical Hazards Through Policy Measures

The possibility that terrorists could turn American industry into weapons that threaten Americans' safety provides policy-makers with a clear imperative to revise existing policy on lethal chemical releases. Furthermore, more action is needed than simply increasing physical security at chemical facilities; more physical security measures do not offset the possibility of terrorists taking extreme measures to orchestrate a chemical release.

Inadequacies of Existing Policies

Industry Cannot Be Left to Voluntary Measures

Since September 11, 2001, our nation has tightened security in a variety of venues. Airports and airlines have numerous new regulations they must follow; airplanes have routinely patrolled our water supply; our government has even established an entire new governmental department, the Department of Homeland Security. Despite the repeated admission by government officials that the chemical industry poses a significant security threat, however, no federal regulations exist that require oil refineries or any chemical facility to reduce their hazards when they are able to do so.

Instead, some facilities subscribe to a voluntary industry program known as Responsible Care. Responsible Care is an initiative developed by the Chemical Manufacturers Association (CMA), now the American Chemistry Council (ACC), in 1988 to respond to the public's lack of confidence in the chemical industry. The CMA needed to act to address the poor public image or "end up in worse shape than the atomic industry," according to John Johnstone, a former chairman of CMA.[65] Fifteen years later, the ACC requires all member companies to comply with this voluntary initiative to improve security. Responsible Care, however, only requires facilities to install physical security and does not address preventing accidents or terrorist attacks by requiring companies to switch to safer technology.

A former member of the Security Committee of the American Chemistry Council, as well as a former Security Manager for Georgia-Pacific Company, has addressed just how inadequate these voluntary guidelines are. In a June 2005 hearing in the House Subcommittee on Economic Security, Infrastructure Protection, and Cybersecurity, Sal DePasquale said, "it may be argued that inner city liquor stores are better protected than are the facilities that manufacture and use highly toxic and lethal chemicals." Despite industry's claims of effective self regulation, he continued, "if the industry will not issue substantive standards, it cannot say that it is self regulating. It is simply a contradiction in terms."[66]

Furthermore, ACC is made up of only around 160 of the largest chemical companies, meaning that, according to Department of Homeland Security statistics, 20% of chemical companies in the United States do not fall under any sort of voluntary chemical security standards.[67]

The Emergency Planning and Community Right-to-Know Act

After the 1984 Union Carbide chemical disaster in Bhopal, India, which killed thousands of people, grassroots pressure convinced Congress to pass the Emergency Planning and Community Right-to-Know Act (EPCRA). This act established a network of

Local Emergency Planning Committees (LEPCs) and required facilities to disclose a baseline of important information on chemical risks, including the amount of hazardous materials stored at particular facilities.

Despite some positive results of EPCRA, including the creation of the Toxics Release Inventory and a general decline in toxic releases, this law did not address the need to prevent chemical releases before they happen. Instead, it only addressed the need to prepare for and respond to them. The LEPCs lack the authority to mandate hazard reduction, and many are inactive.[68]

The Clean Air Act and Risk Management Planning Program

In response to an explosion at a Phillips chemical facility in Texas that killed 23 workers in 1989, Congress took further action by adding amendments to the Clean Air Act in section 112(r). These amendments established the Risk Management Planning (RMP) program that requires industry to develop a hazard assessment that covers various release scenarios, off-site consequences, and a five-year accident history; a prevention program that manages procedural areas such as training and safety audits; and an emergency response program.

Therefore, the Risk Management Planning program addresses the management of chemical risks, but not their prevention. If the RMP program were fully implemented as EPA originally intended, the worst-case scenario estimates of off-site impacts of chemical releases would be available in a national database and could help reduce chemical hazards, much as the Toxics Release Inventory has. Although originally available to the public online, the RMP plans have been removed from public view and are now difficult for individuals to obtain.

The case study of the voluntary changes at the Valero refinery in California shows that when a community is aware of the threat a local refinery poses to its safety, public pressure may successfully persuade the facility to change its chemicals and processes in order to reduce the threat. In order for the RMP program to be successful, risk management plans must be readily available to the public.

EPCRA and RMP Too Specific to Address Threat

In addition to the limitations addressed above, both EPCRA and the RMP program were not designed to address the threat of a terrorist attack at a chemical facility, and consequently pertain only to certain chemicals, industries, and facilities using quantities of chemicals above certain thresholds.

Smaller quantities of chemicals are easier to manage onsite and therefore pose less of a concern for accidental releases. Terrorists, however, could specifically target smaller quantities of chemicals and still cause significant harm. For example, a single one-ton cylinder of chlorine gas can fall below thresholds for the RMP program. The quantity of chlorine gas in this cylinder, however, could result in toxic concentrations nearly two miles off-site.

THE RIGHT-TO-KNOW AS A SAFETY TOOL

The federal government should not limit the public'd access to information about chemical use and releases in the effort to prevent chemical terrorism.

Ensuring a community's right-to-know about chemicals used, stored and released has long been a useful tool in protecting public safety from toxic hazards. In Massachusetts, for example, companies are required to assess and disclose all chemicals used by their facilities as well as complete toxics use reduction plans. Simply by completing these plans, and not necessarily even implementing them,

Massachusetts companies have reduced their overall use of toxic chemicals by 40%, the waste generated by 70%, and their environmental releases by 92% between 1990 and 2003. At the same time, production increased 23% at these facilities.[69]

New Jersey has seen similar success with its Pollution Prevention Act, which requires companies to assess and report chemical use. As a result of these reporting requirements and the Toxic Catastrophe Prevention Act, hundreds of drinking water facilities and sewage treatment facilities have stopped using chlorine gas.

Based on these examples, it is likely that if a petroleum refinery completed an assessment of available inherently safer technologies, and was required to disclose the findings to the public, it would be compelled to switch to a safer technology.

PROTECTING COMMUNITIES THROUGH INHERENTLY SAFER TECHNOLOGY

The most effective means of protecting American communities from the consequence of an act of chemical terrorism is to require facilities to implement inherently safer technologies.

In the case of hydrofluoric acid in alkylation units, inherently safer technology exists and is proven to work. Facilities should change their chemicals and processes to a safer technology as soon as possible, and new facilities should be built using the safest technology available.

Policymakers could require refineries to change their processes or to conduct a technology options analysis.[70] A technology options analysis is a way to address a range of industries and require them to make a concerted effort to identify inherently safer options in their chemical uses and processes. Furthermore, a technology options analysis provides an opportunity for facilities to adopt technologies with acceptable cost and appropriate performance characteristics and to explain why technologically feasible options were not selected. These technology options analyses should be made public, while protecting legitimate confidential business information, in order to inform communities of safety measures at nearby facilities as well as to disseminate information on innovative technologies.

Oil refineries using hydrofluoric acid on site pose an unnecessary risk to surrounding communities. By requiring facilities to switch to inherently safer technologies, as well as requiring them to publicly disclose plans to protect surrounding communities, oil refineries could greatly reduce their risk. As oil refineries are not the only facilities that pose unnecessary risk to surrounding communities, these same policy changes should apply to a variety of industries and facilities. Future policy must focus on significantly reducing or eliminating the potential harm posed by terrorist attacks on industrial facilities, instead of focusing on physical security and the unachievable task of mitigating the consequences of a major chemical release.

METHODOLOGY

The vulnerability zone data in this report were collected from Risk Management Planning reports obtained at public reading rooms operated by the Environmental Protection Agency in Washington DC, in compliance with all rules that currently govern the collection of such data. We collected this data in June and July 2005.

Facilities had to report their latest Risk Management Plans to EPA starting in June 1999 with revised reports due no less often then every five years. The data used in this report are current as of December 2004. Facilities are required to file RMPs if they store hazardous chemicals listed in section 112(r) of the Clean Air Act above a threshold level used in regulated processes. These facilities span a broad spectrum of industries, including chemical manufacturers, petroleum refineries, agricultural wholesalers, drinking water and wastewater treatment systems, electric utilities, and others.

To estimate the total number of people living in the vulnerability zones in each state and nationally, we reviewed the geographic location of each facility, as oil refineries are often grouped together only a few miles apart. In order to avoid double counting, we assumed if the vulnerability zones of two or more facilities overlapped that they overlapped entirely. In these instances, we included the highest at-risk population from the facilities in our calculations.

As a result, the reported totals are a conservative estimate of the total population at risk in each state. In addition, the totals do not reflect the fact that many individuals are at a heightened risk because they live within the vulnerability zone of two or more oil refineries using hydrofluoric acid.

Appendix: Oil Refineries Using or Storing Hydrofluoric Acid On-Site

State	Facility Name	City	Parent Company
CA	ExxonMobil Torrance Refinery	Torrance	ExxonMobil
CA	Ultramar Inc. d/b/a Valero Wilmington Refinery	Wilmington	Valero Energy Corporation
IL	ExxonMobil Oil Corporation Joliet Refinery	Channahon	ExxonMobil
IL	PDV Midwest Refining, LLC	Lemont	CITGO
IL	Marathon Ashland Petroleum LLC IRD	Robinson	Marathon Ashland Petroleum LLC
IN	Countrymark Cooperative, Inc.	Mt. Vernon	Countrymark Cooperative, Inc.
KS	Coffeyville Resources Refining & Marketing	Coffeyville	Coffeyville Resources, LLC
KS	Frontier El Dorado Refining Company	El Dorado	Frontier Oil Corporation
KS	National Cooperative Refinery Association	McPherson	Cenex
KY	Catlettsburg Refining, LLC	Catlettsburg	Marathon Ashland Petroleum LLC
LA	ConocoPhillips Company Alliance Refinery	Belle Chasse	ConocoPhillips
LA	Chalmette Refining, L.L.C.	Chalmette	Chalmette Refining, L.L.C (ExxonMobil and Petroleos de Venezuela S.A.)
LA	Marathon Ashland Petroleum, LA Refining Division	Garyville	Marathon Ashland Petroleum LLC
LA	Murphy Oil USA, Inc. Meraux Refinery	Meraux	Murphy Oil Corporation
LA	Placid Refining Co. L.L.C. -Port Allen Refinery	Port Allen	Placid Refining Co./Petro-Hunt
MN	Marathon Ashland Petroleum, MN Refining Div.	St. Paul Park	Marathon Ashland Petroleum LLC
MT	ConocoPhillips Billings Refinery	Billings	ConocoPhillips
MT	Montana Refining Company	Great Falls	Holly Corporation
MT	CHS Inc. - Laurel Refinery	Laurel	Cenex
MT	ExxonMobil Billings Refinery	near Billings	ExxonMobil
ND	Tesoro Mandan Refinery	Mandan	Tesoro Corporation
NJ	Valero Refining Co. - New Jersey	Paulsboro	Valero Energy Corporation
NM	Navajo Refining Company	Artesia	Holly Corporation
NM	Ciniza Refinery	Jamestown	Giant Industries
OH	Ohio Refining Division	Canton	Marathon Ashland Petroleum LLC
OK	TPI Petroleum Inc.	Ardmore	Valero Energy Corporation
OK	ConocoPhillips Refinery - Ponca City, Oklahoma	Ponca City	ConocoPhillips
OK	Wynnewood Refining Company	Wynnewood	Gary-Williams Energy Corporation
PA	Sunoco Philadelphia Refinery	Philadelphia	Sunoco, Inc.
PA	Trainer Refinery	Trainer	ConocoPhillips
TN	The Premcor Refining Group Inc.	Memphis	Premcor Inc.
TX	Alon USA Big Spring Refinery	Big Spring	Alon USA LP
TX	Borger Refinery and NGL Center	Borger	ConocoPhillips
TX	Flint Hills Resources, L.P. - CC West Refinery	Corpus Christi	Koch Industries
TX	Valero Refining Co. - Texas, L.P. - West Plant	Corpus Christi	Valero Energy Corporation
TX	CITGO Corpus Christi Refinery East Plant	Corpus Christi	CITGO
TX	Valero Refining Co - Texas, L.P. - East Plant	Corpus Christi	Valero Energy Corporation

State	Facility Name	City	Parent Company
TX	Crown Central Petroleum, Houston Refinery	Pasadena	Crown Central LLC
TX	Premcor Port Arthur Refinery	Port Arthur	Premcor Inc.
TX	Marathon Ashland Petroleum Texas Refining	Texas City	Marathon Ashland Petroleum LLC
TX	BP America, BP Texas City Site	Texas City	BP
TX	Valero Refining - Texas, L.P.	Texas City	Valero Energy Corporation
TX	Diamond Shamrock Refinery - Three Rivers	Three Rivers	Valero Energy Corporation
UT	Big West Oil LLC	North Salt Lake	Flying J, Inc.
UT	ChevronTexaco Salt Lake Refinery	Salt Lake City	ChevronTexaco Corporation
UT	Woods Cross Refinery	Woods Cross	Holly Corporation
WA	ConocoPhillips Company	Ferndale	ConocoPhillips
WI	Murphy Oil USA Superior Refinery	Superior	Murphy Oil Corporation
WY	Frontier Refining Inc.	Cheyenne	Frontier Oil Corporation
WY	Wyoming Refining Company	Newcastle	Hermes Consolidated

END NOTES

[1] Congressional Research Service, *RMP Facilities in the United States as of May 2005*, released by Representative Edward J. Markey, July 6, 2005.
[2] U.S. EPA, *Assessment of the Incentives Created by Public Disclosure of Off-Site Consequence Analysis Information for Reduction in the Risk of Accidental Releases*, April 18, 2000: 2.
[3] "Ammonia leaks rousts tourists from Gulf hotels," *Sun Herald*, February 23, 2003.
[4] The Army Surgeon General found chemical plant terrorism to be second only to a major bioterror event. *See* Eric Pianin, "Study Assesses Risk of Attack on Chemical Plant," *Washington Post*, March 12, 2002.
[5] National Infrastructure Protection Center, "Homeland Security Information Update," February 12, 2003.
[6] Agency for Toxic Substances and Disease Registry, "Industrial Chemicals and Terrorism: Human Health Treat Analysis Mitigation and Prevention," 1999.
[7] Carol D. Leonnig and Spence S. Hsu, "Fearing Attack, Blue Plains Ceases Toxic Chemical Use," *Washington Post*, November 10, 2001.
[8] James Belke, U.S. Environmental Protection Agency, "Chemical accident risks in U.S. industry – A preliminary analysis of accident risk data from U.S. hazardous chemical facilities," September 25, 2000. Available at http://www.epa.gov/swercepp/pubs/stockholmpaper.pdf.
[9] Energy Information Administration, Annual Refinery Report, "Petroleum Supply Annual 2004, Volume 1, Table 36. Number and Capacity of Operable Petroleum Refineries by PAD District and State as of January 1, 2005."
[10] Adam Fifield, "In the Shadow of Danger: The Chemical Plant Peril How Safe, How Secure?" *Philadelphia Inquirer*, April 20, 2003.
[11] "Hydrofluoric Acid: Chemical Safety Information," Environment Health and Safety at University of North Carolina. Available at http://ehs.unc.edu/pdf/HydrofluoricAcid.pdf.
[12] Bruce Scott, "Alkylation Process Hazards Management: Does it Matter Which Acid You Use?" Presented at the 1991 Alkylation Seminar. May 1991.
[13] Agency for Toxic Substances and Disease Registry, "Toxicological Profile for Fluorides, Hydrogen Fluoride, and Fluorine." Available at http://www.atsdr.cdc.gov/toxprofiles/tp11.html.
[14] "Occupational Safety and Health Guideline for Hydrofluoric Acid," http://www.osha.gov/SLTC/healthguidelines/hydrogenfluoride/recognition.html.
[15] "Alkylation." *Encyclopedia Britannica Online, 2005.*
[16] Bruce Scott. "Alkylation Process Hazards Management: Does it Matter Which Acid You Use?" Presented at the 1991 Alkylation Seminar. May 1991.
[17] DuPont fact sheet, "H_2SO_4 vs. HF." Available online at http://www.stratco.dupont.com/alk/alkylation_02.html.
[18] Fred Millar, "Too Close For Comfort," *Friends of the Earth*, Winter 1991.
[19] National Response Center, http://www.nrc.uscg.mil/foia.html, accessed July 25, 2005.
[20] Adam Fifield, "In the Shadow of Danger: The Chemical Plant Peril How Safe, How Secure?" *Philadelphia Inquirer*, April 20, 2003.
[21] United States Chemical Safety and Hazard Investigation Board, Chemical Incident Report Center, "Local Refinery Accident Sends 13 to Hospital," March 4, 2003.
[22] Adam Fifield, "In the Shadow of Danger: The Chemical Plant Peril How Safe, How Secure?" *Philadelphia Inquirer*, April 20, 2003.
[23] Adam Fifield, "In the Shadow of Danger: The Chemical Plant Peril How Safe, How Secure?" *Philadelphia Inquirer*, April 20, 2003.
[24] Adam Fifield, "In the Shadow of Danger: The Chemical Plant Peril How Safe, How Secure?" *Philadelphia Inquirer*, April 20, 2003.
[25] Don Steward and Sandi McDaniel, "Danger Adrift: Chaos reigned after Sun Co. Spill," *Tulsa Tribune*, April 28, 1988.
[26] Laura Elder, "Group wants powerful acid out of BP's plant," *The Galveston Daily News*, March 27, 2005.
[27] Adam Fifield, "In the Shadow of Danger: The Chemical Plant Peril How Safe, How Secure?" *Philadelphia Inquirer*, April 20, 2003.
[28] Theodore Karasik, "Toxic Warfare," RAND Project Air Force, 2002.
[29] Theodore Karasik, "Toxic Warfare," RAND Project Air Force, 2002.
[30] Theodore Karasik, "Toxic Warfare," RAND Project Air Force, 2002.
[31] Theodore Karasik, "Toxic Warfare," RAND Project Air Force, 2002.

[32] John Tedesco, "Thief breaks refinery security: Two workers lose wallets," *San Antonio Express-News*, February 16, 2002.
[33] Theodore Karasik, "Toxic Warfare," RAND Project Air Force, 2002.
[34] Carl Prine, "Lax security exposes lethal chemical supplies," *Pittsburgh Tribune-Review*, April 7, 2002, and "Companies respond to infiltration of facilities," *Pittsburgh Tribune-Review*, May 5, 2002.
[35] Bill Walsh, "Toxins Make Local Plants Possible Target for Terrorists," *Times-Picayune*, July 6, 2003.
[36] Alexis Grant, "Chemical plants not secure against attack, officials say," *Houston Chronicle*, June 15, 2005.
[37] Alexis Grant, "Chemical plants not secure against attack, officials say," *Houston Chronicle*, June 15, 2005.
[38] Testimony of Amy E. Smithson, Director, Chemical and Biological Weapons Nonproliferation Project, Henry L. Stimson Center, before the House of Representatives Committee on Transportation and Infrastructure Subcommittee on Water Resources and the Environment, November 8, 2001.
[39] Energy Information Administration, Petroleum Supply Annual 2004, Volume 1, "Refinery Capacity Table No. 36. accessed at www.eia.doe.gov/oil_gas/petroleum/data_publications/petroleum_supply_annual/psa_volume1/psa_volume1.html.
[40] Adam Fifield, "In the Shadow of Danger: The Chemical Plant Peril How Safe, How Secure?" *Philadelphia Inquirer*, April 20, 2003.
[41] Bill Walsh, "Toxins Make Local Plants Possible Target for Terrorists," *New Orleans Times-Picayune*, July 6 2003.
[42] Adapted from N.A. Ashford, J.V. Gobbel, J. Lachman, M. Matthiesen, A. Minzner, and R.F. Stone, *The Encouragement of Technological Change for Prevention Chemical Accidents: Moving Firms from Secondary Prevention and Mitigation to Primary Prevention.* Cambridge, Massachusetts: Center for Technology, Policy, and Industrial Development, Massachusetts Institute of Technology, Boston, 1993.
[43] Milton Lapkin and Sanford Lewis, *Boosting the First Line of Defense: Report of the Good Neighbor Project for Sustainable Industries*, 2003." Available at http://www.safehometowns.org/chapter3.pdf.
[44] National Institute of Standards and Technology, Advanced Technology Program, "Catalysis and Biocatalysis Technologies White Paper, Leveraging Resources and Targeting Performance." Available at http://www.atp.nist.gov/atp/97wp-cat.htm.
[45] ABB Lummus, "New alkylation process ready for commercialization," press release, January 5, 2005.
[46] E.H. van Broekhoven, "A New Solid Acid Isobutane Alkylation Technology AlkyClean," *Catalysts Courier*, September 2001.
[47] Personal communication with Peg Stein, UOP, June 29, 2005.
[48] Nowak, Franz-Marcus, "Advances in Hydrofluoric (HF) Acid Catalyzed Alkylation," http://www.uop.com/objects/NPRASpr2003HFAlky.pdf.
[49] Personal communication with Peg Stein, UOP LLC, June 29, 2005.
[50] "Baku Heydar Aliyev Refinery selects UOP technologies for high-octane gas project," *American International Automobile Dealers*, February 7, 2005, at www.showroommagazine.com.
[51] Exelus, "Exelus Solid Acid Catalyst Technology – ExSact," Accessed July 14, 2005 at http://www.exelusinc.com/exsact.shtml.
[52] Exelus, "Exelus Solid Acid Catalyst Technology – ExSact," Accessed July 14, 2005 at http://www.exelusinc.com/exsact.shtml.
[53] Enviro-Chem, "Spent Acid Regeneration: Sulfuric Acid," accessed July 14, 2005 at http://www.enviro-chem.com/plant-tech/3rdtier/spentacidtop.html.
[54] Personal communication with Randy Peterson, Alkylation Division at STRATCO, July 21, 2003.
[55] Randall Peterson, STRATCO, *The STRATCO® Alkysafe™ Process: Low Cost Conversion/Expansion From HF To* H_2SO_4 *Alkylation*," Available online at http://www.stratco.dupont.com/alk/pdf/STRATCOAlkysafeProcess.pdf.
[56] Personal communication with Randy Peterson, Alkylation Division at STRATCO, July 21, 2003.
[57] Steven Ackerman, Girish K. Chitnis, and David S. McCaffery, Jr. "ExxonMobil Sulfuric Acid Alkylation Process." Presented at the 5th International Topical Conference on Refinery Processing. March 10-14, 2002
[58] ExxonMobil, "Sulfuric Acid Alkylation: Light Ends Upgrading Technology," accessed July 14, 2005 at http://www.prod.exxonmobil.com/refiningtechnologies/fuels/mn_sulfuric.html.
[59] Pam Pryor, "Alkylation Current Events," paper presented at the 2001 Alkylation Seminar in Scottsdale, Arizona, November 2001.
[60] Randall Peterson, STRATCO, *The STRATCO® Alkysafe™ Process: Low Cost Conversion/Expansion From HF To* H_2SO_4 *Alkylation*," Available online at http://www.stratco.dupont.com/alk/pdf/STRATCOAlkysafeProcess.pdf.
[61] ConocoPhillips Technology Solutions, "ReVap in Action," Accessed July 14, 2005 at http://www.coptechnologysolutions.com/alkylation/revap/revap_action/index.htm.

[62] Bill Walsh, "Toxins Make Local Plants Possible Target for Terrorists," *New Orleans Times-Picayune*, July 6, 2003.
[63] South Coast Air Quality Management District, "Highly Toxic Chemical to be Phased Out at Valero Refinery," February 7, 2003, available at http://www.aqmd.gov/news1/2003/hfvalero.html.
[64] Bruce Scott, "Alkylation Process Hazards Management: Does it Matter Which Acid You Use?" Presented at the 1991 Alkylation Seminar, May 1991.
[65] John Holusha, "Chemical Makers Identify a New Hazard," *New York Times*, August 12, 1991.
[66] Sal DePasquale, Testimony before the House Homeland Security Committee hearing, June 15, 2005.
[67] Testimony of Martin Durbin of the American Chemistry Council, Senate Committee on Homeland Security and Governmental Affairs hearing, July 13, 2005, available at http://hsgac.senate.gov/index.cfm?Fuseaction=Hearings.Detail&HearingID=256.
[68] Resources for the Future, *The Future of Local Emergency Planning Committees*, 1993; National Institute for Chemical Studies (Charleston, W.V.), Local Emergency Planning Committees and Risk Management Plans: Encouraging Hazard Reduction, prepared for U.S. EPA, Chemical Emergency Preparedness and Prevention Office (#CX 824095), June 2001.
[69] Toxics Use Reduction Institute, Lowell, Massachusetts, accessed July 14, 2005 at http://www.turi.org/turadata/Success/ResultsToDate.html.
[70] ConocoPhillips Technology Solutions, "ReVap in Action," Accessed July 14, 2005 at http://www.coptechnologysolutions.com/alkylation/revap/revap_action/index.htm.

Survey of Chemical Industry Hazard Reduction to Protect Public Safety

2002 Survey Summary

In February 2002, U.S. Public Interest Research Group and the Working Group on Community Right-to-Know sent a written survey to over 700 chemical plants. All of these facilities had filed Risk Management Plans with the U.S. Environmental Protection Agency because they use or store extremely hazardous substances.

The survey asked whether the facility had taken steps, or planned to take steps, to reduce the size of the off-site area where members of the public could be injured or killed in a worst-case chemical accident or terrorist attack. Such changes include measures such as substituting safer chemicals, reducing hazardous chemical storage and transportation, and using chemicals under less dangerous conditions.

The survey did not ask companies to provide any information about specific site security or safety features.

Only about one percent of the surveyed ACC facilities indicated that they were reducing chemical hazards to protect public safety.

Some 46 specific facilities responded to the survey (by mail or phone). Of these, nine facilities indicated reducing chemical hazard to protect public safety. The nine facilities were:

1. Durez Corporation (Kenton, Ohio)
2. Betzdearborn (Macon, Ga.)
3. BOC Gases (Waycross, Ga.)
4. FMC Corporation (Bessemer City, N.C.)
5. Milliken Chemical (Inman, S.C.)
6. Dow Chemical (Freeport, Texas)
7. ESCO Company (Muskegon, Mich.)
8. Fisher Scientific (Newark, Del.)
9. Rhodia (Baltimore, Md.)

Several additional companies acknowledged corporate interest in inherent safety, but gave no indication of actual changes to use safer chemicals and processes at specific facilities. These companies were: Cytec Industries; Rohm and Haas; Bayer; and, Solutia.

TESTIMONY
OF
THE NATIONAL ASSOCIATION OF CHEMICAL DISTRIBUTORS

BEFORE THE

SENATE COMMITTEE ON HOMELAND SECURITY AND GOVERNMENTAL AFFAIRS

ON

"CHEMICAL FACILITY SECURITY:WHAT IS THE APPROPRIATE FEDERAL ROLE?"

JULY 27, 2005

Introduction
The National Association of Chemical Distributors (NACD) is an international trade association headquartered in Arlington, Virginia. NACD represents 250 chemical distribution companies in the United States and Canada. These companies represent between 80% and 90% of the chemical distribution facilities in the nation and more than 90% of the industry's gross revenue.

NACD member companies process, formulate, blend, re-package, warehouse, transport, and market chemical products for an industrial customer base of approximately 750,000. Approximately $18 billion of U.S. chemical industry sales are through chemical distributors, who are also actively engaged in various phases of import/export trade. Chemical distributors' industrial customers use these materials to produce such everyday items as computers, detergents, cosmetics and toiletries, food flavorings, perfumes, automobile parts, water purifiers, fiberglass, plastics, pharmaceuticals, paints and coatings, and many other products.

To become a member of NACD, chemical distribution companies must take title to product and adhere to management practices related to health, safety, security, and the environment outlined in the Association's industry practice known as the Responsible Distribution Process℠ (RDP).

Before a company is admitted as a member, it must first be approved by successfully completing an independent, third-party verification of its written policies and procedures under RDP. To ensure continued compliance with RDP, every member must undergo an on-site verification by an independent third-party verifier once every three years. This mandatory practice has been in place since 1998, and members are currently undergoing their second on-site verification which will be completed at the end of 2005. NACD's Responsible Distribution Process℠ is the most comprehensive and rigorous industry practice of any in the chemical industry primarily because of its requirement for independent third-party verification of health, safety, security, and environmental practices. Continued compliance with RDP is a condition of membership in NACD.

Although chemical distribution is a sector of the chemical industry, distribution facilities differ in numerous ways from chemical manufacturing facilities. One notable example is the low levels of release of toxic emissions from everyday operations. According to data compiled each year by the Environmental Protection Agency (EPA), chemical distribution is a minor source of environmental releases. Of all industrial sectors required to submit annual toxic release reports, including the chemical industry, chemical distribution is by far the lowest emitter of toxic emissions. The average yearly release per distribution facility is just over 3,000 pounds over a 12-month period. While the possibility of chemical releases exists at chemical distribution facilities, it is minimized because of several factors, not the least of which is adherence to the industry's environmental, health, safety, and security practice – Responsible Distribution Process℠ – among NACD members.

Chemical distribution is also a safe industry in which to work. Industry data show that chemical distribution is a safe industry in terms of employee injuries and fatalities at NACD member companies as well as in transportation-related incidents. For example, last year among 19,000 workers employed by NACD member companies, there were 626 OSHA-reportable employee injuries that occurred within distribution facilities and no fatalities. Among transportation-related incidents, there were 20 injuries among member companies and three fatalities.

what steps must be taken to come into compliance and give them a specified time period to do so. If there is still no or inadequate compliance, DHS should have authority to levy fines and shut facilities down in severe cases.

DHS's authority should be limited to regulating the security of chemical storage and chemical manufacturing. Existing regulations, such as EPA's RMP, the Occupational Safety and Health Administration's (OSHA) Process Safety Management (PSM) rule, and the Drug Enforcement Administration's (DEA) Controlled Substances Act already effectively regulate how facilities manufacture, store, or process their chemicals in the safest manner possible to reduce the risk to employees, the environment, and their local communities.

Sensitive information about facilities, products, vulnerabilities, and security plans must be protected.
Any information companies are required to submit to DHS about their facilities, products, vulnerability assessments, and security plans must be protected from release to the public. Access to this sensitive information must be limited to groups that have a need to know. Measures must also be adopted to protect the confidentiality of sensitive information regarding a facility's vulnerability or security plan in any court proceeding.

Chemical security legislation should not mandate the use of inherently safer technologies.
NACD strongly opposes legislation that would attempt to make determinations about chemical compositions without considering costs, benefits, and safety issues. For example, mandating product substitution may result in needing much larger quantities of new substances, which would increase transportation requirements and the potential for incidents and accidents. A great deal of incentive already exists for companies to use the safest materials and methods possible in their operations. These incentives involve laws and regulations already in existence such as EPA's RMP and OSHA's PSM, the objective of companies to avoid costly incidents and to minimize potential consequences of such incidents, and the desire of companies to do what is in the best interest of their employees, their communities, and their business reputations.

In Closing
NACD appreciates the opportunity to provide testimony on this important issue of chemical security legislation. The Association looks forward to working with the committee to develop legislation that builds upon existing effective laws, regulations, and programs to protect chemical facilities from terrorist attacks.

Addendum
NACD thought it would be helpful to the committee to provide answers to the questions that were asked of the witnesses during the July 13 chemical security hearing as follows:

Q – Explain how NACD monitors compliance with the Responsible Distribution Process[SM] security measures.
A - Before a company is admitted as a member, it must first be approved by successfully completing an independent, third-party verification of its written policies and procedures under the Responsible Distribution Process[SM]. Continued compliance with RDP is a condition of membership

in NACD. To ensure continued compliance, every member must undergo an on-site verification by an independent third-party verifier once every three years. NACD, as the leading association of chemical distributors, was the first chemical industry association to adopt new additional industry practices that address security following 9/11. In April 2002, NACD added security requirements to RDP within key distribution operations, specifically handling and storage of chemical products at facilities, carrier selection for distributing chemical products, and customer qualification for chemical products of concern to various federal agencies. NACD's on-site, third-party verification requirement is currently being conducted to verify implementation of these new security requirements. This cycle will be complete by the end of 2005, and all NACD members will have had their procedures verified.

Q – The MTSA gives the Coast Guard the authority to shut down sites that are not satisfying the requirements of the law. Would NACD support giving DHS the authority to shut down chemical facilities that do not comply with the new chemical security law?
A – NACD supports giving DHS authority necessary to ensure that chemical facilities are secure. Before shutting a site down, it would be beneficial for DHS to work with that facility to let them know what steps must be taken to come into compliance and give them a specified time period to do so. If they do not comply in that time frame, DHS would levy fines or shut them down.

Q – How is the flow of information between DHS and small companies?
A – The flow of information from DHS to NACD members occurs mostly through communication by DHS to NACD. NACD filters information and sends relevant content to our members in an appropriate format. In general, we are pleased with the flow of information from DHS to NACD. Through DHS's ISAC and Homeland Security Information Network (HSIN) systems, NACD receives appropriate information and is working with the Department to improve the information sharing network through our active participation on the Chemical Sector Coordinating Council, which is also actively supported by DHS.

Q – How do you explain the media reports demonstrating how easy it is to access chemical facilities?
A – NACD believes that such unauthorized access to facilities is unacceptable. NACD members, who operate under the Responsible Distribution Process℠, have taken measures such as limiting entry points and requiring all visitors to present identification and wear badges to ensure that unauthorized entry does not occur. NACD supports federal legislation that would require facilities not covered under programs such as RDP to also take such security measures.

Q – It sounds as if under the ACC Security Code/NACD Responsible Distribution Process℠, a third-party verifies that the security plan is being followed, but not that the plan itself is adequate.
A – Before a company is admitted as a member of NACD, it must first be approved by successfully completing an independent, third-party verification of its written policies and procedures under RDP. Many of these policies and procedures are based on existing laws and regulations such as the U.S. Department of Transportation's HM-232 security regulation and the Drug Enforcement Administration's "know your customer" rules.

Q – What are the similarities between the MTSA and the NACD Responsible Distribution Process℠?
RDP bears some similarity to MTSA, namely the requirement to address site and transportation security. However, because the wide majority of NACD members do not ship or receive chemicals via marine terminals or marine vessels, RDP does not include specific considerations regarding marine transportation. A good example of an already-existing federal security requirement is DOT's HM-232 transportation security regulation. Transportation is a critical element of chemical distribution operations, and NACD members have been subject to HM-232 since it took effect in 2003. Compliance with HM-232 was simplified for NACD members because they were already abiding by the rule's requirements through RDP, particularly in the area of carrier selection.

Q – What are the consensus items and differences of opinion on what the federal role should be in chemical security?
A – NACD supports federal legislation. We believe that is the only way to ensure that all facilities adopt adequate security measures, particularly among those companies that follow no recognized management practice like the Responsible Distribution Process℠ or Responsible Care®. NACD further believes that any legislation should be based on risk, provide flexibility to address the needs and different types and sizes of companies, build upon federal programs currently in place, recognize existing effective industry programs, ensure the protection of sensitive security information so that it does not fall into the wrong hands, and avoid mandating ISTs.

Q – Do you see any conflicts between safety and security?
A – Conflicts between safety and security occur when one is compromised to accommodate specific concerns of the other. Removing placards from trucks and rail cars in the name of increasing security presents real safety concerns because it places in jeopardy the safety of individuals handling the shipment as well as first responders in the event of an incident. NACD opposes removal of placards. Mandating product substitution or banning certain products in the name of increasing security can also present a safety issue. For example, banning rail car quantities of chemicals in favor of truck-load quantities increases truck traffic and the potential for incidents and accidents.

Q – Should DHS have the authority to regulate chemical processes and storage?
A – DHS should have the authority to regulate the security of chemical storage and chemical manufacturing. This would require the establishment of a new regulatory program that addresses what the appropriate level of physical security is for chemicals on-site as well as the potential off-site consequence(s). Existing regulations, such as EPA's Risk Management Program, OSHA's Process Safety Management rule, and DEA's Controlled Substances Act, however, effectively regulate how facilities manufacture, store, or process their chemicals in the safest manner possible to reduce the risk to employees, the environment, and their local communities. Any new authority over chemical manufacturing or warehouses must, therefore, be limited to physical security.

Q – What does NACD think of a mandate that safer technologies be considered?
A – We assume you are referring to the composition of products, not the handling and storage of products. NACD members handle, store, and distribute chemicals every day. We do so using the safest methods known and available to us, either by law or by industry practice. We strongly oppose

legislation that would attempt to make determinations about chemical compositions without considering costs and benefit and safety issues.

Q – Does the NACD Responsible Distribution Process[SM] require companies to conduct first responder drills?
A – Communication and interaction with the local communities as well as first responders are large parts of RDP. Member companies are required to communicate and interact with their local communities and first responders about the types of chemicals handled by their facilities and about the environment, health, safety, and security programs they are required to follow under RDP, and the third-party verification process ensures that they do so. RDP has an entire section of its Code of Management Practice dedicated to Emergency Response and Public Preparedness, which requires eight specific actions including communicating with state and/or local emergency planning committees (LEPCs) and response organizations on the potential hazards of the member company's chemicals and participating in the LEPC's process to develop and periodically test the local emergency response plan. RDP also has an entire section of its Code of Management Practice dedicated to Community Outreach. Finally, NACD's Chemical Educational Foundation, on an annual basis, holds a competition among LEPCs and awards those that have the best programs for their communities.

Q – Wouldn't a risk-based approach to chemical security, under which there would be different tiers of regulation according to risk, give companies the incentive to use safer chemicals?
A – Using a risk-based, tiered system is both reasonable and equitable. NACD supports a tiered approach to chemical security that recognizes that more effort must be made by facilities that present the greatest level of security risk. A great deal of incentive already exists for companies to use the safest materials and methods possible in their operations. These incentives involve laws and regulations already in existence, the avoidance and minimization of potential consequences of costly incidents, and the desire to do what is in the best interest of their employees, their communities, and their business reputations. Companies must also meet the needs of their customers, who make purchasing decisions based on factors such as the laws of chemistry, cost, availability, functionality, and safety.

Q – One of the biggest tasks in developing legislation is to define the universe of chemical facilities that should be regulated. What are your suggestions?
A – The RMP list would be a reasonable place to start. Other factors that should be considered include the type and quantity of chemicals on-site and the facility's proximity to municipal settings such as urban areas and high-value buildings such as schools, hospitals, transportation, and neighborhoods. The properties of the substances being stored or handled, such as whether they are flammable, explosive, or toxic-by-inhalation, should also be considered.

AGRICULTURAL
RETAILERS
ASSOCIATION

Written Statement of

AGRICULTURAL RETAILERS ASSOCIATION

On

"Chemical Facility Security: What is the Appropriate Federal Role? (Part II)"

Before

Senate Homeland Security and Governmental Affairs Committee

July 27, 2005

Richard Gupton
ARA Director of Legislative Policy & Counsel
Washington, D.C.

INTRODUCTION

I would like to thank Chairwoman Susan Collins (R-ME) and Senator Joe Lieberman (D-CT) for holding this important hearing today. We appreciate the opportunity to discuss with this committee an issue of importance to all Americans including the agricultural industry. The Agricultural Retailers Association (ARA)[1] represents around two thirds of the nation's retail dealers who provide essential crop input materials to America's farmers. As the only national organization exclusively representing the interests of the Ag retail and distribution industry, ARA is vitally interested in any federal chemical site security laws or regulations affecting the operation of facilities and chemicals utilized in the nation's agricultural sector.

OVERVIEW OF AG RETAIL / DISTRIBUTION INDUSTRY

In 2002, there were an estimated 10,586 retail outlets in the United States.[2] The overall number of retail outlets is lower today and has been declining due to a number of factors taking place within the industry: consolidation, increased domestic and global competition, higher operating costs, and low profit margins. ARA members range in size from family-held businesses to large companies with many outlets located in multiple states. A typical retail outlet may have 3 to 5 year-round employees with additional temporary employees added during the busy planting and harvesting seasons. Many of these facilities are located in small, rural communities.

The retail farm supply dealer provides necessary goods and services to our nation's farmers. Goods and services that include: seed, crop protection chemicals, fertilizer, crop scouting, soil testing, custom application of pesticides and fertilizers and development of comprehensive nutrient management plans. Certified crop advisors (CCA's) are retained on retailer's staff to provide professional guidance and crop input recommendations to farmers and consumers. Retail and distribution facilities are scattered throughout all 50 states and provide important jobs in rural and suburban communities. The food and agriculture production and processing industry contributes substantially to the American economy – accounting for 13 percent of the U.S. gross domestic product and 18 percent of domestic employment.

A HEAVILY REGULATED INDUSTRY

Even before the terrorist attacks on September 11, 2001, Ag retailers have been one of the most heavily regulated industry segments in the country. Many of the products used by the industry are hazardous materials, which are highly regulated and expensive materials. There are countless federal and state laws and regulations related to the safe handling, transportation and storage of agricultural crop inputs. For example, many retail facilities that handle and store a threshold amount of listed substances such as propane, ammonia and / or other chemicals are required to comply with the U.S. Environmental Protection Agency's (EPA) Risk Management Program (RMP)[3]. Under the rule, covered facilities must

---

[1] The Agricultural Retailers Association (ARA) is the national political voice for the retail sector of the agricultural industry in the United States. ARA is an advocate for retailers located throughout the nation, representing their interests before Congress and Federal agencies.

[2] Doane's *Ag Professional* Magazine, Summer 2003, p.40-41

[3] Accidental Release Prevention Requirements: Risk Management Programs Under Clean Air Act Section 112(r)(7); Final Rule; 40 CFR Part 68

develop an RMP that describes their chemical accident prevention programs and submit full updates and resubmissions to EPA at least once every five years. The RMP Rule divides regulated facilities into three program focuses according to the level of potential danger they may present to surrounding communities.

Most retailers fall under the RMP's Program 2 Requirements, which generally are processes of low complexity and do not involve chemical reactions. Program 2 RMP requirements for retailers include the following: Describe how their RMP management systems will be implemented; Conduct hazard assessments, which includes analyses of worst-case and alternative release scenarios; Establish emergency response programs that include plans to inform the public and emergency response organizations about the chemicals onsite and their health effects and strategies to coordinate those plans with the community; and report steps taken to prevent incidents that can release dangerous chemicals. Program 2 RMP reporting requirements are less stringent than Program 3 RMP requirements, which are usually for higher risk chemical facilities and involve complex chemical processing operations. The prevention program requirements for Program 3 are very similar to those of the OSHA Process Safety Management (PSM) requirements. Some of this includes process safety information, process hazard analysis, operating procedures, training, mechanical integrity, management of change, pre-startup review, compliance audits, incident investigation, employee participation, hot work permits, and contractor requirements.

The Federal Insecticide Fungicide and Rodenticide Act[4] (FIFRA) continues to be the basis of EPA regulations covering agricultural pesticides. Sections of the code cover handling, labeling, crop tolerance requirements, precautionary statements, environmental protection issues, worker protection standards, storage requirements, transportation regulations and considerations, product use, and lots more issues designed to protect the public and all workers.

The Department of Transportation's (DOT) Pipeline and Hazardous Materials Safety Administration (PHMSA) formulates, issues and revises Hazardous Materials Regulations (HMR) under the Federal Hazardous Materials Transportation Laws. The HMR issued by the DOT cover hazardous materials definitions and classifications, hazard communications, shipper and carrier operations, training and security requirements, and packaging and container specifications. Ag retailers and distributors are required to comply with many of these DOT regulations.

As an industry, we have done a good job of educating and training employees to judiciously handle hazardous materials and to make sure they remain in the hands of authorized personnel. Employees of Ag retailers and distributors complete numerous training and certification programs that help ensure hazardous materials are being stored and handled with proper care. An employer at an Ag retail facility is responsible that their employees meet requirements such as: 1) A Commercial Applicator Certification Card; 2) Regular training for hauling hazardous materials (i.e. anhydrous ammonia, ammonium nitrate and other certain agricultural chemicals); 3) Trained in Worker Protection Standards; 4) Trained under OSHA Standards (i.e. Worker Right to Know, Lockout / tag out. Confined Space Entry, Personal Protective Equipment, etc.); 5) Subjected to periodic drug and alcohol testing; 5) Trained and tested for a Commercial Drivers License and Hazardous Material Certification; 6) Trained

---

[4] 7 U.S.C. s/s 136 et seq. (1996)

in the use of safety equipment and Hazmat procedures. 7) Trained in handling and record keeping on restricted use materials.

CURRENT SECURITY LAWS AND REGULATIONS

There are a number of federal security laws and regulations in place today that address security at chemical facilities. Some of the existing laws and regulations include:

USA Patriot Act of 2001[5]*:* The USA Patriot Act was enacted by the U.S. Congress and signed into law by President Bush shortly after the September 11, 2001 terrorist attacks. The legislation enhanced the authority of U.S. law enforcement to investigate and preempt potential terrorist attacks and other potential criminal activities that seek to endanger human life. One of the provisions in this new law prohibited states from issuing any Commercial Drivers License (CDL) with hazardous materials endorsement (HME) without first determining whether or not an individual seeking to transport hazardous materials poses a security risk. As a result of this requirement, the Transportation Security Administration (TSA) issued regulations requiring all applicants seeking a state-issued CDL with HME to submit biographical information and fingerprints with their application as of May 31, 2005. In general, the transportation of hazardous materials in commerce comprises 5 percent of all shipments in the United States; however the percentage is significantly higher in the agricultural community.

Maritime Transportation Security Act of 2002[6]*:* The Maritime Transportation Security Act (MTSA) was enacted to provide a framework for ensuring the safety of maritime commerce and the nation's domestic ports. The Coast Guard is the lead federal agency responsible for the implementation of this new law. Vessels and facilities that load/carry cargos deemed dangerous (i.e. flammable, potentially explosive, caustic or environmentally hazardous) must have individual security plans that address fundamental security measures such as access controls, communications, restricted areas, cargo-handling and monitoring, training, and incident reporting. Many ARA member companies have port facilities that have been required to comply with these security regulations since July 1, 2004.

U.S. Department of Transportation's (DOT) Hazardous Materials Transportation Security Regulations[7]: Agriculture is included in DOT regulations designed to better ensure the safe, secure transport of fertilizers and pesticides classified as hazardous materials. These regulations call for security assessments, plans and training for personnel, facility access and en-route transport. We believe that a more secure transportation system creates a safer, reliable food supply.

INDUSTRY SECURITY RELATED EFFORTS

Ag retailers and distributors have and continue to be pro-active in addressing security concerns related to the storage, handling and transportation of agricultural chemicals. It is important for Congress and the Administration to know that our nation's agricultural industry is committed to support effective measures that will prevent terrorists or other criminals from gaining access to these important crop production materials. In fact, DHS has and continues to work with the private sector to identify risks,

[5] Public Law No: 107-56
[6] Public Law No: 107-295
[7] Hazardous Materials: Security Requirements for Offerors and Transporters of Hazardous Materials; Final Rule; Published: March 3, 2003; 68 FR 14509

build systems to communicate those risks, and to prepare plans to keep those risks from becoming terrorist's targets. Industry has taken a very proactive role in dealing with DHS and has participated in the development of the sector working groups. One of the products of our working relationship has been the creation of the Interim National Infrastructure Protection plan, I-NIPP. In the future the I-NIPP will become the working document that all private industry will use as a planning tool for site and product security. ARA has been actively involved with DHS before the current sectors were even established; ARA Vice President of Regulatory Policy & Corporate Relations James D. Thrift currently holds the representative seat on the Food and Agriculture sub-sector Input committee representing some 60 plus agricultural associations.

At the same time, ARA has several concerns related to the new security regulations that could have a long lasting and costly impact on Ag retailers and distributors. ARA has been public in our support of reasonable efforts to protect the security of all Americans, but redundant systems set up by different branches of government to accomplish the same goal serve no useful purpose. For example, the new CDL HME fingerprint and background check regulations are on top of several existing programs that already check the backgrounds of CDL HME drivers. This new effort by TSA unnecessarily slows the security process down and gives the message that government is not organized for security measures.

ARA is a strong supporter of Asmark's Security Vulnerability Assessment (SVA) program. The Asmark SVA tool is licensed to ARA and is currently being utilized by member and non-member companies. ARA is working with CropLife America and The Fertilizer Institute under the "Agri-Business Security Working Group" and state associations to promote security measures and the SVA program. To date this SVA has been utilized by 2,500 retailers. ARA and Asmark recently reached agreement with Clemson University to make the SVA tool available to all Ag retail facilities in the state of South Carolina. This web-based software enables retail facilities to conduct a security vulnerability assessment of their facilities and receive recommendations to improve overall security. A vulnerability assessment is required by DOT for transporters of certain hazardous materials. The SVA satisfies one part of the Department of Transportation's (DOT) new rule governing the shipment of hazardous materials that requires a vulnerability assessment and a transportation security plan. This tool assesses the vulnerabilities of hazardous material transportation and provides countermeasures that can be used to aid in creation of the transportation security plan.

The program, which meets design criteria of the Center for Chemical Process Safety (CCPS) for conducting security vulnerability assessments, is available to retailers nationwide through state agri-business associations. By meeting CCPS design criteria; this SVA tool meets the same criteria that have been required by federal agencies for other chemical industry facilities. As a result, retailers who use this program to assess facility and hazmat transportation security can do so with confidence that their assessment is based on sound risk assessment principles. Even before the release of the SVA, the Agri-Business Security Working Group developed and issued *Guidelines to Help Ensure a Secure Agribusiness* so that retailers, distributors, wholesalers, and farmers and ranchers could begin evaluating and addressing security concerns at their facilities.

The committee should also keep in mind other security related programs being utilized by the industry. For example, there are individual Ag retail / distribution companies that have developed and conducted their own in-house SVA or utilized other service providers to conduct an SVA. ARA urges Congress to give the industry credit for security efforts already being accomplished.

KEY ISSUES RELATED TO ANY CHEMICAL SITE SECURITY PROPOSED LEGISLATION

DHS Lead Agency: As the very basic component of any chemical site security legislation developed by this committee, ARA believes that DHS must be lead agency for all chemical security related issues, as no other agency has the trained personnel, the direct security mission, or the broad security reach over numerous industry disciplines. If future legislation or DHS regulations require additional chemical security activities at our member's locations, ARA would strongly oppose any form of third party security oversight or audits by other federal agencies or private companies, as they are unnecessary and costly.

Tiered Risk-Based Approach: ARA recommends a tiered risk-based approach since not all facilities are equal in risks. Previous chemical security proposals have listed the starting point for regulations using Clean Air Act, Section 112(r) RMP sites. According to the EPA, approximately 15,000 facilities in various industries produce, use or store one of more toxic or flammable chemicals that pose the greatest risk to human health and the environment when present in certain quantities above threshold levels. Agricultural chemical sites of varying sizes represent over 6,000 RMP sites or around 30 percent of the EPA listed RMP sites. We are concerned that eventually Ag retail / distribution facilities and many of their customers are likely to be regulated just like larger facilities. As previously mentioned, our industry falls under the Program 2 RMP requirements, which is viewed to be less of a risk compared to other sites. Whatever the chemical security regulatory scheme that is developed, America's agricultural industry could end up ultimately paying the costs for added security at all sites handling certain crop inputs. A one-size–fits-all regulatory scheme must be avoided, primarily because it will do little to protect Americans while causing severe disruption throughout American agriculture. We agree with DHS that a tiered risk-based approach should be utilized.

Recognition of On-going Industry Efforts: Any legislation should allow DHS to recognize existing and future industry programs developed and used by the Ag retail and distribution industry and other related industries at facilities that are designed to responsibly assess and address security concerns should be recognized by the government as meeting mandated requirements. For example, Ag retailers that conduct the Asmark SVA or a similar vulnerability assessment program and implement recommended security measures should be considered as meeting any DHS security mandates. DHS should provide some flexibility for the industry in relation to what types of security measures are required to be installed.

No Inherently Safer Technologies (ISTs) or Alternative Approaches mandates: ARA does not believe the federal government should mandate the use of inherently safer technologies (ISTs) or alternative approaches for chemical processing, which is extremely complex, and which differs from company to company. Anti-chemical groups have been pushing for an IST mandate long before September 11. Congress should be very careful about how it handles this issue. The General Accounting Office in a March 2003[8] report that found that ISTs could result in shifting, rather than reducing, the risk of terrorist attacks. In that report, GAO stated, "reducing the amount of chemicals stored may shift the risk onto the transportation sector as reliance on rail or truck shipments increases." Availability of lower-cost sources of plant nutrient products or certain pesticides used by farmers could be at risk under an IST or Alternative Approaches mandate. EPA already monitors IST technologies. It also considers agricultural

[8] GAO-03-439 Homeland Security: Voluntary Initiatives Are Under Way at Chemical Facilities, but the Extent of Security Preparedness is Unknown, March 2003, p. 29

pesticide products for fast track registration that it deems safer for use. A new IST approach or mandate would set up a duplicative effort that is not needed and potentially opens the door for anti-chemical groups to file lawsuits against the industry. Ag retailers and their farmer customers cannot afford the loss of essential crop input products, new expensive federal mandates or defending against frivolous lawsuits.

Congress does not have to go to such extremes and should not be in the business of picking winners and losers in the crop protection industry. This an issue best left up to the market place and consumers. If the committee does end up including language related to ISTs or Alternative Approaches it should be based on the threatened impact of a release caused by terrorism or other criminal acts that relate to the site-specific risk (i.e. rural area vs. urban area).

Protect Confidential Information: ARA has serious concerns whether there will be adequate protection of sensitive information submitted to DHS for security purposes or any other government agency. This concern only partly relates to its potential use in civil lawsuits. Public access to this type of secure information would create roadmaps for terrorist's attacks. ARA supports language that will protect the preparation and submission of confidential security information sent to DHS by an owner or operator to help ensure that the information cannot be used by terrorists or as an automatic basis for private civil liability in the event of a terrorist or criminal act. ARA believes that any government-mandated information assessing or addressing facilities' security vulnerabilities must be protected from Freedom of Information Act (FOIA) public open records requests or use in civil lawsuits.

Establishment of Security Grant Program for Agricultural Facilities: ARA requests the committee support authorizing language in any chemical site security bill that establishes a grant program that can provide financial assistance to Ag retailers, distributors and their farmer customers in order to help them meet any new federal security mandates that may be imposed. Installing state of the art security measures is expensive. In a December 12, 2003 GAO letter[9] to former Senator Ernest Hollings (D-SC), then ranking member of the Senate Commerce, Science, and Transportation Committee, its states "one theme we have heard from maritime stakeholders is that the current economic environment makes this a difficult time for the private industry or state or local governments to make security investments. According to industry representatives and experts contacted, most of the transportation industry operates on a very thin profit margin, making it difficult to pay for additional security measures."

Many retailers and their customers are experiencing similar difficulties as maritime stakeholders due to a continuous drought impacting large segments of the United States, higher operating costs, and low profit margins. There is precedent for such as grant program. A grant program for the agricultural industry was included in chemical site security legislation (S. 994)[10] adopted by the Senate Environment & Public Works Committee last Congress. In addition, DHS currently has a Port Security Grant Program that provides resources for security planning and projects to improve dockside and perimeter security. These port security grants contribute to security upgrades such as surveillance equipment, access controls to restricted areas, communications equipment, and the construction of new command and control facilities. According to DHS, $92 million was awarded in June 2002, $168 million in July 2003, and $179 million in December 2003 as part of the Port Security Grant Program, and $75 million from

[9] U.S. General Accounting Office letter to Senator Ernest Hollings (D-SC), December 12, 2003, p.4

[10] The Chemical Facilities Security Act of 2003 (S. 994); Section 11 – Agricultural Business Security Grant Program; Sponsor – Senate Environment & Public Works Committee Chairman James Inhofe (R-OK)

the Urban Area Security Initiative for port security in August 2003. ARA believes this committee should support a similar grant program to help the nation's agricultural industry address security related issues.

SECURITY PROPOSALS INTRODUCED IN 109TH CONGRESS

There are a number of security related proposals that have been introduced in Congress that this committee should take under consideration as you begin to draft security legislation next month:

Agricultural Business Security Tax Credit Act of 2005: Ag retailers as well as distributors, manufacturers, formulators and aerial applicators that store agricultural pesticides and fertilizers sold to farmers in the United States are committed to providing increased security for these products. However, a retailer or distributor could spend tens of thousands of dollars at a single facility on security measures such as fencing, alarms, lights and locks. A security tax credit would allow eligible agricultural businesses to use their own financial resources to take the necessary steps installing state of the art security measures that better protect the U.S. agriculture and food system and the American public from the potential threat of terrorism or other illegal activities.

Rep. Ron Lewis (R-KY) has introduced the "Agricultural Business Security Tax Credit Act of 2005" (H.R. 713). This ARA supported legislation would help Ag retailers, distributors and other eligible agricultural businesses partially offset security costs by providing a tax credit equivalent to 50 percent of the aggregate amount paid on implementing security measures at facilities where crop input materials are stored. Under H.R. 713, a qualified business would be able to take a tax credit of up to $100,000 per site to help install state of the act security measures in order to protect agricultural pesticides and fertilizers that are manufactured, distributed and stored on site. This legislation is a fiscally responsible proposal that will enable agricultural businesses to make the necessary security investments to better protect these facilities.

The bill also has the support of the Chemical Producers & Distributors Association (CPDA), The Fertilizer Institute (TFI), CropLife America (CLA) and the National Agricultural Aviation Association (NAAA). Senator Pat Roberts (R-KS), Chairman of the Senate Intelligence Committee, and Senator Ben Nelson (D-NE), a member of the Senate Agriculture and Commerce Committees, plan to introduce a companion bill to H.R. 713. ARA urges Senators to support this important legislation by being original co-sponsors of the Roberts / Nelson security tax credit bill.

Secure Handling of Ammonium Nitrate Act of 2005: Ammonium nitrate continues to be a key plant nutrient fertilizer product sold by Ag retailers and purchased by their customers for application on agricultural operations. Our industry is committed to providing increased security for ammonium nitrate fertilizer. Several states have enacted registration and record keeping laws for this product with the support of the state agribusiness association. Senators Thad Cochran (R-MS), Mark Pryor (D-AR), Saxby Chambliss (R-GA) and Pat Roberts (R-KS) have introduced the "Secure Handling of Ammonium Nitrate Act of 2005" (S. 1141). Congressmen Curt Weldon (R-PA) and Bennie Thompson (D-MS) have introduced a companion bill (H.R. 3197) in the House. This legislation would regulate the production, storage, sale and distribution of ammonium nitrate fertilizer. Specifically, the bill authorizes DHS to enter into cooperative agreements with state departments of agriculture or other state agencies that regulate plan nutrients to ensure that any person who produces, stores, sells or distributes ammonium nitrate fertilizer registers their facility and maintains records of sale or distribution including the name, address, telephone and registration numbers of purchasers. Also, purchasers would be required to register under this proposal.

ARA supports a common sense, fair and simplified federal registration system for ammonium nitrate fertilizer in order to ensure the product's continued availability for sale, purchase and use by America's agricultural industry. ARA believes it is important for retailers to maintain the ability to sell ammonium nitrate fertilizer if they so desire, as well as ensuring their customers maintain the ability to purchase the product for use on their farming operations.

Combat Meth Act of 2005: The use of illegal drugs constitutes a huge challenge to the health of our nation. It is estimated that illegal drugs cost our health care system almost $15 billion a year. Drug use causes crime and drug money is being used to support terrorists. We agree with the Administration that cutting down drug use in the United States will help the fight against terrorism. An ever-growing problem in rural areas relates to methamphetamine production and use. Ag retailers and others within the fertilizer industry have been particularly hard hit with this nation-wide problem, as drug dealers continue to steal our products for the illegal production of meth. To address this serious crisis, Senators Jim Talent (R-MO), Dianne Feinstein (D-CA) and Chuck Grassley (R-IA) introduced the *Combat Meth Act of 2005* (S. 103). This legislation would regulate the sale of pseudoephedrine by keeping the product behind the counter and by only allowing a pharmacist or state approved sales person to sell such product. Pseodoephedrine is a key ingredient in meth production. Congressman Roy Blunt (R-MO) has introduced a House companion bill (H.R. 314). ARA strongly supports this legislation and is working with TFI to secure its quick enactment.

CONCLUSION

ARA and our members strongly support the war on terrorism and are committed to do our part to address security related concerns. As an industry we have already made great strides, but we believe it is important to have commonsense, workable regulations in effect that do not place unreasonable and unnecessary burdens on Ag retailers and distributors or their customers.

Thank you for considering ARA's views. We welcome the opportunity to work constructively with the committee to address any security gaps that may exist within the industry and on the drafting of any bipartisan legislation. ARA stands ready to work with Congress on the development of a chemical site security proposal that adequately reflects the needs of America's agricultural industry and our rural economy. As we face these challenges, we can only accomplish what needs to be done if we work together.

GUIDELINES TO HELP ENSURE A SECURE AGRIBUSINESS

A special report brought to you by the Agricultural Retailers Association, CropLife America and The Fertilizer Institute.

Security of Chemicals in the Pesticide and Fertilizer Industries:
A Primer for Retailers, Distributors, Wholesalers and End-Users

Agriculture in the United States has faced many difficulties and has always been ready to do its part during times of national emergency. The current war against terrorism is everyone's responsibility and all Americans, including U.S. agricultural retailers, distributors, wholesalers and farmers must do their part along with every other citizen to keep America secure.

Industry Organizations Working for a Secure Agribusiness

CropLife

The Fertilizer Institute

Three Key Security Principles

1 Identification of critical assets; **2** Establishment of layers of protection; **3** Practice deter, detect and delay.

Agribusiness is unique in its use, distribution — and security needs — of materials and finished products because of the diversity of chemicals produced and locations where they are stored. In the days following the Sept. 11, 2001, terrorist attacks on the United States, EPA Administrator Whitman was very clear that all facilities that manufacture, process or store chemicals should conduct a vulnerability assessment and implement security updates.

Agricultural Retailers Association, CropLife America and The Fertilizer Institute have formed an agribusiness security working group of security and environmental health and safety professionals to address security concerns about agricultural chemicals. The working group developed the enclosed security principles and guidelines so you may begin the process of security assessment for your facility.

Other Security Guidelines and Programs

In addition to this checklist, there are a number of other security resources. The American Chemistry Council has published "Site Security Guidelines for the U.S. Chemical Industry" and "Transportation Security Guidelines for the U.S. Chemical Industry." Both are available at www.americanchemistry.com. The North Carolina Department of Agriculture & Consumer Services has developed the "Terrorism Threat Vulnerability Self Assessment Tool," an easy-to-use tool to help you determine site vulnerability. This can be accessed at www.ncagr.com/industry_self-assessment.doc. Additional website resources include the United States Department of Agriculture at www.usda.gov; the Food and Drug Administration at www.fda.gov; and The International Association of Emergency Managers at www.iaem.com. You may locate additional

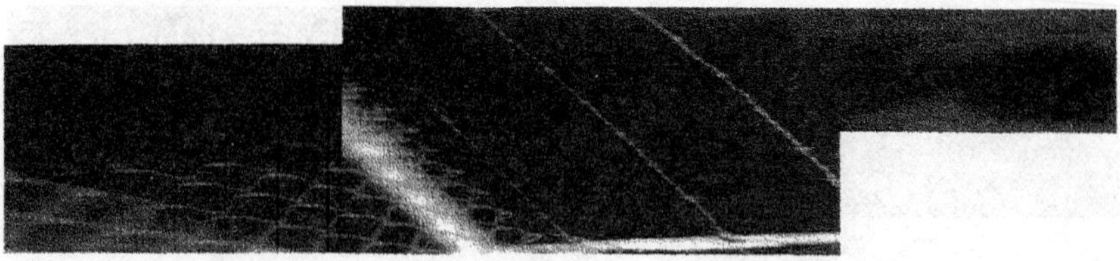

resources by contacting your state Office of Homeland Security at www.whitehouse.gov/deptofhomeland/. As always, it is very important to communicate effectively with your local law enforcement and first responders.

Security of agricultural retail facilities has taken on a whole new meaning since that terrible September morning in 2001, and we will not be returning to pre-9/11 days any time soon. By using the principles and guidelines outlined in this special Ag Retailer section, however, you can play a vital role in securing your facility and agricultural chemicals.

Almost every security situation where agricultural chemicals are handled or used can be addressed by the following three basic security principles:

1. IDENTIFICATION OF CRITICAL ASSETS

Knowledge and identification of your most valuable assets is essential to any security plan. Through such identification, limited resources can be most efficiently utilized. Since 9/11, however, product loss or theft no longer can be simply written off as a financial loss when product use as a weapon of mass destruction is a possibility. To identify critical assets, determine the products you handle that might be illegally used as explosives, chemical weapons or cause harm in other ways. Then, ask yourself these questions:

- What is the threat (theft, sabotage, attack)?
- Is the threat internal or external?
- How might a theft or other illegal action be carried out (overtly or covertly)?
- Products in large, stationary vessels usually are not susceptible to theft, but could a small amount be siphoned off via an accessible valve or other means?
- Are containers of critical products easily accessible?
- Are strangers or visitors allowed to roam the facility unescorted, or have access to critical items?
- Are all employees trustworthy? What about contractors or customers?
- Do you know those with whom you do business?
- Can you protect against an attack from the outside or from the inside (cyber attack on computer information)?

2. LAYERS OF PROTECTION

The use of multiple levels of protection to safeguard your critical assets is sound security. Once such assets have been identified, concentrate security resources to make it difficult for criminals and terrorists to gain access to them. For instance, if a critical asset is portable, focus protective measures on where the product is stored.

Perhaps the first layer of protection would be a fence around the warehouse with a gate locked during off-hours. The next layer might be a locked warehouse door with employee-only access. A final protection layer might be to secure products in a high-value area or cage.

3. DETER, DETECT AND DELAY

The theory behind this principle is simple — deter an unwanted event from happening; detect potential criminal or terrorist activity as early as possible; and, failing all else, delay violators as long as possible until proper authorities arrive. The longer it takes to break into a facility, the greater the chance violators will be caught. Heavy-duty locks and good key control; clear, open zones around property; employee awareness; and solid relations with local law enforcement and first responders are all sound measures that can prevent security breaches.

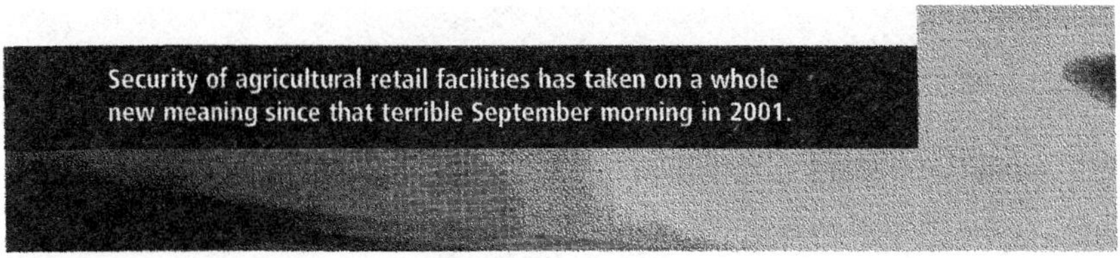

Suggested Facility Security Practices

Awareness
- Conduct a security assessment of your facility.
- Use opening and closing security check lists; note any discrepancies or irregularities.
- Initiate or join your local "crime watchers" program.

Access
- Escort all customers or visitors in storage yards or near loading docks.
- Establish a uniform or ID badge system to distinguish employees.

Alarms
- Install alarms and use a security alarm monitoring service.
- Ensure that phone lines are protected or have a service interruption alarm.
- Locate exterior strobe lights with alarms where neighbors and law enforcement can see them.

Barriers
- Construct structural barriers, including steel doors and barred windows.
- Install fencing as a deterrent where appropriate; fencing should be such that law enforcement and passers-by can view the property.
- Install access gates where fencing is not appropriate.
- Install bollards and chains across driveways or block with trucks and other equipment during off-hours.

Community
- Establish a process for including neighbors and the community as part of facility security and emergency response planning.

Inventory Control
- Know your inventory.
- Establish an ongoing process for inventory control of materials stored at the facility.
- Do not allow unattended, loaded trailers on site.
- Record stored nurse tanks by identification number and weight of remaining product.
- Inspect tanks visually each morning.
- Keep bills of lading, blank forms and all shipping/receiving paperwork secured.

Law Enforcement
- Establish and maintain relationships with local law enforcement and emergency responders. Provide them with your emergency plans and keys to locked gates.
- Provide law enforcement dispatchers with current emergency contact information for the facility. Keep this information current.
- Immediately report unusual or suspicious persons, vehicles or activity to local law enforcement.

Lighting
- Contact your local power company for a lighting assessment and information on leasing lights for your property.
- Install sufficient exterior lighting for law enforcement and passers-by to see your property.
- Discuss your lighting plan with local law enforcement.

Locks
- Establish a procedure and responsibility for locking up at close of business.
- Use high-security locks for doors, enclosures and gates, following local fire code requirements. Keep padlocks locked on hasps while not in use to prevent your lock from being replaced by someone else's.
- Use deadbolt locks on doors with a minimum of 1.5-inch throw.
- Implement key control for locked containers, equipment, hoppers, vehicles and vessels.

Signage
- Post alarm monitoring service signs in highly visible locations. Include signage for:
- No trespassing
- Private property
- Closed circuit TV surveillance
- Patrolled
- No vehicles beyond this point
- All visitors must check-in with front office
- All visitors must be escorted

Surveillance
- Install CCTV surveillance cameras to monitor less visible or high-risk areas.

Training
- Involve employees in security planning.
- Train employees to spot suspicious individuals and behavior.
- Conduct periodic emergency drills, e.g. fire, evacuation and security, with employees.

Vendors
- Know vendors that service your facility.
- Require all vendors to check in.
- Escort vendors.

Visibility
- Assure an open area around the facility, unlimited by shrubs, trees, large signs or other barriers to open sight.

SUGGESTED CUSTOMER TRANSACTION PRACTICES

Awareness
- Heighten employee awareness of what constitutes an unusual customer and sales transaction.
- Heighten customer awareness of potential for criminal misuse of agricultural chemicals.
- Advise customers to contact law enforcement immediately with any concerns about unusual persons, vehicles or activities in the vicinity of your facility or theirs.

Sales Transaction
- Know your customer.
- Follow all requirements for verification when selling restricted use pesticides.
- For all sales, record customer's name, address, telephone number. If in doubt ask for a driver's license.
- Make deliveries only when the customer or agent is available to take custody and sign for the material.
- Do not deliver tanks or other products to empty fields or other unattended locations.
- Make follow-up calls to verify receipt of materials by customer in quantity ordered.
- Be alert to those who:
  - Pay in cash;
  - Won't take delivery;

  - Behave in an unusual manner;
  - Hesitate when asked for ID to complete the sale;
  - Don't know the product;
  - Insist on certain products, such as ammonium nitrate, and will not consider other suggestions;
  - Ask questions about product manufacturing;
  - Aren't familiar with farming, pesticides or fertilizer products.
- If in doubt:
  - Write down vehicle color, make, license number and state and a physical description of the individual;
  - Retain papers the customer may have touched for fingerprints;
  - Save this information in the event that it needs to be provided to law enforcement.

Suggested Special Security Measures

Certain agricultural inputs stored at your facility may warrant special security measures, such as anhydrous ammonia, ammonium nitrate, bulk urea and insecticides.

Alarms

- Install alarms near tanks.
- Install explosion-proof alarm systems near combustible material.

Awareness

- Be alert to those attempting to buy ammonia if they cannot state a legitimate, agronomic need for the product.
- Inspect tank and bulk storage areas daily.
  - Check for fresh tracks in mud or snow or disturbed ground around tanks and bulk storage areas;
  - Check to see if tank valves are closed tightly;
  - Look for suspicious items near tanks such as duct tape, garden hose, bicycle inner tubes, buckets and coolers;
  - Check for broken or missing wire ties or seals that you may have placed on valve wheels as markers.
- Make customers aware of the potential for theft or tampering with tanks and bulk ag chemicals.
- Remove hoses between tool bars and nurse tanks; relieve pressure with the bleed valves when left overnight. Encourage end-users to do the same.

Law Enforcement

- Work with local law enforcement to encourage frequent nighttime patrols.
- Contact local law enforcement immediately if you suspect tampering or theft at your facility or the presence of unusual persons, vehicles or activities.
- Do not disturb a potential crime scene.

Locks for Tanks

- Use brightly colored plastic ties or wire seals between the valve wheel and the roll cage to ease visual checks and to identify tampering.
- Use tamper resistant seals and locks.
- Use high-security locks.
- Use specialized tank locks for nurse tanks containing anhydrous ammonia.
- Paint tank locks red so law enforcement can identify anhydrous ammonia tanks.

Visibility

- Store tanks in well-lit areas with a clear line-of-sight.
- Store tanks with flow valves facing outward to speed visual inspections.
- Do not leave tanks in remote areas.

SUGGESTIONS FOR PARTNERING WITH YOUR CUSTOMERS ON SECURITY AND SAFETY

- Take delivery of tanks as close to time of application as possible.
- Position tanks in open, visible areas.
- Don't take delivery of tanks to unattended locations.
- Don't store tanks and tool bars inside buildings, near the farmhouse or livestock confinement houses.
- Remove hoses between tool bars and nurse tanks and relieve pressure with the bleed valves if tanks are left overnight. Store hoses and tool bars away from tanks.
- Don't leave tanks unattended for long periods of time.
- Inspect tanks every day, especially after a weekend when most thefts occur.
- Return tanks immediately after use.
- Inspect and record the condition of each nurse tank upon delivery and return.
- Store all agricultural chemicals, e.g. bulk, bagged, in a secured area.
- Where appropriate, use alarm systems to protect secured storage areas and chemicals.
- Be aware of and maintain inventory control.
- Lock any containers, equipment, hoppers, tanks and vessels containing product whenever possible.
- Be aware of signs of theft of anhydrous ammonia, ammonium nitrate or bulk urea.

Law Enforcement

- Urge customers to contact local law enforcement immediately if tampering or theft is suspected or suspicious persons or vehicles are seen.
- Do not approach or confront suspicious individuals.
- Do not disturb the area around a possible crime scene.

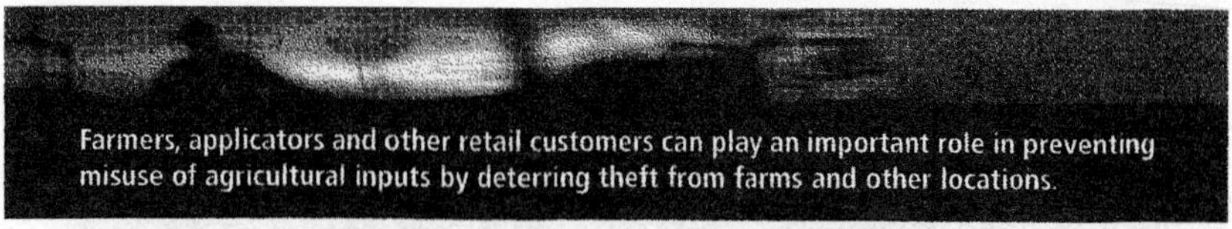

TIPS FOR DEALING WITH THE MEDIA IN AN EMERGENCY SITUATION

Emergencies or criminal activity at your retail location will attract the media. Take a moment to gather your thoughts to ensure that appropriate information is provided without compromising safety or an official investigation.

Contact

- Contact the designated corporate media spokesperson for the product involved and refer media.
- If you have no corporate office to handle media inquiries:

Coordinate

- Keep lines of communication open with law enforcement and emergency responders. They will often have their own spokesperson at the site of a newsworthy event.

Designate

- Designate one media spokesperson and a back up. Have employees direct all media inquiries to these individuals.

Prepare

- When caught off-guard, such as being awakened in the middle of the night by a reporter's call, ask to call them back in a few minutes.
- Be calm, concerned, confident and credible. Stick to your areas of credibility.
- Prepare talking points if the situation allows. Stick to three key points about the situation.

Remember

- There is no such thing as "talking off the record."
- Never lie.

What you say

- Never say "no comment" or something that sounds like " no comment."
- Answer only the question asked.
- Never speculate. If you don't know, say so, but indicate a willingness to find the right person to provide the answer and get back to the reporter.
- Never answer what-if questions.
- You do not have to answer every question but you need to provide a plausible reason if you don't.
- Provide your name and phone number for reporter follow-up and questions.

How you say it

- Show your concern.
- Talk from the public's interest, not the company's. For example, talk in terms of public safety, security or environmental protection.
- Don't be defensive or lose your temper.
- Do not repeat the negative parts of a reporter's words or questions.
- Challenge any incorrect information in a question before answering the question itself.

SUGGESTED RETAIL FACILITY CLOSING CHECKLIST

- Assign patrol for grounds, storage and perimeter areas.
- Be aware of signs of tampering with fencing, locks, doors, windows, equipment, product, etc.
- Secure and seal all containers, equipment, hoppers, vehicles and vessels.
- Verify that all valves on bulk storage tanks are closed and secured with padlocks.
- Secure all pedestrian and vehicle gates and doors or access points along with sliding and overhead doors and windows.

- Secure file cabinets, phone closets and areas with sensitive company and product information, such as bills of lading and customer lists.
- At day's end ensure all computers are logged off and passwords are not visible.
- Ensure all appropriate facility equipment is turned off and all keys are secured in a lock box or are kept with designated personnel.
- Shut off electrical power at inside breaker box for pumps inside and outside of the facility.
- Ensure all seals and product labels are secured.
- Ensure all lighting is operating effectively.
- Arm the alarm system and exit the facility.

Agricultural Retailers Association: Security
Vulnerability Assessment Workshop

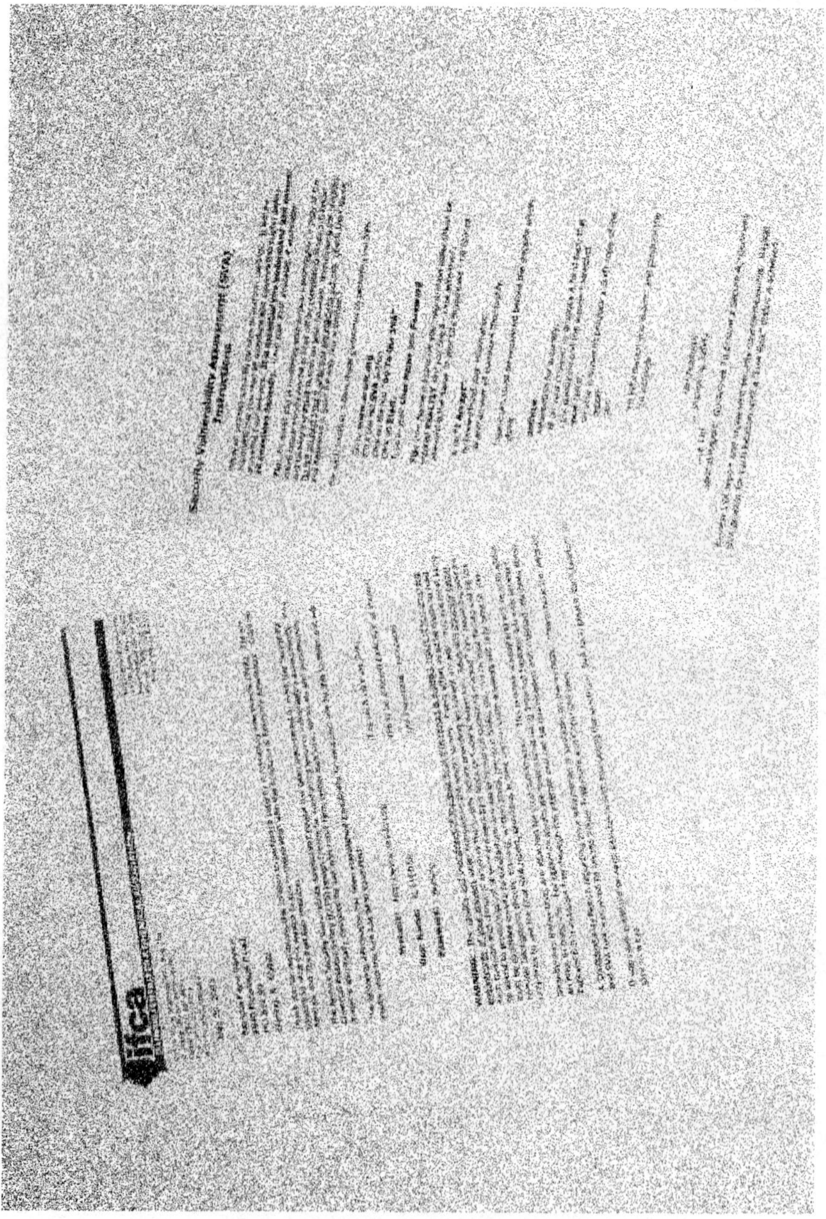
ifca
Security Vulnerability Assessment (SVA)
Instructions
WARNING:

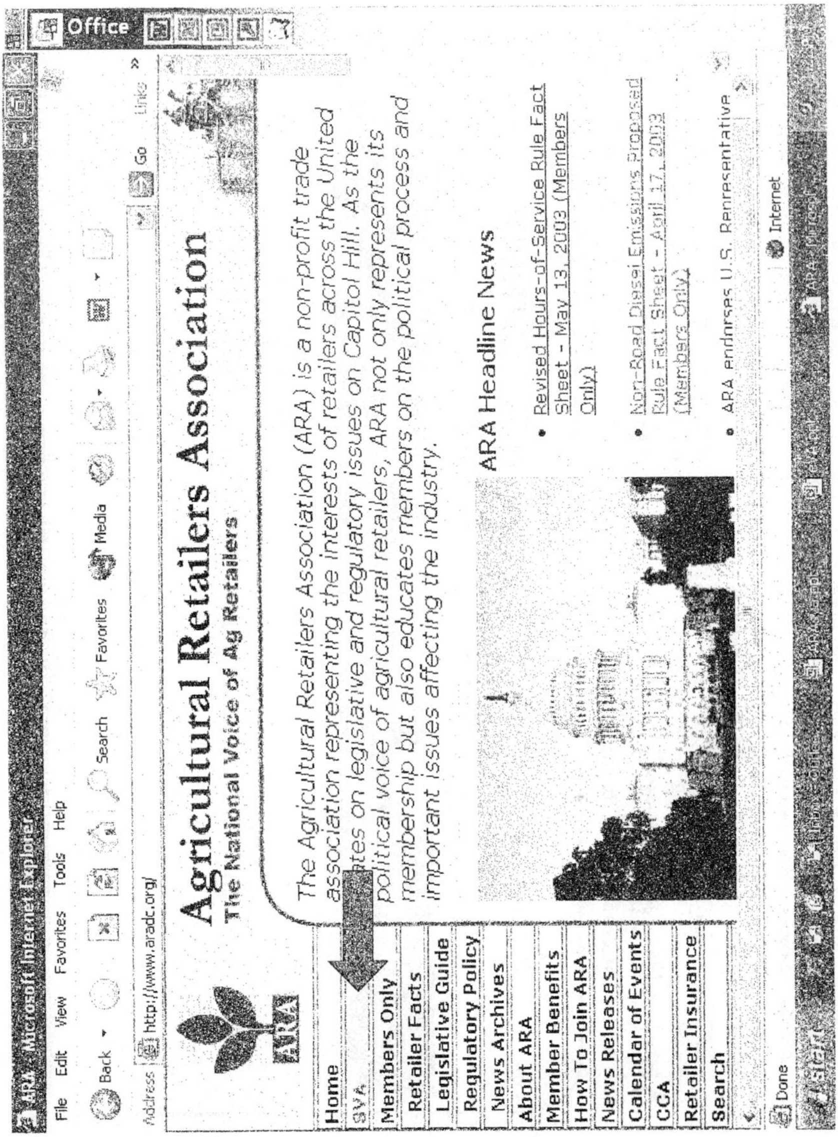
ARA - Microsoft Internet Explorer
File Edit View Favorites Tools Help
Back Search Favorites Media
Address http://www.aradc.org/
Go Links
Office
ARA
Agricultural Retailers Association
The National Voice of Ag Retailers
The Agricultural Retailers Association (ARA) is a non-profit trade association representing the interests of retailers across the United States on legislative and regulatory issues on Capitol Hill. As the political voice of agricultural retailers, ARA not only represents its membership but also educates members on the political process and important issues affecting the industry.
Home
SVA
Members Only
Retailer Facts
Legislative Guide
Regulatory Policy
News Archives
About ARA
Member Benefits
How To Join ARA
News Releases
Calendar of Events
CCA
Retailer Insurance
Search
ARA Headline News
Revised Hours-of-Service Rule Fact Sheet - May 13, 2003 (Members Only)
Non-Road Diesel Emissions Proposed Rule Fact Sheet - April 17, 2003 (Members Only)
ARA endorses U.S. Representative
Done
Internet
start

ARA - Microsoft Internet Explorer
File Edit View Favorites Tools Help
Back Search Favorites Media
Address http://www.aradc.org/svaforara.php
Go Links
Agricultural Retailers Association
The National Voice of Ag Retailers
Security Vulnerability Assessment
for Agricultural Retailers
deter
deter
CLICK HERE TO GO TO SECURITY VULNERABILITY ASSESSMENT PROGRAM
Sponsored by:
Internet
start
Office

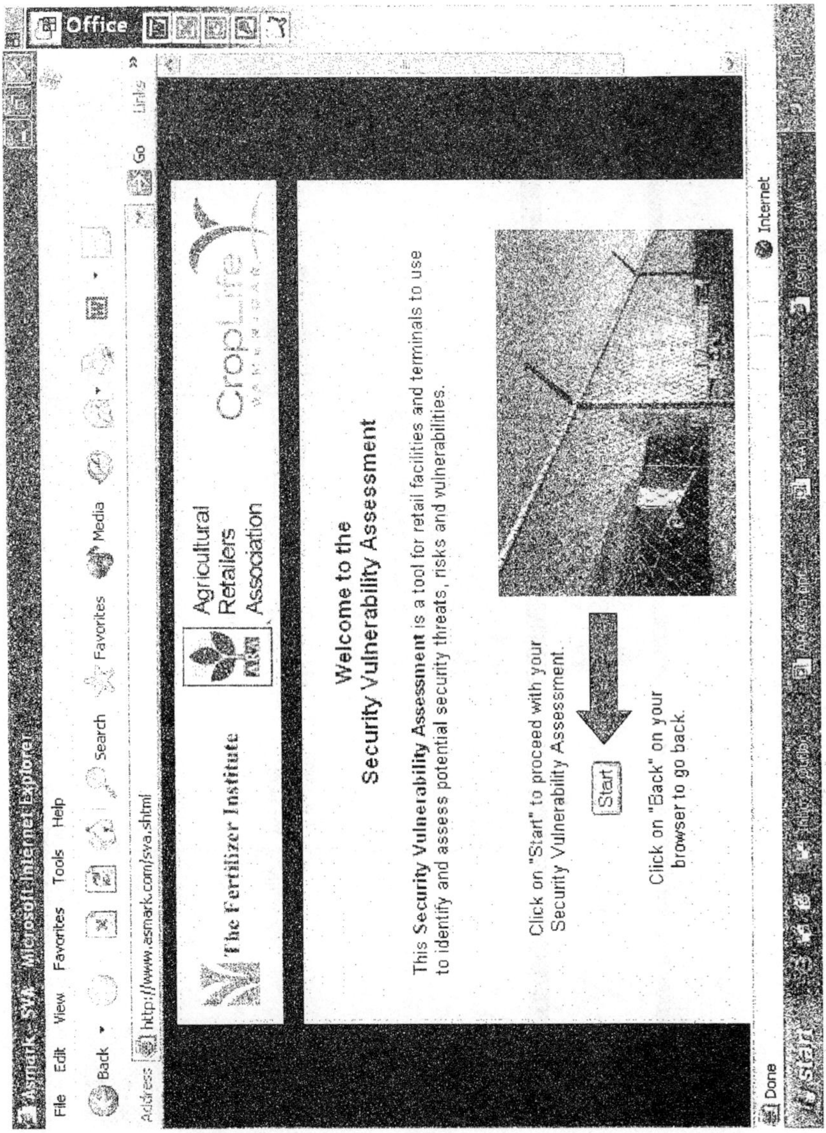
Asmark - SWA - Microsoft Internet Explorer
File Edit View Favorites Tools Help
Back Search Favorites Media
Address http://www.asmark.com/swa.shtml
Go Links
Office
The Fertilizer Institute
Agricultural Retailers Association
CropLife
Welcome to the
Security Vulnerability Assessment
This Security Vulnerability Assessment is a tool for retail facilities and terminals to use to identify and assess potential security threats, risks and vulnerabilities.
Click on "Start" to proceed with your Security Vulnerability Assessment.
Start
Click on "Back" on your browser to go back.
Done
Internet
start

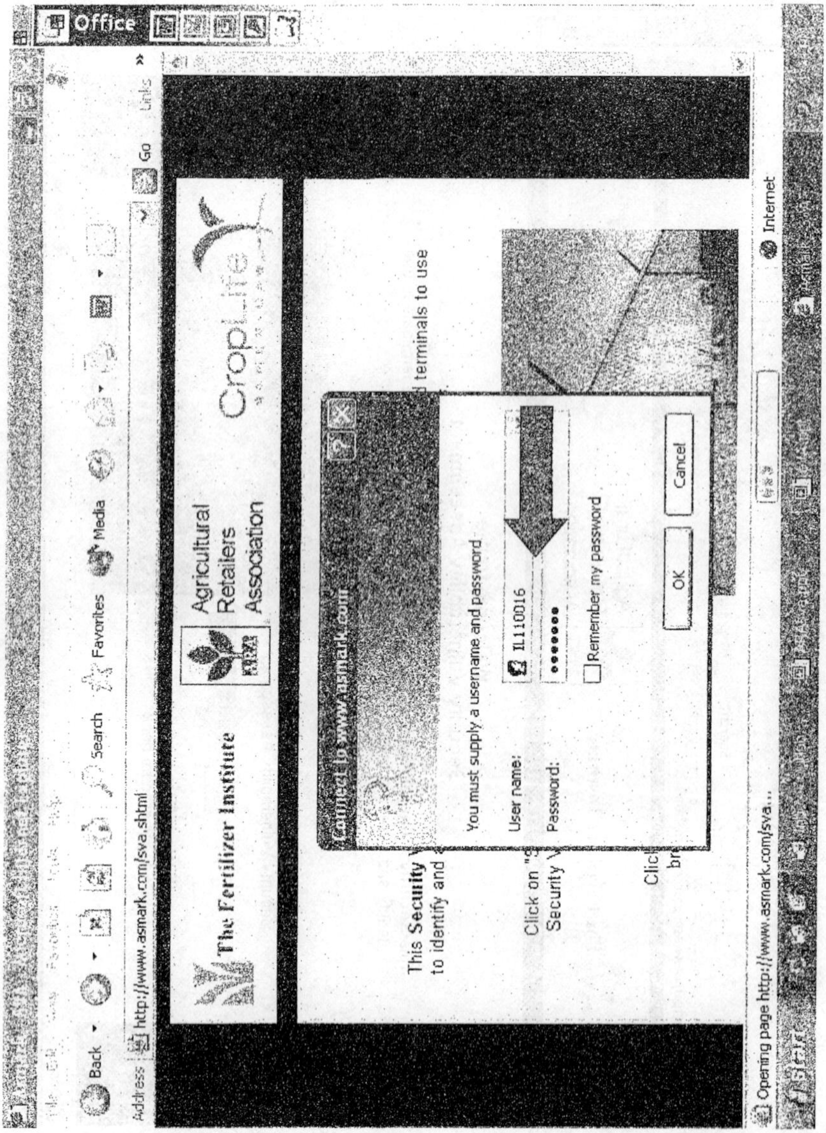
Office
Back
Search
Favorites
Media
Address http://www.asmark.com/sva.shtml
Go
Links
The Fertilizer Institute
Agricultural
Retailers
Association
CropLife
Connect to www.asmark.com
You must supply a username and password
User name:
Password:
IL110016
Remember my password
OK
Cancel
This Security
to identify and
terminals to use
Click on "
Security
Clic
Opening page http://www.asmark.com/sva...
Internet

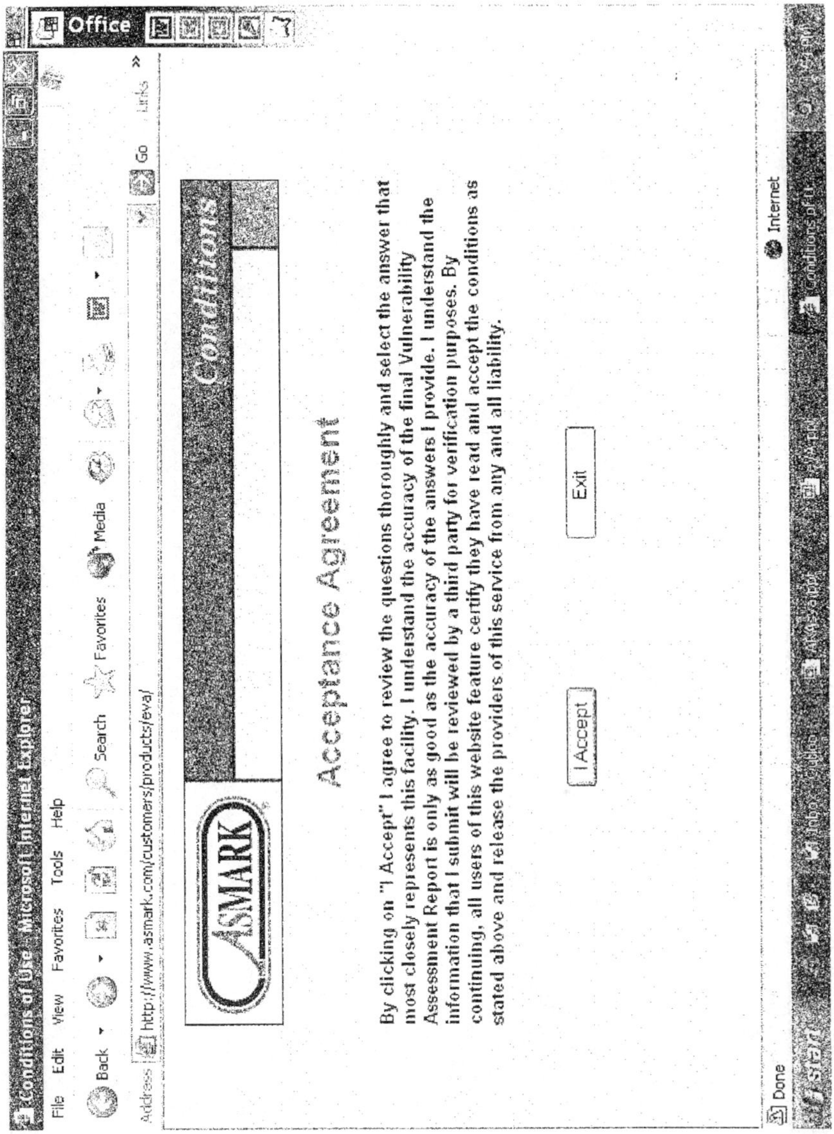
Conditions of Use - Microsoft Internet Explorer
File Edit View Favorites Tools Help
Back Search Favorites Media
Address http://www.asmark.com/customers/products/eva/
Go Links
ASMARK
Conditions
Acceptance Agreement
By clicking on "I Accept" I agree to review the questions thoroughly and select the answer that most closely represents this facility. I understand the accuracy of the final Vulnerability Assessment Report is only as good as the accuracy of the answers I provide. I understand the information that I submit will be reviewed by a third party for verification purposes. By continuing, all users of this website feature certify they have read and accept the conditions as stated above and release the providers of this service from any and all liability.
I Accept
Exit
Done
Internet
start

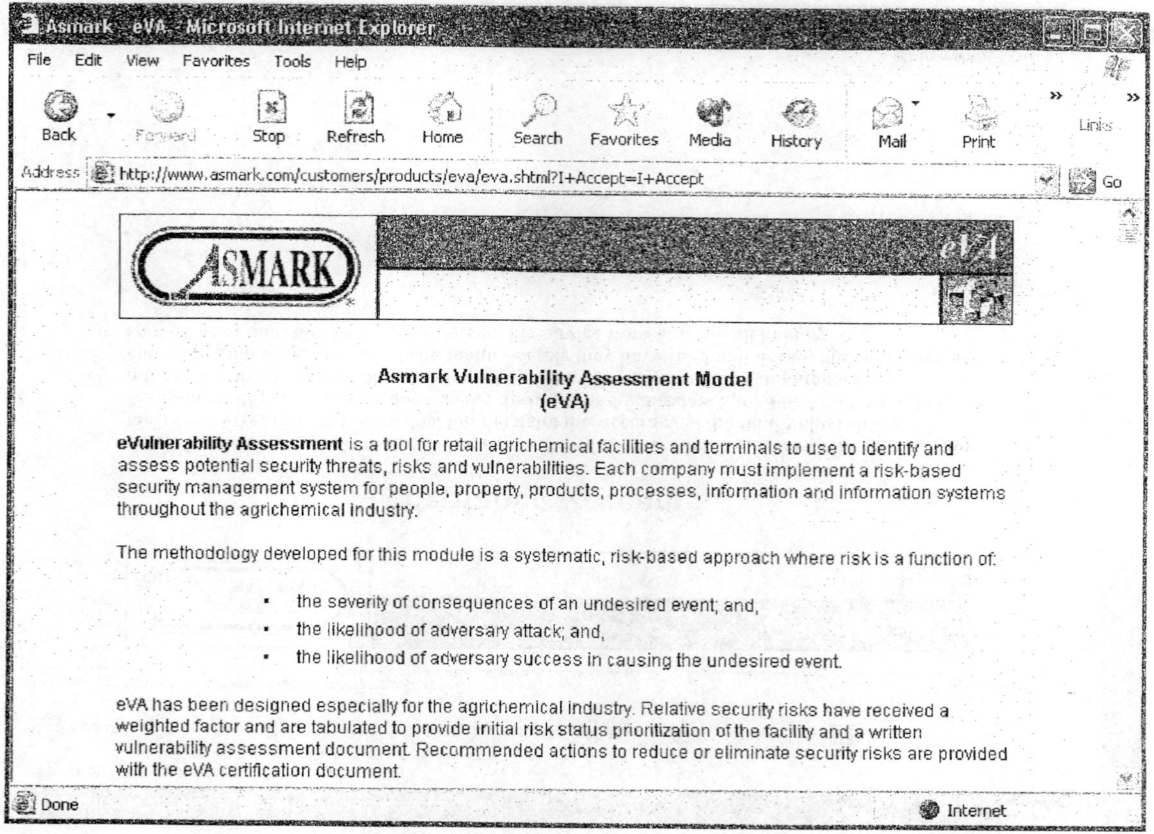

Asmark - eVA - Microsoft Internet Explorer

File Edit View Favorites Tools Help

Back Forward Stop Refresh Home Search Favorites Media History Mail Print Links

Address http://www.asmark.com/customers/products/eva/eva.shtml?I+Accept=I+Accept Go

ASMARK eVA

Asmark Vulnerability Assessment Model (eVA)

eVulnerability Assessment is a tool for retail agrichemical facilities and terminals to use to identify and assess potential security threats, risks and vulnerabilities. Each company must implement a risk-based security management system for people, property, products, processes, information and information systems throughout the agrichemical industry.

The methodology developed for this module is a systematic, risk-based approach where risk is a function of:

- the severity of consequences of an undesired event; and,
- the likelihood of adversary attack; and,
- the likelihood of adversary success in causing the undesired event.

eVA has been designed especially for the agrichemical industry. Relative security risks have received a weighted factor and are tabulated to provide initial risk status prioritization of the facility and a written vulnerability assessment document. Recommended actions to reduce or eliminate security risks are provided with the eVA certification document.

Done Internet

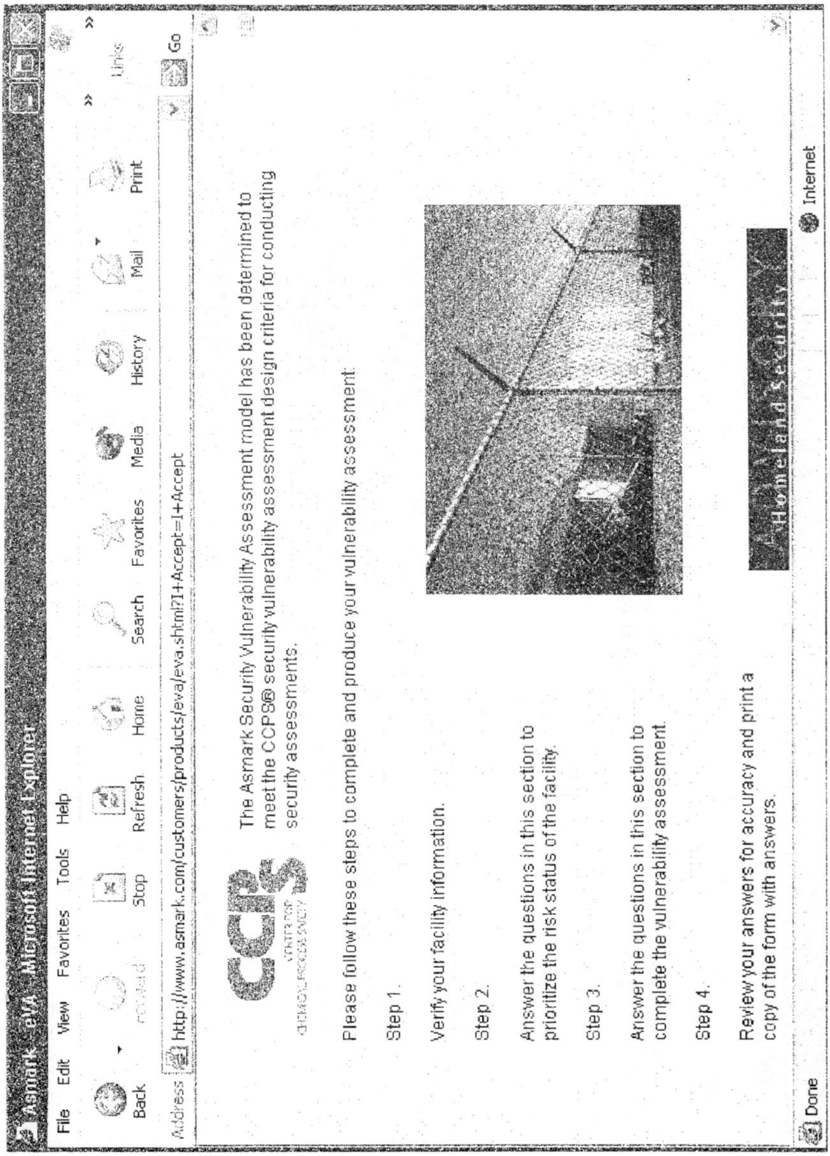
Asmark - eVA - Microsoft Internet Explorer
File Edit View Favorites Tools Help
Back Forward Stop Refresh Home Search Favorites Media History Mail Print
Links
Address http://www.asmark.com/customers/products/eva/eva.shtml?1+Accept=1+Accept
Go
CCPS
The Asmark Security Vulnerability Assessment model has been determined to meet the CCPS® security vulnerability assessment design criteria for conducting security assessments.
Please follow these steps to complete and produce your vulnerability assessment:
Step 1.
Verify your facility information.
Step 2.
Answer the questions in this section to prioritize the risk status of the facility.
Step 3.
Answer the questions in this section to complete the vulnerability assessment.
Step 4.
Review your answers for accuracy and print a copy of the form with answers.
Homeland Security
Done
Internet

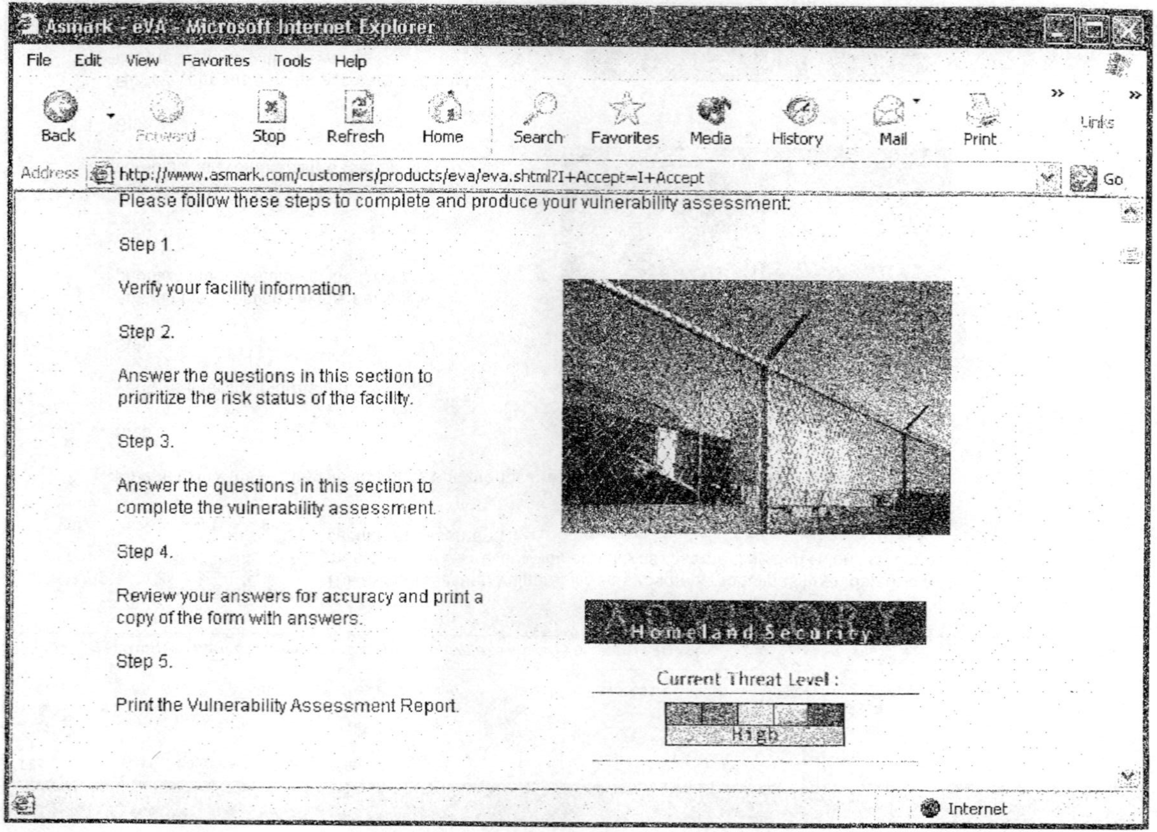
Asmark - eVA - Microsoft Internet Explorer
File Edit View Favorites Tools Help
Back Forward Stop Refresh Home Search Favorites Media History Mail Print
Links
Address http://www.asmark.com/customers/products/eva/eva.shtml?I+Accept=I+Accept
Go
Please follow these steps to complete and produce your vulnerability assessment:
Step 1.
Verify your facility information.
Step 2.
Answer the questions in this section to prioritize the risk status of the facility.
Step 3.
Answer the questions in this section to complete the vulnerability assessment.
Step 4.
Review your answers for accuracy and print a copy of the form with answers.
Step 5.
Print the Vulnerability Assessment Report.
ADVISORY
Homeland Security
Current Threat Level :
High
Internet

Understanding the Questions

- Read the entire question
- Read all choices of answers
- Must understand the question
- Answer honestly
- Answer accurately

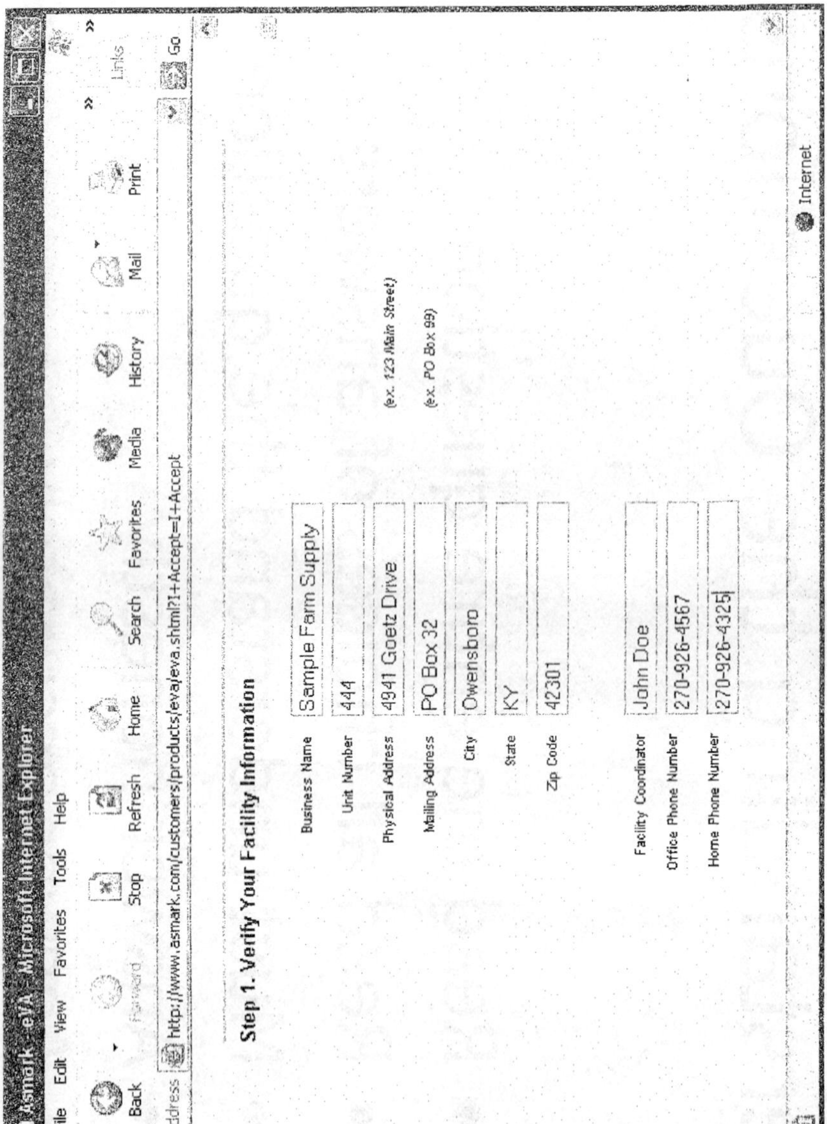
Asmark - eVA - Microsoft Internet Explorer
File Edit View Favorites Tools Help
Back Forward Stop Refresh Home Search Favorites Media History Mail Print
Links
Address http://www.asmark.com/customers/products/eva/eva.shtml?I+Accept=I+Accept
Go
Step 1. Verify Your Facility Information
Business Name
Sample Farm Supply
Unit Number
444
Physical Address
4941 Goetz Drive
(ex. 123 Main Street)
Mailing Address
PO Box 32
(ex. PO Box 99)
City
Owensboro
State
KY
Zip Code
42301
Facility Coordinator
John Doe
Office Phone Number
270-926-4567
Home Phone Number
270-926-4325
Internet

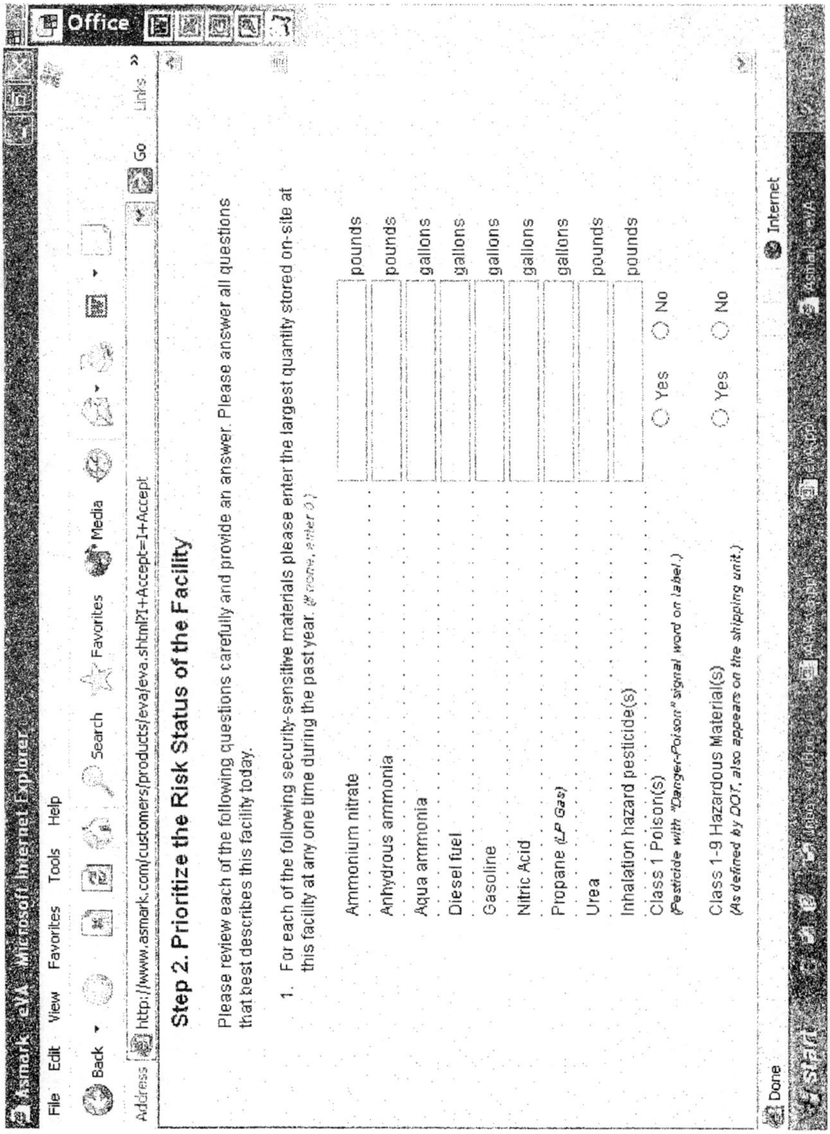
Asmark - eVA - Microsoft Internet Explorer
File Edit View Favorites Tools Help
Back Search Favorites Media
Address http://www.asmark.com/customers/products/eva/eva.shtml?I+Accept=I+Accept
Go Links
Step 2. Prioritize the Risk Status of the Facility
Please review each of the following questions carefully and provide an answer. Please answer all questions that best describes this facility today.
1. For each of the following security-sensitive materials please enter the largest quantity stored on-site at this facility at any one time during the past year. (If none, enter 0.)
Ammonium nitrate pounds
Anhydrous ammonia pounds
Aqua ammonia gallons
Diesel fuel gallons
Gasoline gallons
Nitric Acid gallons
Propane (LP Gas) gallons
Urea pounds
Inhalation hazard pesticide(s) pounds
Class 1 Poison(s) Yes No
(Pesticide with "Danger-Poison" signal word on label.)
Class 1-9 Hazardous Material(s) Yes No
(As defined by DOT, also appears on the shipping unit.)
Done
Internet
start
Office

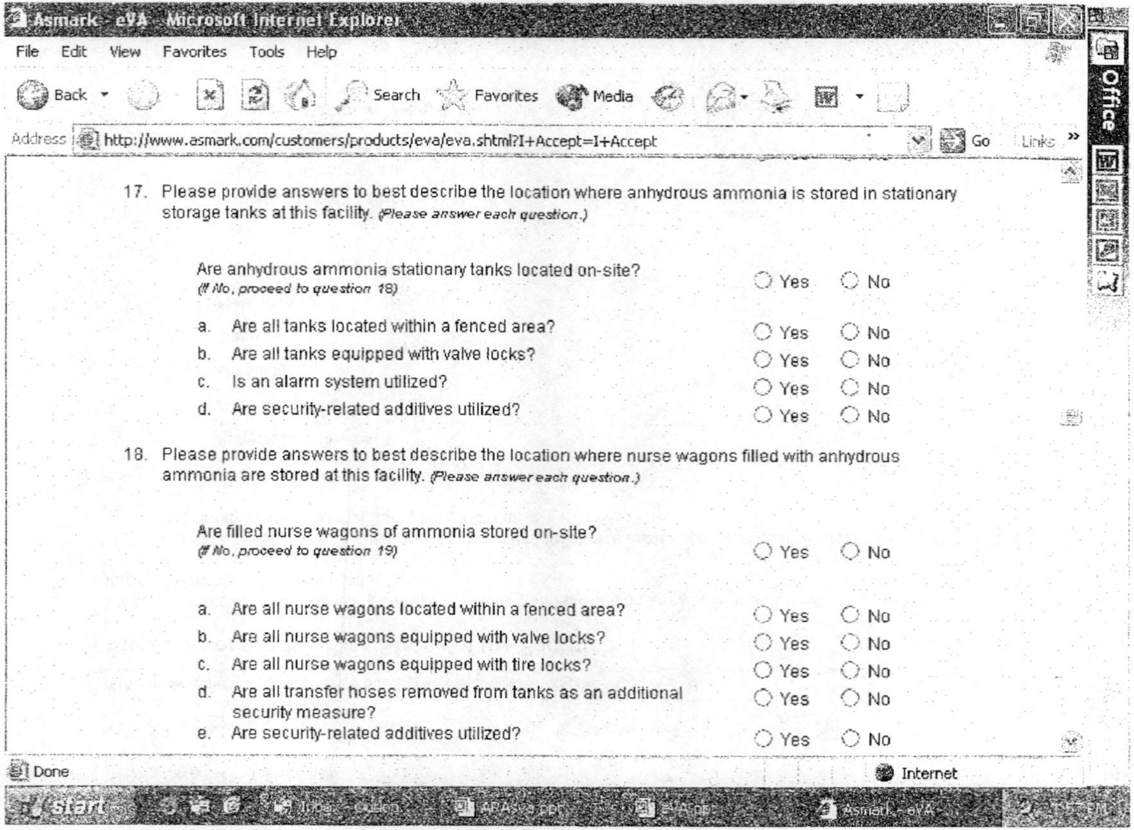

Asmark - eVA - Microsoft Internet Explorer

File Edit View Favorites Tools Help

Address http://www.asmark.com/customers/products/eva/eva.shtml?I+Accept=I+Accept

17. Please provide answers to best describe the location where anhydrous ammonia is stored in stationary storage tanks at this facility. *(Please answer each question.)*

Are anhydrous ammonia stationary tanks located on-site? *(If No, proceed to question 18)*	◯ Yes	◯ No
a. Are all tanks located within a fenced area?	◯ Yes	◯ No
b. Are all tanks equipped with valve locks?	◯ Yes	◯ No
c. Is an alarm system utilized?	◯ Yes	◯ No
d. Are security-related additives utilized?	◯ Yes	◯ No

18. Please provide answers to best describe the location where nurse wagons filled with anhydrous ammonia are stored at this facility. *(Please answer each question.)*

Are filled nurse wagons of ammonia stored on-site? *(If No, proceed to question 19)*	◯ Yes	◯ No
a. Are all nurse wagons located within a fenced area?	◯ Yes	◯ No
b. Are all nurse wagons equipped with valve locks?	◯ Yes	◯ No
c. Are all nurse wagons equipped with tire locks?	◯ Yes	◯ No
d. Are all transfer hoses removed from tanks as an additional security measure?	◯ Yes	◯ No
e. Are security-related additives utilized?	◯ Yes	◯ No

Done Internet

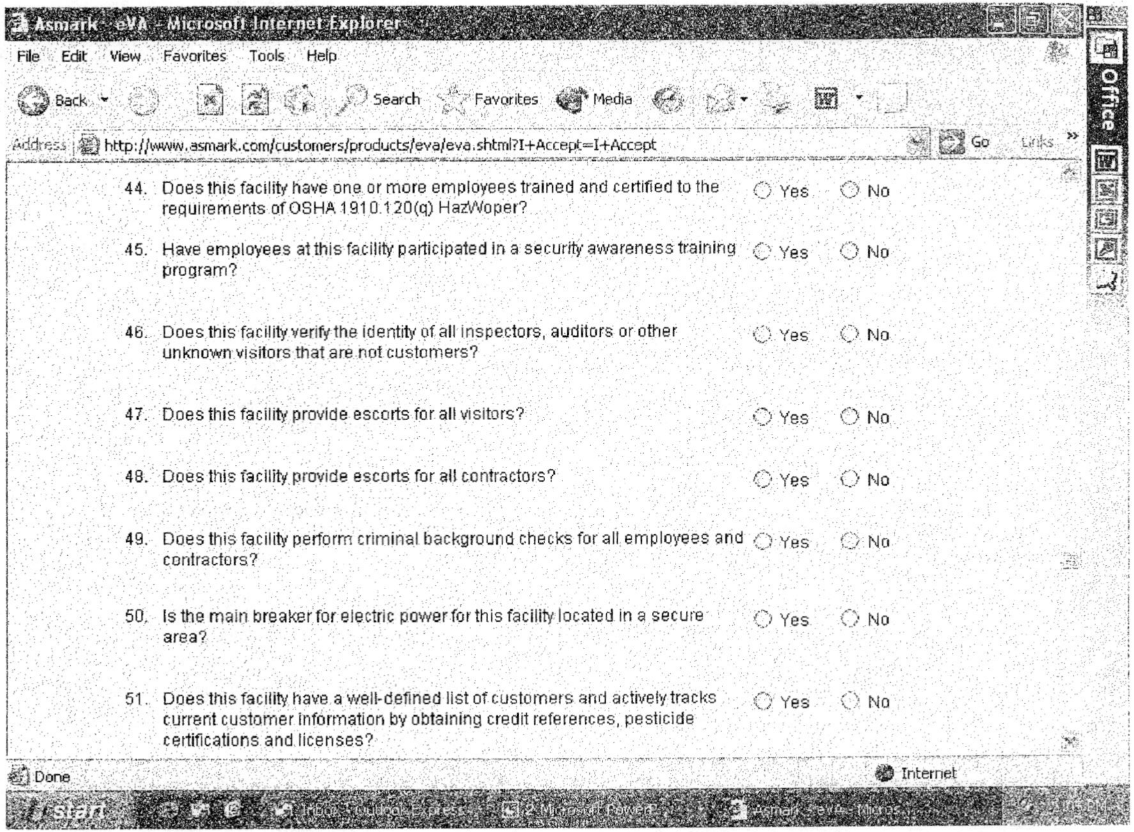
Asmark - eVA - Microsoft Internet Explorer

File Edit View Favorites Tools Help

Address http://www.asmark.com/customers/products/eva/eva.shtml?I+Accept=I+Accept

44. Does this facility have one or more employees trained and certified to the requirements of OSHA 1910.120(q) HazWoper? ○ Yes ○ No

45. Have employees at this facility participated in a security awareness training program? ○ Yes ○ No

46. Does this facility verify the identity of all inspectors, auditors or other unknown visitors that are not customers? ○ Yes ○ No

47. Does this facility provide escorts for all visitors? ○ Yes ○ No

48. Does this facility provide escorts for all contractors? ○ Yes ○ No

49. Does this facility perform criminal background checks for all employees and contractors? ○ Yes ○ No

50. Is the main breaker for electric power for this facility located in a secure area? ○ Yes ○ No

51. Does this facility have a well-defined list of customers and actively tracks current customer information by obtaining credit references, pesticide certifications and licenses? ○ Yes ○ No

Done Internet

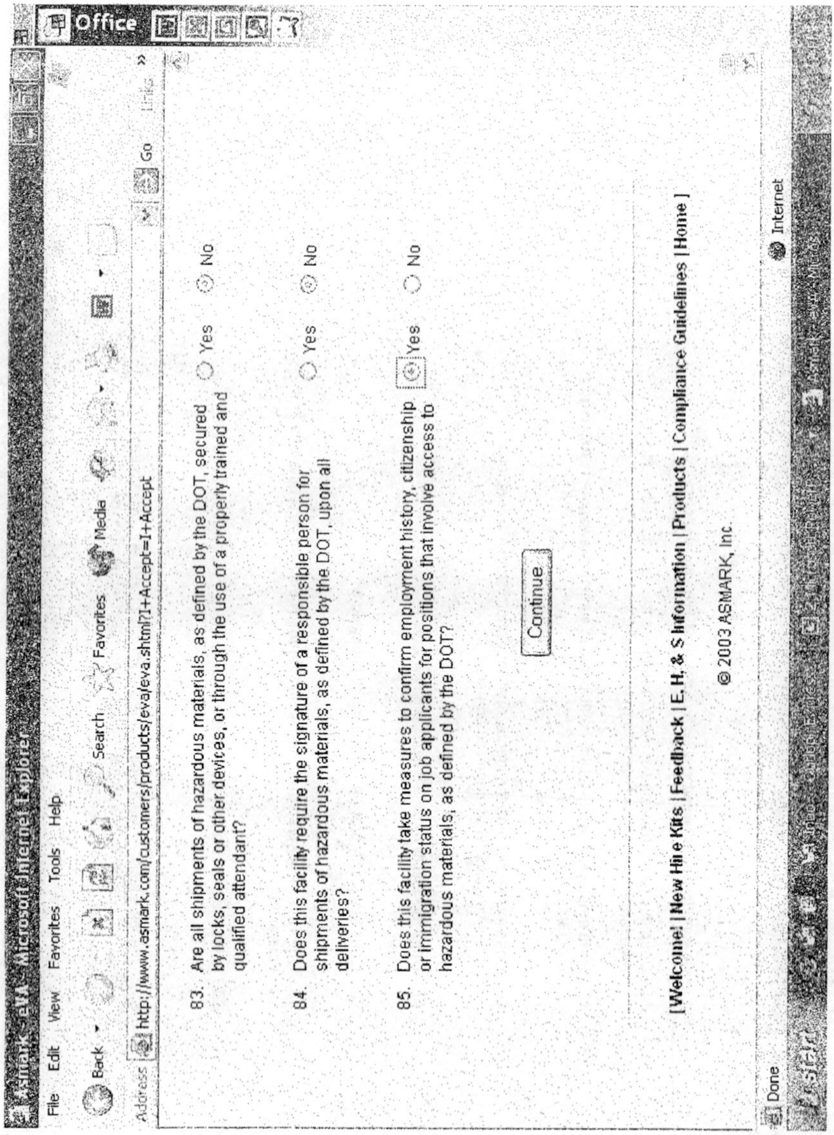
Office
Asmark - eVA - Microsoft Internet Explorer
File Edit View Favorites Tools Help
Back Search Favorites Media
Address http://www.asmark.com/customers/products/eva/eva.shtml?1+Accept=I+Accept
Go
83. Are all shipments of hazardous materials, as defined by the DOT, secured by locks, seals or other devices, or through the use of a properly trained and qualified attendant?
Yes No
84. Does this facility require the signature of a responsible person for shipments of hazardous materials, as defined by the DOT, upon all deliveries?
Yes No
85. Does this facility take measures to confirm employment history, citizenship or immigration status on job applicants for positions that involve access to hazardous materials, as defined by the DOT?
Yes No
Continue
[Welcome! | New Hire Kits | Feedback | E. H. & S Information | Products | Compliance Guidelines | Home]
© 2003 ASMARK, Inc.
Done
Internet

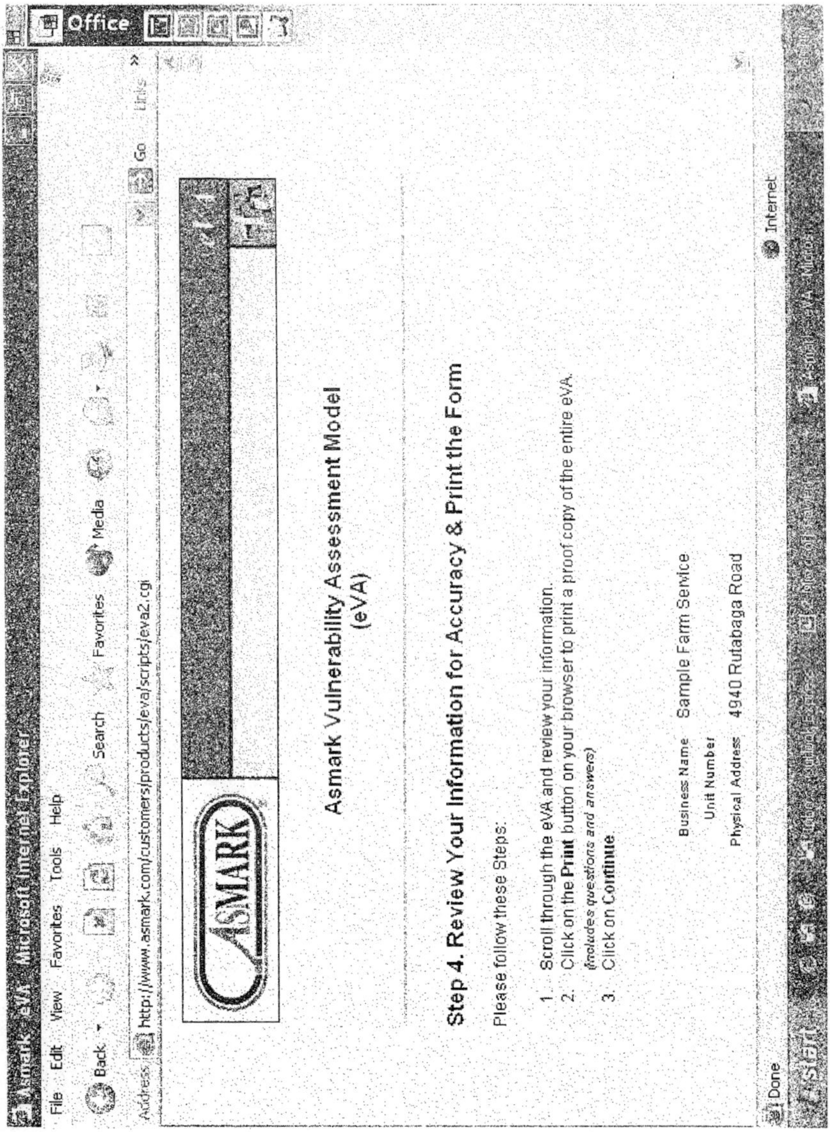
Asmark - eVA - Microsoft Internet Explorer
File Edit View Favorites Tools Help
Back
Search
Favorites
Media
Address http://www.asmark.com/customers/products/eva/scripts/eva2.cgi
Go
Links
Office
ASMARK
Asmark Vulnerability Assessment Model
(eVA)
Step 4. Review Your Information for Accuracy & Print the Form
Please follow these Steps:
1. Scroll through the eVA and review your information.
2. Click on the Print button on your browser to print a proof copy of the entire eVA.
(Includes questions and answers)
3. Click on Continue
Business Name Sample Farm Service
Unit Number
Physical Address 4940 Rutabaga Road
Done
Internet
start

Asmark - eVA - Microsoft Internet Explorer

File Edit View Favorites Tools Help

Address http://www.asmark.com/customers/products/eva/scripts/eva2.cgi

83. Are all shipments of hazardous materials, as defined by the DOT, secured by locks, seals or other devices, or through the use of a properly trained and qualified attendant? No

84. Does this facility require the signature of a responsible person for shipments of hazardous materials, as defined by the DOT, upon all deliveries? No

85. Does this facility take measures to confirm employment history, citizenship or immigration status on job applicants for positions that involve access to hazardous materials, as defined by the DOT? Yes

Continue

[Welcome! | New Hire Kits | Feedback | E, H, & S Information | Products | Compliance Guidelines | Home]

Done Internet

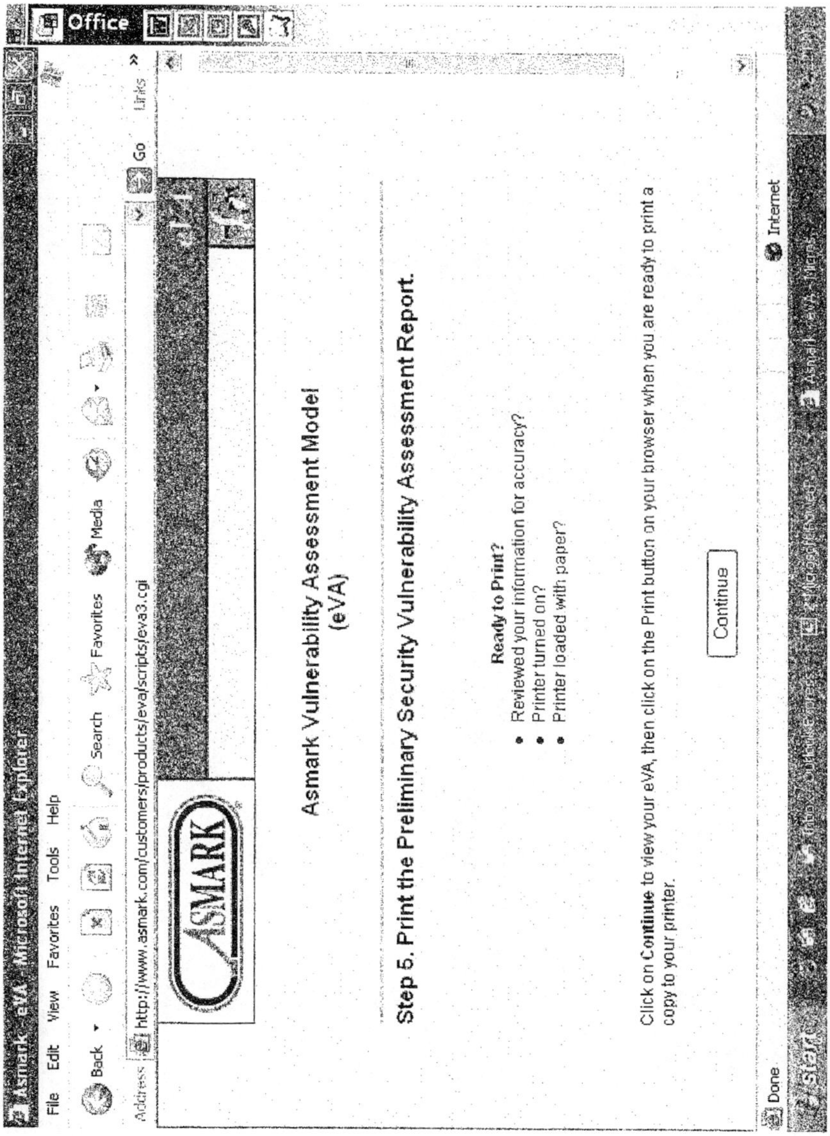
Asmark - eVA - Microsoft Internet Explorer
File Edit View Favorites Tools Help
Back Search Favorites Media
Address http://www.asmark.com/customers/products/eva/scripts/eva3.cgi
Go Links
ASMARK
Asmark Vulnerability Assessment Model
(eVA)
Step 5. Print the Preliminary Security Vulnerability Assessment Report.
Ready to Print?
Reviewed your information for accuracy?
Printer turned on?
Printer loaded with paper?
Continue
Click on Continue to view your eVA, then click on the Print button on your browser when you are ready to print a copy to your printer.
Done
Internet
start
Office

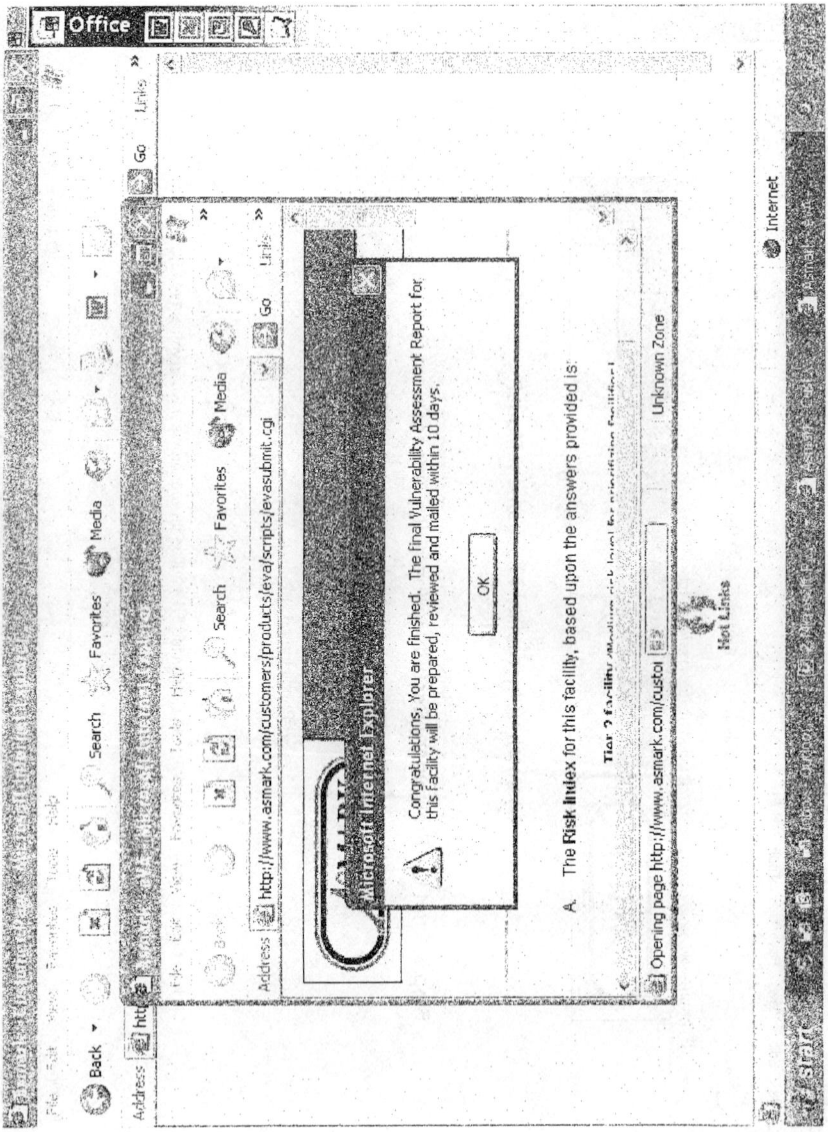
Office
Address http://www.asmark.com/customers/products/eva/scripts/evasubmit.cgi
Microsoft Internet Explorer
Congratulations, You are finished. The final Vulnerability Assessment Report for this facility will be prepared, reviewed and mailed within 10 days.
OK
A. The Risk Index for this facility, based upon the answers provided is:
Opening page http://www.asmark.com/custo
Unknown Zone
Hot Links
Internet

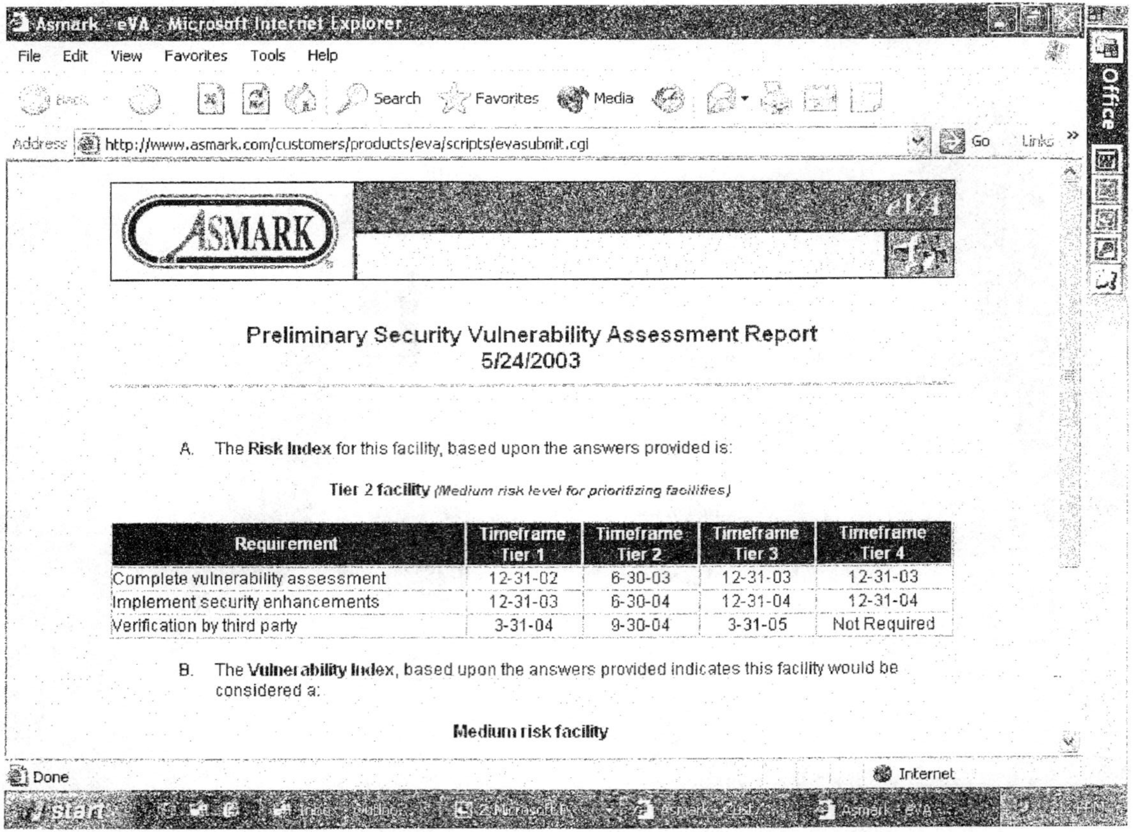

Asmark - eVA - Microsoft Internet Explorer

File Edit View Favorites Tools Help

Address http://www.asmark.com/customers/products/eva/scripts/evasubmit.cgi

ASMARK

Preliminary Security Vulnerability Assessment Report
5/24/2003

A. The **Risk Index** for this facility, based upon the answers provided is:

Tier 2 facility *(Medium risk level for prioritizing facilities)*

Requirement	Timeframe Tier 1	Timeframe Tier 2	Timeframe Tier 3	Timeframe Tier 4
Complete vulnerability assessment	12-31-02	6-30-03	12-31-03	12-31-03
Implement security enhancements	12-31-03	6-30-04	12-31-04	12-31-04
Verification by third party	3-31-04	9-30-04	3-31-05	Not Required

B. The **Vulnerability Index**, based upon the answers provided indicates this facility would be considered a:

Medium risk facility

Done Internet

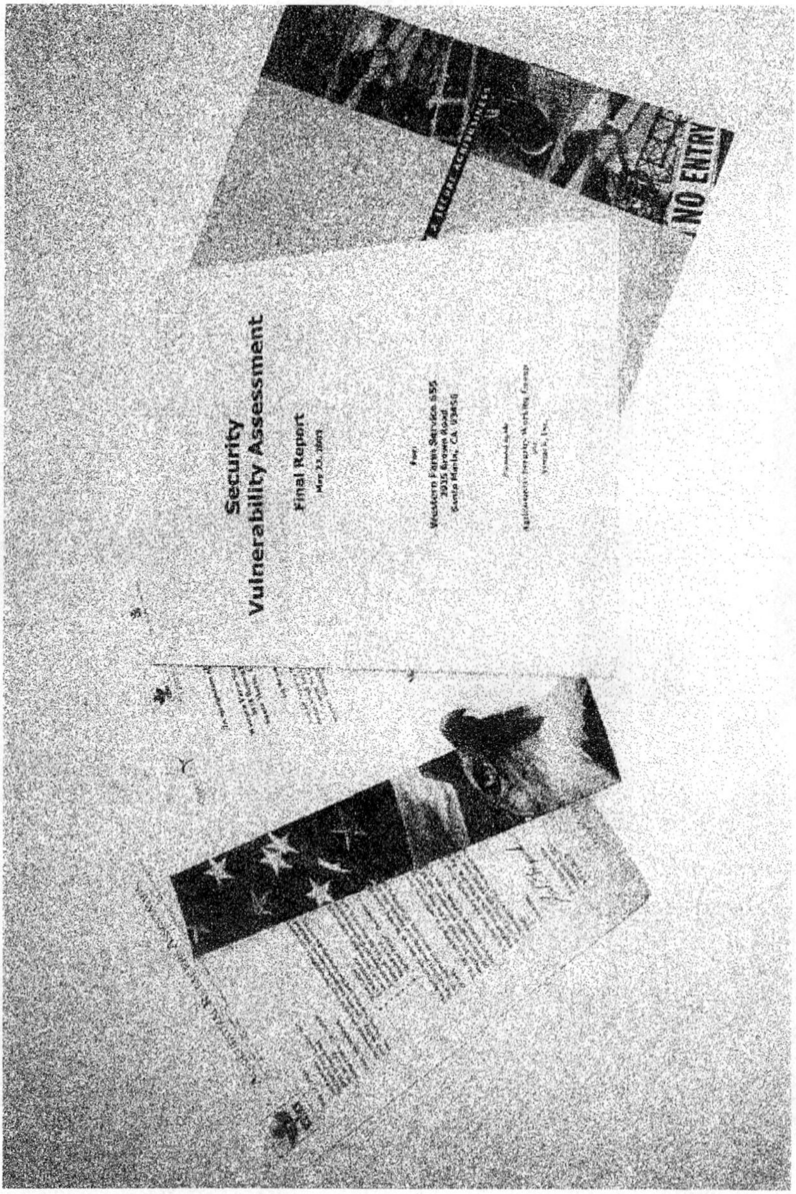
Security
Vulnerability Assessment
Final Report
NO ENTRY